Information Communication Technologies and City Marketing:
Digital Opportunities for Cities Around the World

Mila Gascó-Hernández
Pompeu Fabra University and Estratic, Spain

Teresa Torres-Coronas
Universitat Rovira i Virgili, Spain

Information Science REFERENCE
INFORMATION SCIENCE REFERENCE
Hershey · New York

Director of Editorial Content: Kristin Klinger
Director of Production: Jennifer Neidig
Managing Editor: Jamie Snavely
Assistant Managing Editor: Carole Coulson
Typesetter: Michael Brehm
Cover Design: Lisa Tosheff
Printed at: Yurchak Printing Inc.

Published in the United States of America by
Information Science Reference (an imprint of IGI Global)
701 E. Chocolate Avenue, Suite 200
Hershey PA 17033
Tel: 717-533-8845
Fax: 717-533-8661
E-mail: cust@igi-global.com
Web site: http://www.igi-global.com

and in the United Kingdom by
Information Science Reference (an imprint of IGI Global)
3 Henrietta Street
Covent Garden
London WC2E 8LU
Tel: 44 20 7240 0856
Fax: 44 20 7379 0609
Web site: http://www.eurospanbookstore.com

Library of Congress Cataloging-in-Publication Data

Information communication technologies and city marketing : digital opportunities for cities around the world / Mila Gasco-Hernandez and Teresa Torres-Coronas, editors.
p. cm.
Includes bibliographical references and index.
Summary: "This book promotes understanding of ICT's contribution to the development of city marketing strategies to enhance local socio-economic development"--Provided by publisher.
ISBN 978-1-60566-134-6 (hardcover) -- ISBN 978-1-60566-135-3 (ebook)
1. City promotion--Social aspects. 2. City promotion--Economic aspects. 3. City planning. 4. Information technology--Social aspects. 5. Information technology--Economic aspects. I. Gascó Hernández, Mila. II. Torres-Coronas, Teresa, 1966-
HT325.I45 2009
659.2'930776--dc22
2008035141

British Cataloguing in Publication Data

A Cataloguing in Publication record for this book is available from the British Library.

Editorial Advisory Board

List of Reviewers

Table of Contents

Detailed Table of Contents

Preface

"In a place at La Mancha, which name I do not want to remember, not very long ago lived a country hidalgo, one of those gentlemen or hidalgos who keep a lance in the lance-rack, an ancient shield, a skinny old horse, and a fast greyhound."

In a place at La Mancha El Quijote decided to go out as a knight-errant in search of adventure. In one of his escapades El Quijote met Master Pedro and his divining past-things ape. Master Pedro was a famous puppet-showman, exhibiting a show of the release of Melisendra and thus, for the very first time in our history, he linked marketing and technology. The Master Pedro's show was mainly used to dazzle the audience so later he could get a generous amount of cash with the aid of the divining ape. This is a funny story about the use of "technology in marketing" which remembers us an important and old lesson: Technology is only a part of a solution.

The Master Pedro's show is a story that can be seen from both a technological and a human perspective. The first one allows unlimited options, but the human view is responsible for providing the values and principles that can make technology a tool to be used in benefit of everybody. And this is something that cities have to bear in mind when entering in the fascinating world of city marketing through information and communication technology (ICT).

WHEREIN IS SET DOWN WHAT CITY MARKETING IS

City marketing is defined as the designing of a city to satisfy the needs of its target markets. It succeeds when citizens and businesses are pleased with their community and the expectations of visitors and investors are met. Indeed, city marketing as such is not a new phenomenon. However, as a result of the global transformations that are deeply impacting organizations at the local level, cities have been compelled to actively compete with each other. Now more than ever, cities need to attract tourists, factories, companies, and talented people, as well as find markets for their exports. This requires that cities adopt strategic marketing management tools and conscious city branding. As a result, several city marketing methods, approaches, and instruments have been designed to attract the attention of city stakeholders. Nevertheless, despite the important investments of the last few years, there is still an important tool, brought about by the new era, which remains unexplored: the new ICT and, particularly, the Internet.

As Martinotti states (1999), the boosting of city images, for both political and commercial reasons, can be traced to ancient cities, perhaps to the very origin of the city; the urban dweller has always felt his or her superiority over the rest of the world. However, the unabashed commodification of cities as sales objects has become a matter of course only in very recent years and can be easily dated to nearly the 1990s as the result of the convergence of three macro processes that have been taking place all over the world: the growing urbanization, the technological revolution, and the economy and communications globalization.

These phenomena have made evident the need to rethink the role played by cities since contrary to what many observers asserted, cities have not become obsolete. What's more, besides the decline of several once-great industrial centers in the highly developed countries, a significant number of cities have also seen their concentration of economic power rise (Sassen, 2001). These changes in a city competitive profile have encouraged cities competition, understood as the efforts that cities carry out in order to become competitive and dominate other cities. According to Metaxas (2002), these include common action and different measures for local economic development, as well as strategic thought to implement a development policy concerning the role that the potential city will play in the future. In short, cities compete in order to attract, among other, investments, population, tourists, public funding, students, or international events that can improve their territorial competitive advantage (Budd, 2001).

Most of the research about cities competition has focused on two issues: 1) what influences and facilitates places competitiveness and 2) which competitiveness and cooperation strategies among cities exist. In particular, the latter has to do with the question "how do cities compete" and, therefore, deals with the tools city managers have. Thus, this is the context where the term "city/place marketing" emerges.

According to Kotler, Asplund, Rein and Haider (1999), place marketing refers to a place planning procedure concerning the satisfaction of the needs of target – markets. It could be successful when it satisfies two main parameters: a) the enterprises' and the residents' satisfaction from the purchase of goods and services that the place provides, b) the satisfaction of the expectations of potential target - markets (enterprises and visitors), as long as the goods and the services that the place provides to them are those that they wish to get.

In this respect, a marketing city plan adapts the traditional model of the four marketing "Ps". Metaxas (2002) explicitly states that city marketing includes:

1. The product, which has to do with the production system (that is, the city), the productive good (or the city's image), the good's ingredients (such as the city's character, economic activities, natural environment, services, recreation and leisure, culture, or tourism), and the city's distinctive characteristics (that is, those characteristics that could create city's competitive advantages).
2. The price, which concerns the value of land's use for residence or the value of land's use for setting up new business activities.
3. The place/distribution channels that relate to the network of relationships with channel partners (both internal and external) that the city develops in order to apply its image to the potential target markets or to penetrate into new market areas.
4. The promotion, which includes advertising, public relations, campaign, or slogans strategies.
5. The people, a term which is used in order to satisfy the human resources management process for visitors' attraction and the citizens' contribution to the city's development.

The 21st Century brings about new ways of doing things and also new means to market cities. Although the above five "Ps" remain valid, new tools are available to city managers in order to promote their cities. In particular, ICTs (and specifically, the Internet) are giving rise to the implementation of new strategies and techniques in the city marketing field that are going to change the way cities are managed for, as Deighton (1996) said, "the profession of marketing, its theories, its practices, and even the basic sciences that it draws on are determined by the tools at its disposal at any moment. When the tools change, the discipline adjusts, sometimes quite profoundly and usually quite belatedly. The introduction of television advertising 50 years ago was just such a disruptive event, and marketing theory and practice are still responding, evolving their understanding of how the tool works and how its effects should be measured" (p. 151). If this is so, and the editors believe it is, this book is intended to be one important resource for both researches and practitioners about the potential of the ICT, and particularly of the Internet, in the city marketing field in terms of strategy design and implementation practices.

OF THE DELECTABLE DISCUSSION ABOUT HOW INFORMATION COMMUNITATION TECHNOLOGIES AND CITY MARKETING BRING NEW OPPORTUNITIES FOR CITIES

Cities inevitably, as the economic world does, evolve and change. The socio-economic growth of a city takes time and effort. In their journey to higher development, cities are responsible for the well-being of their targeted groups. At the same time, cities are increasingly competing against each other for attracting sources of wealth. They need to find ways of differentiating themselves thereby making themselves more attractive to gain advantage over their competitors. With no doubt, cities need to work harder. In this hyper-competitive context, fresh ideas are needed for place marketing to give cities a new set of tools. As Rainisto (2003) explains, contemporary place marketing practices have not yet answered the challenges of our information and knowledge society, and there is still plenty of room for improvements.

ICT are one of those tools whose benefits remain still unexplored for both city marketing academicians and city managers. Although the adoption of technology by (local) governments is not new, specific applications in the city marketing field are hard to find except for the use of Web sites and city portals aimed at promoting the city among potential visitors and therefore focusing on tourism strategies.

The Internet is utilized as a system of electronic intercommunication and a way of processing and presenting digital information. With help from people's imagination, it brings together unlimited opportunities for city marketing strategies around the world. The use of Internet for promoting and building a city brand image is very common. But it is not the only thing that can be done. The latest revolution was presented with the Web 2.0 era that encouraged active participation, collectivism and knowledge sharing. The blogging concept is now utilized by private companies as part of their communication and building community strategy. Setting up virtual offices, advertising and promotion for new products, or providing information and news are reasons stated by companies to enter the virtual world of Second Live®. How can cities benefit from the never-ending Internet revolution?

The delectable discussion about how ICT and city marketing have large potential for opening up new areas of opportunities, both in social and business uses has just begun. Now it is time for cities to ask themselves:

- Which ICT features and practices bring value to their activities?
- How to use ICT to better attract the cities' target groups for local development.
- How to identify ICT best practices and trends for their city marketing strategy.
- Which type and amount of investments are needed.
- Which is the cost-benefit for cities to be involved in electronic city marketing?
- Why ICT are now a must in their place marketing activities?

Cities need a general background for understanding the opportunities that ICT can bring to them. Reading this book could be a first step in this adventure, as reading knight-errant books was the first step in the adventures of El Quijote.

IN WHICH IS RELATED THE CONTENT OF THIS BOOK

Information Communication Technologies and City marketing: Digital Opportunities for Cities around the World is a book aimed at enlightening the above concepts and therefore at providing understanding as to how ICT can contribute to the development of city marketing strategies and, therefore, enhance local socio-economic development. In particular, its overall objectives are:

1. To describe the concept of city marketing and to analyze its contribution to both a city's competitiveness and a city economic development.
2. To identify the potential applications of ICT in city marketing, to provide insightful analysis about those factors that contribute to a successful use of ICT by city marketers.
3. To propose strategies to move forward and to address future challenges that involves the whole array of city stakeholders.
4. To identify and describe international successful experiences that explains the previous issues.

The book presents insights gained by leading professionals from the practice, research, academic, and consulting side in the field. This is why it should be useful to a variety of constituencies, who are interested in the interrelationships between information and communication technologies and city marketing strategies and, in particular, to:

1. Politicians and public sector officials (civil servants) who need a convenient source of information on what ICTs can do for city marketing and on what other local governments are doing in this field.
2. Private company executives, leaders, and consultants who frequently liaise with government agencies to design and implement city marketing strategies
3. Academicians, researches and students interested in the field of city marketing and/or the field of the use of ICTs by local governments.

The book is presented in three sections. The first one, "*Discovering connections: Linking City Marketing and ICT*", is a wide-ranging section which contains six chapters focused on the use of information and communication technologies in city marketing strategies from more of a conceptual point of view.

In particular, Muñiz-Martínez and Cervantes-Blanco (Chapter I) analyzes economic development from a phase of industrial production towards a new era that complements management of technology and information with intelligent awareness and creativity. The author concludes by investigating the role of new Internet technology in communicating and promoting the identity of cities with consideration of the cases of a number of cities around the world.

Fernández-Cavia and Huertas-Roig (Chapter II) describe the results of two combined studies (qualitative and quantitative) that analyze a sample of official city Web sites. The results show that official Web sites of cities pay a lot of attention to ease of navigation, but interactivity is much less implemented, especially between users. Furthermore, some lack of attention to the communication aspects of city brands can also be found.

Chapter III, by Mishra and Rolland, proposes a broad approach called strategic stakeholder relationship management (SSRM) that is enabled by information and communication technologies including the Internet in order to help the decision makers succeed in designing the twenty first century city marketing initiatives.

McCready, in Chapter IV, also focuses on strategy and refers to the technology driven aspects of globalization as "techno-globalization". He describes the role of strategic technology planning in the marketing of cities as well as for information and communication technologies and its intersection with marketing planning.

Chapter V, authored by Chudalla and Pousttchi, provides basic knowledge on mobile services, the presentation of restrictions and opportunities of mobile devices, applications and communication techniques. Three case studies are presented to help the reader understand applications from a practical perspective.

Finally, in Chapter VI, Matherly and Jouett present a causal model where ICT is used to not only deliver services to internal stakeholders but also to market a city to external stakeholders. Case study applications are discussed as well as the questions to address in future research.

Section II, "*Unlocking the Power of City Marketing to City Development*", presents new interrelationships and illustrates them with case studies. Thus, it goes beyond the connection between ICT and city marketing and approaches the possibilities of that link for city development strategies.

Therefore, Chapter VII, by Huerta-Carvajal and Luna-Reyes, discusses the strategic scaffolding for ICT as a key component of a city's marketing strategy. Although city marketing efforts and ICT use are still at its initial stages in the city, lessons from current efforts in Puebla (Mexico) are related to the key role of stakeholder networks, ICT interoperability, Geographic Information Systems and government program continuity.

Díaz-Luque, in Chapter VIII, analyzes the possibilities that official Web sites offer from a marketing perspective. It also studies the opportunities to use cities' Web sites to develop a complete marketing mix strategy.

Next, Prats-Planagumà and Camprubí present a case study in order to analyze network configuration through promotional Web sites and determine if the tourist product of a destination is integrated and promoted globally or, on the contrary, each tourist agent acts independently. Conclusions are based on the theoretical model presented.

Chapter X, by McGill and Ensign, discusses the importance of making strategic investments in information communication technologies in order to benefit from globalization and the opportunities created by export-oriented business clusters. Examples of investments made by local governments in India, Jamaica and Hong Kong are presented.

Sigala, in Chapter XI, aims to inform city tourism organizations responsible for the development of city portals about the use of the major Web 2.0 tools and their impact on the tourism demand and supply. It also presents the ways and practices for integrating the use of Web 2.0 into their e-business model and e-marketing practices.

In the last chapter of this second section, Córdoba, Jullien, and Tremembert define three different patterns –idealist, strategic and power-based to understand how city marketing initiatives are designed and implemented, and how information and communication technologies can support their implementation. Experience of using these patterns to understand the situation of Hull and Brest are also presented by the authors.

At last, Section III, "*Stories from the Battlefield: Finding Out the Power of City Marketing*", reviews several initiatives that have taken place all over the world and that illustrate the use of ICT to enhance city marketing strategies.

In Chapter XIII, after introducing the concept of city branding, and a model of how Web site elements communicate brand values and messages, Dobers and Hallin analyze a recent attempt of city managers to promote the brand of Stockholm. This case study illustrates the challenges that city managers face today.

Hallin (Chapter XIV) looks deeply into the Stockholm case by semiotically analyzing a marketing leaflet for the Stockholm-based ICT-project mCity, and two ads for Nokia phones that appeared in Europe at about the same time. Her chapter challenges the traditional cybernetic sender-receiver model of communication.

In Chapter XV, Moffett, McGinnity, Callaghan, Harkin, Woods, and Paris, outline the journey that the city of Londonderry undertook when transforming a traditional walled city to a technology enhanced wireless city. The chapter presents an overview of the three project strands, namely wireless city (civic aspect), wireless walls (tourism aspect), and wireless campus (educational aspect). A detailed case study of the educational element is presented.

To conclude, Morolong looks, in the last chapter of the book, at how information communication technologies can be used to market cities such as Gaborone, the capital city of Botswana. Literature and experiences from other parts of the world are used to underscore city growth as a serious development issue.

More could be said about the relationships between ICTs and city marketing since the possibilities of the former are unknown. This book is only a first approach to this new field. It presents several issues that have to do with the new tools city managers have. It also introduces some interesting aspects about the academic state of the art of the discipline. Both perspectives make the text valuable for researchers and practitioners. But *Information Communication Technologies and City Marketing: Digital Opportunities for Cities Around the World* is only a first stone and the authors hope that the authors' contributions encourage the reader to keep strengthening the way technology can help cities all over the world.

REFERENCES

Budd, L. (2001). Territorial competition and globalisation: Scylla and Charybdis of European Cities. *Urban Studies, 35*(4), 663-685.

Deighton, J. (1996). The future of interactive marketing. *Harvard Business Review, 74*(6), 151-160.

Kotler P., Asplund C., Rein I., & Haider H.D. (1999). *Marketing places Europe: How to attract investments, industries, residents and visitors to cotes, communities, regions, and nations in Europe*. London: Prentice Hall.

Martinotti, G. (1999). A city for whom? Transients and public life in the second-generation metropolis. In R. A. Beauregard & S. Body-Gendrot (Eds.), *The urban moment. Cosmopolitan essays on the late 20th Century city* (pp. 155-184). Thousand Oaks, CA: Sage Publications.

Metaxas, T. (2002, April). *Place/city marketing as a tool for local economic development and city's competitiveness: A comparative evaluation of place marketing policies in European cities*. Paper presented at the EURA Conference Urban and Spatial European Policies: Levels of Territorial Government, Turin, Italy.

Rainisto, S. K. (2003). *Success factors of place marketing: A study of place marketing practices in Northern Europe and the United States*. Unpublished doctoral dissertation, Helsinki University of Technology – Institute of Strategy and International Business, Helsinki.

Sassen, S. (2001). Cities in the global economy. In R. Paddison (Ed.), *Handbook of urban studies* (pp. 256-272). London: Sage Publications.

Seisdedos, G. (2007). *Cómo gestionar las ciudades del siglo XXI. Del city marketing al urban management*. Madrid: Pearson Educación.

Acknowledgment

The editors would like to acknowledge the help of all involved in the collation and review process of the book, without whose support the project could not have been satisfactorily completed.

Deep appreciation and gratitude is due to Kristin Roth, our former development editor and, most important, our friend. Her editorial support, her encouragement and her good advice during this year has turned this project into a worthwhile book. Special thanks also go to all the staff at IGI Global, whose contributions throughout the whole process from inception of the initial idea to final publication have been invaluable. In particular, to Mehdi Khosrow-Pour, whose enthusiasm motivated us to initially accept his invitation for taking on this project.

We also want to sincerely thank all those who offered constructive and comprehensive input for the different chapters. They did an outstanding job. In closing, we wish to thank all the authors for their insights and excellent contributions to this book. They have shared with us their priceless expertise, carrying out a terrific work. Thank you to all of you for helping us to spread your knowledge about how ICT may contribute to the development of city marketing strategies.

Finally, we want to thank our loved ones for their understanding and support throughout this project:

To my wonderful children, Marcos and Hawa, for being the reason that keeps me going every single day. To my parents, Milagros and José Luis, and my sisters, Samantha and Amaya, for their unconditional support during this hard year. To Teresa, for her unlimited patience while preparing this book and for being such a good friend. And to Carlos, for being always there (Mila's special thanks).

To Arnau and Jordi, my two knight-errant children, already surfing the Net in search of unlimited adventures. To my parents, Félix and Engracia, and my husband, Jordi, for all the things they have brought me throughout my life. And of course, to my friends Alegría, Josepa and M. Angel who still answer the phone when I call them (Teresa's special thanks).

Mila Gascó-Hernández and Teresa Torres-Coronas
Barcelona-Tarragona, Spain
September 2008

Section I
Discovering Connections:
Linking City Marketing and ICT

Chapter I
Identity and Marketing of Cities

Norberto Muñiz-Martínez
Universidad de León, Spain

Miguel Cervantes-Blanco
Universidad de León, Spain

ABSTRACT

Cities are acquiring a key geopolitical importance in the shaping of world-wide flows and exchanges, playing a key part in modern socio-economic relations within the framework of the world order termed globalization. Urban areas are the nodes where networks of various types of interchange come together: economic, social, cultural, communications and interpersonal. While having a leading role in these major relations of world-wide exchanges, cities in addition shape their own interchanges between what they can offer and the demands from the various groups within them. These are principally their citizens, but also investors, tourists, and administrative and civil institutions. Strategic marketing and management approaches have been implemented into the field of countries, regions, and especially cities, which are adopting these approaches to sell what they have to offer; to better manage and compete more effectively. Marketing provides a conceptual framework, and tools for managing these exchange relationships between what cities supply and demand. This chapter explores these issues, and examines the evolution of city marketing, from emphasizing infrastructures and urban regeneration towards stressing intangible values, such as multicultural integration, urban quality of life, appreciating aesthetics, the design and beauty of a city, a marketing of cities by means of intangible and emotional elements.

THE IDENTITY OF CITIES IN SOCIO-ECONOMIC GLOBALIZATION

As an outcome of economic globalization and the growing internationalization of markets, numerous relationships are developing that encourage a multiplicity and mixing of identities, in what Castells (1998) calls a *network society.* Cities, as the places or centers where these numerous flows of exchanges converge, gradually form an identity for themselves as a function of how these interconnections take shape within them. In this way, the concept of the identity of cities gains prominence. There are some cities that have clear identities relating to certain more or less unusual features that make them well known. Others, in contrast, especially large cities that constitute centers of power or world standards, become famous as an outcome of multiple facets (Table 1).

These attributes or potentials can be capitalized on by cities taking full advantage of them. In this sense there is a parallel with the approaches of strategic management and marketing in the world of business. These qualities would be equivalent to the concept introduced by M. Porter, *competitive advantage*, a capacity which if developed by an enterprise would allow it to achieve better results than competing businesses. In the same way, when a city is able to capitalize on its resources and potentials, or is remodeled and promoted, as Barcelona and Sydney were through their organization of the Olympics, then it achieves international fame that facilitates attracting investments and visitors.

The profile of a city can be influenced greatly by major public works. These can include new airports, as in Hong Kong, or extensions to existing airports, like Barajas in Madrid; ports as in Yokohama; new museums like the *Guggenheim* in Bilbao, the area of *Arts and Sciences* in Valencia, or the *Picasso Museum* in Malaga, or renovations, such as the *Louvre* in Paris, the *Rijksmuseum* in

Table 1. Cities with international reputation according to some attributes (Source: self compilation)

Events based on cultural singular assets	**Music**	Music festivals: **Bayreuth, Germany**–R. Wagner; **Salzburg, Austria**–W.A. Mozart; **Verona, Italy**–G. Verdi; **Nashville, USA**–North American folk; **New Orleans, USA**–Jazz, Gospel;
		Great philharmonic orchestras: **Berlin, Germany**; **Viena, Austria**
	Theatre	Drama festivals: **Avignon, France**–summer festival; **Edinburgh, UK**–*Fringe* festival; **Stratford-upon-Avon, UK**–festival in W. Shakespeare's natal village
		Classical theatre: **Almagro**, **Mérida, Spain**
		Theatres of international reputation: **Barcelona, Spain**–*Liceu*; **Milan, Italy**–*alla Scala*; **Venice Italy**–*La Fenice*; **Naples, Italy**–*San Carlo*
	Cinema	Cinematographic industry or festivals (and in some cases also festivals of advertising): **Los Angeles, USA**–Hollywood; **Cannes, France**–cinema and advertising festival; **Berlin, Germany**–*Berlinale*; **Venice, Italy**–*Mostra*; **San Sebastián, Spain**–cinema and advertising festival
	Great Museums	**Madrid, Spain**–Museo del Prado, Centre of Art Reina Sofia, Thyssen-Bornemisza; **Bilbao, Spain**–Guggenheim; **Valencia, Spain**–Ciudad de las Artes y las Ciencias; **Paris, France**–Louvre, Centre Pompidou; **Saint Petesburg, Russia**–Hermitage; **London, UK**–British Museum, National Gallery, Tate Gallery; **Berlin, Germany**–Alte National Galeri, Bodes Museum, Altes/Neues Museum, Pergamo; **New York, USA**–Metropolitan, Museum of Modern Art (MOMA), Guggenheim, American Museum of Natural History
	University Cities	**Oxford, UK; Cambridge, UK; St. Andrews, UK; Heidelberg, Germany; Tübingen Germany; Göttingen, Germany; Salamanca, Spain; Coimbra, Portugal; Bolonia, Italy; Princeton, USA; Berkeley, USA**
	Parties and Popular Celebrations	**Rio de Janeiro, Brazil**–*Carnival*; **Salvador da Bahía, Brazil**–*Carnaval*; **New Orleans, USA**–*Mardi Gras*; **Venice, Italy**–*Carnevale Di Venezia*; **Valencia, Spain**–*Fallas*; **Pamplona, Spain**–*Sanfermines*; **Munich, Germany**–*Oktoberfest*, party of the beer; **Siena, Italy**–*Palio*, horse racing; **Calgary, Canada**–*Stampide*, western parade; **Seville, Valladolid, Zamora, Spain**–*Semana Santa*, Holy Week religious parades

continued on following page

Table 1. continued

Cities economic centers	World financial centers	**New York, USA; London, UK; Tokyo, Japan**
	Metropolitan cities that are economic centers of international influence	**Chicago, USA; Los Angeles, USA; Paris, France; Madrid, Spain; São Paulo, Brazil; Shanghai, China; Seoul, South Korea; Toronto, Canada; Dubai, United Arab Emirates; Johannesburg, South Africa; Singapore**
	Cities of medium average size that specialize themselves economically and become more important than they are for their sizes	**Frankfurt, Germany**–stock exchange, banks, European Central Bank; **Geneva, Switzerland**–banks, watches; **Milano, Italy**–economic centre, mode & craft industry; **Rotterdam, The Netherlands**–Mercantile port activity
Industrial cities	Cities of industrial tradition that look for new positioning because of the decline or industrial transformation	**Manchester, UK; Liverpool, UK; Glasgow, UK; Bilbao, Spain; Dortmund-Duisburg-Bochum-Essen-Düsseldorf, Germany**–metropolitan area of the Ruhr region; **Detroit, USA; Pittsburgh, USA**
Cities with research centers	**San Francisco, USA, San Jose, USA**–Silicon Valley; universities: Berkeley, Stanford; **Boston, USA**–universities: Harvard, MIT; **Stuttgart, Germany, Munich, Germany**–high range automobiles Mercedes, Porsche, BMW, Audi; **Kuala Lumpur, Malaysia**–*Multimedia Super Corridor*; **Dublin, Ireland**–integration of the *Digital Knot* in the urban centre; **Bangalore, India**–technological parks	
Vanguard forefront cities, trend setting cities	**Barcelona, Spain**–urbanism, architecture; **Berlin, Germany**–urban renovation, reunification, culture; **San Francisco, USA**–forefront social movements, waterfront bay, cosmopolitanism; **Vancouver, Canada**–multiculturalism, natural beauty, waterfront; **Shanghai, China**–urban, socio-cultural and economic change; **Amsterdam, The Netherlands**–urban singularity, forefront social tolerance; **Curitiba, Brazil**–social and environmental management; **Dubai, United Arab Emirates**–new economic centre, finance and trade, culture, tourism, vanguard architecture	
Natural beauty	Cities, town and villages placed in environments of natural beauty that generate reputation and attract visitors and residents	Sea waterfronts/bays/beaches: **Palma de Mallorca, Spain; San Sebastián, Spain; Santander, Spain; Bay of Naples, Italy; Miami, USA; Mar del Plata, Argentina; Sydney, Australia; Cape Town, South Africa, Durban, South Africa**
		Mountains: **Chamonix, France**–*Mont Blanc*; **Zermatt, Switzerland**–*Matterhorn* or *Cervin*; **Innsbruck, Austria**–the Alps, Tirol;
		Sky: **Calgary, Canada; Aspen, USA; San Carlos de Bariloche, Argentina**–*the Andes*
		Coast and mountains: **Rio de Janeiro, Brazil**–beaches of Copacabana, Ipanema, etc., Mount Corcovado; **Bergen, Norway**–fiords
Touristic cities	Popular destinations of sun and beaches	**Benidorm, Spain; Lloret de Mar, Spain; Rimini, Italy; Acapulco, Mexico; Cancún, Mexico; Varadero, Cuba**
	Cities specializing in a tourism of coast of elitist nature	**Cannes, France; Nice, France; The Principality of Monaco; Punta del Este, Uruguay**
	Health tourism	**Baden-Baden,** Germany; **Evian,** France; **Budapest,** Hungary; **Karlovy Vary, Czech Republic**–*Karslbad*; **Mariánské Lazně, Czech Republic**–*Marienbad*
	Urban tourism	**New York, USA**–architecture of skyscrapers, cosmopolitanism, cultural offer, cinema-tographic evocation; **Paris, France**–urban beauty, monuments, culture, fashion industry; **Rome, Italy**–old Roman civilization, urban beauty, Vatican City; **London, UK**–cultural offer, cosmopolitanism, monuments, Monarchy; **Prague, Czech Republic**–architectural heritage, culture; **Singapore and Kuala Lumpur, Malaysia**–shopping and financial centers
Metropolis or megacities	Big cities of powerful countries; politically, economically and culturally of world reference	**New York USA; Los Angeles, USA; Paris, France; London, UK; Tokyo, Japan; Moscow, Russia; Beijing, China**
	Big cities or urban agglomerations of emerging or developing countries, with over-excited and unplanned growth	**Calcutta India; Bombay, India; Yakarta, Indonesia; Mexico City, Mexico; São Paulo, Brazil; Cairo, Egypt; Lagos, Nigeria; Karachi, Pakistan; Manila, Philippines**
City-states	**The Principality of Monaco**–principality, tax haven, elitist tourism; **Andorra**–winter and mountain tourism; **Vatican City**–Catholicism; **Singapore**–financial centre and communications hub; **Luxembourg**–European crossing of cultures	

continued on following page

Table 1. continued

Cities that host international institutions		**Brussels, Belgium**–European Union, NATO; **New York, USA**–United Nations; **Geneva, Switzerland**–United Nations, International Red Cross; **Paris, France**–UNESCO; **Lausanne, Switzerland**–International Olympic Committee; **Washington D.C., USA**–International Monetary Fund, World Bank, Administration of the USA; **The Hague, The Netherlands**–some international courts of justice
Cities where international treaties have been signed, or international agreements have been celebrated		**Kyoto, Japan**–Protocol on Climate Change; **Maastricht, The Netherlands**–Treaty on European Union; **Davos, Switzerland**–annual meeting of the World Economic Forum
Cities that host big companies		**Eindhoven, The Netherlands**–Philips; **Leverkusen, Germany**–Bayer; **Wolfsburg, Germany**–Volkswagen; **Torino, Italy**–FIAT; **Toulouse, France**–aeronautical industry; **Ulsan, South Korea**–Hyundai; **Seattle, USA**–Boeing (aeronautics), Microsoft (software); **Detroit, USA**–automobile industry
Cities and places with sporting events of international repute	**Annual events**	Car racings: **Monte Carlo, Monaco**–Formula 1, Rally; **Le Mans, France**–*24 hours race*; **Indianapolis, USA**–500 miles
		Tennis: **Wimbledon, UK**; **Paris, France**–*Roland Garros*
		Golf: **St. Andrews, UK**; **Augusta, USA**
		Bicycle racing: **Paris, France**–*Tour de France* final stage
	One–off events with long term returns. They attract investments and boost tourism	Summer Olympic Games: **Sydney, Australia**–presentation like one of the best cities of the world; **Barcelona, Spain**–urban remodeling, world promotion; **Seoul, South Korea**, **Athens, Greece**, **Beijing, China** –presentation and promotion of the cities but also top of their countries
		Winter Olympic Games: **Salt Lake City, USA; Calgary, Canada; Lillehammer, Norway; Albertville, France; Torino, Italy**
		Football World Cup: **Germany** (2006); **South Africa** (2010)
	Sports clubs that provide **international reputation** to the cities that shelter them	European Football: Real **Madrid, Spain;** FC **Barcelona, Spain**; **Milan, Italy** Inter **Milan**; Juventus **Torino, Italy**; Ajax **Amsterdam, The Netherlands**; Bayern **Munich, Germany**; **Manchester** United, **Liverpool, UK**
		Basketball NBA: **Los Angeles, USA**–Lakers, **Boston, USA**–Celtics, **Chicago, USA**–Bulls
Cities with a religious significance		**Jerusalem, Israel/Palestine**–Judaism, Christianity, Islamism; **Rome, Italy (Vatican City)**–Roman Catholicism; **Santiago de Compostela, Spain**–pilgrim way of *Camino de Santiago*; **Mecca, Saudi Arabia**–Islam; **Varanasi, India**–Hinduism
Singular cities		**Venice, Italy**–singular urbanism: canals; **New York, USA**–singular architecture: skyscrapers; **Las Vegas, USA**–city of the game and gambling in casinos, original architecture of the hotels and casinos; **Paris, France**–beauty and monuments

Amsterdam or the *Prado* in Madrid; facilities for artistic activities, like the Sydney Opera House or the new Alexandria Library in Egypt; buildings for conventions; skyscrapers like *Taipei 101*, the tallest in the world in Taiwan or the *Petronas Towers* in Kuala Lumpur in Malaysia or the *Freedom Tower* project at the site of the ill-fated towers of the New York *World Trade Center*; sports facilities, like the Olympic Parks in Munich, or stadiums that are innovatory in their technology and design such as the *Allianz Arena* Munich, or the new Olympic Stadium in Beijing; renovation of waterside areas, as in Barcelona, of riverside zones, as in Bilbao, or of historic centers as in Compostella, all three in Spain; theme parks like *Eurodisney* near Paris. These constructions equate in modern times to what cathedrals represented in mediaeval Europe. Many cities commission major works of this sort from "star" architects who create structures that become new urban icons, symbols advertising the city. Mention may be made of N. Foster, J. Nouvel, R. Koolhaas, D. Libeskind, R. Rogers, F. Gehry, Herzog and de Meuron, A. Siza, R. Moneo, S. Calatrava, R. Bofill.

Figure 1. Cities with international reputation according to some elements

Florence (Italy), the city of the Renaissance

Festival of Wagner's music in Bayreuth. Front page of the influential German magazine *Der Spiegel*

Santiago de Compostela (Spain) is the destination of the pilgrims of the way Camino de Santiago, to that numerous tourists come too

Sao Paulo: over-excited and unplanned growth, *favelas* and skyscraper

Leverkusen (Germany), the city of the Bayer company

Benidorm (Spain), massive popular tourism

The Royal Guard, a tourist attraction of London

Mecca (Saudi Arabia), the sacred city of the islamic religion

Las Vegas Hotel *New York, New York* Hotel that imitates the skyscrapers of New York and the Statue of Liberty

Brussels, the city that host the European Union, European Commission headquarters

Building of the UN headquarters, New York

Figure 2. Big buildings and skyscrapers as urban icons

STRATEGIC MANAGEMENT IN THE RUNNING AND MANAGEMENT OF CITIES

This section looks into the advisability of cities basing their management decisions on a broader system of strategic planning that will allow the steering of urban strategies and policies, and the guiding of marketing strategies. The adoption of principles from economics, business and marketing in a city context may be set in a wider framework (Figure 4). An integrated city study should draw on conceptual contributions from a range of disciplines: town planning and architec-

Figure 3. Thematic buildings that become new urban icons

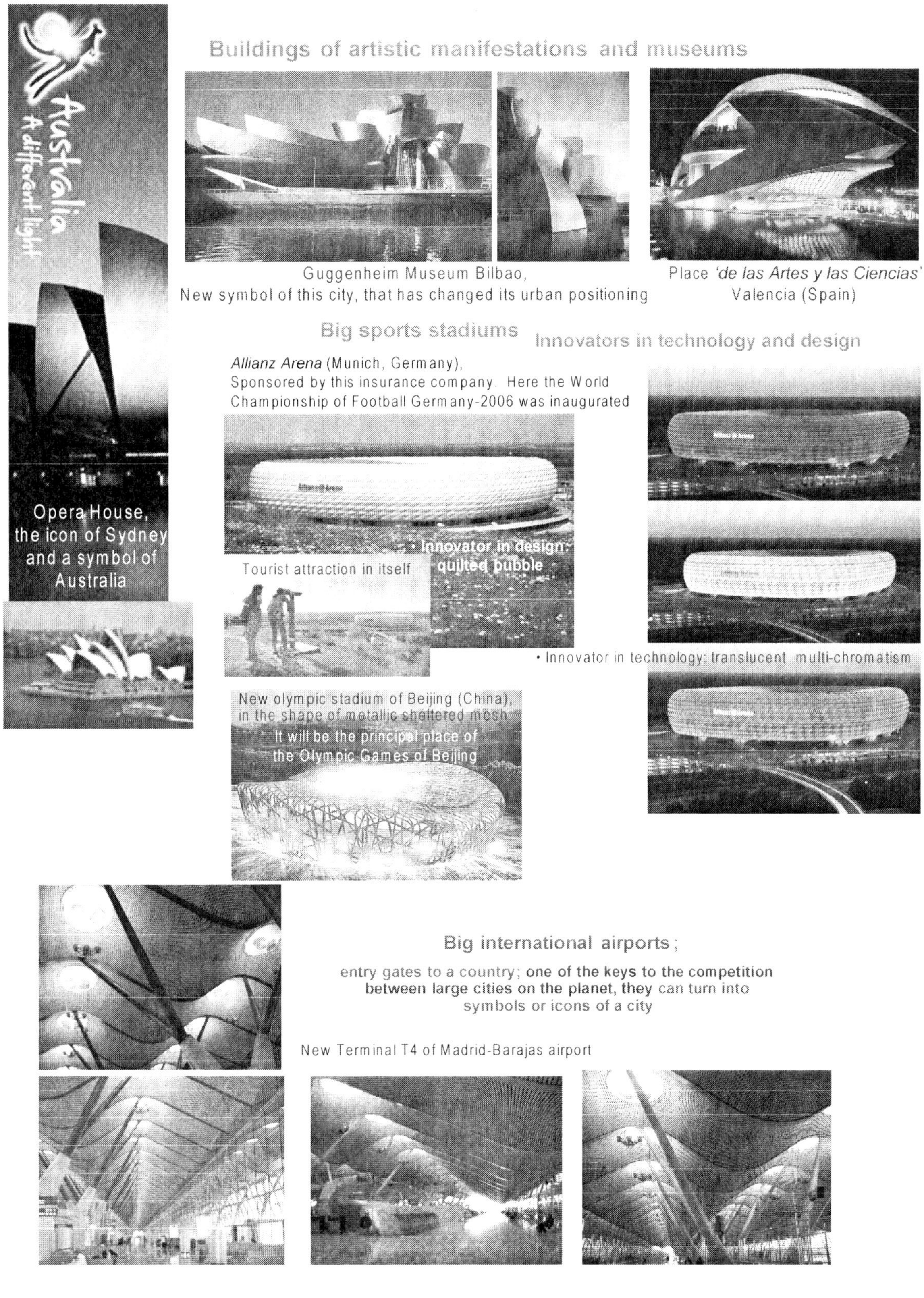

Figure 4. Model for strategic management and marketing of cities

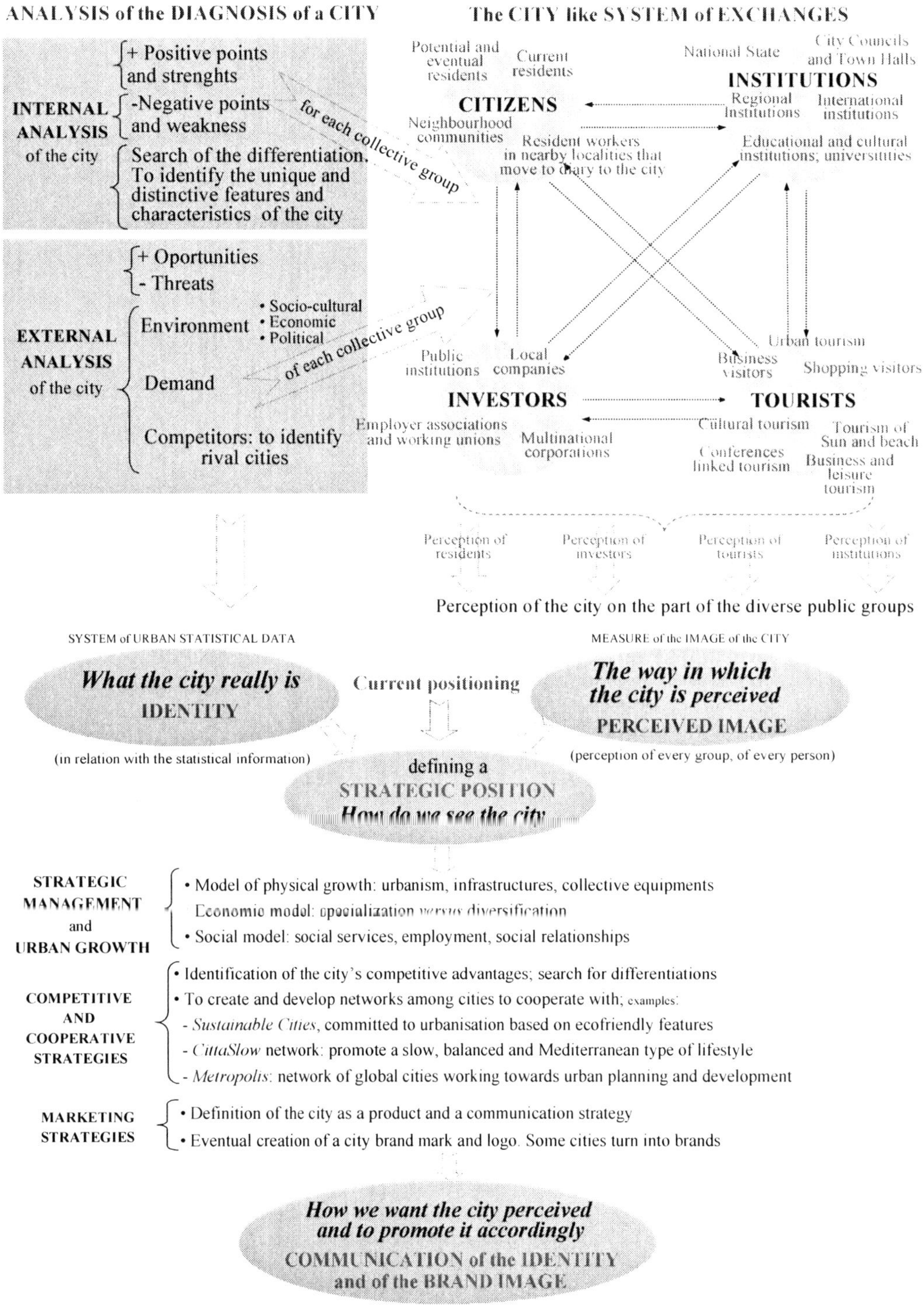

ture, sociology, economics and business, among others. The line being proposed here concentrates on the approaches of Strategic Management and Marketing.

The principles of economics and business studies can be applied to cities because cities are places, but they are also entities and systems of relations. The complexity of the interchanges and relationships between the various groups (citizens, institutions, investors and tourists of various sorts) justifies studying these approaches with an eye to identifying what is the best possible thing the city can offer, whether internally in accordance with the needs of its citizens and other component groups or externally, meeting the demands of outside groups and markets. This would allow appropriate selling of what can be offered by a city in the more and more complex and internationalized markets in which cities increasingly compete.

Defining and developing Strategic Management and Marketing in an urban context, implies making a diagnosis of the situation of the city, and also requires be undertaken and should include an internal and an external analysis. The internal analysis would cover the strengths and weaknesses the city may have for each group, so that those involved may differ over this aspect, in the sense that what for one group may be positive may not be so for another. The external analysis should consider opportunities and threats that might arise from various sources (Fernández Güell, 1997): The surroundings, whether seen from a political, economic or socio-cultural angle, the demands different groups make on the city, and study of possible competitor cities.

Before applying marketing approaches in the area of cities, it is necessary to analyze the concepts of identity and image of the city. The identity has been analyzed in the field of corporate brand (1992). The identity can be defined as the set of traits items and attributes that define the essence of the city, some of which are visible and others are not. The image of a brand is the set of mental representations, both emotional and cognitive an individual as a set of individuals associated with a brand or an organization (Lambin, 1991). Thus, it is very important to link these two concepts, because one will be expressing a defined personality (identity) and another with the perception of that personality (image). In the case of a city, identity is what the city is, and its image is the perception.

The reality of a city can be compared with its image. One step is to investigate *what the city is*, that is the identity, its objective positioning, by means of a set of urban indicators that allow observation of the relative position of a city as a function of a range of variables. Ideally, rather than being restricted to one given moment, this methodological tool should be able to extend over time, so that it would be a question of a permanent observatory permitting analysis of the evolution of the position cities held. This objective analysis should be compared with study of *how the city is perceived*, in other words how it is seen by the various groups related to it. In this case it would be a matter of a subjective positioning obtained by measuring the image of the city. Detecting differences between *what the city is* and *how the city is perceived* would help in determining the current position of the city and serve as a base for the construction of a future positioning, confirming or modifying previous plans and actions. In its turn, the image of a city may differ from one to another of the groups involved in it. Even the image one person has of a city can come from various sources: stereotypes lodged in the collective imagination, personal experiences, what other people say, information drawn from the mass media (radio, television, the press, the cinema, and similar), opinions formed about national and international events, and so forth.

From these steps it would be possible to define what it is desired a city should be, its corporate philosophy and the strategic principles on which to construct it. All of this is strongly determined by aspects that cannot be adjusted, like the physical geography and location, or the climate, history and socio-cultural values of the city. Strategic approaches and marketing may be applied at three levels. Firstly, a definition for the strategic and growth management of a city would include a model for physical development (town planning, infrastructures, socially owned facilities, services in general, and other features); an economic model, in which it would be possible to choose specialization or diversification of economic and business activities (as may be seen in some sections of Table 1); and a social model, which would include social services and employment and should shape the general pattern of social relationships. Secondly, when applying strategies for competition and co-operation, the competitive advantages of the city should be identified; with a view to seeking differentiation both in itself and in relation to the cities that are its most direct competitors. It would be useful at this stage to use the methodological tool of *benchmarking* (Font, 2003), in other words comparison of cities through the key activities that each city does best, in an attempt to achieve urban improvements in the cities under study. An example would be the comparative analysis carried out in the context of the *Eurocities* network covering the cities of Antwerp (Flanders) in Belgium, Bilbao (Basque Country), of Spain, Bologna (Emilia-Romagna), Italy, Eindhoven (North Brabant) in the Netherlands, Lisbon in Portugal, Munich (Bavaria) in Germany, Rotterdam (Mouth of the Rhine zone) in the Netherlands and Turin (Piedmont) in Italy (Van den Berg, Braun & Van der Meer, 1999).

NETWORKS OF CITIES: COMPETITION AND CO-OPERATION

The challenges of globalization imply major economic, social and cultural changes in the flows of exchanges in the modern world. In this context, cities develop numerous networking relations, either horizontal with other cities (whether competitive or co-operative) or vertical with other institutions (the central state, regional authorities, international bodies, or businesses).

Modern approaches in Strategic Management and Marketing stipulate that besides competing there is a need to co-operate and establish strategic networks so as gain a good position in markets. Some authors (Brandenburger & Nalebuff, 1996) call the converging of relationships of competition and co-operation by the portmanteau acronym term *co-opetiton*. Among cities it is also increasingly common for networks to be established. While horizontal strategic alliances and networks for co-operation with other cities are set up to defend common interests and collaborate, vertical connections with national authorities and with regional or international institutions are of importance because of the political and financial support they provide for urban management. Collaboration, associations and networks linking cities are of a range of types, as may be seen in Figure 5.

Collaboration may take many different forms. Sometimes it is economic in nature, as in the case of the *Metropolis* network of 81 cities around the world intended to develop solutions for the problems of great world cities, *Cities Project*, a network for identifying the competitive advantages of the cities forming it, for urban innovations, aiding other cities around the world to spot their strategic opportunities. There is also collaboration of a cultural type, such as, for example, the co-operation on matters of a historical and cultural nature between the UNESCO World Heritage cities in Spain, or the localities that co-operate over the *Pilgrim's Way to Compostella*, promoting tourism and culture. There are networks of cities that concentrate on collaborating to support values, like the group of cities that encourage the creation of music within the framework of cultural diversity, or the network of cities that promote solidarity, backing and refuge for threatened writers from around the world.

Figure 5. Networks of cooperation and competition between cities

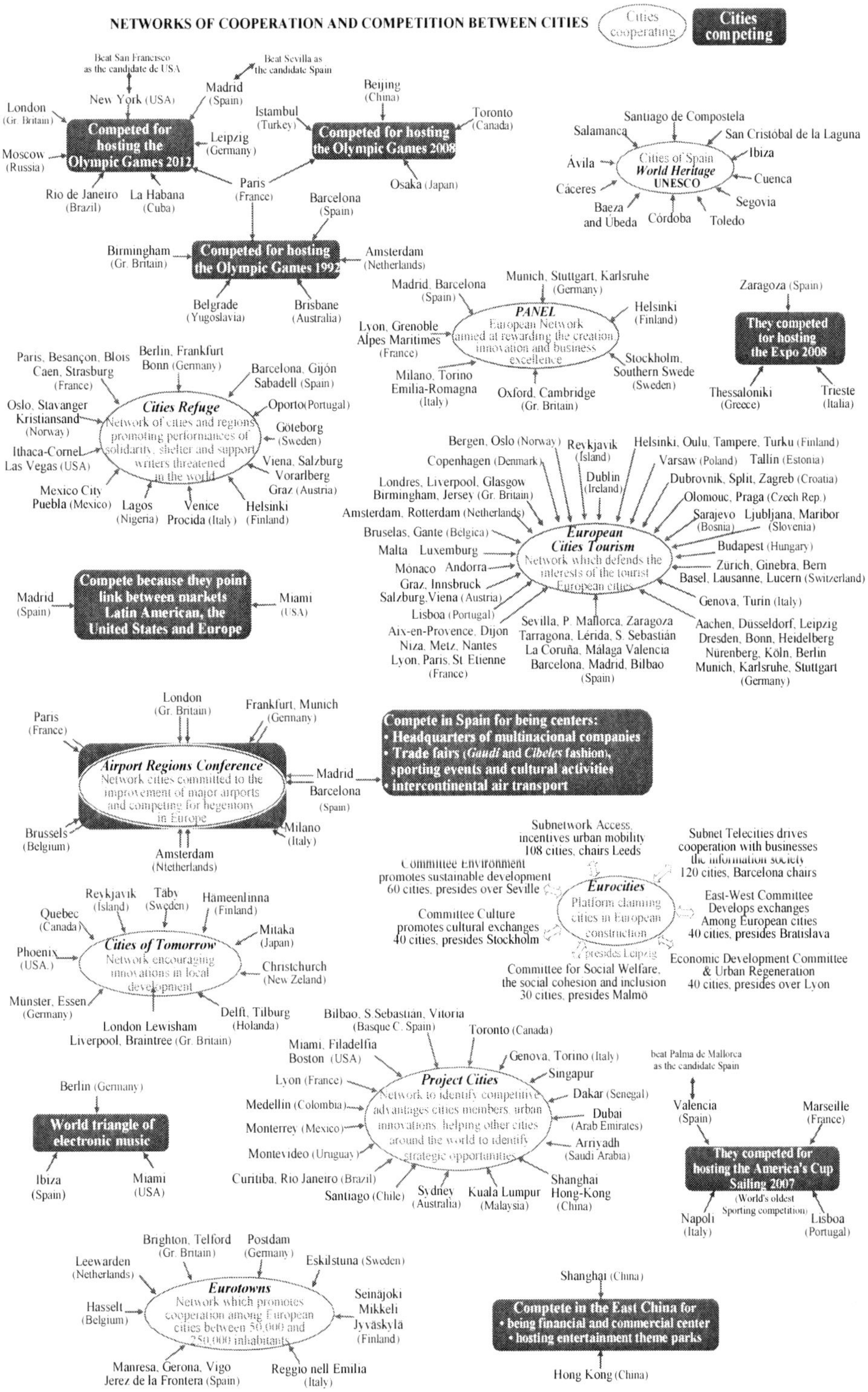

Alongside the dimension of co-operation, cities compete in many ways: to attract investment; to be the headquarters for institutions or businesses; to organize sports events like the Olympics, World Championships of a particular sport (especially football, but also basketball, athletics, swimming and others), various other sports competitions, or events not of a sporting nature, such as Universal Expositions; to be centers for economic activities (for example, the struggle between Hong-Kong and Shanghai in China to be financial and commercial centers in the Far East), for cultural activities (like the competition among the larger cities of Spain through their museums of contemporary art) or for mixed economic and cultural activities (like the competition between cities in the USA to host the *Grammy* music awards).

THE CONTRIBUTION OF MARKETING TO THE MANAGEMENT OF CITIES' IDENTITIES

Marketing is an area of knowledge within economics and business studies that has the basic philosophy of analyzing and managing the relationships of interchange that occur in markets between what firms have to offer and the demand from clients and consumers. Although originally marketing was applied to business sectors alone, its principles later began to be brought into play in non-business environments, non-profit organizations and politics. This was on the premise that in all cases there are relationships involving interchanges between one party that is offering something (products or brands, but also ideas or values) and another party that is seeking something or is likely to do so, allowing the principles of marketing to be extended to cover these contexts. Marketing of an urban nature would thus be a matter of cities developing an ability to offer what will match the demands of the various groups within them: citizens, institutions, tourists and visitors, investors.

The marketing of a city must start from its identity. Every city has its own identity: its name, geography, climate, history, culture, infrastructures, installations, monuments, and people. The great challenge for cities is to get different publics to see these elements as of value. From the point of view of marketing, this represents an opportunity, since these individual characteristics mark out and differentiate each city. Marketing can provide its ideas for improving the management of this identity, for meeting the needs and wants of the various groups. A city is what it is because of what politicians and municipal managers decide over the course of time, but also because of the physical and emotional space that it represents, because of the behavior of residents and firms, and because of what is said about the city, either by those who visit it or by those who do not. All of these contribute to building up and developing the identity of a city.

Towards the 1990s the principles of marketing were extended to the context of places and particularly of cities (Ashworth & Voogd, 1990; Kotler, Haider & Rein, 1994; Noisette & Vallerugo, 1996; Rosemberg, 2000; Elizagárate, 2003). This was because there are also relationships of interchange between what a city has to offer (a place to reside, quality of life, social and economic structure, work, infrastructures and installations, transport and communications, services, leisure and culture, education, events, and so forth) and what is sought by the various publics related to that city: citizens principally, but also tourists, institutions, enterprises and others. Transactional marketing tried to create a supply meeting the profile of the various publics or users; relational marketing attempts to develop stable and satisfactory relationships. Following the latest tendencies in collaborative marketing, cities should concentrate on collaborating with their publics and institutions, building up principles and values. In the modern socio-economic relationships that have been emerging since globalization, cities can also be understood as products competing

Table 2. Parallelisms between the management of the companies and the governance of the cities (Source: Adapted from Fernández Güell, 1997)

	Business companies	Cities or towns
Ownership	Owners, shareholders	Citizens and institutions
Top management	Boards of directors; executive management	City councils
Products	Manufactured goods or services	Public services, diverse offers
Clients	Consumers	Citizens, investors and visitors
Competitors	Other business companies	Other cities or towns

one against another to attract investments and visitors, or to organize events of various sorts. Managing a business and managing a city can have points in common (Table 2).

Despite the application of such a philosophy of business management and marketing, the governance of cities cannot be directed simply by transferring over these principles. This is because of the social and human function that should guide city management with priority over financial profitability. It is also because of the great complexity of the multiple interchanges that take place in a city. Cities have certain attributes that cannot be modified, such as their nature, climate, or geographic situation. They have others which are modifiable, relating to infrastructures, leisure, culture and education, health, economy, businesses and employment, public safety, government and politics, social atmosphere and the development of civil society. Aguilera and Perales (1994) conceive of the city as an open project, developed by public administrations, the private sector and civil society. This view is based on three criteria:

- **What the city has:** natural resources, such as landscapes, or attractive features of the climate, and created items, like green spaces, town planning, historic and artistic treasures, local dishes and so forth.
- **What the city does:** social, organizational and institutional relationships, economic, cultural and educational activities, artistic and sporting events, and the like.
- **What the city represents:** for a given national or cultural identity or for symbols of affiliation (national, iconography, history and legends, famous figures, celebrations and festivals, and similar).

The process of strategic management and marketing should culminate in communication of the urban strategies and policies, which would attempt to broadcast the image of the city among the various groups and markets, whether national or international. A plan for ways and means can be drawn up, which would include publicity, promotion at tourism or business fairs, public relations, and the like. Communication campaigns based on massive publicity reach out a long way and may produce more immediate results. Communication campaigns involving public relations are more direct and credible. Influential figures, such as artists and sportsmen from the city itself or other famous people, transmit a positive view of the city.

One of the principal ways of ensuring external projection and communication of cities is the organizing of events that bring international fame and transmit values. They also have positive effects for the citizens and institutions of the city, since they involve projecting values relating to the pride of belonging to a city or living there, to innovation and change. In a word, they generate collective enthusiasms. Spanish cities are active in this way. Barcelona, Vittoria, Gerona and Santiago de Compostela have received international awards. Besides this, a number of international events have been organized in Spanish cities. These include the World Swimming Championships in Madrid in 1986 and Barcelona in 2003; the Olympics in Barcelona 1992; the Expo in Seville 1992; European Cultural Capital status for Madrid 1992, Compostella 1995 and Salamanca in 2002; the World Athletics Championships in Seville in 1999; the America's Cup sailing contest in Valencia in 2007; and the Expo in Zaragoza in 2008. Marcos Blanco (1993) considers the study carried out for the Barcelona Olympics in 1992[1] to be one of the greatest large-scale applications of city marketing. For this, a great deal of energy was brought into play by citizens and businesses, together with an extensive communication campaign, well programmed in time, for which four types of sector were assigned as the target publics[2], and this led to the designation of Barcelona as the site for the Olympics.

The distribution policy of a city can be conceived of as how it makes available services and brings them closer to citizens and users. For its part, the price of the city product may be seen as the set of costs of using the services offered by the city. These comprise the cost of land for residential or industrial use, the price of services (social and municipal, of various sorts), transport, education and culture, health care, sports, leisure and tourism, and so forth. Some of these will depend on decisions taken at a municipal level, others on the market. Many cities already have cards allowing discounts for residents or tourists using public services, similar to what is also done by retail companies with their payment cards that are intended to generate customer loyalty.

The communication strategy may culminate in the creation and projection of a brand image of the city related directly to its identity. Just as businesses have brands, slogans, logos or emblems can be designed for a city or for an event organized by it. A good slogan or symbol may communicate the identity of an event or of the city itself and transmit values, whether these are current or those it is desired to project as an image.

EVOLUTION IN THE KEY DIMENSION OF CITIES: FROM INFRASTRUCTURES TO VALUES

Once cities reach high levels of development and have largely covered their essential needs in terms of infrastructure, installations, housing, water and power, health, education, transport and similar, they tend to move away from phases building up

Figure 6. Cards for using urban services

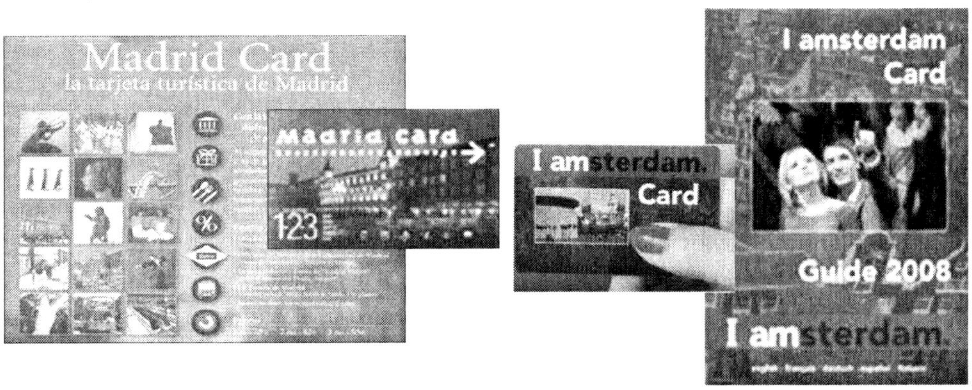

Figure 7. Logos of cities and urban events

projects for investment in such infrastructures, towards an emphasis on values emanating from civil society. This is an aspect of civilization that evolves from the *urbs* towards the *civitas*, as it would have been expressed in Latin. Although the necessary nature of infrastructures is recognized, as cities achieve high standards of installations, both now and in the future the consideration of values will be of growing importance. These include education and culture; quality of life and balance with nature; tolerance and social co-existence; generating energy and enthusiasm among citizens; creativity, creation of an image, differentiation and leadership with respect to a city's own identities, and even a brand image for a city, or, to sum up, an adequate marketing of intangibles and non-material aspects. Thus, while infrastructures are necessary, there comes a time when carrying out further public works will not always improve the quality of life. They can even become counterproductive when there is excessive construction to the detriment of the environment.

Some cases of networks of cities concentrating on values can be quote. One very interesting instance is the network called *Cittàslow* (literally *Slow City*; figure 3), which started with 26 Italian towns that attempted to unite economic progress and cutting-edge technology with a Mediterranean quality of life. They called for a quieter lifestyle than in big cities, one which would protect the environment and preserve the rich aesthetic heritage of Italian architecture. This would be done by encouraging skilled craftsmen and small shops, defending Mediterranean cookery traditions and exploring new ways of drawing benefit from tourism that did not involve massive numbers. This movement has now become a federation of more than 100 cities around the world, which according to the magazine *Newsweek* is laying the foundations for an urban transformation after a century marked by speed and acceleration in metropolitan life (Cueto, 2005). A further example involving city values is the grouping *Sustainable Cities* based on Aalborg in Denmark, subscribed to by 85 cities that are committed to environmentally friendly town planning. They encourage participation by citizens and social equality, plan for energy efficiencies and better public transport, and minimize the generation of waste and emissions of pollutants.

This evolution away from infrastructures towards values highlights a broader paradigm

Figure 8. Evolution and changes of the economy

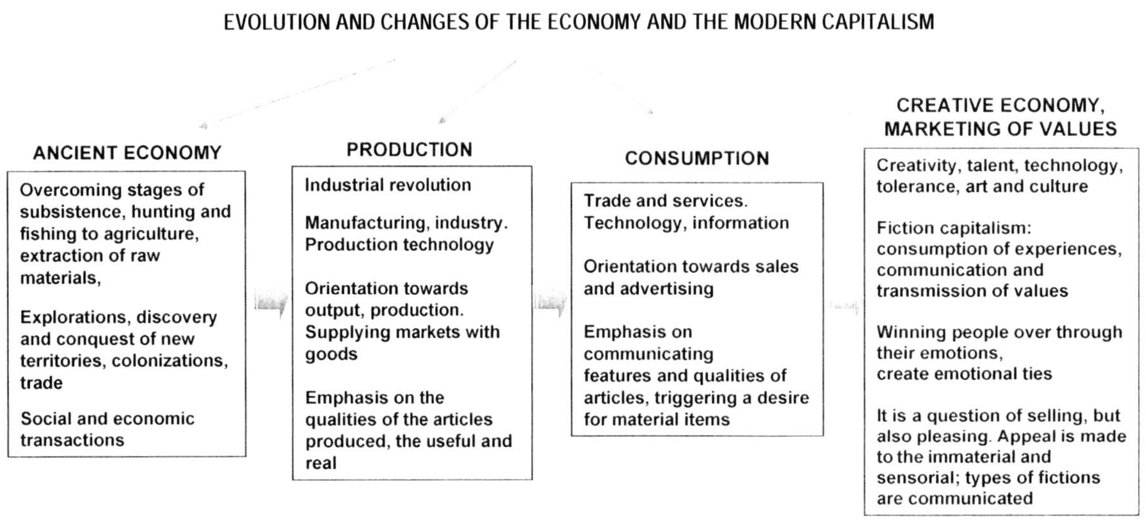

shift, in which the economy evolves away from the primary sector towards the tertiary and fourth phase sectors, this being within the framework of a change from an industrial economy to the information society (Castells, 1998). In effect, the economy has evolved with changes in people's lifestyles, with social developments and in relation to natural resources. In general, stages founded on basic activities like mining, agriculture and farming are termed the primary phase. After that industry and services develop, while in the tertiary stage it is the turn of technology and communications. Trade is present in all these phases.

There have been various interpretations of these changes; one of the most accurate may well be the views put forward by Verdú (2003, 2005). This author calls the final phase of the development fiction capitalism, following production capitalism and consumption capitalism. The first phase of capitalism was oriented towards the production of goods, industrial transformation into manufactured items, and supplying markets with utilitarian products. At a later stage trade and services predominate, with the thrust of marketing being aimed at sales; attempts are made to communicate the quality of products and encourage their consumption. According to Verdú (2003), the most modern stage should be called "fiction capitalism" because it is now no longer just a question of what businesses offer being able to be sold, but also of it pleasing people. An appeal is made to the emotional and the immaterial. This development is illustrated in Figure 8.

It may be asked what implications this socio-economic process has for the evolution of cities, conceived of as places to live and seen also in their relationships of exchange and trade with those who seek things from them: citizens, visitors, investors, institutions. Cities rose up as structures of refuge (Mitchel, 2007), and as sites where trade developed, then later services in modern economies, and throughout the history of urban settlements cities have been meeting points for culture in its various differing manifestations. Many historic cities have survived from ancient times to this day, while many others have disappeared or been transformed.

In the industrial period, a city was a place where industrial production took place; in it there was a predominant presence of capital and labor for production. Cities are where trade and services are sited; they commercialize the wish to consume. At the present day, cities are centers where there are multiple flows and networks of connections generating innovation and economic progress. In a sense, a city may be seen as an object of consumption in itself. It is no longer just the place where exchanges are carried out, but rather is itself an object and subject of them. Additionally, some cities become brands in their own right, objects for consumption to which people travel, where people wish to live, and which of-

Figure 9. Evolution of the economic conception of the city

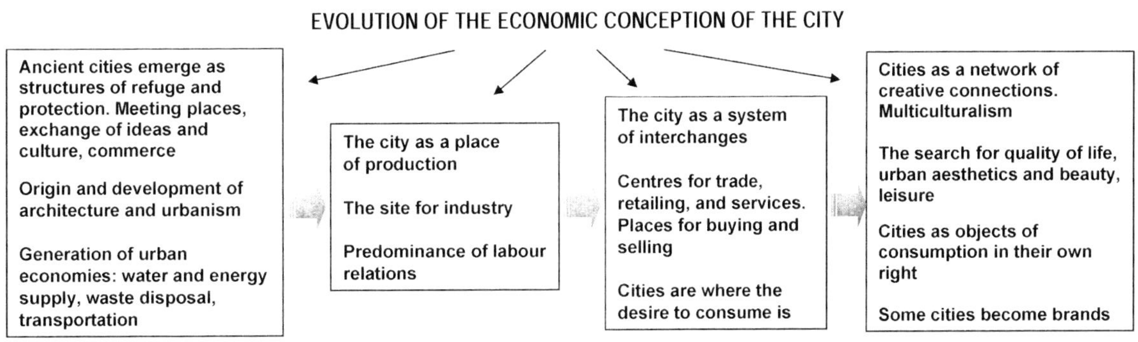

fer experiences or monuments that can become urban consumption icons. This development is shown in Figure 9.

One city that brings together infrastructures and values is Munich in Germany. In this capital of the region and of the Federal State or *Land* of Bavaria, there is a balance and harmony between industries at the cutting edge of technology (such as BMW or Audi nearby in Ingolstadt, and Siemens), institutions (like the Government of Bavaria, or the European Patent Office) and traditions and social life, with a good quality of life, sociability in the beer-gardens, natural surroundings of great beauty (Alpine lakes and mountains). It was the first large German city to be reconstructed as it had been before the Second World War.

Further examples of cities in the vanguard would include Curitiba in Brazil, which has set up participatory environmental management; Vancouver in Canada which is a model multicultural city, with a good quality of life and beautiful surroundings. Others would be Singapore, which is a multicultural Asian City-State, a financial and communications centre, with one the largest port in the world and a major airport serving Southeast Asia and Oceania; or Sydney in Australia, which presented itself through the 2000 Olympics as one of the best and most beautiful cities in the world. Barcelona is cited as a model of town planning; it brought back into use its sea waterfront thanks to the urban revitalization and promotion brought by the Olympics in 1992 and the Universal Forum of Cultures in 2004. The "modernist" or art-deco

Figure 10. Munich (Germany), a city that combines infrastructure and values

Figure 11. Evolution of the key urban dimensions

Infrastructures		Values
Regeneration and urban renewal	➡	To generare and develop social values
Projects equipment	➡	Processes of ideas and principles
Budgets and deadlines	➡	Excitement and collective energies
Management by few people (polititians, administrators, staff officials)	➡	More participatory management (consensus, cooperation, coexistence, commitment)
Planning a city	➡	Exchanges with other cities
Strategic metropolitan plan	➡	City networks

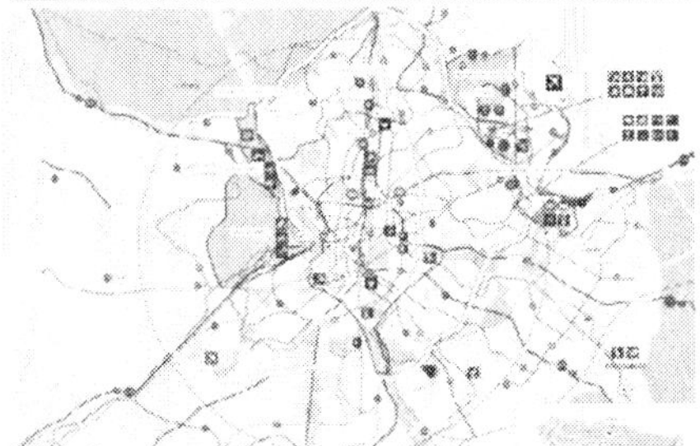

Infrastructures of transport in Madrid to the Olympic Games 2012

The nice welfare of the slow Italian cities.
26 populations join to claim the way of life Mediterranean

style associated with Gaudí lends enchantment to the city; it has been chosen as the best city in the world by a programme that selects the 20 best destinations worldwide under various headings (20 WORLD'S TWENTY BEST 2004).

The evolution of the key dimension of cities from infrastructure to securities is reflected in Figure 11.

If values are becoming the new key factors in the pattern of urban development, creativity, talent and tolerance stand out as generators of economic progress.

THE RISE OF THE CREATIVE ECONOMY. CREATIVE CITIES WORLDWIDE.

One of the most innovative works in the theory of economic development is *The Flight of the Creative Class: The New Global Competition for Talent*, by Richard Florida (2005). This author puts forward the view that the terms of worldwide competition are no longer grounded in natural resources, the direction taken by industrial production, or military domination, but rather that the decisive factors in the creation of wealth and economic progress will lie more and more in the ability of countries, regions and especially cities to generate and attract creative talent. It is on this that leading international economic positions will be based. For this author the keys are three features beginning with "T": *technology, talent* and *tolerance*.

This author argues that *technology* and *talent* have been recognized by economists as key factors in economic development, but that extra synergies in progress and wealth arise when these flows in movement have added to them creativity in places that offer conditions of *tolerance* toward social openness and human diversity. In cities where people of diverse origins come together, they form a hotbed that produces economic growth and technological innovation. When a city is a meeting point for a range of socio-cultural interchanges, arising from a diversity of human types, from people from different places, bringing varying cultural outlooks and lifestyles, if this wealth is brought together in an atmosphere of tolerance, then these forces will combine positively and generate creativity that translates into progress. Generally it is in multicultural cities where there are more sources of creative inspiration. These are places where differences are not just tolerated, but rather people are included by integrating them proactively. Florida states that various studies confirm that places that are open to emigrants, artists, the gay community, or bohemians of all sorts, and where there is racial and socio-economic integration, there is a stronger positive correlation with high-quality economic growth.

The approach adopted to confront problems, creativity and adaptability may vary between the natives of one country and those of others, so that, when they mingle, their contacts may bring about learning on both sides. Florida cites as an example the arrival in the United States of people as different as, for instance, an Italian stylist, a Mexican cook or a Russian dancer, who bring with them differing views on life. This variety adds richness, provides a range of shadings and produces synergies that increase the value of production in the economy. Indeed, the United States historically has welcomed people from many places, immigrant workers but also scientists such as Albert Einstein or Enrico Fermi who fled from Fascism in Europe, raising the standards of research and contributing decisively to the development of the atom bomb by that country. This author muses on the social evolution of the United States and the West as a whole. He states that the original American dream was to have a good job, educate one's children, save, buy a house and attain security. However, once the middle classes gradually achieved these targets in terms of work and material goods, there began to arise artistic movements or creations that symbolized a certain rebelliousness, freedom, attraction, even eroticism. Hence the appearance in the cinema of idols like James Dean (*Rebel without a Cause*), or Marlon Brando (*A Streetcar Named Desire)*, or, in music, *rock and roll* and *pop* with stars like Elvis Presley, the Beatles or the Rolling Stones. These are aspects showing the evolution in social motivations, from mere subsistence and seeking to earn money, towards aspirations to creativity and personal self-realization.

According to the author in question, enterprises have always looked for the best talents; the difference is that nowadays instead of bring these talents to their own bases, businesses are setting up facilities where the talent already exists. There is a small but growing number of highly skilled workers who first seek the city or urban area in which they want to reside and after that look for employment. Florida mentions the revitalization of SoHo in New York or the SoMa zone in San Francisco; areas which after suffering decline took on a new lease of life with the arrival of artists, immigrants and creative people settling into empty factory and warehouse buildings, followed by gay and unmarried folk, then later families, urban professionals (yuppies and similar categories); technological companies and small shops gave a final consolidating touch in these districts. There are firms that gain additional advantages from the fact that they are located in a given place, particularly when this implies attracting talent. It is not just a question of creating employment: people seek work but also keep in mind in what location this work is situated. Successful places take the shape of complex, multisided ecosystems, like those to be found in the natural world. Among them are cities in environments that do not just provide work but offer the crucial conditions for residing with a high quality of life: the ability to find like-minded people and friends, pleasant surroundings, in some cases being in areas of great beauty. These are attractions that permit people to a build a life of the style they wish, taking all these features as a whole.

Economic and business thinking and strategic analyses have gradually evolved. In general the dominant line of thought in the 1980s was to investigate inside companies, gauging strengths and capacities, while studies of nations and regions concentrated essentially on national economic policy. After having introduced the concepts of *strategies* and *competitive advantages* in the 1980s, Michael Porter concluded in the 1990s that a large portion of the competitive advantages enjoyed by firms did not lie within them, or even necessarily in their sector; but was to be found in their geographical sitting or location: nations and regions, but especially cities and urban areas. In other words, a considerable part of a business's chances of success may depend on whether it is located in a particular city or region.

This is a wide-ranging and dynamic theory of the enterprise. It is wide-ranging through not being restricted to the internal features of enterprises and dynamic because it can evolve over time. The globalization of the economy has contributed decisively to expand the internal zone of businesses by increasing the availability of production factors to a larger number of countries. This focuses the capacity for improvement not so much on internal processes or on vertical integration within the firm as on the quality of the economic environment. In certain geographical locations, notably cities and metropolitan zones, accumulations or clusters arise, geographically dense groupings of enterprises and institutions with shared characteristics, networks of creative connections. Porter states that the majority of the members of these clusters or networks of businesses do not compete directly with one another, but target different sectors, in accordance with the strategic groupings that form, and hence share many common characteristics. Geographical proximity brings them closer to production factors like access to suppliers, technology, information and personal communication with other enterprises and institutions. In this way relationships of shared interest and consciousness arise, networks of civic commitment, of capital and social adhesion. A collective reputation grows up and joint marketing can be brought into operation. This was how the technological vanguard formed in the famous Silicon Valley in California and in Austin in Texas. A similar process led to the concentration of the chemical industry in south-western Germany around Freiburg, with its university dating back to the Middle Ages, and in the northern cantons of Switzerland near Basel; the high positioning car industry in Stuttgart and Munich, or the cinema in Hollywood.

While huge countries like China or India are certainly going to become great political or economic powers, if considered from an aggregate or macroeconomic standpoint, their advantages will come more from a capacity for low-cost industrial production than from leadership in core managerial skills. According to R. Florida, world economic leadership will lie in the hands of small European countries, such Finland, Sweden, Denmark, Holland and Ireland, or English-speaking countries like Canada, Australia or New Zealand, which are able to combine a capacity for technological excellence with the ability to attract creative talent from round the world. Hence, although the large countries like India and China are building up technologically advanced cities such as Bangalore in India or great commercial and financial metropolises such as Shanghai in China, as are other emergent states such as Brazil, with its economic centre in Sao Paulo, the true vanguard will be found in the leading-edge cities in the former countries. These will be cities such as Helsinki in Finland, with Nokia leading research into telecommunications, Dublin, concentrating the economic and technological growth of Ireland, Vancouver as a Canadian city that is multicultural and offers a good quality of life, Sydney or Melbourne as the front runners in Australia, or Florianopolis or Curitiba as improved alternatives to the sprawling Sao Paulo or Rio de Janeiro in Brazil. They may also include Singapore and Kuala Lumpur in Malaysia as multicultural Asian cities, centers for trade and finance, but of a more balanced size than the overgrown and chaotic Bombay or Calcutta in India, Karachi in Pakistan, Jakarta in Indonesia, Manila in the Philippines, or Bangkok in Thailand.

The true cause for the flourishing of these cities is their capacity to combine technology with talent and tolerance. According to Florida it is possible to identify in the United States cities that bring all three together. There is San Francisco and its nexus of San Jose, Silicon Valley and the universities at Stanford or Berkeley; Boston with Harvard or the Massachusetts Institute of Technology (MIT); or Seattle con Microsoft and computing, and Boeing the aeronautics industry. All three are open, multicultural, tolerant cities. Cities with one or two of these three variables would include Pittsburgh or Cleveland, which have excellent universities with a good research record, but have little attraction for people of diverse sorts; or Miami, which attracts immigrants, especially South Americans, and is a showcase for a lifestyle related to sunshine and good weather in beautiful coastal surroundings, but has little technological capacity and few people involved in this field and in research.

At a world level, it is possible to distinguish, on the one hand, Asian cities which are becoming outstanding technologically but have relatively little diversity and tolerance. Examples are Bangalore in India, Seoul in South Korea, Taipei in Taiwan, Shanghai in China and Tel-Aviv in Israel with its geopolitical conflicts. Only Singapore would escape this trend. It is a city-state and former British colony where communities of Chinese, Indian and Malay origin live, together with westerners. Kuala Lumpur, the capital of Malaysia, also presents a diversity of cultures, races and religions in an Asian context. On the other hand, mention may be made of avant-garde, multicultural cities, basically English-speaking, in countries formerly colonized by Great Britain, which offer good levels of tolerance and security, without the racial conflicts and inequalities of many cities in the United States. These include

Figure 12. Key dimensions of influence and attraction of the metropolitan cities

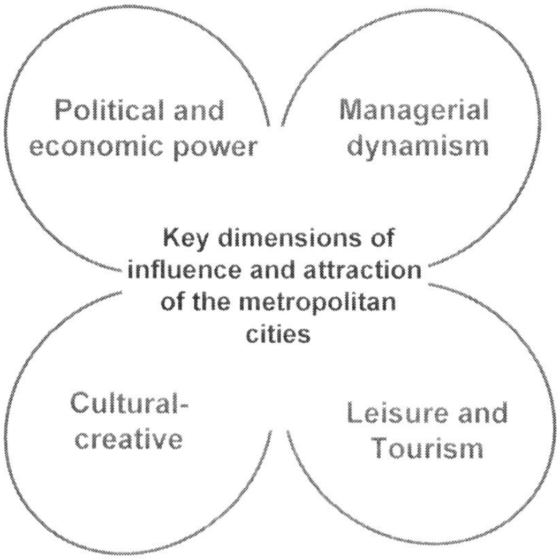

cities with surroundings of natural beauty: Sydney and Melbourne in Australia, Vancouver and Toronto in Canada, Wellington and Auckland in New Zealand, Cape Town in South Africa. Florida explains the curious case of Wellington and how it was the place where the *Lord of the Rings* trilogy was filmed, attracting many people related to the world of the cinema, its special effects, graphic design, and similar. This tendency was strengthened later when the director Peter Jackson brought the production of the new version of *King Kong* to the city. All this encourages the development of creative industries, just as the saga of *Star Wars* by George Lucas gave an impulse to the development of videogames and cinema marketing in California.

The new cities that mark trends in the world are setting their influence and marketing through the following dimensions (International Conference of Metropolitan Capital Regions; Madrid, 2006).

The creative cultural dimension is based on elements of intangible culture, with large interrelationships, as can be seen in Figure 13.

Figure 13. Elements of intangible culture that give international reputation

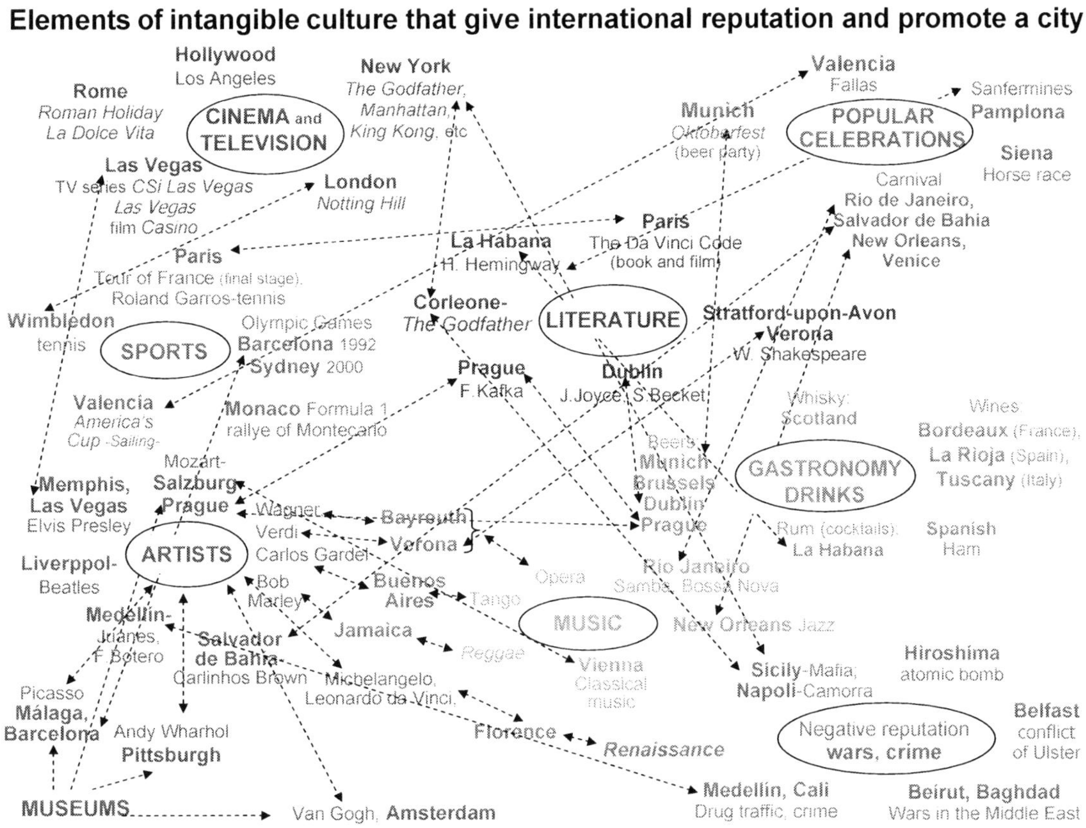

CONCLUSION

It is in cities where the multiple interchanges of the new world socio-economic order flow together. At the same time they are becoming subjects or products for consumption, brand-names with identities and values. To manage the complexity arising from modern relations between what urban areas can supply and the various demands made on them by people and institutions, cities are increasingly adopting marketing approaches. The progress of cities has moved away from the economic aspects of industry and infrastructures towards an emphasis on values, creativity and intangible cultural elements.

Recently successful cities around the world tend to be of a moderate and balanced size, bringing together economic development and quality of life, proactively integrating people of different sorts, accepting a multicultural nature as a treasure leading to cultural and artistic creativity, which translates into social and economic progress. They are cities in which urban life harmonizes with the natural surroundings: indeed, many are set in areas of great beauty. They care about town planning, and urban design and aesthetics.

Marketing as a management philosophy can aid in becoming aware of and orientating the needs of citizens, tourists, and sources of investment. It can develop offers adapted to the needs of these groups and ensure adequate promotion and communication. Marketing in the broader sense can also provide those institutions related to cities, such as councils, businesses, or states, with a conceptual framework, bringing together approaches from various fields, like town planning, sociology, environmental protection and culture, such as to identify the competitive advantages of a city. It can seek out creative ways of differentiating it, create values of tolerance, multicultural integration, balance with nature, and excellence in physical configuration and urban landscapes. At the same time that they need to be better places to live, cities are becoming products competing among themselves, brands and items for consumption in accordance with their identities.

REFERENCES

Aguilera, M., & Perales, A. (1994, March). La Imagen de las ciudades en el marketing urbano. *Marketing y Ventas*, *79*, 10-14.

Ashworth, G., & Vgoodg, H. (1990). *Selling the city. Marketing approaches in public sector urban planning.* London: Bellhaven Press.

Brandenburger, A. M., & Nalebuff, B. J. (1996). *Coopetition.* New York, NY: Currency Doubleday.

Castells, M. (1998). *The information age: Economy, society and culture, I, & The rise of the network society, II.* Cambridge, MA: Blackwll Publishers Inc.

Chias, J. (2005). El negocio de la felicidad. *Desarrollo y marketing turístico de países, regiones, ciudades y lugares.* Madrid: Prentice-Hall.

Cueto, J. (2005, January 23). Ciudades lentas. *El País Semanal*, *1478*, 8.

De Elizagárate, V. (2003). *Marketing de ciudades.* Madrid: ESIC Pirámide.

Fernández Güell, J. M. (1997). *Planificación estratégica de ciudades.* Barcelona: Gustavo Gili.

Florida, R. (2005). The flight of the creative class. *The new global competition for talent.* New York, NY: Harper Collins Business.

Font, J. (2003). *Competencia y colaboración de ciudades: La aportación del benchmarking.* Retrieved July 10, 2008, from http://www.rosario.gov.ar/per/Acti/Ponen2.html

Kapferer, J-N. (1991). *La marca, capital de la empresa.* Bilbao: Deusto.

Kotler, P., Haider, D. H., & Rein, I. (1994). *Marketing places.* New York, NY: Free Press.

Lambi, J.-J. (1995). *Marketing estratégico.* Madrid: McGraw-Hill.

Metrex (2006, May). *International Conference on Planning and Sustainable Development of Metropolitan Capital Regions*, Madrid.

Mitchell, W. (2007). *Ciudades inteligentes.* UOC Papers, num. 5. Retrieved July 10, 2008, from http://www.uoc.edu/uocpapers/5/dt/esp/mitchell.pdf.

MOMA –Museum of Modern Arts- New York (2006). On Site; new arquitecture in Spain.

Newsweek (2006, July 3). *The world's hottest cities, Special Issue*, (pp. 48-79).

Noisette, P., & Vallerugo, F. (1996). *Marketing des villes.* Paris: ESSIC.

Porter, M. (1999). *On competition.* Boston, MA: Harvard Business School Press.

Rosemberg, M. (2000). *Le marketing urbain en question.* Paris: Anthropos.

Vegara, A., & De Las Rivas, J.L. (2004). *Territorios inteligentes.* Madrid: Fundación Metrópoli.

Verdú, V. (2001, April 13). La sexy-ciudad. El País, (p. 20).

Verdú, V. (2003). *El estilo del mundo. La vida en el capitalismo de ficción.* Barcelona: Anagrama.

Verdú, V. (2005). *Yo y tú, objetos de lujo. El personismo: La primera revolución cultural del siglo XXI.* Barcelona: Debate.

Zaera-Polo, A. (2004, September 10). La era urbana. El País, special issue on the World Urban Forum, (p. 12).

20 World's Twenty Best (2004). Television program that selects the top 20 destinations in the World in several sections, including the top 20 cities. RDF Media production for channel broadcasting Ltd.

ENDNOTES

1 Barcelona gained international exposure through the 1992 Olympics. This city is a host to trade fairs, congresses, sports events and urban tourism, as well as being a major stopping point for European and transatlantic cruises (in the Western Mediterranean especially after the 11 September 2001 terrorist attacks). The next major project by Barcelona was the *Universal Forum of Cultures* in 2004, involving further world-wide promotion of the city, and remodelling of urban areas including prolongation of the renovated seafront beyond what had already been done in 1992.

2 Target Sector I "Extraordinarily Important": members of the International Olympic Committee; Target Sector II "Important": members of national Olympic committees, international federations, sports federations for Olympic sports, specialist international press, world and national opinion leaders, Olympic athletes; Target Sector III "Further Interest": ex-Olympic sportspeople, participants in sports, sports clubs, artists, national and local press; Target Sector IV "Associates and participants": professional organizations, businesses, students, citizens of Barcelona, Catalonia and Spain.

Chapter II
City Brands and their Communication through Web Sites:
Identification of Problems and Proposals for Improvement

José Fernández-Cavia
Universitat Pompeu Fabra, Spain

Assumpció Huertas-Roig
Universitat Rovira i Virgili, Spain

ABSTRACT

City marketing tries to position cities in the mind of the public, although the process of creating and communicating city brands is still at an early stage of its development. One of the main tools for the communication of these brands is now the World Wide Web. This chapter describes the results of two combined studies (qualitative and quantitative) that analyzes a sample of official city Web sites. The results show that official Web sites of cities give much attention to ease of navigation, but interactivity is much less implemented, especially between users. Furthermore, some lack of attention to the communication aspects of city brands can also be found. Finally, the chapter submits a number of improvement proposals.

INTRODUCTION

In the current world of cities, competition has increased and the centre of interest has moved to include much broader spheres. Already, cities do not try only to be just significant tourist nuclei, but they also compete in aspects such as quality of life, economic development and sustainability. Aside from tourist interest, cities try to position themselves as comfortable areas to live and important centers of economic development that attract all types of investment.

To achieve this, it is necessary to know the opinions and evaluations of the publics (Prebensen, 2007)[1], to find out what image they have of the city and determine the positioning that it would be desirable to achieve. Therefore, a vision of the city must be formulated and, consequently, a program of identity must be created that is transmitted through a brand and a visual logo and, later, an adequate and effective communication program must be run. With this objective, citymarketing tries to position cities in the minds of the public. In spite of this, the process of creating and disseminating city brands is still in a very early stage of its development.

This chapter tries to show, on one hand, that the concept and application of city brands are still very incipient and, on the other hand, that the official websites, in part due to city brands being underdeveloped, do not pay enough attention to the dissemination of the graphic, functional, and emotional aspects of the brand. With this, there is still a long way to go in the dissemination of cities through their brands on the Internet.

THE CONCEPT OF CITY BRAND

The concept of brand applied to destinations, places or cities is relatively new. It started to spread with the Travel and Tourism Research Association's Annual Conference in 1998 (Blain, Levy & Brent Ritchie, 2005). From that moment, the concept has been developed widely and has been studied from diverse perspectives, especially from the point of view of tourism. All in all, studies on city brands and destinations are still under developed and knowledge on the subject is limited. Some authors consider that it's not correct to talk about branding or place branding in relation to territories, cities or countries. They believe that it is incorrect to associate communicational and marketing terms to realities with their own identity like cities. Nevertheless, the majority or authors appreciate that the territories and the cities do not have the same characteristics as commercial products, but agree that they can apply the same marketing strategies to the territories (Olins, 2002).

The first difficulty we come up against in the study of city brands is the confusion of concepts. Therefore, it is fundamental to distinguish between city brand and brand image (Cai, 2002). Many studies confuse the analysis of the brand image with the brand itself and the branding done by the destination.

The city brand is a new concept and is not very well defined. It is currently very much in fashion, and many people have theories, but few have dared to define it. It is a construct composed of a name, a logo, some symbols and some values that we try to associate with a city, representing its identity, with the objective of creating a position and a vision of the city in the minds of the public. Each city must have its own brand, and each city brand must be the result of a citymarketing plan and a competitive city strategy.

A very complete definition of destination brand, fully applicable to the city brand, which is based on the previous definitions of Aaker (1991) and Ritchie and Ritchie (1998) is that of Blain, Levy and Brent Ritchie (2005), which implies:

The creation of a name, symbol, logo, word mark or other graphic that both identify and differentiate a destination; that convey the promise of a memorable travel experience that is uniquely associated with the destination; and that serve to consolidate and reinforce the emotional connection between the visitor and the destination; that reduce consumer search costs and perceived risk; all with the intent purpose of creating a destination image that positively influences consumer destination choice. (p. 337)

However, the brand image is the result of the branding process, which is the perception created in the minds of individuals. It is "*networks of knowledge elements stored in long-term memory, and the core of such a network is the brand name which is linked to a number of other knowledge elements and/or associations*" (Riezebos, 2003).

As confirmed by Bill Baker:[2] "*A destination without a clear and attractive brand image is like a person without a personality. They blend into the crowd, are seen as uninteresting, and don't get the attention they deserve*".

As consequence of this dichotomy between city brand and brand image, there are two types of studies on the topic. On one hand, those that analyze the brand names themselves, their names, symbols, logos, their identification, purposes and meanings. Our research is included within this first type. And on the other, those that are based on the associations and relationships that the brand names create with the public, that is, the brand image that is created among consumers.

We also consider it is important to distinguish another conceptual aspect. Place branding is not the same as destination branding. Place branding is based on the construction of a global image of the territory that promotes the place in its globality: economical, touristic and as a place of residence. Even in some occasions, tourist branding can be contradictory with place branding. This is why we consider it necessary to give a definition of place branding. It is the sum of beliefs and impressions people hold about places. Images represent a simplification of a large number of associations and pieces of information connected with a place. They are a product of the mind trying to process and pick out essential information from huge amounts of data about a place (Kotler, Haider & Rein, 1993).

Having clarified these concepts, we will concentrate on the city brand, which is the purpose of our analysis. This, like all brands, has its raison d'être, which is based on two basic functions (Aaker, 1991). One of these is the identification of the brand with the town and the attribution of a symbology and some values to the destination. The cities must have new signs of identity, an image and a position. Therefore, the first function of the brand is to attribute functional and emotional values to a city that identify the different cities globally and by consensus.

The second function of the brand is based on differentiating the cities from each other. This has always been the principal mission of all brands. According to the American Marketing Association: "the brand is a name, term, design, symbol, or any other feature that identifies one seller's good or service as distinct from those of other sellers".

Blain, Levy and Brent Ritchie (2005) add three concrete purposes of city brands to the two classics stated previously by Aaker (1991). On one hand, to give the visitors the security of a quality experience at the destination. On the other, to reduce the search costs on the part of the visitors and, finally, to offer a single purchase proposal. But we must be conscious of the difficulties of creating city brands, and the limitations that still exist with respect to their functions.

Moreover, Hankinson (2004) uses the concept of brand networks, in which the destination and city brands have four functions: brands as communicators that represent a differentiation between cities, brands as perceptual entities that appeal to the senses and emotions, brands as values, and brands as relationships.

Having reviewed the theoretical framework of destination and city brands, and defined the concept of city brand, from which the study starts, we will now make a simple classification. As we understand, city brands can be classified according to their degree of evolution and development in the following categories:

1. The graphic brand, which only implies the creation of a symbol and a logo. Logos are the basic element for the creation of a brand and the main vehicle for communicating an image.[3]
2. The functional conceptual brand. This type of brand adds the symbolization of some of the territory's characteristics to the logo,

which are real and tangible, and which are to be promoted, being adopted as strong points of the city. These attributes may be: good climate, beaches, nightlife, quality of life and, level of innovation.

3. The emotional conceptual brand, created by a body, entity or public institution, also trying to transmit abstract, symbolic and personifiable values to the city, such as innovation, multiculturality, modernization, passion, etc. With this combination of values, the aim is to position and distinguish the image of the city from the competition. Previous studies (Ekinci & Hosany, 2006; Hosany, Ekinci & Uysal, 2006)[4] have shown that the emotional and personifiable values of the destination brands have positive influences in the prior choices of purchase and recommendation of these destinations.

Various authors agree that the brand image of a destination has two basic dimensions (Lawson & Band-Bovy, 1977): cognitive and affective, which would correspond with the emotional and functional conceptual brand. The cognitive component would be made up of the beliefs and knowledge of the physical attributes of a city, the functional conceptual brand; meanwhile, the affective component would refer to the feelings about these attributes, the emotional conceptual brand (Baloglu & McCleary, 1999). From the perspective of understanding the brand image as a "cluster of attributes and associations that the consumers connect to a brand", Biel (1997) understands the existence of "hard" associations, which refer to the tangible and functional attributes, and "soft" associations, emotional attributes. Biel acknowledges that the personality of the brand belongs to the emotional aspect of the brand image. Along the same lines, authors such as Etchner and Brent Ritchie (1991), Kapferer (1997) or De Chernatory and Dall' Olmo Riley (1997) confirm that the brand and its image are composed of two attributes: the functional or tangible, and the symbolic or intangible[5].

However, the emotional conceptual brand must be agreed upon by consensus, created jointly by public and private institutions and citizens of the region, which involves both the internal and external public, and is not identified with or property of a single institution, but of the whole region.

The valid significance of a brand is that registered by its public. Certainly, a brand must be adopted by all the public, starting with the city's own residents, companies and institutions; and for this purpose it is fundamental that these are involved in its creation, that they adopt it as their own and thus help with its dissemination. A study by Blain, Levy and Brent Ritchie (2005), based on interviews with the heads of marketing of the destinations, showed that the opinion of the residents and the visitors must be fundamental in the process of creating city brands.

So, in light of the classification above, an existing city brand may be understood as more or less developed according to its degree of preparation. Some city brands are simply logos that do not represent any specific aspect of the destination, while others have elaborate brands agreed by consensus, which come from a prior marketing plan and represent functional and emotional values that can be identified with the city.

QUALITATIVE ANALYSIS OF CITY BRANDS

Having stated the definition, functions and classification of city brands, we performed an initial qualitative study on eight brands of important world cities on their official websites. The study showed the limitations of city brands in their current process of creation, implementation and dissemination.

The objective of the investigation centered on finding out the degree of evolution and development of a sample of city brands, and their dissemination through the websites of their official institutions. The analysis consisted of two stages. The first centered on the prior examination of the city brand itself and its degree of development. The method used in this part of the study was based on the classification of the degree of advance preparation and evolution. In other words, in the analysis of the graphics and logo, the functional and emotional values assigned and the knowledge of the moment of creation, motivation and agents involved in the process of creating the city brand.

The second phase of the investigation centered on the study of how the city brands were treated on the official websites of the destinations. The analytical method used was the BIWAM (Brand Identity Web Analysis Method). This is a technique for qualitative analysis of the establishment of brands on the web, created by Martín Barbero and Sandulli (2005), which includes eight dimensions of analysis, of which we applied six to our study of city brands:

1. Analysis of the Appearance, which corresponds to the strong, real and objective points of the destination (the functional element of the brand) and how it is communicated on the web.
2. Analysis of the Personality. This implies the assessment of the symbolic and emotional elements that are attributed to city brands and their treatment on the websites.
3. Analysis of Humanity. This refers to the interactivity of the page.
4. Analysis of the Style. This analyses the graphic part of the brand, specifically, the relationship of the logo with the colors and the typography of the website.
5. Analysis of the Medium, and the communicative functionality of the websites.
6. Analysis of the Credibility. This refers to errors, slow loading speed, internal coherence, etc.

We only applied six dimensions of analysis created by Martín Barbero and Sandulli (2005) because the other two could be applied to product brands, but no to city brands.

The sample consisted of eight city brands that correspond to international tourism capitals, which have created their city brands, but show different degrees of evolution according to the typology stated above. The city brands selected for the sample were: Barcelona, Madrid, Edinburgh, Amsterdam, Cincinnati, Toronto, Dubai and Hong Kong.

RESULTS OF THE QUALITATIVE STUDY: EVOLUTION AND LIMITATIONS OF CITY BRANDS

The results of the study showed that city brands are a concept of recent creation and still incipient development. Actually, the majority of brands analyzed in the study were created in 2005. Many large world cities still have not created their city brands. And among those that have, the degree of evolution of their brands is still mostly in the initial stages. The vast majority are stuck in the stage of creating logos based on strong points or characteristics of the city that they wish to boost. However, very few cities try to identify themselves with an emotional conceptual brand, based on the appropriation of personifiable values and the creation of a city marketing strategy.

Our results coincide with those of a study by ESADE (2004) on the evolution of the positioning of Spanish tourist destinations and their tourism brands. This stated that the brands, as strategic realities of tourist destinations, are usually fairly general, based on functional values without dealing with emotional aspects. They do not segment their range much and are only transmitted externally, forgetting the internal public, and do not evolve with the passage of time.

Other studies in tourist marketing have shown that, in general, the application of marketing techniques in destinations is still scarcely developed (Gnoth, 1998; Pritchard & Morgan, 2002).

The causes of the lack of evolution of city brands are related to a series of limitations the brands experience in their implementation. The first consists of the complexity of combining a segmentation strategy with the creation of a single brand image. The cities are directed at diverse sectors of the public (citizens, investors, businesspersons, tourists) with whom they wish to communicate. As each sector of the public has certain interests and certain needs, the cities generally create different marketing strategies for each of them. For this reason, it may be complicated, or even contradictory, to create different strategies and integrate them into one single brand positioning.

Another difficulty is found in the existence of more than one brand per city. It often occurs that different institutions create city brands for their websites or for independent use. The result is dispersion, incoherence and the impossibility of creating a single image that is recognizable and adopted throughout the community. For example, the City Council of Madrid has a municipal website (munimadrid) with a brand and a logo that has nothing to do with the Madrid brand on the municipal tourism portal (esmadrid), or with the website of the region (turismomadrid). It is easy to understand that it is absolutely essential to make the effort to coordinate and negotiate when creating a single brand that is not property of one institution in particular, but of all the citizens, and applicable to all the websites related to a destination.

With respect to the second function of the brand, that of differentiating cities from each other, it is possible that different cities try to identify themselves with the same values. The research that we have performed shows that the emotional conceptual brand, that which attributes some personifiable values to a city with the purpose of differentiating it from other cities, is usually created in a very broad and ambiguous way, and this does not fulfill its differentiating function. The majority of cities do not identify with a single value, but with many, some of which are shared by different brands, which, in fact, encourages confusion.

Kotler (1993) already stated that all images of a destination must be simple and distinctive. The main function of a brand must be to differentiate it with respect to the competition. However, in the study it was demonstrated that the majority of cities prefer to be identified with diverse values or characteristics at the same time, perhaps to be attractive to more sectors of the public, maybe because they are values that are currently very attractive in society and which they do not want to renounce although they have been adopted by other destinations. This is shown in the eight city brands analyzed, Amsterdam and Toronto identify themselves as creative; Barcelona and Dubai, adventurous; Barcelona and Edinburgh, friendly; Edinburgh and Toronto, imaginative; Barcelona, Amsterdam and Edinburgh, diverse; Hong Kong, Barcelona and Amsterdam, cosmopolitan; and finally, Barcelona, Amsterdam, Edinburgh and Toronto, innovative. In this sense, the city brand completely loses its distinctive or differentiating function and this limits the creation of a single image for every destination. Similarly, Morgan, Pritchard and Piggott (2002) also showed that the images created of destinations and cities are not different and do not usually contain a single idea or single purchase proposition.

Our research also revealed that the majority of city brands have been created according to a specific event. For example, Madrid's brand was created to promote the destination internationally when the city was selected as a possible candidate for the Olympic games of 2012. Barcelona, however, created its brand to promote the Forum 2004, and Cincinnati, after the results of an economic study, took notice of the need to connect the three States of Ohio, Kentucky and Indiana, and as a result, created its brand with this unifying objective. This fact may also be negative for city brands, which should never be associated with a specific political event, as they must represent a city and not a municipal government, a social or sporting event or any private interest.

The motive that generally moves public bodies to create city brands is mainly based on tourist or economic interests, and thus the brand created usually is identified with these interests. This is an error, a reductionist conception that limits the potential of the city brand. In addition, on occasions, after an election or change of party in the town halls, the city brands and their representation and meanings are changed, precisely with the intention of breaking away from the previous image and promoting a new improved image related to the political party that has formed the municipal government. These changes create dysfunction, as the brands need time to be implemented and require their evolution to be homogenous and coherent. The persistence and durability of a brand is key for its implementation and acceptance by all sectors of the public. Therefore, brand changes only create more confusion in the identities and images of the cities.

Finally, but importantly, difficulty in the creation of city brands lies in coordination, taking into account the umbrella brands of destinations greater than the cities. In tourism, the broader destinations (regions, nations, states or countries) include those within them in their brand for the promotion of tourism. So this presents us with a number of questions: Should the umbrella brands take into account and be coherent with the city brands that they include? Should they all be related? Should they be coherent with each other? Should they have common features? Should the attributes of the umbrella brands be shared by the brands of the respective cities that are represented?

TREATMENT OF CITY BRANDS ON THE WEB

The Internet and new information technologies play a key role in communicating the cities and their brands. They are an important source of information. Destination Management Systems are more than simple websites. In addition to the information, they offer advertising, marketing and sales applications, and have interactive resources that, in an entertaining way, provide services and attract the attention of the users.

Currently, through a city's tourism website, you can get information, make reservations, etc. However, in the promotion of the cities, not only as tourist destinations, but also as business centers and residential areas, portals or broader websites are starting to be created, which offer, in addition to tourist information, business and leisure information for the citizens.

In this sense, and from the field of communication, the brand websites have been marked as the future of marketing communication on the Internet, as they have the potential to provide high levels of information and, in addition, create virtual product experiences (Klein, 2003). Brand websites are capable of combining both of the basic objectives of commercial communication in this channel: to create a brand image and achieve a direct response (Hollis, 2005). As Cho and Cheon (2005) describe, the websites may serve for diverse communication purposes: public relations, sales promotion, advertising or direct marketing.

In the second part of the study, the results of our analysis of city brands on the web showed that the aspects that make up the corporate image, that is, the colors, the lines and the logo of the city brand, in general, are used very little to create graphic coherence and brand image throughout the website. There are some exceptions, such as the sites of Amsterdam or Madrid, that show effective graphic coherence for transmitting the brand, but habitually the typography and the colors of the logo are only used in auxiliary hyperlinked pages, but not the whole official site.

The study showed that what is best transmitted through the web is the functional conceptual brand of the cities, the strong points or the potential that they wish to promote; but in no way the emotional conceptual brand, which ascribes personifiable values to the destination. The exception is a hyperlink that some official websites have, such as Edinburgh or Amsterdam, which links to a page exclusively dedicated to explaining the emotional brand.

Starting with the results of the studies of Hosany, Ekinci and Uysal (2006), where it is demonstrated that the emotional and personifiable values have positive influences on the intention to visit, purchase and recommend destinations on the part of consumers, it is surprising that it is actually the emotional aspect and the personality of the brand that are the least developed aspects on the official websites of the cities analyzed. The heads of marketing of the cities should develop strategies and campaigns that promote the distinctive personality of the destinations, based on the emotional components of these cities, which create better positioning and a more favorable image among users.

Thus, in conclusion, the under use of websites to promote city brands has been verified. The websites centre on the functional conceptual aspects of the brand, that is, all the strong points that are notable in the city, but lack, in general, coherent graphic treatment and the expression of the emotional conceptual brand. The design of the websites tries to be useful and functional to provide the users with the information and services they wish to obtain, but they are not at all creative to disseminate the emotional values attributable to the city through its brand. McMillan (2004) coincides with our statements, arguing that advertising on the Internet and websites must be more creative, that is, better designed, with greater impact, more varied and more entertaining.

Once at this point, and in light of other studies and a bibliography centered on more technical questions of the websites, we decided to analyze other characteristic aspects of the websites that also influence the dissemination and perception of city brands. We considered that dealing with city brands should not be limited to an analysis of the websites' content only, but should take into account aspects such as the interactivity or usability of their pages.

Different research on websites shows that usability is a key aspect in the creation of a good brand image. The sites that seem to be or are easier to open, navigate or use, create a more favorable attitude and image among users (Chen & Wells, 1999; Chen, Gillenson & Sherrell, 2002; Heijden, 2003). Thus, small websites, with very basic iconography and ease of use, such as that of Barcelona, would transmit a good city brand image. However, extensive and complicated sites that are slow to open and that have a confused internal structure, such as that of Toronto, cause less positive or even unfavorable attitudes to be created.

Along the same lines, Jared M. Spool (1996)[6] did a study comparing websites, which demonstrated that the usability of a website considerably and positively affects the brand and the branding process. His results showed that the users that navigate more easily through a website and find the information that they want quickly end up with a better impression of the brand, as it has satisfied their expectations to a greater degree. Contrarily, the obstacles that the users find when navigating negatively and directly affect their perception of the brand. Therefore, usability is essential for effective branding.

Regarding interactivity, Liu (2003) defends the idea that the concept of interactivity unites three correlated but different factors: the active control of information, bidirectional communication and the synchronicity or simultaneity of communication. In a previous article (Liu & Shrum, 2002), the same author classified the brand websites in the maximum range of the three factors stated. This study showed, by the way of bidirectional communication, that the Internet is the only medium that can be used for commercial transactions without the help of other tools, since necessary activities such as showing the product, placing orders, making payments or even, in categories such as music, software or transport titles, distributing the product can happen through the web.

Other authors (Cho & Cheon, 2005) prefer to divide the concept of interactivity into three fields of action or types: consumer-message interactivity, consumer-consumer interactivity and consumer-marketer interactivity. Consumer-message interactivity refers to the capability of the user to personalize his or her relationship with the contents of the page according to his or her interests and motives. Consumer-marketer interactivity centers on the communication between the user of a website and the organizers or those responsible for the content; this relationship may be bidirectional, from user to administrator (questions, suggestions, complaints) or from administrator to user (obtaining personal data, answering questions, etc.). Consumer-consumer interactivity is the relationship that may be created between the people that access a website (virtual communities, chats, forums, etc.).

More recent studies (Sicilia, Ruiz & Munuera, 2005; Ko, Cho & Roberts, 2005) show that interactivity enables the information to be processed better and generates more favorable attitudes towards the website and towards the product and the brand, and greater intention to purchase.

Based on these previous studies on the usability and interactivity of websites we decided to make a broader quantitative analysis about the treatment of city brands on the Internet. In addition to the items related to graphic, functional and emotional aspects of the brand, in the trial we analyzed usability and interactivity characteristics of the websites as elements that also influence the dissemination of a good brand image.

PURPOSE AND METHOD OF QUANTITATIVE STUDY. USABILITY, INTERACTIVITY AND THE CITY BRAND ON THE WEB

The first qualitative study was wider. It analyzed the city brands and their treatment in websites in depth. But later, we decided to carry out a quantitative study analyzing more webs and variables. We had to develop a different questionnaire of analysis that could be measured by quantitative methods.

The main objective of this quantitative study consisted in analyzing the degree of usability, interactivity and treatment of city brands on the official websites of tourist cities.

For this empirical study, we used a quantative method based on a WTO (World Tourism Organization, 1999) analysis model. This was extended with the contribution of recent studies (McMillan, 2003; Liu, 2003; Cho & Cheon, 2005) and adding newly created interactive resources, which appeared as new features on destination websites at the time the study was carried out. At the same time, the model was also extended with aspects to analyze about how city brands are dealt with on the web, the databases that the websites may obtain about their users, and other aspects of website information not considered in the initial model.

The analysis was applied to 40 official websites of important tourist cities of the five continents during 2006. This sample was selected by means of a ranking evaluating the main tourist cities of the world, taking into account the *World's Top Tourism Destinations* of the WTO (World Tourism Organization), the number of visitors and the importance of the city as a tourist destination.

We observed 135 indicators, of which 87 are representative of the three variables analyzed: usability, interactivity and the brand. Interactivity is analyzed in the three factors mentioned previously: consumer-message interactivity, consumer-marketer interactivity and consumer-

consumer interactivity. Its analysis, through the SPSS program, was centered on descriptive statistics and the combination of variables, using Gamma as a correlation index.

RESULTS OF THE QUANTATIVE STUDY. USABILITY

The concept of usability is defined as "the extent to which a product can be used by specified users to achieve specified goals with effectiveness, efficiency and satisfaction in a specified context of use"[7]. This ease of use, nevertheless, is related to very diverse aspects, that go from the page design, to the content quality, the ease of locating information and the simplicity of navigation, all of which are related, in addition, to the subjective perceptions of the user.

The indicators used in the study to measure the usability variable are those that provide the navigability for the website: the access menu for the sections always being visible, the indication of the navigation path, constant links to the home page, the existence of a sitemap and an internal search engine; the possibility of user help by telephone, e-mail, chat or web call; and external links to related sites.

The majority of sites analyzed in the study showed high levels of structure, which means that the websites are, in general, well designed. At a global level, high percentages were seen for the variables that make up the structure, usability and design of the websites.

As seen in Table 1, the majority of the sites analyzed have a list of the site contents that appears on all of the pages, along with a link to the home page from each page. These two resources give the user his or her location on the web, clarify the structure of the site and simplify its use. In addition, almost all the websites have external links with other related sites. Through these links they can provide complementary information and services. In some cases connection to the official sites of the town halls and institutions are provided, in others, the purchase of services, for example the link on Barcelona's website to ServiCaixa, through which you can purchase tickets for entertainment.

With respect to the help that the websites offer their users, 90% of the sites analyzed provided e-mail contact. In addition, 65% provided a telephone helpline. But, however, only one website offered help by chat and none used web calls (calls from marketing staff from the destinations where you request them to call you at a certain time) or human clicks (communication in real time of a visitor to a website with its administrators).

All in all, the results stated up to now show that the majority of websites are well structured and designed. Therefore, they have a high degree of usability.

Table 2 shows the rankings of the websites with best structure, design and usability, along with those that have the least points in these aspects. The structure variable ranges between 5 and 11, with sites with 5 having the worst structure and those with 11 the best.

The initial Hong Kong web page (Figure 1), one of the best in usability of our study, is simple, graphic and structured. It shows only a big image and the list of different languages the user can choose.

Table 1. Items of usability (Source: Authors' compilation)

Indexes of usability	Frequency	Percentage
List of contents on all pages	36	**90.0%**
Link the home page	37	**92.5%**
External links related sites	38	**95.0%**
Up-to-date information	39	**97.5%**
Sitemap	23	**57.5%**
Indication of the navigation path	16	**40.0%**

Table 2. Ranking of websites with the most and least points with respect to usability (Source: Authors' compilation)

Web sites with high scores in structure, design and usability		**Web sites with low scores in structure, design and usability**	
Score	City Web sites	Score	City Web sites
11	London	5	Cairo
11	Hong Kong	5	Beijing
10	Bangkok	6	Lisbon
10	Budapest		
10	Sydney		
10	Amsterdam		
10	Tokyo		
10	Buenos Aires		
10	Mexico City		

When you enter in the initial page of the selected language you can observe that it has the content list in all the pages, visual and graphic links and shows a constant simplicity in the entire site.

Interactivity

The interactivity variable, as explained previously, was analyzed using three aspects: consumer-message interactivity, consumer-marketer interactivity and consumer-consumer interactivity. We measured the concept of interactivity using the three aforementioned typologies of Cho and Cheon (2005).

Consumer-message interactivity is measured with indicators such as: the existence of search engines, user help, the option to customize the display, displaying virtual reality, multimedia presentations, directional maps, virtual leaflets,

Figure 1. Hong Kong Web site

the option of downloads to mobiles, and on-line games, among others. However, consumer-marketer interactivity is based on indicators of relationships with the marketing organizers of the cities. They offer the possibility of questions and complaints from users, opinion surveys, chats with promotional or sales agents, bulletin boards for users, or the possibility of placing orders and making reservations on line. Finally, consumer-consumer interactivity is measured using indicators of the relationship between them, such as the existence of chats or an email service.

Consumer-Message Interactivity

The first result we can see is that, in general, the majority of websites analyzed use many more resources that belong to the consumer-message interaction, than to the consumer-marketer, and consumer-consumer interaction. Therefore, the sites offer the navigator a greater interactivity with the messages that they wish to transmit than with the people that are in charge of marketing the destinations and other consumers. And within the resources of the consumer-message interaction, the most used in all the websites, that is, those that show the highest percentages, are the interactive travel organizers.

So, this shows that the majority of websites analyzed offer services of interactive travel organizers. We refer to organizers without the possibility of purchase. They help users to plan their trips by providing fully personalized information adapted to the needs and interests of the users. The interactive travel organizers allow the users to plan their trips in a fully complete and personalized way, from their transport to the destination to their accommodation and other complementary tourist services.

However, there are other interactive consumer-message resources that are hardly used by the websites analyzed. Certainly, 62.5% have directional maps of the cities to orient the users and 82.5% have geographical markers on the maps with zoom, which offer the possibility of searching for and locating elements on the map in detail. But only three of the sites allow the user to customize the home page, four show virtual excursions or online games for children, five offer virtual flights of the city from the air or allow users to create their own virtual leaflets in folders, which they can save with the personalized information that interests them.

The informative services using optional downloads to mobiles, whether metro maps, information on monuments or audio downloads

Table 3. Resources used in the websites of interactive travel organizers (Source: Authors' compilation)

		Frequency	Percentage
Interactive travel organizers (without possibility of purchase)	How to get there	33	**82.5**
	What to do	39	**97.5**
	Attractions/events	38	**95.0**
	Leisure activities	38	**95.0**
	Cultural activities	37	**92.5**
	Where to stay	38	**95.0**
	Transport	39	**97.5**
	Excursions	33	**82.5**
	Rentals	**16**	40.0

are also resources that are barely used by the websites analyzed.

All of these interactive resources, in addition to offering information, provide a certain entertainment and distraction to the users of the websites, making them more attractive and interesting.

Consumer-Marketer Interactivity

The resources of the consumer-marketer interaction are still less used by the websites analyzed that those of consumer-message interaction. The possibility of placing orders or online reservations is offered by 35% of the sites, and 32.5% allow the user to track the orders placed. For these two resources, the prior registration is usually required of the user, and with this the websites can obtain personal information about their consumers. In 22.5% of the sites, users can fill in surveys or opinion polls and 12.5% offer the option of complaints by consumers.

The rest of the consumer-marketer interactive resources are barely used by the websites analyzed. For example, only Rome's site offers the users the possibility of asking what they want about the city, the services offered or the entity that organizes the marketing of the destinations. Only the websites of Dublin and Istanbul have an electronic bulletin board available to the users. And finally, the sites of Madrid, Amsterdam and Montreal are the only ones that allow the consumers to sponsor the website. This means that any organization can pay some money to sponsor the website. All the conditions are very well explained on the site. In exchange, the sponsor can put its logo on the website.

Not one of the websites analyzed offered the users the possibility to propose new products or services, chats with the marketing agents, or "call me" buttons with time and language selection for those responsible for promoting the city to call the interested users to provide them with the information they want, personally by telephone.

Consumer-Consumer Interactivity

With respect to the consumer-consumer interaction, there are even fewer resources available. The resource of this type that is most used in the sites analyzed is the on-line postcards service, featuring in 40% of the websites, of which 37.5% do not require registration. This resource, which is offered more and more by the destination websites and is mostly used by young users, enables them to send on-line postcards, without the delay or costs of sending them.

Moreover, only Krakow's site has a chat for users, and the possibility for consumers to tell their stories, experiences and summaries of trips is only found on the sites of Hong Kong, Rome and Toronto. Finally, none of the websites analyzed offer the services of a cyber club of users with advantages or a cyber community with common interests. Cyber clubs which offer some advantages of information, discounts and special conditions to the loyal users. The cyber club of users would allow the marketing organizers of the destinations to create loyalty programs for clients using the offer of advantages and discounts. Moreover, the creation of a cyber community would enable the users of this group to build relationships with each other, broaden information in their interest, and create a strong position with respect to these common interests.

So, we can see that there is a great inequality in the use of interactive resources by official websites of destinations. The most used are those that belong to the consumer-message interaction, and especially, the interactive travel organizers. All in all, generally, the percentages of use of interactive resources are still underused, in particular those of the consumer-marketer and consumer-consumer interaction. These results coincide with those of the study by Anton (2004) about the Internet presence of the main tourist destinations of the Spanish Mediterranean coast. In it, he highlights the absence of instruments that promote interactiv-

Table 4. Ranking of websites with the most and least points with respect to interactivity (Source: Authors' compilation)

Most interactive websites		**Least interactive websites**	
Score	City websites	Score	City websites
28	Dublin	4	Moscow
23	Hong Kong	4	Kiev
21	Valencia	7	Sydney
20	Madrid	8	Cairo
18	Berlin	8	Seville
18	Istanbul		
18	Rome		

ity in the websites, such as on-line forms, pages of visitor comments, etc.

The degree of global interactivity of the websites analyzed, which may range between 4 and 28, can be seen in the ranking in Table 4. These numbers represent the number of interactive tools they use.

Visit Dublin's web page it's the most interactive out of all the analyzed sites. If we only observe the first part of the page we can find a searcher and in the graphic frontal there is the option to click and see a video about the city. The whole page is very interactive.

In the accommodation section the user can search information, but can also reserve and buy the products.

Even the maps of the city and the maps of public transport are interactive. They show the routes that the user requires in personalized way.

Treatment of City Brands on the Web

In the analysis of the communication and dissemination of city brands through websites, it is necessary to differentiate the concepts of city brand and brand image. As previously explained, the city brand is a construct composed of a name, a logo, some symbols and values that we try to associate with a city representing its identity, with the objective of creating a positioning and a vision of the city in the minds of the public, which must be based on a citymarketing plan and projected through a communication program. On the other hand, the brand image is that which is created in the minds of the public as a consequence of the communication of the city brand, in conjunction with the perceptions and subjective values of individuals.

In the study, the degree of communication and dissemination of the city brands through the websites was measured using the presence of the brand and the logo, whether there is a description of the city brand, whether the graphic and photographic images represent the brand, whether the colors and the typography of the page are coordinated with the logo, and the functional and emotional brand is disseminated.

The results show that city brands are treated poorly on the websites analyzed. Their treatment is even less developed than that of interactivity.

All in all, the vast majority of websites (97.5%) have the logo on all of the pages, and this is usually situated in the upper left part. However, it must also be taken into account that three of the websites analyzed show more than one logo for the same city brand, which creates dysfunctionality and incoherence.

Of the websites, 92.5% offer a brief description of the destination, but only 5% (two of the sites analyzed) have a minimum explanation of the city brand, of its creation and symbolism. Similarly, only 35% of the websites state the marketing objectives of the tourist institutions of the cities.

The graphic images, in general, also are an underused resource in the dissemination of the city brands. All the websites transmit the functional brand through photographic images, but only half of them (50%) communicate the emotional brand through the photographs. On the other hand, the rest of the graphic images are not used to disseminate the brand at all.

Figure 2. Visit Dublin Web site

Figure 3. Visit Dublin accomodation section

Figure 4. Visit Dublin maps section

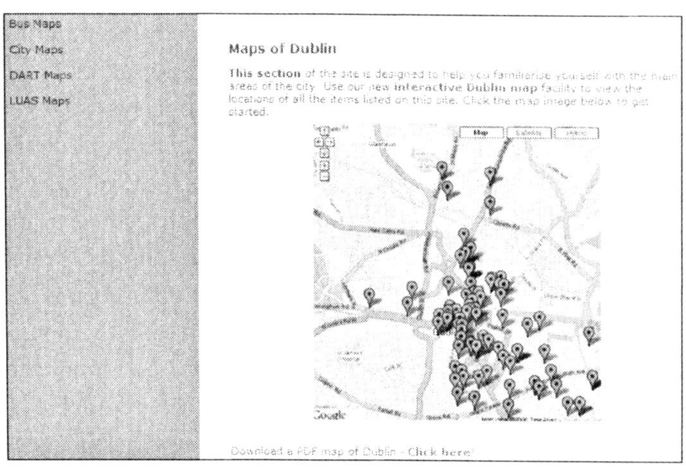

aspects in the construction of the city websites. They seem to be made by information technologists, who pay great attention to usability, and also, to a lesser extent, interactivity, but do not take into account the communication of the brand. Therefore, you can sense the existence of a certain lack of co-ordination between the IT and communication and marketing departments of the entities promoting the cities.

To effectively create websites that disseminate city brands and promote the image of the destinations, it is fundamental to be conscious of what an important tool for image, communication and marketing these websites are for the destinations and cities.

As we go further into the Information Era, the role of websites is changing. They are evolving from being merely sources of information, that is, intermediaries between tourists and destinations, to being involved in tourist transactions. Therefore, tourism and Internet make an ideal combination. The Internet provides the users with a way of obtaining much more varied and detailed information on the destinations and the cities than that which existed before. However, in addition it enables them to, through the same web space, make consultations and reservations quickly and easily.

The websites are considered the future of communication on the Internet, as they offer a large quantity of information and, in addition, create virtual product experiences (Klein 2003). Websites, on one hand, create a brand image and, on the other, can provoke a direct response (Hollis, 2005). As Cho & Cheon (2005) confirm, websites offer different communication possibilities: public relations, sales promotion, advertising, direct marketing and brand image creation. As a consequence, this communicative potential must be promoted and made use of in the field of cities.

With respect to how little city brands are dealt with on the websites analyzed, it is due, firstly, to insufficient conceptualization, creation and development. As we have already stated, city brands are a recently created concept that is still at a very incipient stage. For this reason, the web deal poorly with them. Firstly, it is necessary to create elaborate city brands, based on a strategy of citymarketing, to be communicated later to the public and, if not, disseminated through the websites.

Despite the studies mentioned above (Spool, 1996; Chen & Wells, 1999; Chen & al., 2002; Heijden, 2003) that show the importance of usability in promoting a more positive brand image among users; and the effect of interactivity, which produces more favorable attitudes towards the web, the brand and greater intentions of purchase (Sicilia & al., 2005; Ko & al., 2005), we can confirm that high indexes of usability and moderate indexes of interactivity do little good in the websites analyzed in the promotion of city brands. First it is necessary to correctly create the city brand, to later disseminate it through the websites of the cities. And only after this first step, will usability and interactivity be complementary and effective aspects in the promotion of city brands.

Improvement Proposals

The creation of the city brand must not be an act restricted to the activity of citymarketing. Exactly the opposite, it must be coherent with the whole marketing plan and be derived from a competitive city strategy.

It must start with a prior diagnosis of the current situation of the city image and the competition, to later formulate the vision of the city and the positioning that is desirable to achieve in the world urban system.

Only after the two first stages should an identity program for the city be determined, which must take into account a historical analysis, and the perceptions of the internal and external public. Then the visual identity of the city, its logo, colors, typography should be determined; along with the functional and emotional values of the brand. In this whole process it is essential to identify the internal and external public (citizens, visitors, investors, mass media, public institutions, neighborhood associations), to address them directly and achieve that they take on the identity and the city brand to be promoted as their own and, if possible, this should be by consensus.

Finally, the city brand they must be disseminated by a communication plan to each sector of the public through all types of actions *above* and *below the line,* with the intention of continuity in time, for the values of the brand to catch on coherently in the perceptions of all their sectors of the public and be integrated in the rest of the citymarketing actions.

But city brands must not only comply with a creation process inserted into the citymarketing actions, they must also have certain characteristics:

1. The city brand must be made up of three basic elements: the graphic brand, the creation of a symbol and a logo; the functional brand, based on the real strong and attractive points of the city; and the emotional brand, the symbolic and personalizable values that are associated with the city. Without one of these three elements, the city brand is incomplete, losing its identificative and persuasive power.
2. A single city brand must be created, with a single strategy, that is applicable to the diverse sectors of the public, but that at the same time identifies the brand with unique values, which allow the cities to de identified and differentiated from each other. In addition, this brand must be adopted by all the organizations and entities, avoiding the proliferation of several brands that create dysfunctionality.
3. Finally, and independently of the municipal political changes, city brands must be created to last a long time. If they are agreed upon by consensus when created, and therefore, do not belong to the municipal governments, they will evolve independently of the political channels and have a greater possibility of being consolidated in the minds of the public.

Once the city brand is completely and adequately created, and starting with the importance of websites as tools for communication, promotion and marketing of the cities, attention must be paid to three key aspects in the promotion of city brands through the Internet:

1. The treatment of the city brand through the constant and unique presence of the logo, with an explanation of its symbology and objectives, with the representation of the functional and emotional values that are desirable to associate both textually and as graphically, and by means of the graphic and typographic coherence throughout the website.
2. Maximum development of usability throughout the page, which facilitates user navigation and promotes a positive image increasing the possibilities of recommendation and marketing.
3. Maximum creative use of the interactive resources, as these improve the brand image and the users feel drawn to navigate.

REFERENCES

Aaker, D. (1991). *Managing brand equity.* New York, NY: Free Press.

Anton Claver, S. (2004). La presencia en Internet de los principales destinos turísticos del litoral mediterráneo español. *Actas del Congreso TURITEC 2004.* Universidad de Málaga, 2004. Retrieved June 1, 2008, from http://www.turismo.uma.es/turitec/turitec2004/index.htm

Baker, B. (2007). Places: The new brand frontier. *Total destination management*. Retrieved June 1, 2008, from www.DestinationBranding.com

Baloglu, S., & McCleary, K. (1999). A model of destination image formation. *Annals of Tourism Research, 26*, 868-897.

Biel, A. (1997). Discovering brand magic: The hardness of the softer side of branding. *International Journal of Advertising, 16*, 199-210.

Blain, C., Levy, S. E., & Brent Ritchie, J. R. (2005). Destination branding: Insights and practices from destination management organizations. *Journal of Travel Research, 43*, 328-338.

Buhalis, D. (2002). *eTourism. Information technology for strategic tourism management*. UK: Prentice Hall.

Buhalis, D., & Costa, C. (2006). *Tourism, management dynamics. Trends, management and tools*. Oxford: Elsevier Butterworth Heinemann.

Cai, L. (2002). Cooperative branding for rural destination. *Annals of Tourism Research, 29*, 720-742.

Chaves, N. (2004, April). *La marca destino turístico, cinco estrategias gráficas*. Paper presented at the XIII Simposi Internacional de Turisme i Lleure ESADE-Fira de Barcelona. Retrieved April 30, 2007, from www.esade.es/cedit2004/cat/est_estudios.php

Chen, L. D., & Wells, W. D. (1999). Attitude toward the site. *Journal of Advertising Research, 39*(5), 27-38.

Chen, L. D., Gillenson, M. L., & Sherrell, D. L. (2002). Exciting online consumers: an extended technology acceptance perspective. *Information and Management, 39*, 705-719.

Cho, C-H. & Cheon, H. J. (2005, Summer). Cross-cultural comparisons of interactivity on corporate Web sites. *Journal of Advertising, 43*(2), 99-115.

De Chernatony, L., & Dall'Olmo Riley, F. (1997). Modelling the components of the brand. *European Journal of Marketing, 32*(11/12), 1074-1090.

Ekinci, Y., & Hosany, S. (2006). Destination personality: An application of brand personality to tourism destinations. *Journal of Travel Research, 45*, 127-139.

ESADE, & BDDO Consulting (2004, April). *Evolución del posicionamiento de los destinos turísticos españoles: De lugares de vacación a marcas de turismo*. Paper presented at the XIII Simposi Internacional de Turisme i Lleure ESADE-Fira de Barcelona. Retrieved April 30, 2007, from www.esade.es/cedit2004/cat/est_estudios.php

Etchner, C. M., & Brent Ritchie, F. (1991). The meaning and measurement of destination image. *Journal of Tourism Studies, 2*(2), 2-12.

Gnoth, J. (1998). Conference reports: Branding tourism destinations. *Annals of Tourism Research, 25*, 750-760.

Hankinson, G. (2004). The brand images of tourism destinations: A study of the saliency of organic images. *Journal of Product & Brand Management, 13*(1), 6-14.

Hankinson, G. (2005). Destination brand images: A business tourism perspective. *Journal of Services Marketing, 19*(1), 24-32.

Heijden, H. (2003). Factors influencing the usage of web sites: The case of a generic portal in The Netherlands. *Information and Management, 40*, 541-549.

Hollis, N. (2005, June). Ten years of learning on how online advertising builds brands. *Journal of Advertising Research*, 255-268.

Hosany, S., Ekinci, Y., & Uysal, M. (2006). Destination image and destination personality: An application of branding theories to tourism places. *Journal of Business Research, 59*, 638-642.

Kapferer, J. N. (1997). *Strategic brand management.* London: Kogan Page.

Keller, K. (1993). Conceptualizing, measuring, and managing customer-based brand equity. *Journal of Marketing, 57,* 1-22.

Klein, L. R. (2003). Creating virtual product experiences: The role of telepresence. *Journal of Interactive Marketing, 17*(1), 41-55.

Ko, H., Cho, C-H., & Roberts, M. S. (2005). Internet uses and gratifications. A structural equation model of interactive advertising. *Journal of Advertising, 34*(2), 57-70.

Konecnik, M., & Gartner, W.C. (2006). Customer-based brand equity for a destination. *Annals of Tourism Research, 34*(2), 400-421.

Kotler, P., Haider, D. H., & Rein, I. (1993). *Marketing places: Attracting investment, industry and tourism to cities, states and nations.* New York, NY: Free Press.

Lawson, F., & Band-Bovy, M. (1977). *Tourism and recreational development.* London: Architectural Press.

Liu, Y., & Shrum, L. J. (2002). What is interactivity and is it always such a good thing? Implications of definition, person and situation for the influence of interactivity on advertising effectiveness. *Journal of Advertising, 31*(4), 53-64.

Liu, Y. (2003, June). Developing a scale to measure the interactivity of Web sites. *Journal of Advertising Research,* 207-216.

López Lita, R., & Benlloch, M. T. (2005). La marca territorio. El marketing de ciudad, una herramienta al sevicio de las marcas territorio. *99% com, 2,* 8.

Lynch, P. J., & Horton, S. (2004). *Manual de estilo web. Principios de diseño básico para la creación de sitios Web.* Barcelona: Gustavo Gili, SA.

Martín Barbero, S., & Sandulli, F. (2005). *Marcating en la Web: BIWAM. Identidad desnuda.* Madrid: Dossat 2000.

McMillan, S., Hwang, J-S., & Lee, G. (2003, December). Effects of structural and perceptual factors on attitudes toward the website. *Journal of Advertising Research,* 400-409.

McMillan, S. (2004). Internet advertising: One face or many? In D. Schumann and E. Thorson (Eds.), *Internet advertising: Theory and research (2nd edition).* New York, NY: Lawrence Erlbaum.

Morgan, N., Pritchard, A., & Piggott, R. (2002). New Zeland, 100% Pure: The creation of a powerful niche destination brand. *Journal of Brand Management, 9*(4/5), 335-354.

Morgan, N., Pritchard, A., & Pride, R. (2004). *Destination branding. Creating the unique destination proposition.* Oxford: Elsevier.

Nielsen, J. (2002). *Usabilidad. Diseño de sitios web.* Madrid: Prentice Hall.

World Tourism Organisation (1999). *Promoción de destinos turísticos en el Ciberespacio.* Madrid: WTO.

World Tourism Organisation (2001). *Comercio electrónico y turismo.* Madrid: WTO.

Padín Fabeiro, C. (2004, April). *La formación de la imagen de un nuevo destino.* Paper presented at the XIII Simposi Internacional de Turisme i Lleure ESADE-Fira de Barcelona. Paper presented at the XIII Simposi Internacional de Turisme i Lleure ESADE-Fira de Barcelona. Retrieved April 30, 2007, from www.esade.es/cedit2004/cat/est_estudios.php

Park, C. W., Jaworski, B. J., & MacInnis, D. J. (1986, October). Strategic brand concept management, *Journal of Marketing, 50,* 135-145.

Prebensen, N. K. (2007). Exploring tourists' images of a distant destination. *Tourism Management, 28*, 747-756.

Riezebos, R. (2003). *Brand management. A theoretical and practical approach.* Harlow, UK: Prentice-Hall.

Ritchie, B., & Ritchie, R. (1998). *The branding of tourism destination: Past achievements and future trends.* Paper presented at the 48th Congress, AIEST, St-Gall.

Sicilia, M., Ruiz, S., & Munuera, J. L. (2005). Effects of interactivity in a Web site. *Journal of Advertising, 34*(3), 31-45.

Simon, S. J. (2001). The Impact of culture and gender on web sites: An empirical study, *Database for Advances in Information Systems, 3*(1), 18-37.

Spool, J. M. (1996, January 1). Branding and usability. *User Interface Engineering.* Retrieved June 1, 2008, from www.uie.com/articles/branding_usability.

Tsikriktsis, N. (2002). Does culture influence web site quality expectations? An empirical study. *Journal of Service Research, 5*(2), 101-112.

ENDNOTES

1 According to Nina K. Prebensen (2007), to build a good city brand, with the most suitable elements, that create a positive image, the opinion and knowledge of the visitors or future visitors is fundamental.

2 Bill Baker is the founder and President of Total Destination Management. He is internationally recognised as an expert in building brand recognition for destinations and communities. TDM is a Portland, Oregon based team of destination branding, tourism planning and marketing specialists. www.DestinationBranding.com consulted in July 2007.

3 Blain, Levy and Brent Ritchie (2005) confirm that the logos of city brands globally represent the experience that the visitors expect of a destination or city. Effectively, logos stimulate the communication of the attributes desired among the visitors and influence the tourists' decision to visit a place.

4 Hosany, Ekinci and Uysal (2006) demonstrated that there is a relationship between the image of a destination and its personality. They also demonstrated that the image of a destination's brand is fundamental to position it effectively.

5 Other authors, such as Keller (1993) and Park (1986) add a third category or element to the city brand: experiential attributes, which refer to the experiences, satisfaction and feelings of the destination's visitors.

6 Spool compared two websites: one more usable and informative and the other very graphically meticulous. He showed that the more usable site created a better brand image, as it satisfied the expectations of the users to a greater degree. Therefore, the graphic aspects of a website, such as logos and photographs, have less effect on the branding than expected.

7 International Organization for Standardization (ISO). In: http://www.iso.org/iso/home.htm

APPENDIX: LIST OF WEBSITES ANALYSED

America

México D.F.	www.mexicocity.gob.mx
Los Ángeles	www.lacvb.com
Montreal	www.tourisme-montreal.org
New York	www.nycvisit.com/home/index.cfm
Toronto	www.torontotourism.com
Buenos Aires	www.buenosaires.gov.ar/areas/turismo/home
Río de Janeiro	www.riodejaneiro-turismo.com.br/en/home.php

Africa

El Cairo	www.cairotourist.com
Marrakesh	www.ilove-marrakesh.com

Asia

Bei-jing	english.bjta.gov.cn
Tokyo	www.tourism.metro.tokyo.jp/english/index.html
Hong Kong	www.discoverhongkong.com
Kuala Lumpur	www.kualalumpur.gov.my
Bangkok	www.bangkoktourist.com
Macau	www.macautourism.gov.mo
Bali	www.balitourismauthority.net/home.asp
Delhi	delhitourism.nic.in

Oceania

Sydney	www.cityofsydney.nsw.gov.au

Europe

Amsterdam	www.iamsterdam.nl
Athens	www.cityofathens.gr
Dublin	www.visitdublin.com
Istanbul	english.istanbul.com
Florence	www.firenzeturismo.it/en_default.asp
Helsinki	www.hel2.fi/tourism
Lisbon	www.cm-lisboa.pt/turismo
London	www.visitlondon.com

Paris	www.parisinfo.com
Prague	www.visitprague.cz
Roma	www.romaturismo.com
Vienne	info.wien.at
Berlin	www.berlin-tourist-information.de
Kiev	www.kmv.gov.ua
Krakow	www.krakow.pl
Budapest	www.budapestinfo.hu
Moscow	www.moscowcity.com
Zagreb	www.zagreb-touristinfo.hr

Spain

Barcelona	www.bcn.es/turisme/catala/turisme/welcome.htm
Madrid	www.esmadrid.com
Sevilla	www.turismo.sevilla.org
València	www.turisvalencia.es

Chapter III
A Strategic Framework for City Marketing: The SSRM Approach

Barry Mishra
University of California, USA

Erik Rolland
University of California, USA

ABSTRACT

City marketing in the broadest term can be defined as the strategic design of the city to satisfy the various stakeholders of the city who often have conflicting goals. This frequently requires an integrated approach that aligns and addresses the expectation of various stakeholders to create vibrant communities. The current trend in globalization, formation of regional trade blocks, and the shift in importance of location factors have increased the intensity of competition among regions and cities. More than ever, cities need to compete and cooperate with each other to attract companies, investments, talent, tourists, and create markets for their products and services. This entails that cities embrace strategic marketing management tools and practices, and utilize e-services such as electronic customer relationship management. The authors propose a broad approach, called strategic stakeholder relationship management (SSRM), which is enabled by information and communication technologies, including Internet, to help the decision makers succeed in designing the 21st Century city marketing initiatives.

INTRODUCTION

City marketing in broadest term can be defined as the strategic design of the city to satisfy the various stakeholders of the city who often have conflicting needs and aspirations. This frequently requires an integrated approach of thought and action that aligns and addresses the expectation of various stakeholders to create vibrant communities.

The current trend in globalization, formation of regional trade blocks and the shift in importance of location factors have increased the intensity of competition among regions and cities. More than ever cities need to compete and cooperate with each other to attract companies, institutions, investment, talent, and tourists and create markets for their products and services. Without investing in city marketing, cities will become defunct and may perish in the long run as the stakeholders will leave.

A city, in nutshell, is a microcosm of the modern political state in a socio economic sense. There are multiple stakeholders, and sometimes the boundaries between these stakeholders are blurry. Different stakeholders have different needs and aspirations while together they contribute to the overall well-being of the city. Thus, the cities must bring the various stakeholder views into a congruent strategy that benefits all its stakeholders, or at least, does not negatively affect certain stakeholders. Given the multiple stakeholders and their specific needs, it becomes imperative that any strategic approach to city marketing has to address the common as well as the competing needs in a coherent and logical manner.

The Internet, together with other technologies, has helped organizations re-orient their services as to take advantage of the available information and communications technology (ICT). Such services are often termed e-services, as they are based on ICTs (Rust & Lemon, 2001). Most industries have in the recent years seen a push towards more focus on customer relationship management (CRM)—an integral and important part of e-service. Clearly, the Internet and ICTs have helped many organizations interact better with their customers through electronic CRM (e-CRM). As a result, e-service has become a major area of study. However, it has been stated that government organizations have more trouble than the private sector in successfully applying new technology (Dawes et al., 2004). In this paper we therefore propose a framework to help city governments plan their e-service strategically: that is, align the goals of its stakeholders, as well as frame the perceived dimensions of quality inherent in their stakeholder-base. The framework, referred to as strategic stakeholder relationship management tool (SSRM) can aid cities in identifying, addressing, and managing issues that are important to their various stakeholders within the limited resources the city has.

BACKGROUND

A favorable business climate often perceived as a key factor for local economic development (Blume, 2006). Globalization has effects on regions/cities in that there have been observable shifts in intensity of competition, and shifts towards knowledge factors (Blume, 2006). Even though competitive ability of an organization depends primarily on business-related and knowledge factors (cost-efficiency, an ability to innovate, marketing and other internal factors) (Krugman, 1996), it is clear that local economic policies may enhance or inhibit such competitiveness. For example, unwanted side-effects are seen in areas such as the Inland Empire in Southern California region as well as in and China, where an overly focus on transportation and production respectively has had detrimental effects on air quality and congestion. The locational factors traditionally driving economic development in China has been the availability of cheap labor. For Inland Empire, the locational factor has been availability of cheaper land, located centrally to major distribution routes (air, ship, and road).

In the past decade, arguments have been voiced over concerns of economic growth and its potential negative impact on both local and global environments. An example of this is seen in air quality concerns over the 2008 Beijing Olympics (Der Spiegel, 2007). In addressing the air quality issues, noxious factories and power-plants have been relocated, and experiments are under way to limit car traffic (LA Times, 2007). The recent implementation of variable toll rates and congestion fees (as in Manhattan, NY), are a step in the same direction. Clearly, local economic policies related to the control of pollution drive these concerns. The immediate negative effects, in the case of the Beijing Olympics, are related to image and tourism in China as well as the poor air quality for local residents. What is clear is that there is a strong link between the marketing of Beijing as a city, and local policies for pollution control. One could clearly argue that in these areas, the local governments have not been aware of some of the issues that are important to many of its stakeholders (for example, the residents). At best, the local government has not been cognizant of one stakeholder view: the quality of life factor. Indeed, Wong (2001) found that when traditional economic factors already are in place (such as infrastructure and workforce), quality of life issues become crucial in the competitive process between cities. That is, as competition matures, the notion of being in touch with ones customers becomes even more important.

The concept of customer relationship management (CRM) is based on value maximization for the organization. The core concept of CRM is that value creation between the firm and the customer is related in such a way that an increase in customer value leads to an increase in firm value (Mithas, Krishnan & Fornell, 2005). In the words of Payne and Frow (2005):

CRM is a strategic approach that is concerned with creating improved shareholder value through the development of appropriate relationships with key customers and customer segments. CRM unites the potential of relationship marketing strategies and IT to create profitable, long-term relationships with customers and other key stakeholders. CRM provides enhanced opportunities to use data and information to both understand customers and co-create value with them. This requires a cross-functional integration of processes, people, operations, and marketing capabilities that is enabled through information, technology, and applications.

The customer, in the city government setting, is a stakeholder: such as a resident, a business, a visitor, etc. It is clear from the discussion above that one should go to great efforts to clearly identify and understand the needs of the stakeholders (or the customers, as they might be). Once these issues are known, one must then align the stakeholder goals as to create a congruent plan for marketing the city. The framework proposed below provides a semi-structured way of identifying, understanding, and aligning stakeholder goals within a resource-based, e-service context.

THE FRAMEWORK

The resource-based view of the firm (RBV) argues that an above average return may be generated when the firm develops or obtains a resource that—when successfully applied in a business strategy—provides competitive advantage. The competitive advantage typically comes about after the firm discovers opportunities for change, executes on its strategy paradigms, and implements its strategy (see for ex. Barney & Hesterly, 2006). Since cities compete for customers—employers, residents, and visitors—it is clear that they must take an approach to competing similar to that of the RBV, but where the returns are to be defined by their stakeholders. In section 3.1, we will illustrate a framework for achieving this, and outline the illustration with a specific example in section 3.2.

Construction of the Framework

A city's resources are similar to most organizations' resources, yet perhaps more encompassing.

Organizational theorists typically categorize the assets of the organization into logical groupings (Schermerhorn, 1999). These groupings include personnel, facilities and operational resources, and organizational knowledge as reflected in operational processes. The city's customers and specific external environmental elements should be included with the other organizational assets from the perspective of an extended resource-based view of the organization. This extended classification is often apparent in service industries where the customer has an extensive relationship with the firm—much like that of a city and its customers. The city's stakeholders should be viewed as its customers. Thus, in the discussions below, we use the term stakeholder and customer interchangeably.

Rolland, Patterson and Ward (2008) proposed a 2-dimensional framework for linking organizational resources with the customers' perception of quality (in other words, the things customers may care about). In the spirit of their suggestions, and based on the above discussion, we propose that a city possess organizational assets that can be broadly categorized into:

1. **Personnel:** this organizational asset includes all city employees with whom the stakeholders may interact.
2. **Operational Processes:** this asset includes all major organizational processes, such as city planning, issuing permits, infrastructure development, etc.
3. **Facilities and Operational Assets:** this asset includes infrastructure, land, buildings, parks, equipment, and all other non-personnel related resources.
4. **Customers:** this asset includes the stakeholder, and his/her abilities, restrictions, etc.
5. **External (or environmental) components:** this asset includes regulation, media, competition, and other external factors that may impact the customer/stakeholder.

The first three categories of organizational assets are those that are under very direct control of the city government.

In a CRM system, we are seeking opportunities to understand and improve any interaction (or potential interaction) that takes place between the organization and its customers. Burr, Patterson, Rolland and Ward (2007) as well as Rolland, Patterson and Ward (2008) proposed that such interactions be categorized into a set of "quality dimensions". These quality dimensions are related collections of factors that express (or explain) what customers care about—similar to customer satisfaction factors from studies such as ServQual (Parasuraman, Zeithaml & Berry, 1988). For city marketing, we hypothesize that a reasonable set of example stakeholders would be:

1. **Residents:** This group includes current and potential residents.
2. **Employers/Local businesses:** This group also includes current and potential businesses.
3. **Employees:** includes both people who are residents, and those work, but who do not live, in the city.
4. **Visitors:** This group primarily includes tourists and conference attendees.

Clearly, the stakeholders may have incongruent views as to their own "quality dimensions" that they care about. For example, the resident would typically value low traffic congestion and good air quality, whereas the local businesses would value lower taxes and availability of an appropriate workforce. These quality dimensions could be congruent, but typically will not be as (for example) higher taxes might be needed to reduce traffic congestion.

The dimensions of perceived quality (benefits) are typically determined through in-depth knowledge about the stakeholders, and are measured through the use of surveys (see for example Parasuraman, Zeithaml & Berry, 1988; Ward, Rolland

& Patterson, 2005; or Burr et al., 2007, Rolland, Patterson & Ward, 2008). Indeed, from a quality of life and business perspective, such surveys are often found in the public domain already. For example, CNNMoney.com (Fortune, Money, and Business 2.0 magazine) publishes annual ratings for best cities for both living and business (CNNMoney, 2007; Forbes, 2007)—effectively addressing 2 of the city stakeholders: residents and businesses. Although the "quality dimensions" used in such articles sometimes are rather vague, the variables for best places to live are typically centered on variables such as job, income and cost-of-living data; housing affordability; school quality and education scores; crime rates; arts and leisure opportunities; ease of living; access to airports or teaching hospitals, and others. For best places for business, the variables often used are business costs, living costs, education, crime rates, job growth, and income growth. It should be noted that these quality dimensions differ from those often proposed in the literature. For example, ServQual (Parasuraman, Zeithaml & Berry, 1988) proposes 10 original quality dimensions for evaluating service quality: Access, Communication, Competence, Courtesy, Credibility, Reliability, Responsiveness, Security, Tangibles, and Understanding/Knowing the Customer. Sullivan and Estes (2007) implemented ServQual for local governments, and used all the 10 quality dimensions proposed in the original ServQual instrument, albeit many studies of applications of ServQual across various industries do not find support for all 10 quality dimensions. Indeed, a smaller subset of 5 ServQual dimensions is often used: Tangibles, Reliability, Responsiveness, Assurance, and Empathy. Ward, Rolland and Patterson (2005) found 4 quality dimensions to be dominant for healthcare organizations: Interaction & Communication, Access, Tangibles, and Outcome. The dimensions were found by analyzing more than 60,000 surveys of healthcare patients over a seven-year period. It would be expected that perceived service quality dimensions for cities would most likely show 3-10 dominant dimensions, where the variables listed for popular domain surveys would group into several more generic service quality variables. For the sake of the examples in this paper, we will use the following sample quality dimensions:

1. **Interaction and Communication:** This includes all interaction and communication with city resources.
2. **Access:** This could include access to city facilities and systems, schools, hospitals, airports, and others.
3. **Tangibles:** This includes cost of doing business/living, school performance scores, quality of life, quality of labor, crime rates, and other things that can be measured or observed, such as city equipment, personnel, and communication materials.
4. **Reliability:** A city's ability to perform the promised services dependably and accurately.

Once the stakeholders' perceived quality dimensions are known, we need to construct an interaction-space matrix to relate the quality dimensions to the organization resources that the city controls. Thus, for each stakeholder, we generate interaction-space matrices as shown in the examples in Figures 1 and 2. For each interaction between an organizational asset and a quality dimension, we find an interaction-space. Each interaction-space can hold zero or more interaction-points. These interaction-points constitute an opportunity for the city to interact with the stakeholder in a manner that matters positively to the stakeholder. Particularly, each interaction-point may constitute an opportunity for the city to interact with the stakeholder using electronic means (such as the internet). By definition, the interaction-points may constitute a crucial part of a CRM system, since such systems are seeking opportunities to understand and improve any interaction (or potential interaction) that takes place between the organization and its customers. Figure 3 shows the collection of interaction point matrices for 3 city stakeholders: Residents, Employers, and Visitors.

The measurement and analysis of the interaction-points can be done by surveying the customers. Survey instruments need to be developed for each stakeholder, and data collected on a regular basis to monitor and analyze interaction-points. For example, a survey instrument should be developed that measures the customers' perceptions of the e-services. That is, the surveys are measuring the stakeholders' perceived value and their priority in relation to the organizational assets. A collection of statistically significant questions would be needed to assess each interaction-point. Such surveys typically consist of 5 point, Likert-scale questions, and must be tested and statistically validated for the survey to be deemed useful. City management can use these surveys as a basis for allocation of resources to different interaction points based on their priority and perceived value to stake holder so as to get the best return on their resources. For example, in the healthcare industry medical providers and hospitals conduct annual surveys that may be very useful in the context of healthcare; in the US these surveys are often mandated by national quality

Figure 1. The SSRM framework for city residents

Organizational Assets	Quality Dimensions for City Residents			
	Interaction & Communication	Access	Tangibles	Reliability
Personnel	Interaction point	Interaction point	Interaction point	Interaction point
	Interaction point	Interaction point	Interaction point	Interaction point
Operational Processes	Interaction point	Interaction point	Interaction point	Interaction point
	Interaction point	Interaction point	Interaction point	Interaction point
	Interaction point	Interaction point	Interaction point	Interaction point
Facilities & Operational Assets	Interaction point	Interaction point	Interaction point	Interaction point
	Interaction point	Interaction point	Interaction point	Interaction point
Customer	Interaction point	Interaction point	Interaction point	Interaction point
	Interaction point	Interaction point	Interaction point	Interaction point
External	Interaction point	Interaction point	Interaction point	Interaction point
	Interaction point	Interaction point	Interaction point	Interaction point

Figure 2. The SSRM Framework for city businesses

Organizational Assets	Quality Dimensions for Employers/City Businesses			
	Interaction & Communication	Access	Tangibles	Reliability
Personnel	Interaction point	Interaction point	Interaction point	Interaction point
	Interaction point	Interaction point	Interaction point	Interaction point
Operational Processes	Interaction point	Interaction point	Interaction point	Interaction point
	Interaction point	Interaction point	Interaction point	Interaction point
	Interaction point	Interaction point	Interaction point	Interaction point
Facilities & Operational Assets	Interaction point	Interaction point	Interaction point	Interaction point
	Interaction point	Interaction point	Interaction point	Interaction point
Customer	Interaction point	Interaction point	Interaction point	Interaction point
	Interaction point	Interaction point	Interaction point	Interaction point
External	Interaction point	Interaction point	Interaction point	Interaction point
	Interaction point	Interaction point	Interaction point	Interaction point

Figure 3. The SSRM framework showing 3 stakeholders

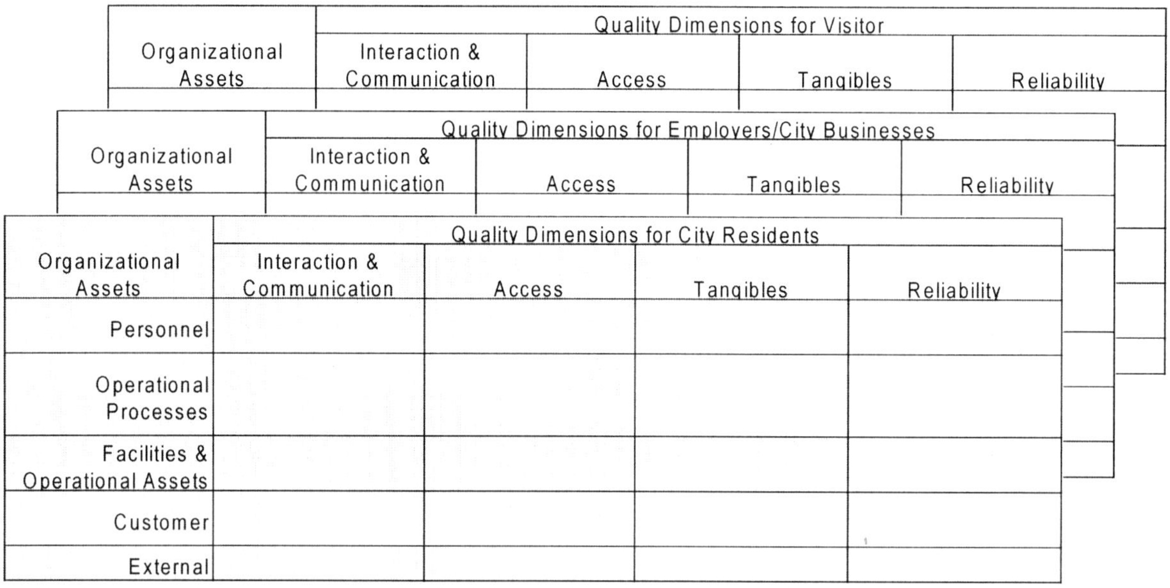

organizations such as the Agency for Healthcare Research and Quality (AHRQ). AHRQ has an online site with copies of their questions and surveys used for an annual survey where consumers assess their healthcare providers (AHRQ, 2008). J.D. Power uses similar surveys in their highly publicized annual quality measures (J.D. Power, 2008). For a brief introduction to survey use in government planning, complete with resource links, see Dawes et al. (2004)

Provided that cities regularly use and measure the performance of their interaction-points through such surveys, the organizational assets can then in turn be changed/removed/improved to promote necessary improvements in the interaction-point measures as needed. The relative importance of the interaction-points can be measured by factor analysis (Burr et al., 2007; Rolland et al., 2008), or by administrating direct questionnaires as to what the stakeholder opinions are regarding importance of interaction-points.

One major task in making the framework functional is to align stakeholder goals: that is, ensure congruency in the stakeholders' perceived quality views of the city. Rolland and Maghroori (1997) proposed that efficient organizations must be responsive to all its stakeholders. They propose a set of actions (mission realignment strategies) to correct for situations where this is not the case. The interesting thing to note about this, is that the when the stakeholder goals are aligned with the organization's mission, there is by definition congruency in the stakeholder views (Rolland & Maghroori, 1997). Thus, aligning stakeholder views may entail some tuning of the organization's mission. That is, the exercise of identifying stakeholders, perceived quality dimensions for the stakeholders, and interaction-points for the stakeholders vis-à-vis organizational assets may indeed help define or improve the goals of the city.

An Illustrative Example

In our example, we will develop the some interaction-points related to the 3 stakeholder views found in Figure 3 above. For the sake of this example, we assume that the city managers will be addressing 3 stakeholders: city residents, city businesses, and visitors. We further assume, based on surveys of the stakeholders as well as on findings from the research literature, that the 4 perceived quality dimensions dominant for servicing the 3 stakeholder groups are: Interaction & Communications, Access, Tangibles, and Reliability.

Figure 4 shows potential interaction-points for city residents. Under the *Interaction & Communication* column we see that potential interaction points between residents and city employees are through phone and online chat systems (other possibilities do of course exist). The operational processes that the city residents seeks to interact with are billing systems, appointment systems, a reminder systems (for appointments and bills due etc.) They also seek to communicate with a reservation system for use of public space/facilities.

Under the *Access* column, we see that city residents seek to have access to the systems by which they sought to communicate (in the previous column), as well as have access to other systems, such as pet licensing and emergency room services. Access to emergency room services could for example be via live chat or telephone, or web scheduling, etc. We also note that the residents want access to training as to how to use the city's e-service systems.

The *Tangibles* dimension primarily lists factual information about issues related to the city's organizational assets. For example, information about: the city employees' qualifications, school rankings, crime rates, water and air quality, and tax issues. Additionally, the residents seek to know about the telecommunications infrastructure (for example, what broadband options are available), and if a free wireless network is available. The latter two issues are related to resources external to the city's organizational assets, but show that it may be important for cities to attract such services from the external industry.

Reliability would include adequate city staffing around the clock—particularly of police services and other emergency services. It also would include access to city e-service 24 hours per day, including holidays, and also may include statistics of city-operated facilities (police, fire, electric, water, etc.) in terms of service reliability. Of course, residents expect that all city services that are conducted electronically can be performed securely (IT security).

In Figure 5, we list some potential interaction-points for the employers/city businesses. This

Figure 4. Potential interaction-points for city residents

Organizational Assets	Quality Dimensions for City Residents			
	Interaction & Communication	Access	Tangibles	Reliability
Personnel	1. Phone system, including live operator 2. Live chat system	1. Live phone operator 2. Live chat with qualified employees	1. Information about employee qualification	1. Live person available 24 hours 2. Adequate police/fire staffing
Operational Processes	1. Billing system 2. Appointment system 3. Reminders - status reports	1. Appointment scheduling 2. Maps and directions 3. Status of permits/licenese/.. Applications 4. Utility services (electric, water, trash, gas) 5. Pet licensing	1. School rankings 2. Crime rates 3. Water and air quality 4. Tax rates, property taxes	1. Automated phone system available 24/7 2. Web system available 24/7 3. Provide information during non-working hours 4. IT security
Facilities & Operational Assets	1. Reservation system for use of public space	1. Available web services for all city issues 2. Emergency room services	1. Pools, parks, tennis courts, golf courses	1. Statistics of faciltities
Customer		1. Training for use of city e-Services		
External			1. Telecommunications infrastructure for the residents 2. Free wireless web access	

Figure 5. Potential interaction-points for city businesses/employers

Organizational Assets	Quality Dimensions for Employers/City Businesses			
	Interaction & Communication	Access	Tangibles	Reliability
Personnel	1.Phone system, including live operator 2. Live chat system	1. Live phone operator 2. Live chat with qualified employees	1. Information about employee qualification	1. Live person available 24 hours 2. Adequate police/fire staffing
Operational Processes	1. Billing system 2. Appointment system 3. Reminders - status reports	1. Appointment scheduling 2. Maps and directions 3. Status of permits/licenese/.. Applications 4. Processing times	1. School rankings 2. Crime rates 3. Water and air quality 4. Business costs and times (starting new business)	1. Automated phone system available 24/7 2. Web system available 24/7 3. Provide information during non-working hours 4. IT security
Facilities & Operational Assets	1. Reservation system for use of public space	1. Available web services for all city issues	1. Conference centers, pools, parks, tennis courts, golf courses	1. Statistics of faciltities
Customer/Employer		1. Training for use of city e-Services	1. Business growth rates and profiles	
External		1. Network of similar businesses	1. Telecommunications infrastructure for the businesses 2. Economic growth rates	

Figure 6. Potential interaction-points for visitors

Organizational Assets	Quality Dimensions for Visitors			
	Interaction & Communication	Access	Tangibles	Reliability
Personnel	1. Dedicated tourist service personnel and info	1. Availability of personnel with knowledge of regional tourism	1. Quality of personnel with knowledge of regional tourism	1. Live person available 24 hours 2. Police staffing
Operational Processes		1. Maps and directions 2. Ticket and transit pass purchases 3. Transit system, transferability	1. Quality of Hotels 2. Airports 3. Weather and air quality	1. Automated phone system available 24/7 2. Web system available 24/7 3. Provide information during non-working hours
Facilities & Operational Assets			4. Quality of [illegible] recreational areas 5. Crime rates	[illegible]ability between transit systems
Custolmer			1. Visitor profiles	
External			1. Available recreational activities 2. Free wireless web access	

interaction-space matrix differs only slightly from the resident's matrix. For example, an important *tangible* related to city facilities could be the existence and size of conference centers for business conferences. Also, businesses would be interested in business growth rates, and profiles of local industry (both variables related to the employers themselves), as well as economic growth rates for the region/state (an issue external to the city). Of course, businesses would like to have access to processing times for business license issues, but they normally would not care about pet licensing as residents do. Also, businesses may care about access to a network of similar businesses for purposes of professional exchange (for a Silicon Valley-like environment), skilled labor pool, university and vocational institutes.

Figure 6 shows some potential interaction-points that visitors may deem as important. For example, potential visitors may want to have access to and interact with city personnel who are dedicated to tourist services. They also want access to maps and directions, as well as purchasing tickets for public transportation systems (all access issues). Visitors are also interested in tangibles (mostly information) related to air quality, weather, hotels, airports, beaches, mountains, and other recreational areas. On the *reliability* side, visitors would be interested in the reliability and interoperability of public transit systems, police staffing in the city, and perhaps reliable availability of city tourism experts even on holidays and nights.

Links to Organizational Strategy

Strategic planning in an organization must take into consideration an assessment of the firm, planning for the future, and implementation of the plan (see Figure 7). Bryson et al. (2007) proposed a resource-based approach to strategy formulation and implementation that is relevant to the framework proposed in this paper. In their paper, Bryson, Ackermann and Eden (2007) presented mapping methods for identifying distinctive organizational competencies, as well as a livelihood scheme that shows what the organization desires, as supported by the organizational competencies (Eden & Ackermann, 2001).

The SSRM method, as proposed in this paper, can certainly build upon (and is informed by) the approach by Bryson et al. (2007), but also is more specific in its treatment and identification of interaction-points—the latter being crucial in an e-service context. For example, Bryson, Ackermann and Eden (2007) identify distinct organizational

Figure 7. SSRM and organizational strategy

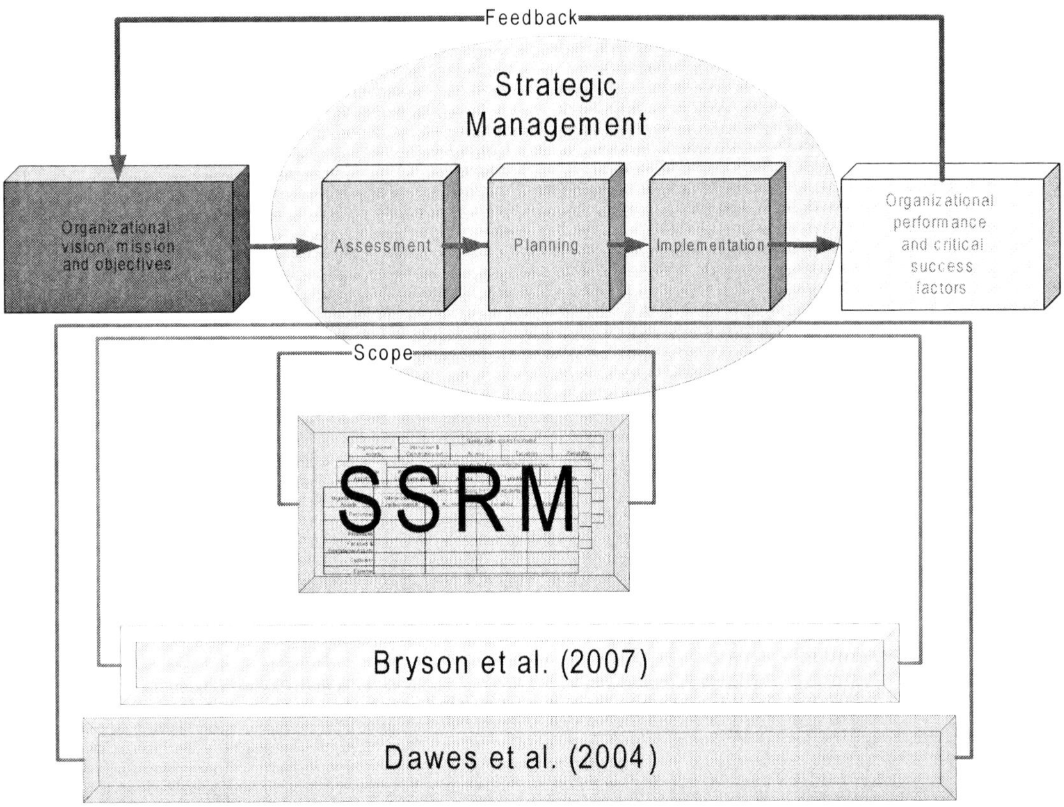

competencies. These competencies are typically thought of as things the organization do especially well in comparison with other organizations. Such items may be thought of as subsets of the organizational resources, and would typically fall under Operational Processes in the SSRM framework. Thus, the e-service relevant capabilities found by using the method proposed by Bryson et al. (2007) can be thought of as various Operational Processes in the SSRM framework.

The SSRM framework discussed above is designed to allow for some guidance in the strategic planning process, without limiting the creativity of the city managers. That is, the goal is to generate e-service interaction-points that tie firm resources to customer-perceived issues. This planning process is clearly a creative one, and it would be counter-productive to specify in detail what must take place in such a process. However, one could clearly integrate the SSRM with existing planning methods, such as the one suggested by Bryson, Ackermann and Eden (2007). The SSRM framework by itself simply enables the managers' creative planning to take place within the context of firm resources and customer-perceived quality issues.

FUTURE TRENDS

It is clear that the development of e-services for cities is progressing rapidly. A large number of cities have implemented many of the services discussed in the examples above. However, it is not clear what tools have been available to city managers to conduct studies to frame proper analyses of stakeholders and interaction-points between organizational assets and perceived quality in the e-service context. Dawes et al. (2004) have presented and summarized an excellent collection of general tools and tips for improved use of ICT in government organizations, and these tools and tips are applicable across the strategic management processes. Bryson, Ackermann and Eden (2007) have proposed specific and important methods to better help government organizations to plan strategically. The SSRM method complements this prior work by specifically aiding the development of e-services in the strategic planning process. The potential strategic management scope of the research by Dawes et al. (2004), Bryson, Ackermann and Eden (2007) are shown in relationship to the SSRM in Figure 7. The rectangles emanating from each method shows the potential use of the method in strategic management. We assume that the proposed SSRM framework, along with the works of Bryson, Ackermann and Eden (2007) and Dawes et al. (2004) should be both useful and of crucial importance to city managers. This should be true whether or not one believes that cities compete with one another: at any rate, all cities are charged with providing its stakeholders with the best services in the most cost efficient manner.

Numerous future research opportunities exist in fine-tuning the organizational assets in the light of the perceived quality dimensions. That is, organizational assets can perhaps be broken into smaller chunks in order to see a finer picture of how the organization (the city) should relate its resources to the perceived quality dimensions. For example, one might want to identify the actual organizational processes, and compare those to existing or new processes that might serve the stakeholders' perceived quality better.

One limitation of this proposed method is clearly that while parts of the framework have been well tested in the healthcare setting (Burr et al., 2007; Rolland, Patterson & Ward, 2008), it is otherwise unproven in the city setting. That is, the perceived quality dimensions used herein must be proven significant for the purposes of city marketing in its e-service context. As such, another fruitful avenue for future research lies in verifying or finding the perceived quality dimensions for city marketing. Even with such a limitation, this framework constitutes the first attempt to align city stakeholders and their perceived quality dimensions to organizational assets through identifying and measuring/monitoring interaction-points in order to improve a city's e-service efforts.

CONCLUSION

In this paper we have proposed a framework to help cities plan their e-service strategically. The approach includes identifying the city stakeholders, aligning the goals of these stakeholders, as well as framing the perceived dimensions of quality inherent to each stakeholder. Each stakeholder's perceived dimension of quality can now be linked to the city organizational assets by identifying one or more interaction-point between the city's organizational asset and the stakeholder's perceived quality dimension. These interaction-points are touch-points between the city and its stakeholders.

By measuring and monitoring the interaction-points on a regular basis, using for example annual stakeholder surveys, the city can adjust the "functionality" of its organizational assets, or adjust the use of interaction points, in manners which lead to higher stakeholder-perceived service quality. For example, the city can change its assets by changing its operational processes, training its personnel, acquire or improve its facilities and operational assets, or influence the stakeholders or external entities to cooperate with the city. Also, the framework enables the city to monitor the importance of the interaction-points (by using statistical techniques - factor analysis and regression) in order to de/emphasize or delete/change interaction points. The framework also allows city managers to think about potential new interaction-points that their stakeholders may value highly—particularly such interaction points that may already be in use at "competing" cities.

The impacts of using the framework are in improved service to all of the city's stakeholders. From a management perspective, the application of the method proposed herein requires deep local knowledge along several dimensions: First, the managers must clearly understand who the stakeholders are, and what these stakeholders care about in terms of perceived quality. The strength of the proposed method is that the final interaction-points outline a broad map for the design of the city's e-services (that is, direct things to put into the web site design related to the interface for each stakeholder). Further, the framework allows city managers to think about interaction points in a semi-structured way, which does not inhibit creative thinking or discussion about the interaction-points. Use of the framework for strategic purposes (setting direction both for goals and processes) may contribute strongly to the city's dynamic capabilities, and in turn their competitiveness as a city (Rolland et al., 2008). While it has been shown that increases in the customers' perceived service quality leads to an increase in firm value (Mithas et al., 2005), we argue that this relationship should persist even if, as is normally the case, the city is a non-profit entity.

REFERENCES

AHRQ (2008). Consumer assessment of healthcare providers and systems. Retrieved February 21, 2008, from https://www.cahps.ahrq.gov/default.asp

Barney, J., & Hesterly, W.S. (2006). *Strategic management and competitive advantage*. Upper Saddle River, NJ: Pearson Prentice Hall.

Blume, L. (2006). Economic policies as determinants of the local business climate: Empirical results from a cross-section analysis among East German municipalities. *Regional Studies, 40*(4), 321–333.

Bryson, J.M., Ackermann, F., & Eden, C. (2007). Putting the resource-based view of strategy and distinctive competencies to work in public organizations. *Public Administration Review, 67*(4), 702-717.

Burr, C., Patterson, R., Rolland, E., & Ward, K. (2007). Integration of E-CRM in healthcare services: A framework for analysis. *International Journal of E-Business Research, 3*(2), 1-12.

CNNMoney (2007). Money best places to live 2007. Retrieved August 1, 2007 from http://money.cnn.com/magazines/moneymag/bplive/2007

Dawes, S.S., Pardo, T.A., Simon, S., Cresswell, A.M., LaVigne, M.F., Andersen, D.F., & Bloniarz, P. A. (2004). *Making smart IT choices—Understanding value and risk in government IT investments*, 2nd edition. Center for Technology in Government, University at Albany, SUNY.

Der Spiegel (2007). Bad Air in Beijing? Pollution dangers cast shadow over 2008 Olympics. Retrieved June 28, 2007, from Der Spiegel Online at http://www.spiegel.de/international/world/0,1518,491184,00.html

Eden, C., & Ackermann, F. (2001). A mapping framework for strategy making. In A. S. Huff and M. Jenkins (Eds.), *Mapping strategic knowledge* (pp. 173-195). London: Wiley.

Forbes (2007). *Best places for business and careers*. Retrieved August 1, 2007, from http://www.forbes.com/lists/2007/1/07bestplaces_Best-Places-For-Business-And-Careers_land.html

J.D. Power (2008). *Business ratings*. Retrieved February 23, 2008, from http://www.jdpower.com/business

Krugman, P. (1996). Making sense of the competitiveness debate. *Oxford Review of Economic Policy, 12*, 17–25.

Krugman, P., & Venables, A. J. (1993). *Integration, specialization and adjustment*. Discussion Paper No. 886. Centre for Economic Policy Research (CEPR), London.

LA Times (2007). *Clearing the air for Beijing Olympics—China is moving factories, reducing traffic and even changing the weather to limit pollution at the '08 Games*. Retrieved August 11, 2007, from http://www.latimes.com/news/printedition/asection/la-fg-air11aug11,1,2200892.story?coll=la-news-a_section&ctrack=4&cset=true

Maghroori, R. &, Rolland, E. (1997). Strategic leadership: The art of balancing vision with policy, procedures, and external environment. *The Journal of Leadership Studies, 4*(2), 62-81.

Mithas, S., Krishnan, M. S., & Fornell, C. (2005). Why do customer relationship management applications affect customer satisfaction? *Journal of Marketing, 69*, 155-166.

Parasuraman, A., Zeithaml, V. A., &, Berry, L. L. (1988). A multiple-item scale for measuring consumer perceptions of service quality. *Journal of Retailing, 64*(1), 12-40.

Payne, A., & Frow, P. (2005). A strategic framework for customer relationship management. *Journal of Marketing, 69*, 167-176.

Rolland, E., Patterson, R. A. &, Ward, K. F. (2008). Dynamic capabilities and E-Service. Working paper, University of California at Riverside.

Rust, R. T., & Lemon, K. N. (2001). E-Service and the consumer. *International Journal of Electronic Commerce, 5*(3), 85-101.

Schermerhorn, J. S. (1999). *Management (2nd* edition), New York, NY: Wiley.

Sullivan, B., & Estes, C. (2007). Measuring customer service quality in local government: Fulton County Human Services shares its experience with surveys and recommends ways to use them to improve customer satisfaction. *The Public Manager, 36*(1), 37-39.

Ward, K., Rolland, E., & Patterson, R. (2005). Improving outpatient healthcare quality: Understanding the quality dimensions. *Health Care Management Review, 30*(4), 361-371.

Wong, C. (2001). The relationship between quality of life and local economic development: An empirical study of local authority areas in England. *Cities, 18*(1), 25–32.

Chapter IV
Strategic Technology Planning for the Techno-Global Economy: Cities in the Market

Al D. McCready
McCready Manigold Ray & Co., Inc., USA

ABSTRACT

Indications are strong that globalization is an irresistible force, fomented by, or at the very least, enabled by technology. This chapter refers to the technology driven aspects of globalization as "techno-globalization" and describes the role of strategic technology planning in the marketing of cities in this global economy. It describes strategic technology planning for information and communication technologies and its intersection with marketing planning. It is intended to guide managers through the technology planning aspects of ICTs and city marketing. In addition to providing practical guidelines for preparing a technology plan that supports the organization's strategic and marketing objectives, the chapter explains many of the nuances of the preparation and alignment of organizational strategic plans using current information systems and organizational theory concepts.

INTRODUCTION

Strategic technology planning is a specialized instance of strategic planning. Because of the particular nature of technology strategy, it embraces many elements of strategic planning, but necessarily includes technology planning considerations and technology evolution forecasting that are not normally part of standard strategic planning. Strategic technology planning can be thought of as a strategic view of the entity's technology infrastructure and requirements that results in the identification of "best fit" technology initiatives in support of the entity's overall strategic direction. In parallel, marketing planning takes the marketing perspective at both a strategic and tactical level. When viewed in relation to the marketing plan, the strategic technology plan moderates and supports the marketing plan at the tactical level and in its transition from a strategic perspective to a tactical focus. The chapter provides specific guidelines regarding the development and content of the strategic technology plan, but leaves the details of the actual plan preparation to each municipality's information technology governance and planning process. The chapter describes strategic technology planning, what it is, and how it properly integrates with the organizational strategic plan and peer strategic planning components.

Naturally, municipal organizations have many active plans in a multitude of areas. Some of these plans are strategic, some are financial and some are operational. The following chart, Figure 1, depicts some of the elements of the overall strategic planning process. This diagram shows the segments of the overall system of planning that are addressed by this chapter. The shaded areas are directly addressed and the areas without shading are not addressed or are mentioned only tangentially. The organizational strategic plan and the strategic technology plan are discussed. The role of strategic planning in marketing, the role of information and communication technologies (ICTs) in marketing, and the potential interactions with ICT providers are addressed.

This chapter is intended to provide useful information to people who have oversight or direct responsibility for a municipality's marketing planning and/or technology planning activities. The purpose of this chapter is to describe strategic technology planning and its intersection with ICTs, and their joint intersection with marketing planning as it relates to municipal marketing. Strategic technology planning is oriented to the specific requirements of planning technology to support the planning entity's overall strategic and tactical plans. Therefore, this chapter is a little less of a "what is" description and more of a "how to" explanation. The next sections will discuss strategic technology planning, the role of strategic technology planning in marketing planning,

Figure 1. The overall system of planning

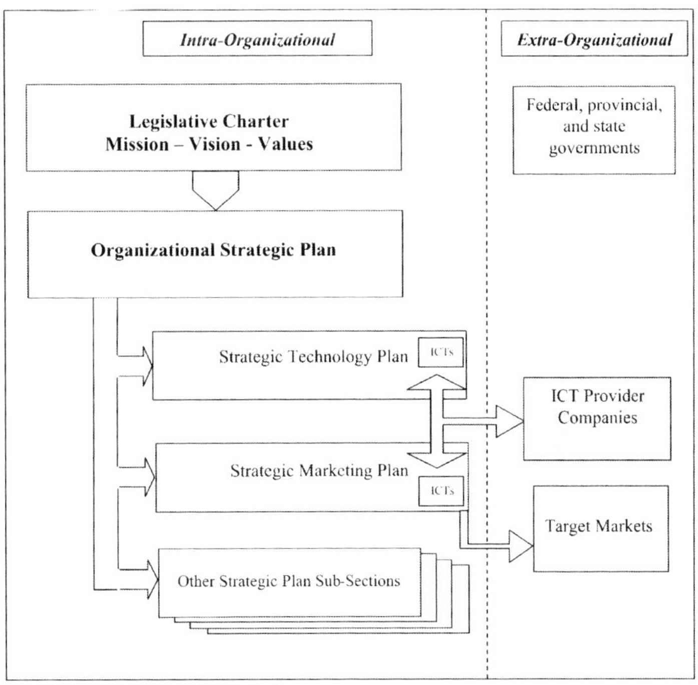

and, in parallel, the role of ICTs in marketing. This chapter also discusses the process of constructing the technology segment of your overall city marketing plan, and reviews certain tactical considerations related to successful implementation. It should be noted that, by their very nature, governmental entities are monopolies. However, when their considerations move beyond their geographic boundaries, they move from being the only one to being one of many. Planning, and specifically various types of strategic planning, is often couched in a "competitive" context. Within the monopoly, the competitive frame of reference has limited application. But outside the monopoly, in the "marketing as one of many" framework, the competitive frame of reference applies. It is from this perspective that commercial and quasi-governmental planning references are used in the following discussion.

STRATEGIC TECHNOLOGY PLANNING

The Techno-Globalization Planning Context

In many nations of the world, people are increasingly migrating into a technology laced way of being. The proliferation of Information and Communication Technology (ICT) is having an impact around the world, and the impact is not limited to the "industrialized nations" that are the usual suspects. Recent scholarly literature provides examples of the role of ICTs in many a far-flung place. Trinidad and Tobago have a national ICT plan (Warnaby & Bennison, 2006). There are examples in Mongolia (Amarsaikhan, Lkhagvasuren, Oyun, & Batchuluun, 2007; Uyanga, 2005), in Tunisia (Hulm, 2007), and the Philippines (Ramos, Nangit, Ranga, & Triñona, 2007). In his 2006 Nobel Prize acceptance speech, Muhammad Yunis (2007) describes how the success of his Grameen Bank in providing micro-credit to poor women in Bangladesh morphed into Grameen Phone. Grameen Phone, in what Yunis describes as the synergistic confluence of micro-credit and ICT, provides 300,000 women with mobile phones which they use to sell telephone services to villagers around Bangladesh.

Certainly many people in the world do not currently have access to technology. However, in many of the more industrialized nations of the world it appears to be mostly those who actively choose to avoid email and cellular phones that do not have them in their lives. For many people, the ubiquitous nature of ICTs makes them so much a part of daily life they are essentially invisible.

These contextual changes manifest themselves in international communications. The proliferation of ICTs worldwide now makes it possible to have a substantive virtually instantaneous document or data exchange with a person on the other side of the globe. This sort of instantaneous global communication is now many people's expectation. It is now of note when someone separates themselves from ICT connectedness by temporarily abandoning their technology devices or traveling to locations without connectivity. Techno-globalization is the proliferation of ICTs and the resulting information and communication exchange that has become institutionalized in much of the global society.

This institutionalization is the new personal existence paradigm for an increasing number of people. These are the digital aboriginals. There is a book (Tarlow & Tarlow, 2002) with these words in the title, but here we will use our own definition. Digital aboriginals are here defined as people who have grown up in a quasi-digital environment, and who consider solid state portable music/video players, computerized games, and handheld computing/telephone devices part of the wallpaper of their lives. These digital natives have ICT expectations that appear to exceed the capabilities of the extant ICT infrastructure and their numbers are growing throughout the world. Naturally, there really is no definitive line between digital aboriginals and those who are not. This is more of a continuum and we all exist somewhere along that scale. Furthermore, this is a multi-dimensional issue and we may be at different points on the continuum in our personal lives than in our professional lives. However, it is fair to extrapolate this line of thinking to the conclusion that increasingly, people move through their lives in a personal technological environment that might be referred to as a "techno-bubble," which is, of course, highly dependent on ICTs. A person's "techno-bubble" can be defined as that assemblage of technology that ranges from our mobile phones and mobile computing devices to automobiles with "brains" and their own mobile data connections. For many people this is the technological environment or even the techno-ecosystem in which they live, and, in many cases, move about. A question to consider is how this evolution and migration of peoples' expectations will impact geographically oriented marketing efforts.

The Planning Paradigm

Plans ranges from the very broad and high level mission statements that may be applicable over decades, down to very tactical task lists related to that day's events. There are an enormous number of documented approaches to sub-dividing and structuring every step along this planning continuum. In general, any planning that is prefaced with the word "strategic" differentiates itself by virtue of the fact that it relates more to strategy than it does to tactical or financial considerations. In different words, it relates to what are we intend to do and why are we doing it, more than it relates to exactly how we will do it or how much will it cost or how we will pay for it. The strategic planning process was explained well by Porter (1980; 1985) and has since been suitably elaborated by many other authors. When strategic planning is done in a governmental organization, public policy considerations often supplant, or at least supplement, the typical mission and purpose guidelines used in commercial strategic planning work. Naturally, this moderates the applicability of commercial sector planning approaches in governmental planning.

An interesting example of this difference is described by Davison, Wagner and Ma (2005) in their analysis of various paths that governments can take in the transition to e-government. They specifically address the somewhat pregnant pause between the adoption of web-based technologies for commercial B2B (business to business) and B2C (business to consumer) interactions and governmental adoption of the same for G2B (government to business) and G2C (government to citizen) interactions. These authors note that there usually is a significant difference between a commercial B2C website's interest in keeping the consumer on the site as long as possible and the governmental G2B or G2C website's interest in providing information or service quickly and allowing the citizen to go on their way as quickly as possible. It is a difference that is typical of the difference between commercial and governmental planning requirements. Beyond their useful explication of this differentiation, Davison, Wagner and Ma (2005) do a good job of describing the most successful paths to be taken in planning the transition from government to e-government. In a similarly insightful way Mason (2007) does a good job of using complexity theory to describe the external environment's effect on the development of strategic plans. Overall, the planning process often has multiple mediating, moderating and sometimes intervening factors that will influence the direction established by the plan. These influencing factors, in turn, are appropriately expressed in the follow-on tactical and operational plans.

Tactical plans address the implementation and operational issues, and financial plans address costs and funding issues. In order to avoid deadlocks or an excessive number of feedback cycles and iterations, many strategic plans include some high level revenue and expense projections thereby setting the direction and magnitude expected. Of course, one of the traps here is that operations oriented managers who have trouble dealing at the true strategic level will often turn a strategic planning process into a tactical or financial planning process. The balancing act of keeping the strategic planning strategic while also keeping it feasible and affordable benefits from the involvement of experienced executives whose accumulated experience helps guide the strategic with a practical and tactical perspective.

> **Guideline 1:** *Keep the plan strategic, but include high level revenue and expense projections.*

The organization's strategic plan addresses the entire organization and properly sets the direction and overall bounds for all other organizational planning. Each significant organizational function or department could reasonably develop their own sub-strategy and prepare their own strategic plan. If a city is interested in marketing itself, for example, then a strategic marketing plan that is supportive of the direction established in the super-ordinate strategic plan is appropriate. Another area that very often warrants the development of a supporting strategic plan is information technology or information systems.

The Intersection of Strategic Planning and Strategic Technology Planning

A great deal has been written about strategic technology planning. Commercial organizations regularly prepare strategic information systems plans (SISPs) or strategic information technology plans (SITPs) as part of their strategic planning process. Information technology researchers, theoreticians and practitioners regularly write about the SISP/SITP process. From a taxonomical perspective, the scope of information systems (IS), information technology (IT), and information and communication technology (ICT) are, technically, all different, though they describe substantially overlapping domains. When these overlapping domains are raised to a strategic planning level, the differences begin to diminish. The choice between them is often driven by organizational history and the orientation of those managing the planning effort. The IS domain often includes data and processes not entirely computer systems based, where IT is usually essentially a technology oriented domain. The ICT domain is very closely aligned with the IT domain, but the term is more popular in countries other than the U.S.A. Since, at the organizational planning level, information systems, information technology, and communication technology all need to be addressed, this chapter will use the terms interchangeably in their more generic sense and abbreviate to SISP/SITP to refer to that strategic plan. The following paragraphs summarize some noteworthy thinking on these topics that is relevant to the planning process discussed in this chapter.

Malnight and Keys (2007) provide useful perspective on tuning the planning process to global change in their article entitled *Surfing the storm: Translating long-term global trends into today's decisions.* They discuss the advantages of taking a continuous and systematic approach to scanning the environment and collecting information about what it means to the organization. Applied here, their ideas indicate that there needs to be a continuous effort at gathering the relevant data, constructing shared insights, understanding the implications for the city and making appropriate decisions. This advice is particularly salient to the topic at hand given the fast paced changes in ICTs and the challenges of incorporating technology planning into the marketing planning and implementation process. It can be done. Wonglimpiyarat (2007) provides an example of how a similar process was successfully used by the country of Thailand. Wen and Shih (2006) describe how another large Asian country proactively managed strategic IT prioritization.

Guideline 2: *Make a practice of systematically and continuously scanning the environment.*

One aspect of the intersection of strategic planning and strategic systems planning is the need for alignment of technology planning with the organizational strategy. A great deal has been written about the need, and how it can be effectively accomplished. For example, in 1999 Henderson and Venkatraman wrote an excellent monograph presenting their "Strategic Alignment Model" (1999, p. 472). They parse the process into four domains, business (organizational) strategy, IT strategy, organizational infrastructure and processes, and IT infrastructure and processes. As you can see in their breakdown, organizational strategy and IT strategy are the first two essential components. They also take their alignment model to the next step and insightfully incorporate the organizational dynamics and the technology infrastructure. These second two are an important insight and are highly recommended for inclusion in the city marketing planning process.

Guideline 3: *Beyond alignment, also incorporate organizational and infrastructure considerations.*

The question of alignment is not always intuitively obvious. Even when the overall plan and various sub-parts are in place, they can have substantially different orientations. This can occur in various degrees and in some, or all, components. Figure 2 represents this concept diagrammatically. The plans need to be aligned under a single overall strategy. They need to be directionally aligned meaning that the route to achieve the organizational goals is coordinate. They need to be aligned in terms of scope; a narrow scope sub-section plan may not be out of alignment in any way other than the fact that it simply doesn't cover enough ground, or perhaps is overly ambitious. They also need to be aligned

Figure 2. Alignment of plans

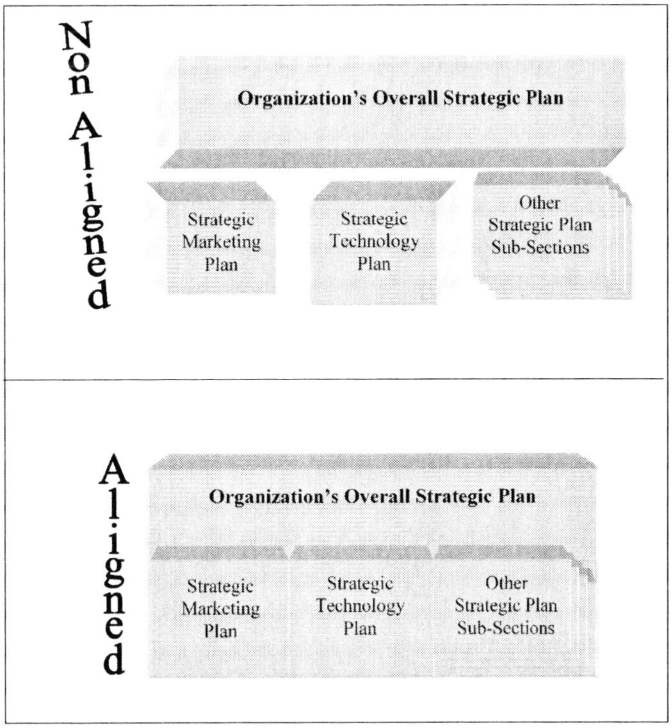

horizontally, meaning the peer-to-peer sub-sections are aligned.

Many other authors, researchers, and theoreticians have written about IT alignment. Peak and Guynes (2003) describe a successful IT alignment implementation at the Omaha Public Power District. Chan, Sabherwal and Thatcher (2006) report on their research into factors affecting the IT alignment process at both business and non-profit organizations. Huang and Hu (2007) describe how to use the balanced scorecard system to achieve IT alignment. Kearns and Sabherwal (2007) present a knowledge based view of achieving IT alignment. These research articles and reports are a sampling of the broad range of available literature on IT alignment. Two observations are immediately evident. First, the need for, and benefits of, strategic technology planning alignment with the organizational plans are widely discussed and accepted as a necessary part of successful planning. Second, if your own organization has a unique need or a preferred approach, it is very likely you will be able to find specifically applicable literature to study.

It may also be helpful to note how other governmental organizations view the SISP/SITP question, and here are some examples. Thorogood, Yetton, Vlasic and Spiller (2004) describe an instance at a Southern Australian governmental organization where a public policy decision was successfully implemented through an SITP incorporating IT alignment. Another interesting example is the work by Holley, Dufner and Reed (2002; 2004) who have written two research articles. One addresses SISPs in U.S. state governments and the other addresses SISPs in county governments. Morton (2006) takes a social sciences perspective and uses critical realism to analyze IT alignment in a multi-department governmental organization.

And finally, Roge and Chakrabarty (2003) have written about their research into IT's integration with marketing strategy and operations. These authors indicate that their research shows the evolution and development of the integration or alignment of IT with marketing operations has progressed to the point that IT is now regularly integrated with marketing strategy. Taking the concept of alignment to the next logical step, we can now look at not only the vertical alignment of IT strategy with organizational strategy, but also the horizontal or peer-to-peer alignment of IT strategy with marketing strategy.

Guideline 4: *Be sure the SISP/SITP is fully aligned with the organization's strategy and the marketing strategy.*

How Strategic Technology Planning Integrates with Other Types of Planning

As was mentioned above, strategic technology planning is oriented to the specific requirements of planning technology in support of the overall strategic plan. One of the more challenging aspects of this process is lack of clarity, and sometimes the out-right absence (non-existence), of the overall strategic plan. The organizational strategic plan is often an emergent design, and even when brought to a firm conclusion, the best of them are immediately open to revision and improvement.

Consequently, a successful approach to strategic technology planning requires an element of simultaneous development with the overall plan and any concurrent peer-to-peer sub-section development. This process requires the technology planners to be ready to facilitate and enable the development of the grand plan. The questions asked, the ideas developed, and the documented progress can all help move forward a deficient or incomplete overall plan.

Guideline 5: *These plans are best done in a cooperative, simultaneous co-development mode.*

In addition, the technology planners must actively interact with any concurrent planning efforts, including the marketing plan develop-

Figure 3. Planning components and relationships

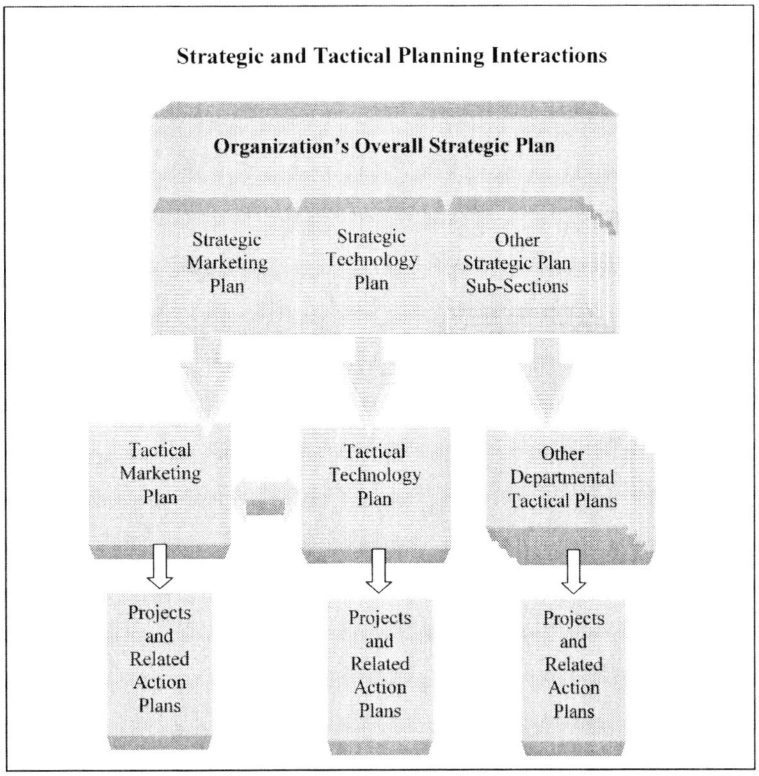

ment. As is depicted in Figure 3, below, there is a top level strategic planning component that defines organizational mission, purpose and overall direction. The supporting strategic plan sub-sections address strategic planning for the various major sub-sections or departments, such as the strategic marketing plan and the strategic technology plan.

Key Elements of Strategic Technology Planning

Strategic technology plans usually incorporate the conceptual development of enterprise software, hardware, networking, and personnel plans. The plan typically includes identification of the preferred technologies, and a staged program for the acquisition, implementation, and support of various technological solutions. Technology planning also frequently incorporates consideration of external entities and their technological direction and potential for success. The reasons for this can range from a simple reliance on an external provider up to a tight lockup with a true technology partner. Unlike strategic technology planning for an enterprise with a more limited scope of activities, the strategic technology planning in support of a city, including its marketing planning can have a rather broad perspective and naturally it should address internal and external considerations of all major ICTs.

Guideline 6: *Setting the scope of the plan is very important and should receive careful consideration.*

The strategic information technology/systems planning process. Although there are many acceptable variations, the SISP/SITP processes generally have broad elements of similarity. It usually starts with an assessment that answers the question "where are we right now?" One aspect of this current assessment is the evaluation

Figure 4. The SISP/SITP planning process (© 2007 McCready Manigold Ray & Co., Inc.)

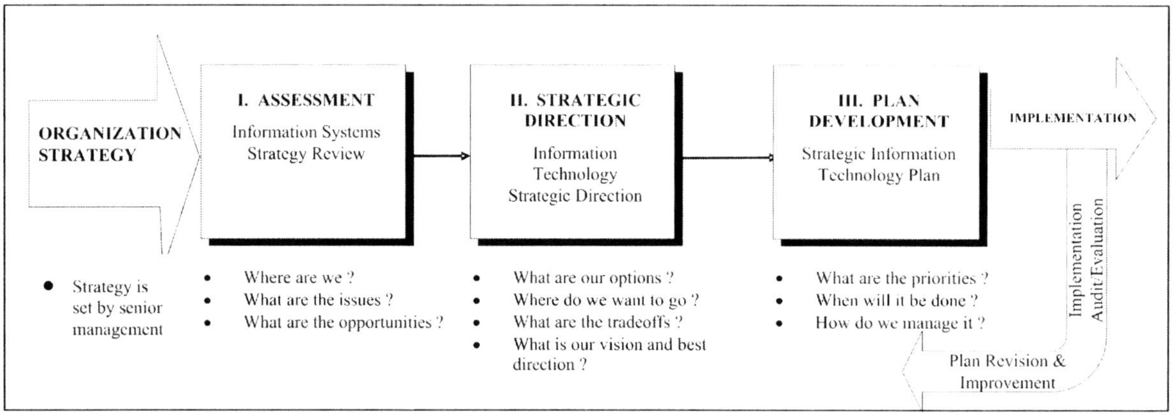

of both the state of the implemented technology and the organizational environment, including the status of the organization's strategic plan. The next phase of the planning process answers the question "where do we want to go?" This phase is typically the development of one or several systems and technology strategic initiatives that maintain alignment with the organizational strategy. The third phase, the "how will we get there?" phase is the plan development that structures the specifics and establishes priorities within the strategic direction and approach established in the second phase. The Figure 5 depicts the flow of the process.

The written plan combines the products of all three phases of the planning process. Naturally there are many variations on this planning process, but processes that do not incorporate all of these elements should be used with caution. Once the implementation process has been initiated, the quality of future planning efforts can be improved by a plan evaluation process. This process is based on the plan evaluation criteria established at the outset plus a practical ex post facto perspective. Following the principle of commensurate complexity (Thorngate, 1976; Weick, 1979), larger organizations require more detailed and complex plans, and typically will have multiple large subsections in each phase that deal with individual organizational subdivisions and major initiatives.

> **Guideline 7:** *Establish the audit and evaluation criteria for the plan during Phase One and Two.*

The components of the plan document. Typically most SITPs or SISPs will have an outline that is reasonably similar. Once again, the size, complexity, and level of technology dependence of the organization will influence the number of sections and the complexity of the plan. An internet search for "SITP" or "SISP" or "ICT plan" will usually generate links to a few copies of governmental or quasi-governmental organizations' strategic plans. Acquiring some of these plans will provide some live examples to review and compare. Because many users of the SISP/SITP are not interested in reading all of the detail, the documents usually contain a fairly comprehensive executive or management summary.

Other key planning elements. A strategic technology plan in support of city marketing includes elements such as the technology infrastructure and governmental facilitations such as tax incentives and regulatory concessions. More specifically, a complete strategic technology planning process will address the target technological environment, the specific ICTs incorporated, the facilitating regulatory and taxation environment, the budgetary/funding strategy—including the revenue strategy, if any.

One significant point that should be brought up early in this discussion is a very real risk raised by Grenny, Maxfield and Shimberg (2007). Their research reports a problem noted by 85% of their participants in technology planning projects. It is referred to as "fact-free planning" by these researchers. In summary, it is the phenomenon of the people responsible for developing the plan acceding to the pressures and expectations of various managers and organizational stakeholders and thereby ignoring contravening facts. Based on the authors' explanations, the key to managing this risk is the ability of those conducting the planning functions to "confront, discuss and manage the phenomenon effectively" (p. 47). This point is further supported by Holley, Dufner and Reed. They make the following point: "When elected officials negotiate to set objectives, feasibility issues may not be considered fully. Setting an organizational objective without consideration of its information systems (IS) feasibility precludes SISP" (2002, p. 399).

Guideline 8: *Be alert to and attentively manage the "fact-free" planning phenomenon.*

The Connection Between Technology Planning and Marketing

The strategic technology plan and the strategic marketing plan can intersect in several ways. The marketing plan may anticipate the organization having a particular technological capability, like an internet website or an on-line system for employees to use in servicing customer/citizen requirements. If the marketing plan calls for advertising or marketing a particular technological capability to potential short term or permanent arrivals, the technology plan will need to assess the capacity and readiness of that capability if it exists, or how to put it in place if it does not.

For city marketing, where the marketing plan addresses a target market segment that is known to, or is likely to, desire or require a specific technological environment or capacity, inevitably that should become part of the technology plan. If the marketing plans target teenagers with iPod devices that play videos, then planning the capability to deliver iPod video broadcasts is important. Similarly, if the marketing plan calls for attracting a certain size or type of business that requires ICTs, planning the appropriate type and level of ICT infrastructure support is required. In a municipality, IT is a service provided to the organization. The SITP will need to be aligned with the organizational strategy in a vertical sense, and will also need to be aligned with the marketing plan from a horizontal or peer-to-peer perspective. The following section addresses that concept.

THE ROLE OF STRATEGIC TECHNOLOGY PLANNING IN MARKETING

Although Kavaratzis and Ashworth suggest that it is time to "transition from city marketing to city branding" (2006, p. 184), others appear to be looking even farther ahead. In an interesting article about technology marketing in this new era, Singer (2006) notes that "the paradigm of brand is worn out. Marketing and strategy must now be about shaping the competitive ecosystem" (p. 96). This raises a new standard that applies to the issue at hand. Through skilful representations of a city's advantages, interests, and facilities that substantially exceeds that of "competing" cities, the city marketing competitive ecosystem can evolve and be shaped to the advantage of the market leaders. The question to be addressed is that of an appropriate method to shape this particular competitive ecosystem.

Defining the Marketing Objective

The marketing objective explains the target and intentions of the marketing plan. It is a risky venture to undertake the development of a technology strategy to support an unknown or poorly defined objective. That applies to the objective and strategic direction provided by the superordinate organizational plan and the objective and strategic direction provided by peer plans like the marketing plan. This risk assessment is based on the obvious logic, recognized strategic planning theory, and the practical experience of having seen unsuccessful attempts to circumvent the inevitability of this requirement.

> **Guideline 9:** *Be sure that the objectives of related plans have been developed and clearly set forward.*

In parallel with the strategic planning misfires mentioned above, there are times when those whose role properly includes the responsibility for planning the marketing objective don't fully understand or know what is required. When the strategic technology planner runs into that circumstance and it is exhibited as a lack of the recognition of the need for a clear marketing objective, the strategic technology planner can work to facilitate the development of enough of a marketing objective to support the strategic technology planning process. This can be done in several ways, and one successful approach is to lead others to the objective through a process of asking questions as Marquardt (2005) so effectively explains in his book *Leading with Questions*. Without a clearly defined marketing objective that has been properly vetted within the organization, the strategic technology planner is substantially constrained in producing a successful technology plan to support the marketing effort. In short, the marketing objective is a necessary, but not sufficient, condition for a successful technology plan.

> **Guideline 10:** *The technology planners should support other organizational strategic planning teams in developing actionable objective statements.*

Assessing Strengths and Weaknesses

With an understanding of the marketing objective in hand, the strategic technology planner can move through a traditional strategic planning strengths and weaknesses analysis of the organization and the region's technological position relative to the marketing objective. If these concepts are a little rusty or perhaps somewhat new to the reader, a good place to start are the well-known books on the topic written by Porter (1980; 1985). In a subsequent section we'll describe the more tactical process of creating the ICT inventory and the related evaluation and planning. This assessment process is a review of the entity's general ICT strengths and weaknesses. Naturally it is important that this strength and weakness assessment be honest, without any boosterism. The extant weaknesses are not necessarily a liability, simply a fact that should be acknowledged and dealt with in the strategic technology planning process. As is often the case, properly identifying a problem here is some substantial part of the solution.

Technological Market Positioning

An important issue is how to position your organization from a technological perspective. For example, a city trying to attract technology businesses could promote both the availability of ICT infrastructure and the potential employees being trained by the city's educational system. If a marketing objective is to attract conventions and trade shows, having a convention center with a high capacity wireless network installed and ready for use can be a significant support. In instances like these the role of strategic technology planning is, obviously, to map out the ICT strategy to support the strategic marketing objective.

Another part of the marketing research the technology strategist will need to undertake is to develop an understanding of the technologies being offered or marketed by the likely municipal competitors. It is rare for a city to have such a unique offering to market that no competition exists. In this era of techno-globalization, the competition may be much more geographically distant than was historically the case. Outsourcing has crystallized many people's understanding of how a city a great distance away can be a viable competitor.

The strategic technologist's contribution to the development of the city marketing plan can be substantial. In the ideal situation, the overall strategic plan is completed first and then the marketing plan and the strategic technology plan are completed concomitantly and in concert. Most likely this will be done in an iterative fashion, as the two plans are taken to increasing levels of detail. In the implementation of this cooperative, iterative approach, the relevant sections of the draft marketing plan are provided to the IT planners who do the necessary data collection and evaluation. The IT planners, in turn, provide the marketing planners with their assessments that provide an initial reading on technical feasibility and perhaps order-of-magnitude costs. It certainly is possible to have a completed marketing plan, and then to undertake the supporting strategic technology planning process. However, experience indicates that the marketing plan will be richer as a result of an interactive process, and it is more likely to avoid revision detours generated by technology roadblocks. These occur when the technology required to support a marketing concept turns out to be unavailable, impossibly expensive, or otherwise untenable.

Guideline 11: *Ideally, the SISP/SITP and the marketing plan are developed cooperatively and iteratively.*

In this collaborative, iterative process, the technology planner will benefit from the creative and expansive thinking of a good marketing planning person or group. The inspiration generated can lead the technology planner to new insights into the marketing/branding process and perhaps contribute to "shaping the competitive ecosystem" as Singer suggests (2006, p. 96). Of course the same benefits can apply in the reverse direction when the channels of dialog are open and flowing. An exceptional job of market positioning can create new competitive dynamics in a world where strong and early market position can translate into competitive advantage. As was noted earlier, this is not necessarily based on unique capabilities, simply on unique marketing. However, when the marketing program lays claim to a capability, the capacity to deliver on that promise is essential in order to maintain both organizational credibility and any advantage gained. This next section discusses the role of strategic technology planning in setting forward the implementation framework.

Framing the Structure for Implementation

There is a reasonable possibility that the marketing objective and/or marketing plan will not explicitly call out specific technologies. It is entirely possible it will be a generic requirement. The strategic technology planners serve best when they apply their knowledge of the technological environment and offer up the leading alternatives with their advantages and disadvantages.

For example, if, as in our earlier hypothetical, the concept of a wireless network in a convention center is in play for attracting convention business, it would be the technology planner's role to provide the advantages and disadvantages of an quasi-open network with limits on internet browsing and email attachments, versus a fully open network free to everyone within range of its signal. The technology planner might contrast these to other options such as an array of "hotspots" or even a "walled garden" solution. Hotspots are small area open networks of the type common in the technological market positioning of some coffee shops. Walled garden networks are those that apply strict limits to the websites and content permitted on the network. These are common at kiosks and at public terminals in some hotels and airports. The user may be accessing the sponsor's selected website(s) via the internet, but is not able to get over the wall to the rest of the internet world.

All of these are viable options for our hypothetical convention center wireless network market positioning example. All of these options can reasonably be implemented, and the decision most properly is not left to the technology planner. When the marketing function eschews this sort of evaluation and interaction as "too technical" or not essential, the organization misses an opportunity to obtain significant advantage. Without input, the technology planner makes the decisions based on technology factors and/or their intuition about what the marketing planning group intends or can use to the city's advantage. The interaction described is an example of an effective healthy organizational interaction. The strategic technology planner, the marketing planner, and the senior municipal executives who oversee them are jointly responsible for enabling and ensuring this process.

After the alternatives are presented and discussed, and the choices or preferential rankings are done, the implementation framework can be completed. Naturally, even though we only discussed one example, that of the convention center wireless network, there will very likely be three to five, or occasionally as many as seven, technology elements included. If there are more than seven, it is likely that this stage of the planning is being done at too low or detailed a level. Larger more complex organizations may have more total elements in their planning, but their *strategic* technology planning and marketing planning will initially be done at a more general level, and subsequent efforts will take the plans to a more detailed level. A smaller municipality may have a complete plan with only three technology elements.

The strategic technology framework should lay out the elements required to achieve the technological market positioning described in the previous section, the ways that these elements interact and support each other, and should clearly state how they will integrate (or not) with the enterprise technology architecture in the final implementation. In addition, if any acquisition, development, or construction (hard or software) is required, the timing, sequence, and responsible agency should be set forward as part of the strategy. At this level of planning, the timeframes are broad estimates that are often stated in calendar quarters, and occasionally in months for smaller, short term, efforts. The strategic technology plan is a high level direction-setting plan and the particular technology being planned is one of its most tactical elements. In this current era, it is almost certain that every complete strategic technology plan will include consideration of communication and networking technologies. The next section reviews ICTs and their role in marketing and consequently strategic marketing plans.

THE ROLE OF INFORMATION AND COMMUNICATION TECHNOLOGIES IN MARKETING

As was mentioned in the previous section, writers and thought leaders in the field of "place marketing" or geography oriented marketing propose moving forward from marketing, to branding (Kavaratzis & Ashworth, 2006), to "shaping the competitive ecosystem" (Singer, 2006, p. 96). Many aspects of marketing and branding revolve around differentiation, and a city's ICT infrastructure certainly can be a differentiator. Furthermore, since perception is the most salient aspect of many of these efforts, promoting an ICT that competitors have, but do not present well, can result in a perceptual differentiation that provides a distinct advantage. This can be accomplished in part by removing an element of uncertainty in the marketing target's evaluation and assessment. If they have explicit assurance that a capability exists in one location, but no explicit assurance that it exists in another, the location that successfully marketed the capability will have the advantage. This can be as simple as making specific note of the geographic coverage of mobile telephone service, or as complex as promoting the existence of high capacity "backbone" data transmission capability.

When this concept is combined with the "techno-bubble" discussed earlier, the intersection is the opportunity to assure, or remove some uncertainty, regarding the ability to maintain connectivity when they come to the geographical area being marketed. So, one of the roles of ICTs in marketing is connectivity assurance for this increasingly technology dependent way of being.

Recognizing the Ubiquitous Nature of ICTs

Regarding the ubiquity of ICTs, think for a moment of a street corner in a moderately affluent larger city. Some percentage of the cars going by have a mobile phone based computer data connection. The people in the cars are likely to be carrying at least one mobile communication device on their person. A phone installed in the car may serve a dual purpose of allowing either the passenger or the vehicle to make a call. Some cars are programmed to automatically call for assistance when the crash-safety air-bags are triggered. The municipal bus passing by may have one or more communication devices installed, including a global positioning system and an emergency communication system. The police car passing by may have multiple instances of active ICT devices including a computer with an intranet and/or internet online research capability. The teenager passing on the sidewalk may be armed with multiple devices, but likely has a mobile phone at a minimum, and that phone provides voice, text messaging and probably internet access. In the apartment above, a grandmother may have her high-speed internet connection open on her computer while talking on her land-line telephone with her mobile phone in her purse. The utility meter at the rear may have a passive or active data connection. In the retail store across the street there may be corporate data connections via either satellite, fiber-optic cable, land lines (cable or telephone), and in a few cases, mobile phone technology.

This snapshot inventory of the pervasive presence of ICTs in many people's lives simply serves to remind that this connectivity is both fixed and mobile, transient and persistent, and personal and employment related. When a person or business is considering coming to your city, what role will the availability of these ICTs play in their decision-making process? Naturally that translates into the role of ICTs in city marketing. The following four sub-sections offer some practical suggestions for the next step is thinking through this aspect of information and communication technology planning.

A Framework for Analyzing the Role of ICTs in Marketing

When considering the role of ICTs in marketing planning and the related technology planning process, it may be helpful to parse the concepts to achieve greater specificity. Some ICT capabilities are simply basic infrastructure, like paved roads, traffic lights, water, and sewer. The data network at city hall is an example of basic ICT infrastructure. If it's not in place and functioning it gets attention, but otherwise is simply expected to be maintained in an adequate manner. On the other hand, a new initiative to put free public wireless internet service, like that in place in St. Cloud, Florida or downtown Hermosa Beach, California rises above the level of basic infrastructure. It is likely that residents of St. Cloud and Hermosa Beach will someday see the free wireless internet the same way the rest of us see our basic infrastructure, but until it is in place in a significant percentage of municipalities, it will be viewed

Figure 5. ICT infrastructure analysis matrix

Infrastructure Type	ICT Infrastructure Ownership		
	Municipal	Partnership	Commercial
Special Initiatives			
Basic			

as a special initiative. The continuum described here is one that ranges from very basic essential infrastructure to special technology initiatives that differentiate the municipality.

A second continuum to consider is the ownership and control of the infrastructure, in this case the ICTs. This range extends from municipally owned telephone equipment in city hall, through joint ventures with technology partners, to commercial ICT companies resident in or providing services to the community.

In the first continuum, concept of basic ICT infrastructure to special ICT initiative can be somewhat artificially divided into two halves, basic and special initiatives. In parallel, the second continuum can be divided into municipal ownership, partnership ownership, and commercial ownership. If these two continua are arrayed against each other in a matrix as in Figure 5, we get a view of some of the aspects of this topic that need to be assessed in planning the role of ICTs in city marketing. The St. Cloud, Florida and Hermosa Beach, California examples of free public internet access running on the municipal high speed internet access is a special initiative municipally owned ICT infrastructure from the top left box on the matrix. Commercial telephone company or cable fee-based internet access is basic commercially owned ICT infrastructure. This classification is very much context dependent. For example, the Grameen Phone project in Bangladesh that was described earlier is a special initiative in Bangladesh, but basic ICT infrastructure in other places. This extreme comparison is drawn to make the point that each city's environmental circumstances will be a little different and placement on this matrix is relative.

The value of the matrix is that it can be an aid to help parse your city's ICT infrastructure so you can assess, inventory, plan, and partner using an appropriate referential framework. Your city marketing plans may include elements from all six divisions in this framework. If your marketing plan includes partnership and commercially owned ICT infrastructure, you may find that you will need to move beyond the standard introspective model for municipal planning. This is not, in and of itself, problematic. It is simply a different requirement that enters into the equation as a result of the role of ICTs in city marketing.

Guideline 12: *An infrastructure ownership/type matrix helps analyze ICT's relationships to marketing.*

Assessing Your ICT Infrastructure

In an earlier section, we noted the importance of conducting an assessment of installed or current technological capabilities. The pace of change is too rapid to here provide an authoritative list of the ICT technologies to be included in your municipality's ICT infrastructure assessment and inventory. Therefore, the recommendations here regard the process more than they provide a sample inventory. As an example, the elements of the communications technology infrastructure assessment can be broken into three broad segments: high speed backbone, commercial grade, and consumer level.

High speed backbone capacity is the fiber optic, satellite, micro-wave, radio frequency, or other high-speed/high-capacity wholesale transmission capacity. Inevitably, this process requires some level of partnership information exchange with private enterprise organizations that own or manage this level of "backbone" transmission capacity. If marketing to businesses that have significant ICT demands is part of your marketing strategy, a thorough review of the availability of commercial grade service is essential. A marketing plan that can describe the area's overall high speed backbone capacity and also list multiple commercial grade service providers is a stronger presentation to today's technology savvy commercial decision-makers. The provision of consumer level service generally falls to private enterprise organizations, and thinking of the purpose here, city marketing at the consumer level might fall into three categories. Cities that are trying to attract new residents. Cities that trying to attract individual tourists. Cities that are trying to attract conference/convention/trade show business.

The assessment work product. Applying the results of the assessment of your ICT infrastructure in the framework suggested in Figure 5 should result in a reasonably complete inventory of the matrix of ICT capabilities and their capacity. As was described above, the first phase of the IT planning process is the assessment of the current environment. In order to develop the portion of that strategic plan which relates to ICTs and marketing, you will need to have an inventory the addresses both quality and quantity (or capacity) of your ICT infrastructure. This assessment should produce that inventory for both the governmental infrastructure and the relevant commercial capabilities.

Mapping Out Your Position

Once you have assessed the ICT infrastructure you will be able to engage in a strategic review of your city's ICT strengths and weaknesses. If the organization's strategic plan clearly sets an ICT direction or the marketing plan is sufficiently developed to indicate a planned marketing direction, you can use the inventory to analyze your strengths and weaknesses in light of those intended directions. This is often done using a "gap analysis." The gap analysis is exactly what the name implies. A list of requirements is compared to the existing capabilities and the magnitude of the "gap" between existing and required is documented. The obvious benefit of the gap analysis is that it results in the identification of the areas where current (and perhaps planned) capabilities don't meet expectations.

Guideline 13: *Conduct an assessment using gap analysis as a tool.*

When a municipality is planning to use ICTs as part of its marketing strategy, the ICTs fall into two categories. Those which the city owns or controls, and all others. Since the marketing plan may include the promotion of ICTs controlled by others, this process of assessment, inventory and analysis may require coordination or a partnership with private enterprise. The next section describes some elements of the tactical approach related to the role of ICTs in city marketing.

Planning Your Tactical Approach

When the cooperative or iterative process of reconciling and aligning the strategic technology plan and strategic marketing plan within the bounds and direction of the organizational strategic plan have been completed, the tactical action plan will be prepared. When you have completed the assessment, the inventory and the gap analysis described above, you may want to quickly assess whether there is any "low hanging fruit" in the results. These can be taken as an immediate tactical win.

CONSTRUCTING A TECHNOLOGY PLAN TO SUPPORT THE CITY MARKETING PLAN

This section addresses the amalgamation of these planning and strategy elements. The strategic technology plan and its coordination with the organizational strategic plan, support the city marketing plan. Knowing what the current and planned technology capabilities are, and the organization's strategic intentions as described in the city's overall strategic plan provides part of the framework for the city marketing plan. With these perspectives clearly in hand, the marketing plan can market to these strengths and intentions.

Multiple Audiences

As the reader can clearly see, the segmentation and construction of each city's plan will, of necessity, be different. In any event, there is a need to address multiple audiences. The choices made in this regard will depend on the city's planning orientation and the prioritization of the various audiences. To facilitate that process, some of the various audiences and typical reasons for interest are described here.

The first, and most obvious, audience is that of the elected officials. Generally the agenda here is fairly well known, and consequently a little easier to address and accommodate. It is an error to ignore this reality since many of these politicians are acutely conscious of the public positions they have taken during and after the election cycle and will apply that litmus test to their evaluation of the technology and marketing plans.

A second audience is that of the professional/career municipal managers. Support from the strongest and/or most influential members of this group is virtually a necessary condition for successful implementation of the plan, consequently the planning process and its interim results need to be fully vetted within this group.

Cities usually have business oriented groups, like a Chamber of Commerce or downtown business group that combines boosterism and perhaps some quasi-governmental role. Since some amount of boosterism is a natural part of their role, they have great potential to support the city marketing plan, but need to be included in the plan development.

Another audience that needs to be considered when ICTs are part of the city marketing plan is the ICT companies. Using a current example, if free internet service is to be provided through the city's broadband connection, the contractual provider may see this as an advantage in marketing to other municipalities, or may have concerns about the plan. This audience should be included in the plan development in an appropriate manner.

Of course a significant audience for the plan is the citizenry and their common proxy, the local mass media. It is unlikely that citizens will want to read the entire plan, and this leads to a specific requirement for dealing with this audience. The plan will need a clear, complete and succinct summary that both sells and informs. Individual communities will have different numbers and types of media that can be offered access to the plan summary (and the detailed plan). Local television stations may be receptive to a well prepared position piece. This quickly becomes an opportunity to demonstrate your city has examined the "competition" and is ready to move forward.

Guideline 14: *Be sure that your plan successfully addresses the many important audiences.*

Formulating Multiple Orientations of the Plan

In the discussion of the multiple audiences for the strategic technology plan, we reviewed various groups inside the municipal community who are mostly potential users of the plan, a somewhat introspective view. This section addresses a parallel, more outward looking perspective that may appropriately be implemented in summaries or abstracts of the plan for those who are more *readers* of the plan than they are *users* of the plan. In a sense, this is about marketing the strategic technology marketing plan.

Local stakeholders. The local stakeholders are primarily those addressed in the multiple audiences section above. However, local stakeholders also include the individual businesses in the hospitality or services businesses that may benefit from increased business driven by some aspect of the city marketing program. In some cases, these companies may have a marketing plan that could be adjusted to coincide with, or be supportive of the city's plan. Attention to these opportunities can strengthen both public support for the plan and a successful implementation. In addition, you should include in your considerations any local stakeholders that may see the plan as disadvantageous. As is usually the case, foresight and a specific plan of action is more likely to result in a better conclusion.

National targets. When a city's marketing plan includes an international orientation, there is a possibility that there will be an audience for the plan at the national level. When a municipal marketing plan is coordinate with national marketing plans, there is an opportunity for synergy and national level support. Whether the aim of the city marketing plan is tourist, conference/trade show, or commercial business, there may be potential support at the national level. This should be considered and where the opportunity exists, an appropriate nationally oriented summary of the plan should be prepared.

Consumer targets. At the consumer level, the city marketing plan can address tourists, transitory, or permanent visitors. Generally targets of this type will be offered some specific set of benefits in the marketing plan. As was discussed in the techno-globalization section at the beginning of this chapter, in this day and age, assurance of the opportunity to stay "connected" while visiting the city is an attractive feature and should be included in your ICT planning considerations. Overall, the consumer orientation of the plan, if any, may have the broadest range of potential messages. Once again, careful consideration of how and what message will be delivered is better than acceptance of a random or default choice in this matter.

Industry and commercial targets. This segment can be targeted with considerable precision. If the orientation of the city marketing plan includes commercial or industrial targets, writing an appropriate summary or applicable section of the plan can be a benefit. As was mentioned earlier, removing an element of uncertainty for a decision-maker can be advantageous. For example, if the plan is able to say that your city is interested in attracting businesses of a certain type and is willing to offer certain advantages or specific concessions, these give the corporate decision-maker both an assurance of a welcoming municipal climate and a referential authority to include in her/his explanation and justification of a decision to come to your city. This may be simply to hold a conference or trade show, or it may be to locate an office or other facility. If it is a trade association or travel/hospitality industry company, they may be interested in both the business climate and the tourist orientation and, as described above, the marketing plan should address both of these aspects.

Supporting Research

Because the field of technology is perennially fast moving and evolutionary, current research is inevitably required. As you might expect, the internet is an excellent place to get both current information and investigate what your city's "competitors" are doing and the state of the art. As an example, an internet search of the words "municipal wireless" will generate a significant list of results. It isn't possible to provide a comprehensive list of useful sites here, however, one example is a site named www.muniwireless.com founded by an attorney named Esme Vos. This is a commercial site with abundant advertising. However the advertising supports useful research. You can find lists of municipal wireless implementations, articles on current requests for proposals for municipal wireless networks, and articles on specific orientations like tourism and wireless municipal networks. In short, this website and other similar ones are a valuable source of information and research data. Naturally you will want to tailor your own searches to your specific needs, but your planning process will be richer and more complete with this important and useful internet based scan of the environment.

Once the strategic plan is complete, begin the tactical and operational planning process. One of the keys to the transition from a strategic plan to a tactical plan is having a skilled tactical planner working with the approved strategic plans and working in conjunction with the financial planners. The reality in governmental entities is often that the financial plan is produced through a combination of administrative recommendations/requests and a legislative/political process. The more closely the tactical, financial and implementation plans are aligned with a properly vetted organizational strategy, the more likely it is that they will be approved and successful.

CONCLUSION

The intersection of strategic technology planning and marketing is, inevitably, different for each municipality. This chapter surveyed the strategic planning paradigm, the role of strategic technology planning in marketing, and the role of ICTs in marketing. In addition, the chapter described the process of constructing and organizing both the strategic technology plan and the ICT marketing sub-plan.

What Can We Conclude?

A successful strategic plan depends on the existence of a clear organizational strategy, which is best captured in a written strategic plan document. If a written plan document doesn't exist, the strategic technology planner needs to invest the time and effort required to get a statement of strategic direction approved by top management. As we address the topic of the interaction of marketing and technology, the same thing applies to a strategic marketing plan. If one doesn't exist, then the technology planner needs to help guide those responsible for marketing the city to the development of a definitive statement of the marketing strategy. The strategic technology plan needs to be aligned with both the organizational strategic plan and the strategic marketing plan (and any other appropriate peer plans). Successful strategic technology planning generally involves a process of simultaneous development with peer strategic plans like the strategic marketing plan. The iterative co-development method is usually the shortest path to successful completion of both planning processes.

Lee and Bai (2003) captured these concepts well in their article discussing organizational mechanisms necessary for successful SISP/SITP plan preparation in the digital era. They trace the evolution of the SISP/SITP planning process over several decades. They explain the progression from "technology mode" to the "align mode" to the "impact mode" to the most developed, the current "fit mode" (p. 33). These four phases were driven by technology, business issues, competition, and organizational issues respectively. The approach described in this chapter intends to reflect the full range described by Lee and Bai (2003). Successful participation in the techno-global economy virtually requires this level of sophistication.

Organizational strategic technology plans incorporate conceptual development of enterprise wide software, hardware, networking, and personnel plans. They often incorporate consideration of external entities and their technology direction. This is especially true when marketing plans will include non-owned ICT capabilities. The objective of the intra-organizational aspect of the plan is to set the strategy and direction of the information and communications technology for the municipality and its delivery of service and support to its constituencies. The objective of the extra-organizational aspects of the plan will include identifying and assessing the technology and capabilities that are required to support the objectives of the strategic marketing plan.

The three phases of the strategic technology planning process are the assessment (Where are we right now?), setting the strategic direction (Where do we want to go?), and developing the plan (How will we get there?). In the discussion of this process, we noted the need to manage the fact-free planning phenomenon. In conclusion, with the organizational strategic direction clearly understood and the marketing objective documented and agreed upon, the scope and requirements for the intra-organizational and extra-organizational assessment should be very clear. The gap analysis technique is recommended as a useful tool to collect and present the differences between what is in place, and what will be needed. This level of sophistication is also essential to success.

What can we conclude the attributes of a successful final product are? The well-done strategic technology plan will address the multiple targets and audiences discussed in the previous sections. It will be aligned with both the organizational strategy and the marketing strategy. It will lay out the elements required to achieve the technological market positioning set forward in the marketing plan. It will clearly set forward the organization's technology strategy and plans for both internal and outward looking initiatives. It will exhibit a broad organizational perspective, result from good group interaction, capture the required organizational knowledge and incorporate organizational change management requirements.

Planning of this sort is almost always an iterative organizational learning process. The review, revision, and renewal of the plan in subsequent years is a natural part of the process and should, as was noted in Figure 4 and Guideline 7, be anticipated at the beginning of the effort. Wherever possible, the initial planning process should include measurement and evaluation metrics that can be used to assess how well your plan has succeeded in meeting your objectives.

With or without specific measurements, a frank assessment of the previous plan should be done before the next iteration of the planning process is undertaken. This can be done very successfully by convening a group of appropriately knowledgeable people. The charge given to this group should be to bring to the meeting their own list of the successes and failures of the prior plan. When these are discussed in the group setting, the individual perspectives get incorporated into the group understanding of the plan. When this group learning is recorded, it becomes a valuable organizational knowledge tool to guide those responsible for the development of the next planning iteration.

Guideline 15: *Consider using a peer-review group to help assess the success of the prior plan and recommend future improvements.*

FUTURE TRENDS

From 2000 through 2007 there was much discussion about municipalities providing free internet access to citizens. In spite of the existence of some free municipal networks, the current trend appears to be moving away from broad, free access networks. There are several factors which militate against free municipal networks. Free wireless service is not currently an essential municipal service, but more of a luxury. In the final analysis, it may be provided free to the users, but there are costs to building and maintaining the network. Some service providers who agreed to partner with cities and provide free access to citizens have found that advertising and other anticipated sources of revenue are really not sufficient to make it worthwhile at this current time. Some rather localized free municipal networks, like those at some airports seem more likely to survive and be replicated.

There is clear potential to capitalize on technology in addressing the city marketing challenge. The techno-global economy fosters an increasing reliance on and demand for technology infrastructure. In the future, technology infrastructure may come to hold the same position as an essential utility that water supply, sewer, land-line telephone, and electric utilities currently hold. As people come to see various aspects of technology infrastructure, especially communications capabilities, as increasingly essential and assumed, the future city marketing potential grows. Farsighted city marketing and technology plans can anticipate this future state and market toward it.

REFERENCES

Amarsaikhan, D., Lkhagvasuren, T., Oyun, S., & Batchuluun, B. (2007). Online medical diagnosis and training in rural Mongolia. *Distance Education, 28*(2), 195.

Chan, Y. E., Sabherwal, R., & Thatcher, J. B. (2006). Antecedents and outcomes of strategic IS alignment: An empirical investigation. *IEEE Transactions on Engineering Management, 53*(1), 27-47.

Davison, R. M., Wagner, C., & Ma, L. C. K. (2005). From government to e-government: A transition model. *Information Technology & People, 18*(3), 280-299.

Grenny, J., Maxfield, D., & Shimberg, A. (2007). How project leaders can overcome the crisis of silence. *MIT Sloan Management Review, 48*(4), 46-52.

Henderson, J. C., & Venkatraman, H. (1999). Strategic alignment: Leveraging information technology for transforming organizations. *IBM Systems Journal, 38*(2/3), 472-484.

Holley, L. M., Dufner, D., & Reed, B. J. (2002). Got SISP? Strategic information systems planning in U.S. state governments. *Public Performance & Management Review, 25*(4), 398-412.

Holley, L. M., Dufner, D., & Reed, B. J. (2004). Strategic information systems planning in U.S. county governments: Will the real SISP model please stand up? *Public Performance & Management Review, 27*(3), 102-126.

Huang, C. D., & Hu, Q. (2007). Achieving IT-business strategic alignment via enterprise-wide implementation of balanced scorecards. *Information Systems Management, 24*(2), 173-184.

Hulm, P. (2007). Tunisia's boom in ICT. *International Trade Forum, 1*, 24.

Kavaratzis, M., & Ashworth, G. J. (2006). City branding: An effective assertion of identity or a transitory marketing trick? *Place Branding, 2*(3), 183-194.

Kearns, G. S., & Sabherwal, R. (2007). Strategic Alignment between business and information technology: A knowledge-based view of behaviors, outcome, and consequences. *Journal of Management Information Systems, 23*(3), 129-162.

Lee, G.-G., & Bai, R.-J. (2003). Organizational mechanisms for successful IS/IT strategic planning in the digital era. *Management Decision, 41*(1/2), 32-42.

Malnight, T., & Keys, T. (2007). Surfing the storm: Translating long-term global trends into today's decisions. *Perspectives for Managers, 145*, 1.

Marquardt, M. J. (2005). *Leading with questions.* San Francisco, CA: Jossey-Bass.

Mason, R. B. (2007). The external environment's effect on management and strategy. *Management Decision, 45*(1), 10-28.

Morton, P. (2006). Using critical realism to explain strategic information systems planning. *Journal of Information Technology Theory and Application, 8*(1), 1-48.

Peak, D., & Guynes, C. S. (2003). The IT alignment planning process. *The Journal of Computer Information Systems, 44*(1), 9.

Porter, M. E. (1980). *Competitive strategy: Techniques for analyzing industries and competitors.* New York, NY: Free Press.

Porter, M. E. (1985). *Competitive advantage: Creating and sustaining superior performance.* New York, NY: Free Press.

Ramos, A. J., Nangit, G., Ranga, A. I., & Triñona, J. (2007). ICT-enabled distance education in community development in the Philippines. *Distance Education, 28*(2), 213.

Roge, J. N., & Chakrabarty, S. (2003). Waiting for the other shoe to drop: Has information technology integrated marketing operations with marketing strategy? *The Journal of Computer Information Systems, 43*(2), 16.

Singer, J. G. (2006). Systems marketing for the information age. *MIT Sloan Management Review, 48*(1), 95.

Tarlow, M., & Tarlow, P. (2002). *Digital aboriginal / the direction of business now: Instinctive, nomadic and ever-changing.* New York, NY: Warner Books, Inc.

Thorngate, W. (1976). "In general" vs. "it depends": Some comments on the Gergen-Schlenker debate. *Personality and Social Psychology Bulletin, 2*, 404-410.

Thorogood, A., Yetton, P., Vlasic, A., & Spiller, J. (2004). Raise your glasses - the water's magic! Strategic IT at SA Water: a case study in alignment, outsourcing and governance. *Journal of Information Technology, 19*(2), 130-139.

Uyanga, S. (2005). The usage of ICT for secondary education in Mongolia. *International Journal of Education and Development using Information and Communication Technology, 1*(4), 101-118.

Warnaby, G., & Bennison, D. (2006). Reciprocal urban place marketing and co-branding? Retail applications. *Place Branding, 2*(4), 297-310.

Weick, K. E. (1979). *The social psychology of organizing (2nd ed.).* New York, NY: McGraw-Hill, Inc.

Wen, H. J., & Shih, S. C. (2006). Strategic information technology prioritization. *The Journal of Computer Information Systems, 46*(4), 54-63.

Wonglimpiyarat, J. (2007). National foresight in science and technology strategy development. *Futures, 39*(6), 718-728.

Yunus, M. (2007). The Nobel Peace Prize 2006 - Nobel Lecture. *Law and Business Review of the Americas, 13*(2), 267.

Chapter V
City Marketing Goes Mobile:
Use of Mobile Commerce Techniques for City Marketing

Juliane Chudalla
University of Augsburg, Germany

Key Pousttchi
University of Augsburg, Germany

ABSTRACT

Mobile services have great potentials in different fields, so it is interesting to have a closer look of them, and about the way they can be used sensiblebly in the present; in the future for city marketing too. This chapter provides basic knowledge on mobile services, the presentation of restrictions, and opportunities of mobile devices, applications, and communication techniques, to help to understand what advantages mobile services have and for what they could be used for. The desriptions of present and possible mobile services and the three case studies inform the readers of the design of such services and help practitioners to design and implement their own successful mobile services for city marketing purposes.

CITY MARKETING AND THE MOBILE CHANNEL

City marketing is an emerging market everywhere as more and more cities market themselves to get a better reputation. People are getting overwhelmed by information, and cities have to prevail against each other so that enterprises settle there, people intend to live and work there, and tourists want to spend their holidays in that city and not in another. Consequently, cities have to take special positions and need special fields of attention to contrast with other cities. On the one hand the growth in city marketing originates from this competition between the cities and on the other hand also stems from the fact that local economy and citizens expect factors like a high value of leisure and culture and the proximity of public administration.

As in a number of other contexts the use of mobile phones or other mobile devices is a very promising option to city marketers. By using these devices for their services, cities enter a more or less new field—they can make forecasts but there will be no guarantee for the acceptance of said services. Still, they have to compete with each other. If a city offers mobile services it will look innovative. In most countries the penetration of mobile devices is higher than the internet penetration and mobile devices are cheaper than computers. Thus, mobile services have a great chance to become a popular method to increase the benefit of city marketing.

We define *mobile city marketing* as any marketing activity in a city which involves the transmission of data via mobile communication techniques in conjunction with mobile devices to increase the attractiveness of the city for different target groups. As *mobile devices* we refer to mobile phones, smartphones, personal digital assistants (PDA), and—where appropriate—tablet PC.

With mobile services cities can provide information their stakeholders can retrieve at any place or time. Thus cities have a new opportunity to get in contact with their (potential) stakeholders. These measures regard the city, and its surroundings in the domains economic, habitation, shopping, infrastructure, transportation, spare time, environment and culture. While interaction based on the stationary Internet requires forward planning, mobile services enable its extension to the mobile situation where the customer either directly needs to assess information or is potentially very responsive to context-sensitive offers that are made to him. On the other hand, the opportunities of mobile devices in presentation, interaction and bandwidth for data transfer are very restricted (Turowski & Pousttchi, 2004, p. 61).

In this chapter we provide basic knowledge on mobile services and their use. Based on this we give an overview of present mobile services that could be or are already used for city marketing and their prospective chances. This will enable city marketers to generally assess the value of mobile services for their work, to have an overview over existing types of services and to understand the development of new applications for mobile city marketing.

The chapter is organized as follows: In section two we introduce the relevant types of mobile devices, applications and communication techniques in order to examine their potential for city marketing purposes. In section three we will examine existing mobile applications in the area of city marketing, beginning with an itemization of single applications and concluding with case studies of m-city projects in Tartu, Stockholm and Bregenz. In section four we will identify future trends and give an outlook. Section five will provide a conclusion with recommendations to city marketing practitioners about how to use the mobile channel.

MOBILE SERVICES

Mobile services add value in contrast to electronic or classic services. They can be used everywhere and anytime without temporal or spatial restrictions *(ubiquity)*, the information can be adapted to the general preferences and even to the current location and situation of the particular user *(context-sensitivity)*. Furthermore the user can be system-inherently identified through his telecommunication account over the network *(identifying functions)* and the mobile device can interact with and control other devices *(command and control functions)*. (Turowski & Pousttchi, 2004, p. 157) We do not want to explore the technical aspects in too much detail but we will provide background information on how the attributes of mobile services differ from other communication techniques and provide added values. As an example for proper use of the mobile channel, Koelmel and

Porak (2003) worked out that the user should only receive information that is relevant for him and brings an additional service. Any kind of service or advertising should be adapted for different types of mobile devices.

Mobile Devices and Application Types

Any end device that is enabled for wireless telecommunication and is typically used in situations where the user is mobile is generally considered a mobile device. First of all, this definition excludes laptop or notebook computers due to the fact that, admittedly, they are easily portable, but their typical use still remains stationary. Furthermore, usage patterns follow the rules of stationary computing rather than those for mobile applications. There are different types of mobile devices that are currently used:

Standard mobile phones are to be considered as the most likely target device for city marketing applications as they are owned by a vast majority. The main function here is voice communication, including (1) the ability to access voice-controlled applications *(Interactive Voice Response, IVR).* Other relevant functions of contemporary mobile phones are (2) text and multimedia messaging, (3) browsing special mobile-enabled Internet microsites following the *WAP (Wireless Application Protocol)* standard and (4) execution of small phone applications following the *J2ME (Java 2 Micro Edition)* or the *BREW (Binary Runtime Environment for Wireless)* standard. As a general rule, the operating systems of these phones are proprietary to the device manufacturer or even to the product line. Thus, although possible, applications should not be specially developed for single handsets but should rely on standard functionality as indicated from (1) to (4). Following, we will provide some details:

1. **IVR applications** represent the first generation of m-commerce applications. They announce information via text components that have been pre-recorded or are automatically generated by voice computers. The user is able to provide information or to navigate through simple menu structures either by keystroke ("for public transport information, press 3") or by simple voice commands ("which sight do you want to see?"). IVR applications are completely device-independent and easy to use even for people with little affinity for technology, e.g. the elderly. On the other hand, handling is very slow and generally perceived as uncomfortable.
2. **Text and multimedia messaging** (in most countries known as *short message service, SMS,* and *multimedia messaging service, MMS*) can be used either for initiation or delivery of services. A typical combination of both could be an advertisement "Send GOAL to 12345 to receive the winning goal of today's FC Bayern Munich soccer final!" For an answer, customers receive a multimedia message with pictures or a short video of the event. It can be assumed that virtually all mobile phone users are able to cope with text messaging, and most are also able to receive multimedia messages. For target groups containing mostly younger people, the ability of sending and receiving both types can be taken for granted.
3. **WAP sites** are websites which are specially adapted for small screens and small transmission bandwidth. They may contain text and/or pictures. However, the length of texts as well as the size and resolution of pictures should also be chosen carefully in order not to annoy users. It is also possible to use forms, although it should be considered that users of standard mobile phones have to enter any text with the use of their number keypad which is extremely uncomfortable. A typical use of a WAP site could be to provide brief news or descriptions of events where reservations could be made over the

air. However, acceptance of WAP offers is not only determined by technical issues. As all providers of WAP sites, city marketers should carefully analyze their target group with regard to the data plan in their mobile phone contract. This will allow detecting reluctance to mobile Internet access which may still be common in some markets, especially among prepaid users.

4. **Java technology** was first introduced in mobile phones in 2001/2002. Today, most up-to-date mobile phones in developed markets such as Europe, Japan or North America can be assumed to be Java-enabled. With this technology, applications can either be pre-installed on the devices (which is only relevant in co-operation with a mobile operator) or can be downloaded over the air (which is more relevant for city marketing). A typical download size will not exceed 30–100 kilobyte and take from a couple of seconds up to some 3 minutes. After this procedure, the user will have to install the application. Although this is quite obvious and only lasts some additional seconds, it will require a certain knowledge and perseverance on the side of the user. Anyway, once he has downloaded the application he could use it offline without any additional downloads (and costs). Optionally, the data in the application could be updated over the air regularly. A typical example could be an event calendar which is updated with new content monthly. Roughly similar functionality can be realized with BREW. Although each standard has its own advantages and disadvantages, city marketers will rarely find themselves in a situation where they have to balance these. Instead, the decision is typically determined by the dominance of one of these standards in the respective market. As a rule of thumb, BREW is primarily widespread in markets dominated by North American radio standards, especially *CDMA IS-95* and *CDMA-2000*. Java is typically leading in most other markets, especially those relying on the European *GSM* and *UMTS/W-CDMA* standards.

Personal digital assistants (PDA) are handheld devices which dispose of a bigger screen and are equipped with a PC-like operating system, especially Windows Mobile or Palm OS. While these mostly rely on a handling via pen and touch screen, devices with an alphanumeric keyboard such as Blackberry or Nokia Communicator are also considered a PDA. In addition to voice and messaging services, PDA are typically equipped with e-mail and full Internet access to regular websites (although these should be generally adapted to the mobile use). Compared with mobile phones, PDA may run complex, large applications in higher programming languages. Especially relevant is the Microsoft .NET compact framework for the Windows-enabled devices which can use their full multimedia capabilities. However, a higher circulation of PDA can only be assumed among business users or in special target groups.

Smartphones are a crossbreed between a standard mobile phone and the PDA. Their constitutive characteristic is that they are primarily used and considered a phone while they also dispose of a PDA-like operating system. The display is smaller than that of a PDA and the handling is limited. The functionality of smartphones combines elements of both standard phone and PDA and differs greatly between different manufacturers and product lines. It is not recommendable to target city marketing services especially to smartphones. If the target group is business users, an offer which is adequate to PDA and smartphones should be chosen, potentially a scalable approach can be useful.

An additional category is build-in navigation and entertainment systems of cars. These are often equipped with cellular mobile radio and thus, generally addressable for mobile city marketing applications. However, proprietary standards and often the necessity of special contracts with car manufacturers make this option difficult.

As outlined above, laptop or notebook computers are not considered mobile devices. Contrary to that, tablet PC are, but they play no major role outside of special enterprise applications within mobile-integrated business process concepts. For B2C applications in general and city marketing in particular, they can be nearly neglected.

There are two general options to deliver information to the mobile commerce user, pull services and push services. With a *pull service* the communication is initiated by the user. The most prominent example is the provision of information on websites, regular as well as WAP sites which can be accessed by the user. Another typical type of service is initiated by the user asking for information by sending an SMS with his request to a special number and getting back an answer message with the requested information, e.g. the address and opening hours of souvenir shops via SMS, a small map with special information or some pictures/videos via MMS. With a *push service* the communication is initiated by the service provider and the user gets information, typically as SMS, MMS or e-mail, without individual request. That could mean when a user comes to a special place in the city he will get a message with information regarding the nearest restaurants and their menu, the opening hours of the nearest museums, or information on evening entertainment, combined with a link for ticket sales. Generally, push services are activated either because the user requests an information service or because the service provider gets information about the user from another source. A typical pattern is a user registering for an SMS raffle specially offered for tourists (e.g., via a placard at the airport, main station or hotel) and hence receiving city information specially tailored to his interest profile. Push services underlie different rules as respective laws differ from country to country, supporting either the principle of *opt-out* which means that messages can be send to a user without his registration as long as he does not complain or the principle of *opt-in* which means that sending messages is forbidden until a user registers for a service.

COMMUNICATION TECHNIQUES

Besides mobile devices and application types, mobile city marketing options are influenced by the applied communication techniques. The relevant types of these are the use of wide area networks via mobile radio telephone service, the use of wireless local area networks via Wi-Fi/802.11 (known as "WLAN" in many parts of the world) and the use of so-called personal area networks such as Bluetooth or infrared (IrDA). Another option is to not use any communication technique and provide offline applications.

The use of mobile radio telephone service relies on the respective cellular networks in a country. These networks may differ between GSM-based standards (that address roughly 80% of mobile users worldwide) and other standards (e.g. US standards based on TDMA IS-135 or CDMA IS-95). However, after the latter introduced messaging services alike SMS and MMS the main difference relevant to mobile city marketers is the difference in data transfer rates the users are able to receive. Third generation (3G) networks such as UMTS/W-CDMA, CDMA-2000 and TD-SCDMA allow for high data rates that enable for instance the transmission of high-resolution images such as detailed photos or large maps and even mobile video streaming. The decisive question, however, remains the percentage of users in the target group that dispose of 3G devices. City marketers should be careful and ask for expert know-how when obtaining these numbers as mobile network operators usually publish numbers that are much higher than those relevant for the city marketing target groups. Apart of some high-developed Asian markets such as Japan, today's standard user disposes of 2G or 2.5G devices that enable only limited data rates. For instance, GPRS networks theoretically allow data rates up to 171 Kbit/s. In reality, networks and standard user devices will allow for typical data rates of 50–60 Kbit/s. EDGE networks could roughly double this rate.

Mobile radio telephone service is characterized by relatively low bandwidth and high data transmission cost but can be used virtually everywhere with existing commercial networks (i.e., without being obliged to care for network infrastructure) and mobile users. All other types of communication techniques offer higher data rates at lower or no costs but cannot be used with all users and remain spatially restricted to the point or area where network coverage is provided.

Wireless local area networks (WLAN) may reach users around up to 300 meters from the access point and provide very high data rates, allowing all kinds of applications. While a lot of newer smartphones and PDA support WLAN, standard mobile phones typically do not. Unlike it is the case for mobile radio telephony, a handover of moving users between different cells is not supported in a standard WLAN. Thus, it mostly remains a technology for stationary wireless Internet connections of laptop computers. However, the allocation of WLAN access points is an option for city marketing activities. Additionally—although not a mobile strategy—it is promising for city marketers to use the welcome page of commercial WLAN service providers in a city for promotion of activities; especially business travelers using laptop computers can be reached this way. An interesting option in this regard could be the provision of free WLAN Internet access to visitors at a central point of the city.

Bluetooth and infrared technology is not only included in most Smartphones and PDA but also in newer standard mobile phones. The primary purpose of this technology is the transmission of data between mobile devices or to peripheral devices in close distance (1 meter for infrared, 10 meters for Bluetooth). For city marketing purposes, this technology basically allows for download of applications to mobile devices at a central point. These applications may later be used either offline (e. g., a directory of services, a commented map or a city guide) or connected via cellular mobile radio for transmitting small data amounts or use positioning over it. In the same way, applications could be provided for download over the stationary Internet, without using any mobile communication techniques (i.e., the user downloads the application at home and installs it on the mobile device before traveling), maybe even involving other devices such as the MP3 player in order to download a city podcast.

Location-Based Services

A major added value of mobile applications is context-sensitivity. Besides other elements, such as interactivity or preference profiles, this considers the spatial context of the user. According applications are termed *location-based services* and require positioning. City marketing applications can use four major ways to position the user. For easy applications, the user may indicate his position manually (e. g., text street name and number in order to get back an advice how to reach the nearest subway station). Application of WLAN, Bluetooth or infrared technology system-inherently delivers the position of the user as he has to be within the range of an access point. A third option is the readout of the mobile radio cell number and the derivation of the user's position in co-operation with the mobile network operator, in a city center that typically delivers the position within a radius of 100–300 meters. In some countries, the operators even offer exact positioning via their networks. Finally, a number of PDA and smartphones as well as car navigation systems are equipped with satellite positioning systems such as the US-owned Global Positioning System (GPS) or the European System GALILEO which enable exact positioning (with an accuracy of approximately 10m) as well. A categorization of location-based services can be done with the help of the "Location-L" framework (Turowski & Pousttchi, 2004, p. 77) which is shown in Figure 1.

Location-based services either depend on the position of the triggering person or the position of the target or both. If a position is relevant it may be known and already stored (such as the parking spaces in a city) or it has to be located. The combination leads to a distinction of the five cases in the "Location-L" which we will now explain, beginning at the top of the "L".

In the first case the position of the triggering person is irrelevant, but the position of a certain moving target is relevant and has to be determined by positioning. Applications of this type are rather uncommon in city marketing, although it may be useful to locate certain services, e.g. for a tourist who tries to find a person that acts as a volunteer tourist guide. The second case relies on predetermined information, for instance a service that is activated when the mobile device comes near to a special set place. A museum that sends information about a special exposition or other information in connection with a link to buy tickets when the tourist passes the museum would be such a service. A third case requires that both, the triggering person and the target are to be located. An example could be a service for inhabitants and tourists for a real life game like the well-known board game "Scotland Yard". The forth case requires locating the trigger and uses the known position of a target. That could for instance be a service showing tourists the way from their current position to some objects of interests or their hotel. The last case also requires locating the trigger but does not involve the position of the target, such as a service to locate

Figure 1. The "Location-L" framework for categorization of location-based services (Source: Turowski and Pousttchi, 2004, p. 78)

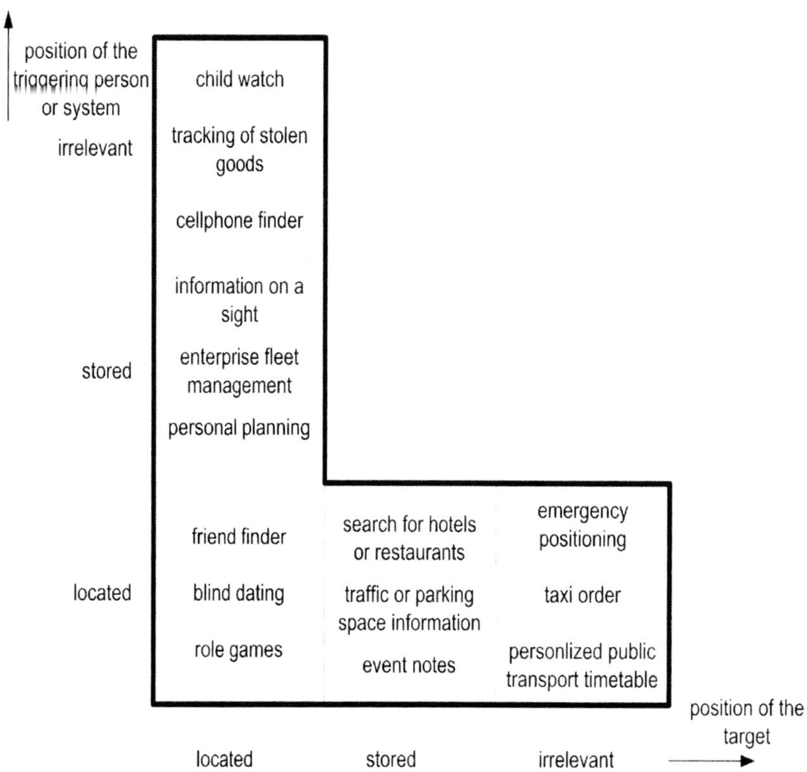

a user in an emergency situation and reporting the position to a public safety answering point (whose position within the city is irrelevant for the completion of the task).

The importance of such services in the city marketing especially for tourists is growing due to the fact that the capabilities of mobile devices expand techniques for positioning get accessible for more and more devices and so the presentation of complex maps including different positioning services and other media services become possible. (Baeumerth, 2002, p. 228)

Considerations for Mobile City Marketing

Mobile city marketing, in order to be successful, must pay attention to the particularities of mobile devices, applications and communication techniques. The restrictions, the smaller display, the limited way of inputs, the limited memory space, the limited computing power and other requirements must be noticed when a mobile city marketing application is established. On the other hand, the mobile channel offers special opportunities which have to be exploited, above all the anywhere-anytime capability and context-sensitivity of applications. Applications shall not be economy versions of PC websites or applications but must be "designed to mobile" and fit the special needs and expectations of the mobile user. Therefore, it is extremely important to know the target group and its preferences.

The latter is important for another problem as well. Unlike for Internet PC, the mobile channel must address a large variety of devices differing in a lot of major features. On the other hand, the patience of the mobile user is extremely limited and his expectations to the user experience on his particular device are high. If this is not considered, users will ignore the mobile applications instead of using them.

APPLICATION OF MOBILE TECHNOLOGIES FOR CITY MARKETING

After explaining the technical determinants and other particularities of the mobile channel we will now describe present and possible mobile services for city marketing and also some case studies of cities which have already introduced mobile services for city marketing purposes.

Involving the Mobile Channel

With city marketing the city intends to transmit information to propagate opinions, expectations and attitudes. Marketing for inhabitants and resident enterprises intends to inform, disclose suggestions and problems and intends to interfere. The city intends to communicate its offers to the different target groups just as enterprises do—but not in an advertising way. For this communication mobile technology is just another channel to deliver information. In this section we will provide mobile applications that exist in literature and are relevant for city marketing and also some ideas what services could be still developed. The itemization in this section will be followed by three complete case studies in section 3.2.

Garbow and Hollbach-Groemig (1998) worked out different elements of city marketing which are now transferred to mobile city marketing. First the city should be regarded as a whole to show the quality of life and work in the city, not only the city centre or the industrial areas. For enterprises the city should be interesting concerning areas, work forces and service orientation. Mobile city marketing can support this service for instance by offering maps with free real estate areas on mobile devices, a mobile newsletter with important events or services like reminding important tax deadlines. The problem is that business people may get too much information onto their mobile equipment and might ignore such messages. So these services should be handled carefully.

A city cannot work alone; it needs help from inhabitants, local enterprises and also of tourists to match their expectations. Some countries try to increase the explanation of crime by sending out messages to police officers, bus and taxi drivers when searching for a missing person, a wanted criminal or a stolen car. (Rannu & Semevsky, 2005, p. 7) The mobile channel allows for fast reactions and also stores the query, a major improvement as compared to voice radio.

The improving of services offered by the city is another element of city marketing. Services a city can offer especially for inhabitants could be typical m-government functions such as opening times of city institutions, order new identity cards or passports and receive a message when their documents have arrived or order tax forms. Inhabitants could inform the city about broken traffic lights or street lamps with their mobile device so that the city can react faster, voting with their mobile device would be a good way to make the administration cheaper, faster and user-friendlier. People do not have to go out for voting and do not have to make a motion for voting via letter. Meanwhile the documents for the postal voting can be ordered online. With this the government hopes to simplify the voting so that the election turnout is higher. When mobile voting works this will much easier than postal voting and the data could not get lost on the way and can not be so easily modified or destroyed. Some countries like Switzerland try to implement a service with which people can vote by using their mobile phone. (Voutsis & Zimmermann, 2005, p. 434) City marketing should meet the interests of all customers of a city who live, work and spend their spare time there. Thus, the number of special target groups which could be addressed with mobile commerce applications is immense. Two examples: A university could send out information about cancelled lectures combined with suggestions what to do instead in the city; commuters could receive information about train delays or cancellations combined with a carpool offer, allowing them quickly organize a new opportunity to get to work.

Another service which is sometimes used at shopping malls or theatres but could also be provided by cities or museums is a childcare station. Parents can leave their children here while they go shopping, make a city tour or visit events while they keep contact with the help of a mobile application. A labor office could provide job offers (with an option to make a direct appointment with the potential employer) via the mobile channel.

The targets which have to be achieved are varying like improving the economy, the image of a city, the residential surroundings, the infrastructure and the numbers of tourism. Examples will be shown later. The option for citizens to communicate opinions and expectations to the administration and the communication of public services to the public is also an element of city marketing. Although mobile options for this are limited so far (as the entry of text to mobile devices is cumbersome) and the stationary Internet may seem to be the predominant alternative, the opportunity of giving feedback directly from the spot could be very valuable. Also viable could be brief surveys about the citizens' satisfaction with different topics of the city administration. As the surveys via the mobile channel should be (very) short, more citizens might even more often participate. But the number of surveys should be small so that they are not stressed out.

Advertisement on mobile devices could be enabled by offerings downloads of background graphics, ringtones or screensavers of the city; also mobile postcards that can be send as MMS are an option. Especially when someone does not know the address of his friend, he has got the mobile phone number, can send a mobile postcard and show his friend he thought of him.

Citizen should benefit from this communication channel in different ways. M-government functions could e. g. provide opportunities to declare a new domicile, order written confirmations or ask for a new passport. This is not only relevant if the according functions may be used while people are away from their fixed-line Internet access but also for those people who do not own one—as most of them own a mobile phone.

Most people do not like to go to the administration as it takes lot of time and nerves; a first step to improve the services was to provide online services. The city can save costs and time and the whole system is transparent. That means all civil services can check the status of a process and can handle the problem of the citizen and give him the information he looks for. So even if the responsible contact person is not available he can get information and does not have to call once more. Such services have also an positive effect on the attractiveness of the city administration, which is not only important for citizen but also for enterprises as they can track how far their concession or request is processed. If the administration uses XML-based forms, they could also be filled in with a mobile equipment. If the form was filled out wrong, the administration sends back a request. Furthermore civil servants can work at one file at the same time; there are no more waiting times. (Arnold, 2007, p. 41)

The services of the government should be available to all citizens. Electronic services which are already offered as tax services have improved the service of the government. But electronic services are not accessible by all population groups as not everyone has internet access. The penetration of mobile phones is much higher and so more people are able to use mobile services. But these options should only be another opportunity and not replace previous services as some people do not want to or can not use mobile or online services.

In some countries the city administration sends reminders about payments of fees and the opportunity to pay the fee directly with the mobile device with a mobile payment system. (MobiSolutions, n. d.) So the city lowers the costs of reminders and people have the opportunity to react faster. An option where the user can save this reminder in his calendar would complete this service so that he will not forget it so easily.

The city should advice its enterprises and carry out interrogation so that the economic conditions can be improved. To become well-known and accounted for good location in a further environment the city in the classic city marketing uses advertisings, detailed location booklets, personal consultants and fairs. (Guenther, 2002, p. 68) Especially at fairs there are good options for mobile services. When a ticket for the fair is booked via mobile, the city can offer cheaper tickets for public transportation or reducing parking tickets. As people move only in a limited area, special services can be offered to exhibitors, for example free mobile marketing opportunities where they can invite participants to their stand when they come near. For this service participants first have to register for that service but that could be handled with the reception. Or they can pay with their mobile equipment so that they do not have to join the queue.

Local enterprises like restaurants, coffee shops and theatres could be able to make an advantage out of a mobile tourist guide, including mobile marketing to their services. When tourists ask for restaurants they could receive a list of restaurants with a link to the menu and a coupon, e. g. with a 5 % discount or a free drink. (Cheverst et al., 2000)

Another major area is m-health. As in most cases the local authority is responsible for the health sector, this is an interesting field for mobile city marketing. As the mobile device is always carried with a person it could be used for storing emergency data about the state of health of the person. Blood group, allergies, if the person is a blood or organ donor could be such data. In case of an accident this data should be available for the ambulance and doctors could demand with mobile devices important information about patients even if the mobile phone is off so that they could react faster and easier. Patients who need to check physical functions, like blood pressure and heartbeat will like the opportunities a mobile device could provide. For example if they have problems with the heart and the blood pressure attains a critical value or the heartbeat drops under a critical value the doctor or ambulance could be automatically called. (Partnership for the Heart, 2006) But most of the time these people can live a normal life in their familiar environment. Ill people who need to take medicine at regular intervals could be reminded to take it or are reminded to come to a check up or appointment. A reminder of appointments is especially useful in countries where people have to wait weeks or even months for the appointment. Using the mobile channel, Great Britain intends to reduce cancellations of appointments by 30 per cent. (Rannu & Semevsky, 2005, p. 17) Finally, the family of a patient may receive periodic information on the state of health of their relative. Therefore the patient should have given an approval so that his privacy is guaranteed.

An example where the health or elderly care institutions can include mobile communication technique is in the organization of working hours for people working in these institutions like it is used in the elderly care services in Stockholm like it is explained in the next section. (Lundeval & Ozan, 2006, p. 13)

A service that is already used in some European cities is mobile ticketing where people can pay for the tickets for the public transportation with their mobile device and also get the ticket directly on it after their demand. An advantage of this service is that people can get an aggregated bill at the end of the month, charged with the cheapest customs tariff. Experiences, e. g. in Vienna, show a decline of fare dodger numbers by m-ticketing services. Transportation tickets may be included in other mobile tickets, e. g. for an event, or a restaurant may offer free public transport in case of a table reservation over the mobile channel. Not only could the tickets for public transportation be paid mobile but also parking tickets. This could be easily handled by sending an SMS or calling a service number at the beginning and at the end of the parking session. So the users can be charged only for the time they were parking and so could perhaps save money and do not have to look for change.

But mobile services can also be used by people who do not use their own cars but go by public transport. Customers of the local traffic enterprises could register to be informed if the bus or tramway is belated or cancelled, along with opportunities for rerouting. In some cases, (personalized) mobile schedules for busses, tram- and subways are already very helpful applications, especially in the case of irregular depart dependant to the daytime or for tourists that are not familiar with the public transport system.

The most-examined domain in mobile city marketing is tourism. The services of human tourist guides may be disadvantageous in some cases: some tourists are rather inflexible for starting time, duration, or program of their visit. Tourists also must accept the breaks the guide proposes; they are not able to go for shopping or for a coffee during a tour. (Cheverest, Davis, Mitchell, Friday & Efstration, 2000) While this is a luxury (but any how relevant) problem to standard tourists, it may pose serious difficulties to handicapped or elderly persons. When a tourist comes the first time to a foreign city he will need a map for the first orientation or even to find his hotel. An opportunity to ask for the location of restaurants with additional information like special meals of the restaurant should be integrated. In their paper Rakkolainen and Vainios (2000) tested the usability of a 3D map. They worked out that these maps are intuitive and user-friendly due to the visualization and it is easier for the users to follow a tour with a 3D map as they recognize the environment. But these maps have high rendering times and high transfer costs when downloaded or used online.

Luley et al. (2004, p. 34) worked out that mobile maps are especially useful in areas where no other information sources are available like rural areas. These applications should provide the following functions:

- Tour description and guide, general information, description of interests,
- Infrastructure of the route, natural environment,
- Opening times of the facilities like museums and restaurants,
- Weather report, forecast and warnings (especially in rural areas for hikers),
- Emergency information,
- Positioning, location based services,
- Events,
- Hotels, restaurants and their menus,
- Transportation options and time schedules.

Different alternatives of guided tours which the user can pursue or which even allow him to arrange his own tour by selecting objects of interest have been developed. Objects are explained with texts or multimedia applications (audio/video). The explanations may be variable, scalable and theme-focused due to the user preferences and device capabilities. This could, for instance, allow a person whose main interest is history and who owns a PDA to see multimedia scenes of historical views or events while a person whose main interest is architecture and who owns an older mobile phone only gets texted advices where to go and what to see.

Cheverest et al. (2000) suggest a mobile tourist guide should notice these topics.

- Offer flexibility with reference to starting time, duration, breaks, route and objects of interests,
- Incurring context-sensitive information, as the user profile, his interests, preferences which are saved on his mobile phone and location, and also the time of the day and the opening hours of attractions,
- Offer support of dynamic information, as changes in opening and closing times of the attractions,
- Offer support for interactive services, for example booking of accommodation or tickets as an additional service of the tourist information.

The latter function could also integrate a system so that the tickets are paid directly, as it is included in the m-toguide project (Schneider & Schroeder, 2003). A map of the city could also include a map with possible trips in the environment or the opening and closing times of shops. The challenge is to provide a card that is big enough and where the user does not have to scroll too much.

The voice of the electronic guide should be polite and friendly; the guide should provide further information if the user demands it and a map of the city should be available. (Cheverest et al., 2000) A printout of the tour at the hotel or at home could not only compensate for the small screen but also serve as a memory and souvenir for the tourists; they may know even months or years after their holiday where they have been and can advise their relatives to also make a trip to that city. Experiences with electronic or mobile ticketing (e. g., for a concert or a sports event) show that the function of paper tickets as a souvenir is highly appreciated. A tour guide should also support user profiles, allowing saving preferences and completed tours. (Schneider &Schroeder, 2003) As most cities or museums only offer tours in selected languages another advantage of a mobile tourist guide could be that the tour or the information is provided to the user in his own language so that he will understand all of it. But that means that the cities will have to put all these information into their database in different languages. That all will take time and might cost some more money so that some cities are not able to afford it.

Cities sometimes offer a ticket with which you can visit several museums with one ticket that is cheaper than the single entrance to these museums. This system can be enlarged by a mobile service. When the user visits a museum he could be granted a mobile coupon for cheaper entrance to other museums. One idea is mobile city information, where people can inform themselves about events, weather or sport activities. Another opportunity is a mobile tourist guide, with maps, information about the objects of interests and their opening hours, so tourists can make a tour through the city on their own without a guide at anytime they like to. That service can be combined with location based services to show tourists the next best cafes and restaurants they can go to or other places they can visit.

To get the information they need, tourists can use WLAN which is provided for free in the city centers. The advantage for them is that they do not have to bear transfer costs which are much higher abroad. (Heidelberg-Mobile, n. d.).

A service that a university can offer to its students and which is already used, is to inform them about the places and times of their exams with a reminder. Also, the results may be sent as a message to their mobile device. Some offer a wake-up service for students who tend to be late or wish them "Good luck" for their exams. Or students are informed when a lecture is cancelled. Additionally information could be included, such as offers of coffee shops in the city, exhibitions in museums or in the summer information about the frequency at lakes. In order to promote the university, local pupils could be invited for an open day in order to receive information on study courses and other important facts. As today young people tend to perceive the mobile phone as a central part of their life, this allows universities to address them directly and fast on their most relevant channel, generating fast and positive responses and a modern corporate image in this target group. Another service that can be offered to young people in the field of m-education is a vocabulary trainer for the improvement of their language skills.

Mobile communication techniques could even be used for real-life games like the famous "Scotland Yard" where group one must find group two in the city and gets the position of the other group every ten minutes or like a treasure search at unusual places in the city—accidentally leading tourists to places that city marketers would want them to visit. Mobile devices with a GPS receiver may use an according service called "geocatching". (Bellmann, 2007) That service can be used for school trips or offered to pupils in their holidays. The underlying idea is to enhance their identification with the city already in early stages and influence them to develop localism.

CASE STUDIES

Most of the cities might not yet trust in the chances mobile services can offer and do only offer very limited or no such services. In this section we present three case studies of cities that already have started coordinated projects with different mobile services: Tartu, Stockholm and Bregenz. They might be still at the beginning, but their ideas and approaches may serve as valuable examples. Other cities do not have the combination of different sorts of applications these cities try to cover. Furthermore these cities have already finished the trial and established the mobile services. The services implemented by these pioneers comprise domains like m-tourism, m-education, m-parking, m-ticketing, m-health and some others. What is more, they are worth to observe them as they are working to enlarge their services.

Case Study 1: The m-city project in Tartu (Estonia)

Tartu is the second largest city in Estonia. The penetration with mobile phones is bigger than the Internet penetration, which means that mobile services are available to a greater number of people. Therefore it is not astonishing that Tartu launched the first m-city project in 2004. It comprises several services with different topics which Rannu and Semevsky (2005) have collected in their study.

Estonian people can pay for their parking space or their bus ticket with their mobile phone. To use the m-parking service it is necessary to be a customer of EMT, the major Estonian mobile network operator. The service has to be activated and the user has to pay an initiation fee. It can only be used in special parking places and a dedicated prepaid account has to be established. There are three options to use it: an SMS message by EMT, SurfPort and by phone call to the service number 1901. (AS EMT, n. d.) The user has to send the number of the license plate and the location code of the parking lot. When he leaves from the parking lot he sends over the same information. There should also be a sticker on each car using m-parking so that controllers are able to check if the parking fees have been paid by SMS or a built in menu in their mobile device. The controllers check with their mobile device if the user is logged in the system. (Rannu, 2003, p. 29; Rannu, 2003, p. 10) The tickets for the public transportation can be purchased by calling the service number 1312 or with the EMT SurfPort. In return the user gets an SMS with a serial number and the expiration that could be 1 hour, 1 day or 30 days. The user pays for this service via mobile phone bill. (Rannu, 2003, p. 30) This service can also be used in the cities Tallinn and Harju. (AS EMT, n. d.) Citizen can also get information about traffic on their mobile device so that they are able to avoid traffic jams. (MobiSolutions, n. d.)

The inhabitants may use a mobile payment procedure for bus tickets as well as for other products and services. The merchant and the customer must both register first. The payment process is executed via phone call to the service number 1214. The customer must enter the merchant's code and the amount of the purchase (1214*shopcode*kroons*cents) and the merchant gets an SMS message as receipt. (Rannu, 2003, p. 11) The customers do not have to pay a monthly fee but a transaction fee of 3 per cent. Banks support m-payment as cash is reduced and merchants search for a cheaper alternative to credit cards as the terminals for the credit cards are too expensive for smaller merchants. (Rannu, 2003, p. 31)

There is also a mobile service for tourism called T-number. Every object of interest has its own code which can be found on the official city maps of the tourist information, on the website of Tartu, on signs next to these sights or can be asked by sending an SMS to the short code 17120. If the customer's device supports automatic positioning, he may receive the codes of the closest sights. When the tourist dials the code he receives an audio clip that explains him the history and significance of the sight so far only in English. There is also a service with which the inhabitants can help to improve the service of the city administration: the city short code 1789. Inhabitants can send their ideas or policy recommendations to the short code. If they see broken traffic lights, street lamps or other damages in public places they can also send this information to the short code by SMS, make a phone call to the number or send the information by email (1789@tartu.ee). Another service where special occupational groups can help to increase the security in Tartu is called m-neighbourhoodwatch. The police, taxi and bus drivers and security companies get SMS with issues like missing persons or stolen cars. So this information is spread faster and wider, helps to react faster and so make the city safer.

The library has also a service for customers. Customers who want to borrow a book, movie or audiotape which is not available can register that they get an SMS message when it will be available.

With another service Tartu expects to get closer to the users. Users can download logos, background images, graphics and ring tones to their mobile phone. So inhabitants can show their relationship to the city and provide good publicity for the city of Tartu. (Rannu & Semevsky, 2005, p. 8)

A mobile device that allows for positioning can be used via SMS, WAP or IVR by inhabitants as "friend finder", for finding locations or even child acquaintances. The location with an exactness of about 500 meters in a city and a few kilometers in rural areas, will be shown on a map or is explained in an text message. (AS EMT, n. d.) If a person shall be located this person must accept the positioning via SMS message or IVR via an EMT service number.

Case Study 2: The mcity project in Stockholm (Sweden)

The mcity project started 2001. The pilot was planned to expire 2004 but due to its success it was retained. New services are worked out with the respective target groups to guarantee that service meet the needs of these. (Rannu, 2004, p. 37)

A service that focuses on the young target group is a mobile vocabulary trainer. Language schools offer a service with which students can test their knowledge on their mobile device. They can learn new vocabulary and repeat older words. (Lundeval & Ozan, 2006, p. 5)

The coordination of schedules in the elderly care is handled with an SMS management system. When a nurse gets ill and she needs a substitute, SMS are sent to a group of other nurses who can answer via SMS. With this system a lot of time is saved as the whole staff can be informed at once. (Lundeval & Ozan, 2006, p. 7) A pilot is tested where not only schedules are coordinated mobile but also the working hours are documented mobile. Additionally, nurses documented the work they carried out for each case with their mobile device. They do not have to go back to the office and can make their documentation instantly after their work everywhere they are. Another service used in the elderly care is called "safety TV". Elderly people can communicate with their relatives or their care staff through messages which are send with mobile devices. These messages—text, pictures or videos—are received with a small box and can be seen over the TV. Elderly people do not feel alone but safer and they can easily interact with their environment this way, for example when their family lives far away. (Lundeval & Ozan, 2006, p. 13) The main problem may be to persuade the elder ones of this system.

The employees of the city of Stockholm can work mobile everywhere they want. That is possible because the city administration works together with TeliaSonera, the major mobile network operator in Sweden, offering the service "St Erik Connect Pro". It enables users to synchronize their calendar, to-do-lists, contacts and e-mails of their notebooks and their mobile devices. (Lundeval & Ozan, 2006, p. 7)

Information on the city can be obtained via explorestockholm.com, an Internet-based service that can be adapted through different terminals like mobile phones, PDAs, notebooks and desktop computers if these devices have an Internet browser. For a scalable presentation of the information the approximate screen size for the different devices is automatically estimated. (Ahlstrand, Asperen, Gardlund & Leandri, 2004, pp. 27) The service provides guided tours with information about objects of interest in Stockholm and is also connected to a database that provides information about buildings, places and statues in Stockholm. The tourist can get as much information as he wants and he has the time to visit the sight as long as he wants to. In addition it has a "mail to a friend" function so that the user can send the information provided on that website to other users, perhaps as substitute for a postcard. (Ahlstrand et al., 2004, p. 20)

Designated opportunities to extend this service are (Ahlstrand et al., 2004, p. 47):

- maps and location based services could be included,
- data transfer (which is only executed via GSM/GPRS, WLAN or a fixed-line Internet connection) could include Bluetooth at different objects of interest in the city,

- the provided information could be augmented,
- a technical solution for monitoring usage and generating statistics could be developed in order to analyze and improve the solution.

In the early phases of this project an SMS service for tourists was also planned. Each object of interest had its own number on a sign. The user sends an SMS to the number at the signs and receives an SMS with information about the sight. Actually, this service was never carried out because of the limited capacity of text messages which presupposes lots of them, imposing the costs on the city. (Ahlstrand et al., 2004, p. 51-57)

Another project is a website where users can get information about the current traffic situation in Stockholm. The service was improved so that it can now be used via WAP with mobile devices or a service number can be called. Users who always take the same way, for example to work or back home, can subscribe a traffic news service. If there is a traffic jam or a building site the user will be informed with an SMS so that he can take another way. (Lundeval & Ozan, 2006, p. 11) The service "mystockholm" enables tourists and inhabitants to tell their own stories about Stockholm. They can add texts, pictures, audio or video clips to a common database. With these information they can compose a city tour and download the tour on their mobile device (if there is enough memory space. But files for mobile devices are usually small). (Lundeval & Ozan, 2006, p. 15) During two weeks in the summer of 2006 a trial offered free WLAN access in a public park in Stockholm. (Lundeval & Ozan, 2006, p. 15) Finally, USK, the Stockholm Office of Research and Statistics, and mCity tested a service with which field surveys are evaluated directly with a mobile device so that paperwork could be reduced. Another advantage is that there is no media disruption, which means no data is typed in wrong and the evaluation can be carried out faster. (Lundeval & Ozan, 2006, p. 17)

Case Study 3: The "mobile city Bregenz" project in Bregenz (Austria)

This project started under the name of "modern city Bregenz" in 2004. It originates from the m-commerce activities of Mobilkom Austria, the major Austrian mobile network operator who successfully convinced the city of Bregenz to begin a co-operation. The intention was to provide inhabitants and tourists innovative and modern services on their mobile devices. For the use of the service, a user must register only once through the website www.bregenz.at or via SMS to use all services. Customers pay for the services via telephone bill or by paybox, a mobile payment service provider majoritarianly owned by Mobilkom. (Kuehnberger, 2004)

All services can be used by calling the service number 0664 660 6900. In present there are five services: m-parking, m-train, m-ticketing, m-shopping and m-cityinfo. (Mobilkom, 2004; Bregenz, n. d.) M-parking can be used in all cities which joined the system of Mobilkom Austria by sending an SMS to the service number. The newest service is m-ticket where the user can buy a ticket for the train via wap or SMS even at the last minute and the schedules can be checked mobile. Following information must be send to the service number: Train, departure station, arrival station, first name, last name, optionally date of departure, number of passengers, class. Users can order products if they are walking from shop window to shop window or when they see an advertisement in the newspaper. They send an SMS to the service number including the product number and the product will be delivered to their home address. Delivery charges have to be paid additionally. Tickets for various events can also be ordered via SMS. The customer sends the title of the event or the name of the artist to the service number and gets in return a mobile ticket for the event on his mobile device. Also tourists can use the service through the tourism agency of Bregenz.

Finally, a service makes it possible for the user to get information about the city on their mobile device. The user sends key words to the service number and gets up-to-date news about the topic like events or information especially for the citizens. During the registration the user can sign a regular newsletter with selected topics like culture, sport, leisure, children, seniors and information for citizen. (Mobilkom, 2004)

Overview

Now we want to provide a short overview, what the case studies have in common and / or where they differ. So far only Stockholm provides services in the health sector, Tartu offers his own friend finder and Bregenz gives the opportunity to buy tickets for events mobile. The only thing all three cities offer is information on objects of interest.

FUTURE TRENDS

As most of the services presented in this chapter are only pilot projects or are not implemented yet, the whole chapter could be seen as a future trend in city marketing that has got good potential. Mobile services are a new opportunity. So far there are not so many services in all domains which are already implemented, but numbers are growing every day. As city marketing itself is emerging in more and more cities, so will mobile services for city marketing. However, upcoming technology will allow for more and better services. It depends on how the enterprises will use these services and how much they will cost the user

Figure 2. Comparison of the case studies

	Tartu (Estonia)	Stockholm (Sweden)	Bregenz (Austria)
M-City	citizen send information to the city download of images of Tartu	mobile working employees of the city	
M-Health		coordination of schedules in elderly care mobile work reports	
M-Transportation			
tickets for public transportation	x		x
parking-tickets	x		x
traffic information	x	x	
M-Tourism			
city-map		x	
information on sights	x	x	x
M-Education		vocabulary trainer	
other topics	m-payment in shops library friendfinder	safety TV online platform for the exchange of experience free WLAN access was tested surveys carried out mobile	m-payment in shops m-payment for event tickets

(or the citizen, if they have to pay for the services offered by the city).

Location-based services all over the world could benefit from GALILEO, the European satellite navigation program. Unlike its US counterpart GPS, GALILEO is a civilian system which will provide enduring location services for public and commercial use. It is more sophisticated, efficient and secure than the GPS. Customer will even have the option to use both systems to increase security and reliability when relying on position data. (Europaeische Kommission, 2002, p. 8) The satellites have an atomic clock to provide an exact temporal and local positioning up to an accuracy of one meter. Basic usage of the services for the public will be free. For commercial or professional applications there will be surplus values with costs for the offerer. (Europaeische Kommission, 2002, p. 4) With this new technology a lot of new services can be developed. The location of people and of things will be easier, faster, exacter and more reliable as it is possible today. And as the standard of mobile devices will also progress, they can be connected with a GALILEO receiver, providing an easy opportunity to present maps on mobile devices. But perhaps some people may feel as if they live in a kind of surveillance state where every step is watched and recorded. It should be guaranteed that the people have the control by whom and when they are located by others and can deny the action. Security topics should be always regarded so that no third party could take an advantage of the position of the user, e.g. thieves checking the time when the house-owners are coming back home.

There is another alternative for positioning: via WLAN. Most cities have a very good WLAN infrastructure. So people can be located in cities over a special software even more precisely than with GPS. The problem of GPS in cities is that it is too imprecisely to help tourist finding their way through the city. So the tourist does not have to be afraid of their language skills. But this sort of positioning is so far only tested in Nürnberg in Germany since January 2008. When the first testing phase is over and the experience is good, the system will be established in other cities like Berlin, Munich, Frankfurt, London, Paris und Mailand. (Fraunhofer IIS, 2008)

An upcoming technology that is expected to be included in mobile devices within the next years is *near field communication (NFC)*. NFC is a derivate of radio frequency identification (RFID) and will enable handsets to read amounts of information from a smart tag when the user touches it with his mobile phone. Two relevant applications are to automate the initiation of data transmission over for Bluetooth (which is very prohibitive for a widespread use) or to redirect the Internet browser of a mobile device to a certain page. The first would allow cost-free and very easy download of applications at a spot, the latter enables to directly go to any information that is provided over the web.

As already described, mobile communication networks are cellular. This allows for sending a message to every user located in a cell or entering it. In GSM, this service is called *cell broadcast* and already part of the network specification—however, in most markets due to legal and commercial restrictions it is only used by the mobile network operators themselves to address their customers. Some used it to send business news or weather forecasts to their clients. ("SMS Cell Broadcast", n.d.) To receive cell broadcast messages users have to activate the service in their menu on the phone. An example where cities could use this service is for thunderstorm or security threat warnings. The mobile channel could be another option to inform the population but should never be the only one. Unfortunately, the days of this service seem to be counted before it comes into commercial use—up to now, it is not supported by subsequent mobile radio standards. ("Cell Broadcast", n.d.)

So far, the benefits of third generation mobile networks are not really used by mobile city marketing applications. This is supposed to change as 3G penetration reaches the majority of standard users in a country or the majority of tourists visiting a city. Even if the previous services will not be replaced by the mobile technology, mobile services are a good opportunity to enlarge the services offered by a city. The major change will be increasing bandwidth, allowing better use of graphics (such as complex maps) and multimedia with online applications. Handsets features will increase accordingly. Here again, careful target group analysis is to be carried out in order to provide city marketing applications that meet the users' needs and device capabilities.

Not only are the mobile networks improved but also the mobile equipments. New features are implemented and new control functions. For example the new iPhone is now handled over touch screen so that the user can use a whole keyboard when he types messages. So the handling gets easier as some users had problems using the small keypads of previous mobile equipments especially mobile phones with phone keypads. Further on the newest generation of mobile equipment is HTML-enabled, what means they access the "normal" Internet with all its features. The producer of mobile equipment should improve the technology so that the high-tech equipment is not prone to error and work fast. A progress in the performance of the battery should also be made so that the user has always got the opportunity to use their mobile equipment without being afraid of having no power left.

In some countries, e. g. Germany, mobile data transmission is still very expensive as it is still a very new technology and is still under construction. So the enterprises did not have the time to amortize their investment. This results in a major reluctance with regard to mobile services. Two developments may be beneficial here. First, there is a clear tendency in virtually all mobile markets leading to lower prices for voice and data, especially when it comes to flat rate data pricing. The second development could be even more interesting for city marketers: The more intelligent mobile network operators' billing systems become, the more probable will be the support of sponsoring models. In other words: Innovative operators will enable third parties to pay for the data transmission costs when the user downloads information on the city—for instance, a mobile video stream could begin with a short commercial and thus be downloaded free of any charge.

CONCLUSION

In most countries, mobile handsets are more widespread than stationary Internet access, television or any other technical device. Many people tend to perceive the mobile phone as a central part of their life—the younger the people are, the higher is their affinity to this technology. In this chapter we examined how this trend can be employed for city marketing purposes and provided an overview on mobile applications for city marketing.

Most people that are part of the target group of city marketing carry their device with them all the time. This leads to the mobile added value of ubiquity: On one hand, the user is almost anytime and anywhere reachable via this channel. On the other hand, he is able to use this channel almost anywhere and anytime to access information. In this topic context sensitivity is very important. That means that the special needs of a person in his present situation are solved with a mobile service immediately. If the user has to send out information requests this should be easy and short for him. Another topic that should be regarded is that the user does not want to get too much information on his mobile phone as he might not have the time to read it all and does not want to get the information. For example if he goes through the city and every second step he takes he gets an mobile marketing message of shops near him.

Also questions of security should be regarded when such services are implemented, e. g.:

- What happens to my data when my mobile equipment is stolen or I lost it?

- Can I lock my whole mobile equipment with all the services included and my SIM-card? How fast?
- What happens when my mobile phone is attacked by viruses?
- What about data protection?
- Can my data transfer be watched by unauthorized people?
- But also questions like "What happens when I change my mobile equipment or my mobile phone number? Have I to do the whole registration again?" should be regarded when mobile services are offered.

As can be seen in other parts of this book, city marketing covers a wide range of activities. In many of these, the mobile channel can play a role. In sections 2 and 3 we introduced a variety of different applications: mobile city info, m-tourism, m-commerce, m-parking, m-health, m-payment, m-ticketing, mobile marketing, mobile learning/m-education, m-government and others. We also saw three case studies of European cities that already have started coordinated projects with different mobile services: Tartu in Estonia, Stockholm in Sweden and Bregenz in Austria. They might be still at the beginning, but their ideas and approaches may serve as valuable examples for successful projects. They are worth to be observed and they are working to enlarge their services.

Mobile services cannot replace classic city marketing as there will always be people not using them. They might even own a mobile phone, but just not intend to use for other things than phone calls. They like their old mobile phone even if it does not have so much functions but it works well and the battery works a long time. But lots of others might be glad that such services exist and will use them. And if it's as easy as sending an SMS to receive location and timetable of the nearest subway station—maybe it is even worth a try...

Upcoming new technology will enable more and richer applications by better location-based services, higher bandwidth and better handling. Existing cost barriers will lower and sponsoring concepts will become feasible. Multimedia applications will become widespread. But the danger of a mobile virus may grow.

Nowadays it is not possible to imagine a world without mobile phones and they get more and more features. It is no more only for making phone calls and sending messages, but also cameras, radios, audio-players, etc. are included. So the importance of mobile phones is growing. There is one thing so far only mobile equipments can offer: that a personalized message can be send directly to a particular person. Services offered from different suppliers have advantages as the user can chose the best offer. But the problem is that the user has perhaps to register with all of these service providers and the services can not be combined.

Mobile applications offer the utmost direct access to the customer and thus a great potential for city marketing in the near future. The success in this domain is determined by customer acceptance. To achieve it, marketers need expert knowledge on both their target group and the particularities of the mobile channel—in order to provide applications that are useful, easy (and fun) to use, and overcome the hurdle of high cost.

REFERENCES

Ahlstrand, K. I., Asperen, A., Gardlund, M., & Leandri, P. (2004). *explorestockholm.com—A tourist service from .tourism.* Retrieved June 20, 2008, from www.tslab.ssvl.kth.se/csd/projects/0412/tourism%20final%20report.pdf

Arnold, M. (2007, Spring). *Die elektronische Behörde.* Pictures of the Future—The Magazine for Research and Innovation, Siemens, (pp. 41-42).

AS EMT (n. d.). Retrieved August 13, 2007, from http://www.emt.ee/wwwmain?screenId=mainpage.private&language=ENG

Baeumerth, (2002). *Handbuch mobile commerce: technische Grundlagen, Marktchancen und Einsatzmoeglichkeiten* (pp. 225-230). Berlin/Heidelberg, Germany: Springer.

Bellmann, B. (2007). *Schicken Sie ihre Gaeste auf Schatzsuche*. Retrieved August 17, 2007, from http://www.bjoernbellmann.de/pdf/pdf_infotrail-geocaching.pdf

Bregenz (n/d.). *Herzlich Willkommen in der Mobile City Bregenz!* Retrieved February 14, 2008, from http://www.bregenz.ws/mcity/

Cell Broadcast ist bald Geschichte - GSM-"Textrundfunk" beim UMTS-Mobilfunk nicht vorgesehen (n/d.). Retrieved August 25, 2007, from http://www.teltarif.de/i/cellbroadcast.html

Cheverst, K., Davies, N., Mitchell, K., Friday, A., & Efstratiou, C. (2000). *Developing a context-aware electronic tourist guide: Some issues and experiences*. Paper presented at the SIGCHI Conference on Human Factors in Computing Systems. Retrieved June 20, 2008, from http://www.guide.lancs.ac.uk/CHIpaper.pdf

Europaeische Kommission (2002). *GALILEO—Das europaeische Satellitennavigationsprojekt. Information zu dem Projekt*. Retrieved June 20, 2008, from http://ec.europa.eu/dgs/energy_transport/galileo/doc/galileo_info_note_2002_03_26_de.pdf

Fraunhofer IIS (2008). *Testumgebung WLAN-Lokalisierung Nürnberg: Vor-Ort-Information und ortsabhängige Dienste mit starken Partnern*. Retrieved June 20, 2008, from http://www.iis.fraunhofer.de/pr/Presse/Presseinformationen_2008/WLAN_Testumgebung.jsp

Grabow, B., & Hollbach-Groemig, B. (1998). *Stadtmarketing—eine kritische Zwischenbilanz*. Berlin, Germany: Deutsches Institut fuer Urbanistik.

Guenther, K. (2002). *Stadtmarketing—Grundlagen und konzeptionelle Gestaltungsansaetze fuer eine raeumlich ganzheitliche Versorgungspolitik unter besonderer Beruecksichtigung des stationaeren Einzelhandels*. Goettingen, Germany: GHS.

Heidelberg-Mobile (n/d). Retrieved May 23, 2007, from http://www.heidelberg-mobil.de

Heise Mobil (2008). WLAN-Navigationssystem leitet Nürnberger Fußgänger. Retrieved February 20, 2008, from http://www.heise.de/mobil/newsticker/meldung/print/101765

Koelmel, B., & Porak, A. (2003). *Real Life Scenarios of Location Based Advertising*. Eurescom Summit 2003.

Kuehnberger, P. (2004). *Mobile City Bregenz*. Retrieved June 20, 2008, from http://wko.at/bshandel/Zahlungsverkehr/Veranstaltung-Bezahlen-im-Internet/Praes-Filme-Unterl/mobilkom/BREGENZ.ppt#336,6,Umsetzung

Mobilkom (2004). *Mobilkom austria macht Bregenz mobil - Neue Massstaebe fuer Buerger- und Kundennaehe*. Retrieved June 20, 2008, from http://www.pte.at/pte.mc?pte=040708010

MobiSolutions (n. d.). *M-Governance Solutions*. Retrieved June 20, 2008, from http://www.mobisolutions.com/en/mgovt.php

Luley, P. M., Almer, A., Schnabel, T, Massimo, R., & Herpolsheimer, W. (2004). Geo-Data presentation on mobile devices for tourism application. *Symposium on Location Based Services & TeleCartography, University of Technology*, Wien, Austria. Retrieved June 20, 2008, from http://www.joanneum.at/uploads/tx_publicationlibrary/img1999.pdf

Lundeval, K., & Ozan, H. (2006). *Report Mcity. Mobile projects in the city of Stockholm to make internal work more efficient and to increase the use of self service by city citizen.*

Partnership for the Heart (2006). Retrieved August 13, 2007, from http://www.partnership-for-the-heart.de

Rakkolainen, I., & Vainios, T. (2000, November). *A 3-D city info for mobile users.* Paper presented at the 3rd International Workshop in Intelligent Interactive Assistance and Mobile Multimedia Computing, Rostock, Deutschland, S.115-212.

Rannu, R. (2003a). M-services in Estonia. *Baltic IT&T Review, 32*, 28-33.

Rannu, R. (2003b). *Mobile serivces in Estonia.* Praxis Working Paper No. 8.

Rannu, R. (2004). M-Governance from vision to reality. *Baltic IT&T Review, 35*, 35-40.

Rannu, R., & Semevsky, M. (2005). *Mobile services in Tartu - Existing services, best practices, methodology and suggestions for future development.* Retrieved June 20, 2008, from http://www.ega.ee/public/Mobile_services_in_Tartu_FINAL1.pdf

Roettger-Gerigk, S. (2002). *Handbuch Mobile Commerce: technische Grundlagen, Marktchancen und Einsatzmoeglichkeite (pp. 419-426).* Berlin/Heidelberg, Germany: Springer.

Schneider, J., & Schroeder, F. (2003). *The M-Toguide Project—Development and deployment of an European Mobile Tourism Tool.* Eurescom Summit 2003.

SMS Cell Broadcast (n. d.). Retrieved May 23, 2007, from http://www.gsm-modem.de/sms-cell-broadcast.html

Turowski, K., & Pousttchi, K. (2004). *Mobile Commerce—Grundlagen und Techniken.* Berlin/Heidelberg, Germany: Springer.

Voutsis, N., & Zimmermann, F. (2005, July). *Anonymous code lists for secure electronic voting over insecure mobile channels.* Paper presented at the International Conference on Mobile Government, Sydney, Australia, (pp. 434-444).

Chapter VI
Strategic Management in City Government:
Integrating Information Communication Technologies and Marketing in a Causal Model to Drive Stakeholder Satisfaction and Economic Development

Laura L. Matherly
Tarleton State University, Central Texas, USA

Maureen Jouett
Tarleton State University, Central Texas, USA

ABSTRACT

Integrating information communication technologies (ICTs) and marketing in strategic management of city government is critical to achieving stakeholder satisfaction and economic development. As a result of the rapid growth in computer networks and access to online services, the use of ICTs, for example, Internet and Intranet, as a communication and marketing platform can provide a city with a global advantage. City marketing focuses on promoting the attributes of a location to prospective stakeholders so that these individuals, businesses, and investors are attracted to visit, locate, or invest in the city. A causal model is presented where ICT is used to not only to deliver services to internal stakeholders, but also to market a city to external stakeholders. To be successful, managers need to be skilled in current technologies and marketing practices. Case study applications are discussed as well as the questions to address in future research to most effectively integrate ICTs and marketing in city management.

INTRODUCTION

Organizations are increasingly faced with the global challenges of international competition and as a result, the adoption of information communication technologies (ICT) as a marketing competency has become a strategic imperative (Gummesson, 2002; Lapierre & Medeiros, 2006; Brady, Fellenz & Brookes, 2008). City governments are utilizing emerging technologies and especially, ICTs, as a major catalyst to market and communicate with stakeholders to ultimately attain strategic objectives such as growth and economic expansion. The purpose of this chapter is to provide a model for the use of marketing and ICTs in strategic planning to create a distinctive competency and address the challenges that result in achieving strategic performance objectives. The causal model developed in this chapter proposes that ICT initiatives and investment in the ICT infrastructure can be used to improve internal stakeholder services and satisfaction as well as with a marketing plan to promote the attractiveness of the city image to external stakeholders. Marketing and ICTs are presented as activities in a strategic map of causal linkages: by developing and recruiting employees with the skill sets required to effectively integrate ICT initiatives and marketing plans that develop customer relationships and improve service quality, a city's image to both internal and external stakeholders will be enhanced. Excellent internal processes and service levels can advance community outcomes such as health, safety and welfare as well as a city's overall reputation which ultimately impact the attainment of strategic goals such as economic development.

First, a review of the developments in performance management, balanced scorecards, and strategic management is discussed with an eye toward developing a generic causal model for city management that starts with developing the employee and the information capital of the organization which in turn improves internal processes, services and eventually, stakeholder satisfaction. The strategic group map (Kaplan & Norton, 1996) was chosen as a framework for the model because it challenges strategic planners to go beyond measuring lagging performance indicators such as economic development and growth to specify the drivers of economic development and growth. For example, if a city's image drives economic growth and marketing positively impacts a city's image, by measuring and investing in marketing initiatives, corresponding improvements in image and then economic growth will occur.

The role of ICTs in contemporary marketing practices (CMP) over the last decade is examined as an effective solution to meet the challenges brought about by a connected, global economy and expanded and empowered customer base. Given the importance of technology in developing a strategic marketing plan, a number of ICT applications are presented. Several case study applications are discussed to illustrate the application and benefits of ICTs to city marketing. Last, the future role of ICTs in city government is examined by posing the fundamental issues in the form of research questions to be answered that will illuminate how to effectively integrate specific ICT and marketing practices into strategic management of city governments. The management mindset that is required to be successful and competitive, both in determining the strategic direction and implementing the technological changes required by the new marketplace is addressed.

PERFORMANCE MANAGEMENT AND A CITY STRATEGIC GROUP MAP

Over the last decade, developments in performance measurement have signaled the need for organizations to monitor performance dimensions that go beyond traditional financial measures to include measuring and improving those factors that ultimately impact financial performance such as stakeholder satisfaction. In 1992, Kaplan and Norton introduced the balanced scorecard as an overall framework for establishing a performance measurement system that predicts financial results. They provided a framework for capturing metrics at the executive level based on four categories: (1) customer satisfaction, (2) financial performance, (3) internal processes and (4) employee innovation and growth. In brief, nonfinancial measures provided the balance needed to supplement financial measures and align employees with strategy. Although many of the balanced scorecard applications are in industry, the City of Charlotte, North Carolina applied the concepts to city management and developed a city scorecard (Syfert & Elliot, 1998 ; Eagle, 2004).

Kaplan and Norton (1996, p. 149) define strategy as "a set of hypotheses about cause and effect" and maintain that every measure should be identified in a chain of outcome measures and corresponding performance drivers of the outcome measures. They introduced the concept of strategy maps--there is a causal path among the four perspectives. Hence, the categories and measures within the categories are linked in a causal path of leading and lagging indicators. Essentially, improving employee innovation and growth (which includes technological and organizational capabilities) will result in continuous improvement of key internal processes. Improved internal processes that deliver value to the customer will lead to subsequent improvements in customer satisfaction, which in turn will result in improved financial performance. The value of using this approach is that managers are more likely to be successful in meeting/improving the performance of lagging indicators, such as economic growth, by identifying and verifying the factors that drive economic growth.

The four categories (and performance measures) in Kaplan and Norton's (1996; 2004) strategic group map are typically adapted or customized when applied to a particular organization. In Figure 1, an example of a strategy map for city government is proposed that illustrates the central role that marketing and ICTs play as leading indicators to customer and stakeholder satisfaction. The model reads from the bottom up--the leading indicators at the base drive the next level of indicators until the uppermost level of stakeholder outcomes is reached. First, employee development is essential to drive effective resource management. Next, investment in ICT initiatives and infrastructure as part of a marketing communications plan helps manage resources efficiently as well as market a city. Using ICT as a marketing tool can improve branding/image perceptions, customer service and customer satisfaction. Enhancing a city's identity or image, services and customer satisfaction will ultimately impact key performance indicators such as overall reputation, economic development and growth. In the following sections, the major categories and factors within each category are discussed with an emphasis on the central role that ICT and marketing play in eventually producing positive stakeholder outcomes.

DEVELOP EMPLOYEES

The catalyst for successful resource management is a committed and skilled workforce which is supported by the organization's culture and leadership to continue to learn and grow (Matherly, 2007). Figure 1 is an example of a strategic group map for city management. The category at the bottom of the model includes three areas to address in order to develop employees; an institution needs to achieve a high level of employee commitment, recruit and retain a skilled workforce, and promote learning

Figure 1. City strategic map

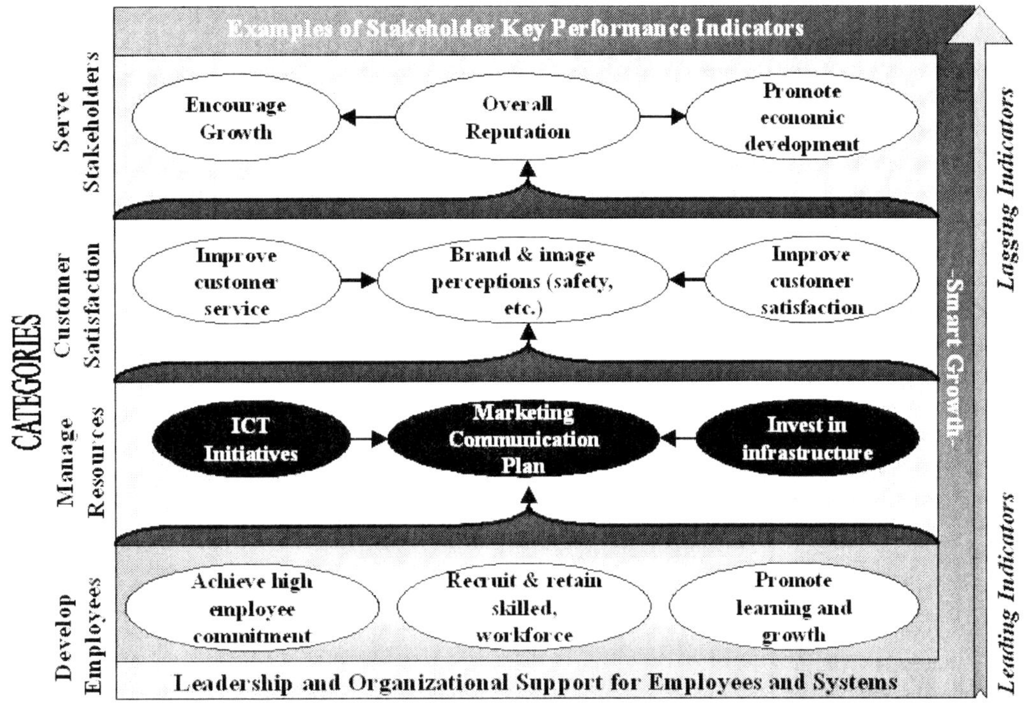

and growth of employees. Both Baldrige (2008) and Kaplan and Norton (1992) recognize the essential role that the people, i.e., employees and managers, play in driving process improvements and performance excellence. In a study of ICT deployment in marketing applications, Brady, Fellenz and Brookes (2008) conclude that there is a "need to study and expand the skill set of marketers into technological, managerial, and organizational areas to more fully enable the use of ICT within contemporary marketing practice" (p. 108). Not only do employees need the skills to work effectively within an organization's existing technology infrastructure, they also must be familiar with new technologies to expand ICT initiatives to create competitive advantage. Only by training and developing existing employees or recruiting and hiring employees/contractors with the required ICT skill set can a city effectively implement new applications and forms of ICT.

With the rapid changes in and complexity of ICTs, there is a corresponding increase in the information processing and analytical requirements of managers creating a challenge in integrating technology changes into strategic planning and marketing (Fisher, Raman & McClelland, 2000; Holland & Naude, 2004). Executives must choose technologies that are integrated with the needs of the organization *and* customers and manage any resistance and barriers to implementation (Leverick, Littler, Bruce & Wilson, 1998; McAfee, 2006). Without a sufficient skill set in technology, planners will be unable to take advantage of emerging opportunities and maximize the value of ICTs (Brady, 2003). The practical reality is that managers are being asked to understand, develop and implement ICT applications that efficiently and effectively market their services to stakeholders—the customer interface—as well as integrate the programs across operations. For example, an organization's ability to assimilate, store and retrieve new information and apply this learning affects successful customer relationship management (CRM). CRM systems capture and integrate customer data from across the city, consolidating the data, analyzing the data, and then distributing the results to various systems and stakeholders across the organization. Effective organizations are those that invest in both the technology infrastructure and organizational learning to support the building of their marketing intelligence (Chen & Ching, 2004). A learning organization that fosters employee commitment and growth, for example, by investing in training and development, will result in more effective and efficient resource management.

MANAGE RESOURCES

Employees who are committed to achieving a city's performance objectives and skilled in technology applications and marketing can drive the development of the appropriate technology infrastructure that can effectively market a city to stakeholders. In Figure 1, the ability of a city to manage resources effectively and positively impact customer satisfaction is determined by how effectively ICT initiatives are deployed which requires an investment in the ICT infrastructure. Further, the ICT infrastructure needs to be integrated into the marketing plan both to deliver services to internal stakeholders and to market the city to external stakeholders. ICTs provide greater access to information, dramatically increase the speed of transmitting information and allow for wide-spread broadcast and use of information. By providing services through an ICT interface, improvements in service, quality, accuracy and operations will result in a subsequent increase in customer satisfaction.

Marketing, through the effective use of ICTs, can improve a city's image, attract new businesses or help retain and expand existing businesses, as well as accentuate a city's attributes such as its workforce, talent pool and services offered. Similarly, an effective ICT marketing program can counteract negative images through the positive interaction of city employees with citizens and visitors. ICTs can enhance communication and improve relationships between the city and its citizens and help secure community support for changes in service delivery. For example, ICTs can be used to increase citizen participation in local government and through collaborative problem solving and communication help reverse an economic decline of an abandoned downtown area. Moreover, the potential for ICTs to alleviate poverty and promote economic growth in developing countries justifies greater attention and systematic analysis (Eggleston, Jensen & Zeckhauser, 2002).

Expand ICT Initiatives and Invest in the Technology Infrastructure

Expanding ICT initiatives and investing in the organization's technology infrastructure are essential to success and a growing assortment of technology tools are available for cities to employ. Technological devises and software are primarily related to obtaining and sending information at unprecedented speeds. Wireless networking affords a city with greater mobility and in many cases affords smaller cities the ability to maintain the same presence and competitive advantage that many larger cities have on the internet. The business community argues that wireless internet access is good for economic development and a handy tool for business travelers (Swope, 2007). Mobile devices and software allow remote connection to desktops and office networks for access to information. Electronic transfer of information is not only more efficient and timely but it can improve accuracy.

Develop and Implement City Marketing Plan

Marketing is recognized as a body of professional skills and techniques for bringing the producers of products and services together with people that need or can use them. A review of strategic group maps in peer reviewed journals as well as those reported in the public domain reveals that although hundreds of organizations have developed strategic group maps, a key factor absent in these applications is marketing as a leading indicator and driving force in developing community perceptions of the organization's image which ultimately impacts lagging indicators such as economic growth and financial performance. More and more, marketing principles are being applied to city management and city websites deliver brand information that can create value, a positive image, recognition and recall that attracts external stakeholder groups. The qualities which make a city unique create an identity or image that can be used to market a location (Erickson & Roberts, 1997). Based on a communication-based marketing model, Duncan and Moriarty (1998) point out the importance of managing consistent messages across all stakeholder groups so that communications are managed strategically to build brands and external relationships. Tschirhart (2008) demonstrates that while many city government websites deliver brand information, there are substantial inconsistencies in whether cities follow this recommended branding practice.

Marketing can play a central role in establishing a particular city as an attractive location and ICT is an effective and efficient method for communicating and facilitating a host of marketing activities. In Figure 2, the four P's (product, price, place and promotion) are integrated in a marketing and ICT model that emphasizes the importance of target market relationships. ICT is place, i.e., ICT is the mechanism/channel used to promote a favorable city image; conduct services (product); and improve operations and efficiency (price). Both internal and external stakeholders can be targeted by marketing communications efforts of cities. These stakeholders include the citizens, city council and mayor, city manager, department heads, employees, bonding agencies, investors, economic development prospects, visitors, shoppers, tourists and current businesses. Moreover, ICT can be designed to be interactive, e.g., by conducting services and communications online.

Having an excellent offering or product does not ensure that prospective buyers will know about it. ICTs are fast becoming the main medium for proactive, targeted marketing – they can be customized, personalized, shared, two-way, interactive and cost-effective. In Figure 2, targeted marketing can be used to improve the image of a city with external stakeholders. ICTs can facilitate marketing and the exchange of mutually beneficial information between a city and its stakeholders. The use of ICTs provides a viable resource to cities, facilitates competitive advantage and opens opportunities for future success. CRM systems can be used to organize, store and retrieve relevant information required for tailored, multi-channel marketing. Pinpoint marketing – getting the right message to the right people at the right time – and data mining capabilities are enhancing marketing efforts for seasonal activities so that previous users are solicited when similar activities are planned. Online community networks are natural marketing channels. Marketing packages can be quickly produced to prepare professional presentations that are available with user friendly software programs to effectively communicate a city's financial condition, for example, to bond rating agencies. Ease of access to financial information helps these agencies determine bond ratings that can lead to substantial monetary savings as investors evaluate the risk involved in purchasing city bond offerings.

Figure 2. Integrated marketing and ICT communication strategy

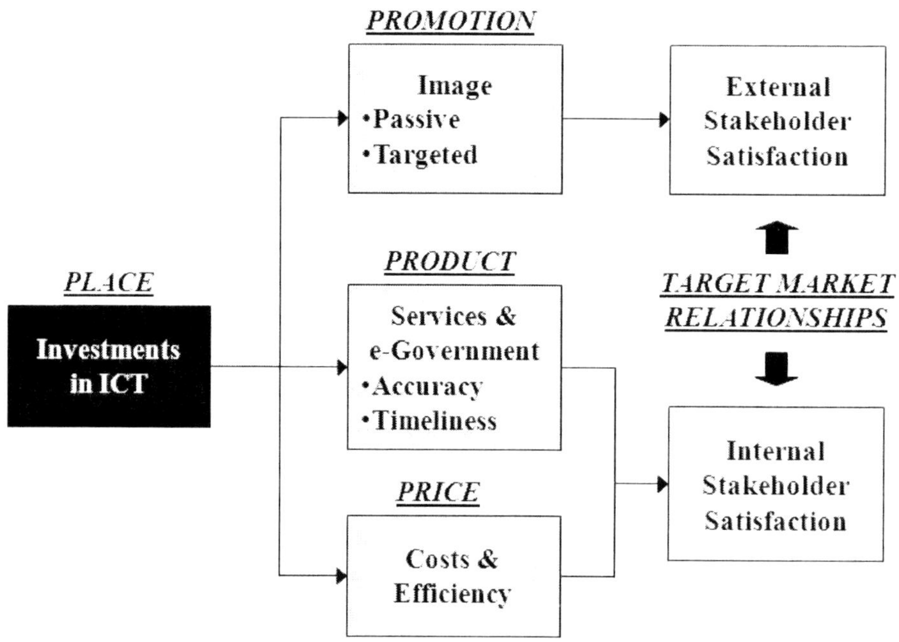

Integrating ICTs with the Marketing Plan

Over the last decade, new directions for marketing and management have been identified in response to changes in factors that influence organizational success. Numerous researchers are calling for a multidisciplinary framework that integrates managing ICTs uses, forms and infrastructure with contemporary marketing practices (e.g., Achrol & Kotler, 1999; Coviello, Miller & Marcolin, 2001; Gommans, Krishnan & Scheffold, 2001; Brady, Saren & Tzokas, 2002; Brady, 2003; Brookes, Brodie, Coviello & Palmer, 2004; Chen & Ching, 2004; Barwise & Farley, 2005; Brodie, Coviello & Winklhofer, 2008). Gummesson (2002) defines the new economy as a network society that focuses on services, emerging customer roles, information technology, globalization and alliances between countries and these forces have ushered in a shift from the traditional marketing paradigm (the four P's-product, price, place and promotion) to relationship marketing and interactive marketing (Webster, 1996; Deighton, 1996; Webster, 1998; Day & Montgomery, 1999; Webster, 2005). Relationship marketing is a necessary perspective requiring a new mindset to address the networks and interactions that are embedded in organizations, markets and society. The network organization includes all stakeholders that interact with the organization—both intra and internet interactions. Similarly, in a survey of marketing research professionals, Struse (2000) identified the top influences of the 21st century as the internet, globalization of business, culture and research and one-to-one (relationship) segmented marketing. Increasingly, technological expertise and investment will be required as part of strategic planning and marketing.

Generally, ICTs can be used as a marketing communication tool to promote a city's image in either a passive or targeted mode (see Figure 2). In the passive form, they can provide: citizen access to information, government functions and services; locations and directions to parks and community centers; calendars of city-sponsored events and activities; access to property information and citizen services; job postings; phone and staff directories; meeting notices; statistics about the city; and tourist information. Stakeholders recognize that the Web is a good place to conduct a site search (Bastain, 2007). Having information readily available and accessible can provide a competitive advantage as a city can showcase strengths and opportunities to potential stakeholders. For example, having accurate, referenced data allows businesses and cities to conduct systematic and realistic evaluations of prospective locations. Increased levels of information access enable a city to differentiate incentive packages to increase competitiveness and improve success rates in attracting new residents and businesses. Stakeholders can conduct research using community and regional websites as primary information sources and avoid the high costs of external consultants (Bastain, 2007).

E-government is the application of the Internet and related technologies to digitally enable government and public sector agencies' relationships with citizens, businesses, and other arms of government. In addition to improving the delivery of government services, e-government can make government operations more efficient and also empower citizens by giving them easier access to information and the ability to network electronically with other citizens. Governments use internet technology across many operations, agencies and levels to deliver information and services to citizens, employees, and businesses with which they work (Laudon & Laudon, 2006). The Organization for Economic Co-operation and Development (OECD) recognizes e-government as a major enabler of good government practices (OECD, 2003). The OECD's charter is to promote policies designed to achieve high economic growth and expansion for over 30 member countries and recognizes that e-government helps improve efficiency and services as well as achieve policy outcomes and economic objectives. To illustrate, ICTs can be transactional providing for: license renewal and payment; remittance of parking citations and court fines; registration for programs and other functions; interactive job applications; online permits, business licenses, court documents; sales tax collection; distance learning; web casting of city/county meetings; communications with local leaders; and links to other key sites (airport, transportation, hotels). ICT and access to digital democracy can enhance the community's democratic participation in public affairs; for example, with e-voting and e-ballots, and citizens can express and exchange views with elected representatives (Bozinis, 2007).

ICTs can be used to facilitate compliance with governmental regulations. For example, in some countries, citizens are entitled to information about the affairs of government and the official acts of public officials and employees. Interactive technology expedites the delivery of information, and provides a conduit for continuous feedback that can significantly impact responsiveness to stakeholders. Timeliness of information and availability allows cities to respond quickly to community needs. However, this requires the monitoring and coordination of numerous communications from multiple media.

In summary, ICTs can be used to communicate service information to internal stakeholders, improve operations and efficiency and promote city programs that advance community well being, e.g., health, safety and welfare. The success of e-government in the provision and delivery of public goods and services as well as marketing communications is well documented (e.g., Mellor, 2006; Kim, 2007; Halaris, Magoutas, Papadomichelaki & Mentzas, 2007). In order to implement ICT changes effectively and efficiently, Gummesson (2002) stresses the importance of focusing on internal networks and interactions.

ICTs allow for employee interaction and shared information that can improve communications, operations and efficiency. On a global level and especially where stakeholders are geographically dispersed, applications such as video communication technology are more cost efficient and timely than face-to-face interactions. Moreover, interactive technology provides a mechanism for supplying information and interfacing with stakeholders on a personal basis such as after hour's contact which is especially effective at facilitating transactions and dealings across international time zones. ICTs expedite the communication and exchange of information which can facilitate consensus building and support for decisions and plans, for example, to invest or locate in a given community (West & Berman, 2001).

CUSTOMER SATISFACTION

The reputation of an organization, defined as the set of attitudes and beliefs that an individual or market holds about an institution, can provide a competitive advantage that enhances an institution's long-term ability to create value (Goldman, Gates & Brewer, 2001; Suh & Amine, 2007). In Figure 1, customer satisfaction results from how well resources are managed, i.e., integrating ICTs and marketing applications will (1) improve operations and efficiency, (2) promote the city image, brand identity and relationships, and (3) improve service and quality. Since ICTs can provide immediate access to a wealth of information and allow for a tailored response to inquiries, cities can improve services and lower costs, cultivate direct relationships and further facilitate customized communication. To promote image and brand identity in order to remain competitive in a rapidly changing global environment, city managers can use ICT to communicate, market and interface with stakeholder groups. City managers and elected officials typically set specific objectives to focus their marketing, financial and operational efforts. These activities include: establishing the image a community seeks to portray; target marketing to national and international investors; providing data for global economic development opportunities; promoting city activities; and enhancing city services.

Research in marketing points to the pivotal role that image marketing targeted to identified market segments has in impacting economic development and growth (Suh & Amine, 2007). An organization's reputation is an important intangible resource, can provide a competitive advantage and is particularly important in global markets (Gardberg & Fombrun, 2002). How stakeholders develop their image of an organization is based on the information they receive (Schuler, 2004) and an effective and efficient vehicle for communicating and facilitating image marketing is the use of ICTs.

Two-way marketing and communications techniques in local government can establish stronger links between citizens and government, build community, and improve the local business climate, while boosting a municipality's image and civic pride (Kellogg & Lillquist, 1999). This can also provide cities with a competitive advantage and opportunities for an increased rate of return on marketing investments. Cities that capture and analyze the demographics of their communities can use the information to market to potential retail and economic development stakeholders, i.e., targeted promotions. For example, education, occupation and income are all measures of purchasing power, but they also reflect other attributes that are vital to economic development, such as the labor pool and intellectual capital. Since it is important to recognize that all businesses are not equally interested in all segments or demographic information, cities should design strategies that will best serve their different target markets. Providing regular demographic updates that are available to requestors in a downloadable format gives a city the opportunity to provide data, analysis and reports to prospective investors.

Providing easy access to municipal government information that can be used, for example, to estimate traffic flow and demand, can enhance the bottom line for private business. Sharing and showcasing the creative tools and innovations local governments are using results in more effective communication with citizens, businesses, visitors and other audiences. To illustrate, Geographic Information Systems (GIS) are tools for storing, retrieving and manipulating data to solve complex problems, e.g., operations such as the school districts can display a map with features associated with data. By selecting a given school, test scores, student demographics and financial expenditures are available. Similarly, police departments can profile crime statistics and characteristics by geographic areas which can help lower crime rates. Hence, using GIS results in more efficient and effective responses to community needs (Nedovic-Budic & Godschalk, 1996; Brown & Brudney, 1998).

The accumulation and circulation of community information in electronic databases can have a broad impact on city marketing and brand identity. By compiling a city's consumer characteristics, potential customers or prospects can be selectively targeted with promotional material through various media. This information is valuable to marketers and retail decision makers. Optimal selection of locations for stores and showrooms or franchises, regional shopping malls, etc., is facilitated through the use of this information. Target marketing to potential stakeholders offers an opportunity for city growth and economic development.

Websites with up-to-date and salient information will attract more traffic from location consultants and companies seeking site locations. Additionally, cities can target the distribution of information to increase marketing effectiveness and reduce media planning expense by resource allocation in areas that provide the greatest rates of return. Cities can focus their marketing efforts to the economic development segments they are interested in developing, e.g., industry, investors, etc. Software is available to track information requests and management can use this tracking technology to further enhance their web-based information offerings.

Some of the additional benefits of e-government are that it: is always open (i.e., 24-hour city hall); expedites transactions; reduces costs; improves citizen and business satisfaction; improves security application and accessibility; allows language options; and facilitates greater citizen participation (Shark, 2007). Also, the advantages of ICTs include timely compilation and dissemination of information and increased efficiency, e.g., lower transaction costs. ICTs facilitate participatory democracy/public decision making and enable wide, efficient and transparent participation between citizens and organizations to arrive at better and faster decisions through negotiation (Benyoucef & Verrons, 2008; Kim 2008). Such proactive problem solving helps cities to compete in an increasingly changing and global environment.

Presently, city administrators and leaders are leveraging resources and forming partnerships to expand and improve services, communications and infrastructures. They are striving to make their communities more attractive places to live, work or vacation – places where people want to raise children and perhaps retire (Kellogg & Lillquist, 1999). Cities must perform cost/benefit analyses to offer incentives that will contribute to the long-term success of the community. Successful implementation of ICTs can result in the acquisition and delivery of community information as well as individual consumer information. This information can then facilitate the design of new effective marketing strategies. Investing in ICT initiatives that result in better efficiency and cost savings ensures a city's ability to remain competitive. To illustrate, with rising fuel costs, there are significant cost savings in making data accessible and coordinating interactions such as telecommunications and video conferencing that do not require stakeholders to travel and be available on site.

Cities employing ICTs can create a competitive advantage when they are able to integrate information regarding factors such as building permits issued, market characteristics, constituent data and infrastructure condition, within a jurisdiction. Tracking building permits issued provides a competitive advantage for cities as they are an indicator of the economic vitality of a city, reflecting the amount of private investment being made in the local community. This is advantageous to a city given that bond agencies desire non-governmental investment in the local community. Examining trends in long-term growth helps prospective businesses/investors determine the likelihood that their investments will be profitable.

By improving operations, service and quality, community (internal customers) satisfaction will increase. In addition, by integrating ICT and marketing initiatives to proactively promote the city's image and build customer relations, improvements in external stakeholders will result. Cities that can more effectively promote the qualities of their locale to prospective stakeholders seeking information for tourism, relocation, economic development, etc., have a competitive advantage over other cities in attracting visitors, residents, and investors.

STAKEHOLDER OUTCOMES

In Figure 1, the final, top category is stakeholder outcomes and includes lagging performance indicators that are a result of improvements in customer satisfaction. In brief, if a city promotes its image and brand identity, builds relationships with customers and has high levels of customer service, its overall reputation will be enhanced which will also encourage economic growth and development as well as advance community outcomes such as health, safety and welfare.

The role of economic development is to increase the per capita income of a city's residents. For example, community information about activities, attractions and events can be communicated through ICTs resulting in increased tourism which brings in financial resources vital to community prosperity. Local government communications can encourage citizen responsibility, create citizen ownership and support and empower people to solve problems on their own, without having to rely solely on government (Kellogg & Lillquist, 1999).

More and more, stakeholders are sophisticated participants requiring substantial flows of information from a more organized but complex market; therefore, ICTs are essential for attracting and retaining business activity in an intensely competitive market (Perryman, 2006). The ultimate goal is for cities to meet their stakeholder needs, to provide information to prospects and to improve overall productivity. Cities seek to respond to an increasing demand from users for access to information that does not require the use of intermediaries. ICTs allow for service and resource comparisons with other jurisdictions which are increasingly sought after by prospective investors and economic developers on a regional, national and global basis. User-friendly interactive software can eliminate the need for users to have extensive technical knowledge in order to access information.

Further, cities can partner in a collaborative environment with other stakeholders to support common goals of economic growth and development and enhance global competitiveness. By leveraging critical expertise and financial resources, the community benefits. ICTs can facilitate the development of complementary interests between cities and private corporations to support city marketing. Government agencies may realize the commercial value of the data that is gathered for the purpose of public administration and offer that data to provide economic development opportunities in their communities. ICTs are changing the speed and efficiency of many industries, including cities. ICTs enable cities to personalize marketing media to individual organizations seeking to invest/relocate to a city.

Investing in the technology infrastructure can improve a city's services and operations and, therefore, enable it to effectively meet important stakeholder outcomes such as advancing health, safety and welfare. To illustrate, in the police and emergency areas, systems that monitor key performance indicators can significantly improve services such as response time. Computer assisted dispatch and mobile data terminals combined with global positioning systems enable reduced response times by providing the geographic information and route in the field. Similarly, field reporting systems that allow officers to generate reports on offenses and incidents real time improve description accuracy and efficiency.

The strategy map example developed in Figure 1 is a strategic tool that provides an illustration of how city planners can drive improvements in key performance indicators. Developing a strategic group map is part of strategic management and should reflect a city's mission, vision and values and be based on an analysis of the internal and external environment (strengths, weaknesses, opportunities and threats) as well as key success factors for a given city (Matherly, El-Saidi & Martin, 2008) . For instance, the vision articulates the image a city seeks to attain in the future and a community survey could determine what the overall image/reputation of the city is with key stakeholders. Collecting and analyzing data on each of the performance indicators in each category will reveal whether the causal paths are correct. For example, the implementation of specific ICT applications and marketing initiatives should result in improved operations. Improvements in operations, such as timelier turnaround for service and accuracy in transactions, are predicted to result in a better reputation and image for a city with internal stakeholders. Similarly, investments in customer relationship management and external marketing should result in improvements in external stakeholder reputation and therefore, more targeted economic development.

CASE STUDY APPLICATIONS

There are numerous examples of successful applications of ICTs in city management and marketing. In Killeen, Texas, U.S.A., a city of 110,000+ in central Texas, over 89% of the population uses ICTs. Citizens perform a myriad of activities on the city's website such as:

- obtain permits and licenses
- view council meetings or review past minutes
- check city documents and/or regulations
- obtain police accident reports
- verify book availability at the library
- view pets available for adoption
- look up and pay their utility bills
- obtain demographic information
- check airline flight schedule/status
- check for upcoming events
- register for park and recreation events
- obtain records requests

Using CRM, the city can track and evaluate the effectiveness of solicitation campaigns. By integrating ICT and marketing applications, the CRM system provides more effective outreach, tourism, and recruiting campaigns, resulting in targeted economic development. GIS capabilities allow citizens to determine the location of utilities, schools, churches, etc. Other forms of ICT include a government television channel, automated telephone systems, email and mass email services, text messaging and internet streaming. Websites bring a host of information to the citizens, promote city activities, enhance citizen involvement, aid in the adoption of pets using online files and result in greater citizen participation in city programs.

Thousands of internal and external users access web-based services daily affording the city new opportunities to market itself. This marketing outreach would not have been possible before the advent of ICTs. GIS fosters collaboration among service providers. Any street address can be displayed in GIS including work orders, building permits, code enforcement cases, occupational licenses and garage sales. New development websites target developers, homebuilders and prospective property buyers. These websites become portals of information for prospects and display sites for development review files including, for example, subdivision master plans, plats and commercial site plans. They give the city a competitive advantage by saving customers a trip to city hall.

The city of Charlotte, North Carolina uses the balanced scorecard and strategic group maps to facilitate communication and strategic planning as well as drive improvements in key performance indicators (Kaplan, 1999). One of the objectives on the city's corporate balanced scorecard is to promote growth in targeted types of business. By using GIS, the police department was able to improve services and reduce the crime rate, i.e., advance the health, safety and welfare of the community and encourage economic growth. The ICT application facilitated the mobilization of municipal resources to address specific local problems and to track the condition of neighborhoods. Ultimately, the police department was able to determine the reasons for higher incidents of crime in targeted areas. Computer aided dispatching (CAD) and the use of lap-top computers enabled police to query information while traveling to their destination or before questioning the driver of a car stopped for a traffic violation.

The National League of Cities (NLC) of the United States (2007) cites several examples of stakeholder benefits from ICTs in its highlighted city practices. Albany, New York has an interactive map of the downtown area on its web site with links that provide more information about Albany such as investment opportunities, living in Albany, current news about Albany as well as links to City of Albany resources, economic development, government, media, professional affiliations and travel and tourism. The Downtown Albany Business Relocation & Development Package provides comprehensive information for stakeholders who want to invest in the future downtown area. Topics include: Empire Zone Benefits, Tech Valley Investments, Quality of Life, Parking/Transportation, Market Data, Financing Programs, Headlines/Testimonials, Entertainment District, Downtown Living, and Downtown Development. Clay County, Kansas, seven cities, and a water conservation district collaborated to create a suite of web sites with a shared events database. The web sites provide expanded opportunities for marketing and economic growth. Rock Hill Rocks is a sports marketing campaign to showcase the city of Rock Hill, South Carolina, to promote its world-class venues and to attract visitors from across the country. The campaign includes a sports web site and sports marketing brochure. Adel, Georgia invested in the technology infrastructure to provide high-speed wireless Internet services to its citizens. The city used the existence of the service, as well as its reputation for having a high level of service, as a marketing tool to attract new businesses and encourage economic growth.

Additionally, Fort Wayne, Indiana implemented lean manufacturing and six sigma (quality improvement) by bringing business applications and philosophies to city management (George, 2003). Claiming that providing excellent services to its citizens is part of "e-City," the city launched a total of 60 projects over a 3 year period that saved almost $3 million. Fort Wayne's free wireless network covers 85 percent of its citizens, including 87 schools, 2000 teachers, 54,000 students, and all libraries and airports. The broadband network facilitates online learning for teacher training and makes it easier for parents and teachers to communicate. Innovative programs undertaken include virtual diabetes diagnosis with retinal scans and the "Net Literacy" program, in which young people teach seniors about computers.

The NLC in 2008 provides additional examples of creative ICT and marketing applications that resulted in improvement in strategic performance indicators such as economic growth--examples from Colorado, Texas and California are discussed in the following paragraphs. Englewood, Colorado's Citywide Computer Coaching Network is a cost-effective approach for developing employees and meeting the internal training needs of city employees. Computer training is delivered by peer-trainers through traditional classroom instruction, brown bag sessions, and one-on-one coaching. Coaches are not technology professionals, but city employees who volunteer their time to teach the practical skills they use on an everyday basis. By leveraging the knowledge and skills of internal resources, the City is able to deliver effective training at a fraction of the cost typically associated with technical training. This unique approach to internal training has dramatically reduced the costs associated with technical training because expenses are limited only to the costs of copying handouts and materials.

In response to a low unemployment rate and surplus of new jobs and companies, Austin, Texas shifted the focus of its economic development strategy from attracting new businesses (a lagging indicator) to recruiting and training workers (a leading indicator). AustinAtWork.com is an online recruitment tool which gives a job applicant Web access to search for jobs and submit a confidential skills profile. Job postings and profiles are matched based on pertinent criteria, and both employer and job seeker are notified. The software also allows employers to conduct on-line interviews and skill testing with job candidates. In its first year, over 1,000 jobs were posted and over 20,000 candidates (15 percent from outside the area) visited AustinAtWork.com.

Sunnyvale, California's Patent and Depository Library, which had been in operation since 1963, was outmoded and expensive to operate. To maintain and update the service, the city partnered with the U.S. Patent and Trademark Office to create the Sunnyvale Center for Innovation, Invention and Ideas (SCI3). The Center (1) assists entrepreneurs and start-up businesses to use new technologies and (2) helps established firms maintain their competitiveness through better access to technological information which increased the number of technological jobs in the region (economic growth). SCI3 uses new technology to provide timely, accurate, and cost-effective patent, trademark and intellectual property information. It also offers direct on-line image access to the patent database, and has video conference capabilities.

Littleton, Colorado's New Economy Project uses the idea of "economic gardening," or growing jobs locally by creating a nurturing environment for entrepreneurs. The program provides sophisticated information services as well as tracking of best ideas, best practices, and best technology of select high growth companies. Some examples of services are tracking construction bids for architects, developing profiles for multimedia companies on their competitors, and finding investors for foreign companies. In addition, the city subscribes to GIS and database research services, keeping track of statistics on local businesses, running focus groups, and monitoring legislation, new product releases and trends in order to share marketing and other information with businesses. The project also seeks to increase networks and connections between industry, research facilities, universities, competitors, suppliers, and customers. As part of the New Economy Project, the city sponsors the "econ-dev" Internet mail list of 300 economic developers worldwide interested in economic gardening. The investments in the technology infrastructure significantly led to more growth in the region.

The Glasgow City Marketing Bureau has been using leading-edge technology to promote Glasgow nationally and internationally. The prestigious Institute of Electrical and Electronic Engineers (IEEE) International Conference on Communications was broadcast live on Glasgow's Convention Bureau Website and viewed by more than 400 technology enthusiasts from around the world. The number of monthly page views on www.seeglasgow.com has grown from 800,000 to 1.8 million in 2007 (Glasgow, 2007).

The OECD (2003) provides numerous examples of international case study applications. Germany has implemented e-government, and in order to facilitate the smooth flow of information between citizens, business and the federal government has developed uniform, government wide standards, procedures and architecture for e-government applications. Given the complexity and coordination challenges across agencies, they recognized the importance of using standard procedures for a multitude of electronic services. Italy provides e-literacy training aimed at providing managers with the skills necessary to meet the organizational needs relating to e-government and innovation plans. IT literacy and technical skills are necessary to implement e-government, which results in better government. Delays in implementing new technologies will penalize economic development in a competitive, global and rapidly changing market.

These case study applications represent only a small portion of successful implementations of ICT and marketing initiatives but depict the importance and potential of marrying technology with marketing to improve internal processes, promote a more favorable image and meet stakeholder expectations. The effectiveness of specific ICT and marketing initiatives may vary depending on the municipality. So as to achieve improvements in lagging strategic performance indicators, each city should hypothesize and verify the major categories and variables in a strategic group map. By integrating the appropriate ICT and marketing initiatives in a causal path, a city will be better positioned to meet their performance excellence targets.

RESEARCH ON THE FUTURE ROLE OF ICTS AND CITY MARKETING

In the new electronic marketplace, selecting the appropriate technology requires taking a calculated risk and success is contingent on senior management support. Competitive advantage comes from integrating and aligning ICTs with the overall strategy and marketing plan—and doing so better and faster than the competition. However, implementing ICT initiatives can be expensive and difficult when there are project cost overruns and service failures (OECD, 2003). Descriptive research can determine what ICT and marketing applications are used most frequently across different municipalities. For example, a survey conducted in Mecklenburg County, North Carolina indicated that 44% of respondents in a community survey had accessed the County's website and were familiar with e-government services (2007). For those respondents who had used e-government services in the previous twelve months, the survey assessed their experience in four areas: convenience, navigation, accuracy and up-to-date information, and the likeliness to use the service again. Overall ratings averaged ninety percent customer satisfaction (Mecklenburg Community Survey, 2007).

Whether a city is successful in driving positive stakeholder outcomes needs to be verified by collecting data/research on the factors identified in the strategic group map. Causal research should investigate which applications are "best practices" and drive improvements in lagging indicators such as operations, service, or image and in the end, economic growth, e.g., those specified in the strategic group map in Figure 1. Research that uses path analysis which allows variables to act as both independent and dependent variables will be required to test a given strategic group map model (Byrne, 2001). For example, in a study on city management of cities > 50,000, West and Berman (2001) used a nonrecursive structure model and found that the use information technology was a significant predictor of management practices (e.g., management that promoted openness, support and risk taking) which, in turn, increased organizational effectiveness. However, controlling for management practices, information technology was not positively related to organizational effectiveness, nor was there a direct relationship between management practices and information technology.

Some fundamental issues to guide future descriptive and causal research posed as thought-provoking questions follow:

1. How do different cities use ICTs to improve service and market/ communicate their image? What are the most frequently used forms of ICTs for internal services/operations and external marketing?
2. What do we really know about how stakeholders make choices? How suitable is a city's website for meeting the divergent needs and wants of multiple stakeholders?
3. What do target stakeholders want from the city and an ICT? Working back from the customer, how well does the current system meet those needs?
4. How well does a given city's ICT capability compare to rivals—both now and in the future? Who are the "best in class" organizations worth benchmarking against?
5. What are the most effective ways for cities to reach and serve their markets? How effective are different uses and forms of ICTs in ultimately affecting stakeholder outcomes? What are the implications for technical expertise requirements and education?
6. How do stakeholders, e.g., potential investors, use information from the website? Can exchanges that create mutual value expand global and long-term stakeholder relationships that benefit the city? What is really known about traffic on the internet and it's potential value to the organization?
7. Who is accessing the city's website? How can that information be used to develop interactive relationships with stakeholders?
8. What specialist expertise and knowledge is necessary to incorporate rapid developments in ICTs? How are processes and services coordinated and aligned with shifting market requirements? How can an organization effectively adapt to changing customers, relationships, and markets?
9. What is the added value of and return on ICT applications? What determines the success or failure of an ICT initiative?
10. Do managers have the education and expertise to effectively integrate ICTs in strategic planning and marketing? How can this intellectual capital be increased?
11. How can stakeholder needs and priorities be incorporated into the latest communications, networking and interactive software and hardware?

CONCLUSION

The overall benefits for a city in integrating the use of ICTs and marketing as part of a strategic plan are effective direction of scarce resources to key business drivers that impact stakeholder satisfaction. In general, a strategy map provides a process for the city to articulate and execute its mission with regard to its key stakeholders by identification of key issues and focus on priorities, strategic activities, budgets and projects. By linking key performance indicators such as marketing and ICT initiatives to strategic areas of focus, management can provide the impetus for proactive, focused process improvement in internal processes and ultimately, stakeholder satisfaction. Future studies should examine the impact of investment in marketing and ICT initiatives on brand and image perceptions of a city. The results of stakeholder surveys on customer service, satisfaction and perceptions of the city identity are key performance indicators that can be reported on city scorecards.

Once the primary goals of a marketing plan are established, ICTs can be used to help improve a city's image and positively position a city in the evaluation of visitors, residents and businesses. Organizations will need an internet strategy and expertise to prosper in the new environment. It is incumbent upon management to ascertain if the objective and value are cost effective or to find the most effective communication mix at the least possible cost (Kerin and Peterson, 2007). The ability of the internet and ICT to disseminate large amounts of information to large numbers of stakeholders in an effective and efficient manner across the globe often results in a high value/cost ratio or a reasonable return for the investment.

Cities and countries differ in the extent of access to ICTs—wealthier countries have more investments in technology and are more accessible to external marketing communications (Shih, Kraemer & Dedrick, 2008). Improvements in a country's ICT infrastructure, human capital, institutional environment, e.g., economic freedoms and entrepreneurial activities determine the readiness of an emerging market to adopt the ICTs. Known as the digital divide that separates countries by the extent of access to ICTs (Gregorio, Kassicieh & De Gouvea Neto, 2005), cities on the less developed side of ICTs may find it difficult to market their cities on a regional or global basis while cities on the more developed side may be able to utilize first mover advantages that result in early adoption and improvement of key performance indicators such as economic growth. Recognizing the challenge of global connectivity and regional differences in ICT services, the International Telecommunication Union and European Commission committed to (1) support the development of human and organizational learning capital in technology, (2) standardize market and information policies across regions as well as (3) build the technology infrastructure for Africa, Asia-Pacific and the Caribbean (2008).

The development of ICTs in city management has profound effects on image marketing, communication and services while opening the paths to improve customer relationship management on an international scale. Successful deployment of ICT initiatives and investments in the technology infrastructure requires integration across organizational functions and relationships so that an organization's strategic approach reflects what is valued by the stakeholders.

REFERENCES

Achrol, R. S., & Kotler, P. (1999). Marketing in the network economy. *Journal of Marketing, Special Issue*, *63*, 146-163.

Ashling, J. (2008). Market growth in Asia, Africa, and Middle East set to tip the balance. *Information Today, 25*(2), 24-25.

Baldrige National Quality Program. (2008). Criteria for performance excellence. *National Institute of Standards and Technology, Department of Commerce*. Retrieved May 31, 2008, from http://baldrige.nist.gov/PDF_files/2008_Business_Nonprofit_Criteria.pdf.

Barwise, P., & Farley, J. U. (2005). The state of interactive marketing in seven countries: Interactive marketing comes of age. *Journal of Interactive Marketing, 19*(3), 67-80.

Bastain, L. (2007). Web surfing for great locations. *Area Development Magazine*. Retrieved September 5, 2006, from http://www.angeloueconomics.com/areadevelopment_websearching.html

Benyoucef, M., & Verrons, M. H. (2008). Configurable e-negotiation systems for large scale and transparent decision making. *Group Decis Negot, 17*, 211-224.

Bozinis, A. I. (2007). Internet politics and digital divide issues: The rising of a new electronic aristocrats and electronic meticians. *Journal of Social Sciences, 3*(1), 24-26.

Brady, M. (2003). Managing information technology assimilation: A marketing perspective. *Irish Journal of Management, 24*(1), 125-138.

Brady, M., Fellenz, M. R., & Brookes, R. (2008). Researching the role of information and communications technology (ICT) in contemporary marketing practices. *Journal of Business & Industrial Marketing, 23*(2), 108-114.

Brady, M., Saren, M., & Tzokas, N. (2002). Integrating information technology into marketing practice—The IT reality of contemporary marketing practice. *Journal of Marketing Management, 18*, 555-577.

Brodie, R.J., Coviello, N.E., & Winklhofer, H. (2008). Contemporary marketing practices research program: A review of the first decade. *Journal of Business & Industrial Marketing, 23*(2), 84-94.

Brookes, R.W., Brodie, R.J., Coviello, N.E., & Palmer, R.A. (2004). How managers perceive the impacts of information technologies on contemporary marketing practices: Reinforcing, enhancing or transforming? *Journal of Relationship Marketing, 3*(4), 7-26.

Brown, M. & Brudney, J. L. (1998). A "smarter, better, faster, & cheaper" government: Contracting and geographic information systems. *Public Administration Review, 58*(4), 335-345.

Byrne, B. M. (2001). Structural equation modeling with AMOS, EQS, and LISREL: Comparative approaches to testing for factorial validity of a measuring instrument. *International Journal of Testing, 1*(1), 55-86.

Chen, J., & Ching, R. K. H. (2004). An empirical study of the relationship of IT intensity and organizational absorptive capacity on CRM performance. *Journal of Global Information Management, 12*(1), 1-17.

Coviello, N., Miller, R., & Marcolin, B. (2001). Understanding IT-enabled interactivity in contemporary marketing. *Journal of Interactive Marketing, 15*(4), 18-33.

Day, G. S., & Montgomery, D. B. (1999). Charting new directions for marketing. *Journal of Marketing, 63*, 3-13.

Deighton, J. (1996). The future of interactive marketing, *Harvard Business Review, 74*(6), 151-152.

Duncan, T., & Moriarty, S. E. (1998). A communication-based marketing model for managing relationships. *Journal of Marketing, 62*, 1-13.

Eagle, K. (2004). Translating strategy: Public sector applications of the balanced scorecard. *Government Finance Review*, 16-27.

Eggleston, K., Jensen, R., & Zeckhauser, R. (2002). Information and communication technologies, markets, & economic development. *Discussion Papers Series, Department of Economics, Tufts University 0203, Department of Economics, Tufts University,* 62-74.

Erickson, B., & Roberts, M. (1997). Marketing local identity. *Journal of Urban Design, 2*(1), 35-59.

Fisher, M., Raman, A., & McClelland, A. (2000). Rocket science retailing is almost here—are you ready? *Harvard Business Review, 78*(4), 115-24.

Gardberg, N. A., & Fombrun, C.J. (2002). The global reputation quotient project: First steps towards a cross-nationally valid measure of corporate reputation. *Corporate Reputation Review, 4*(4), 303-307.

George, M. L. (2003). *Lean six sigma for service: How to use lean speed and six sigma quality to improve services and transactions.* New York, NY: McGraw-Hill.

Glasgow (2007). *Scotland with style.* Retrieved August 16, 2007, from http://www.seeglasgow.com

Goldman, C. A., Gates, S. M., & Brewer, D. J. (2001). Prestige or reputation: Which is a sound investment? *Chronicle of Higher Education, 48,* 13-15.

Gommans, M., Krishnan, K.S., & Scheffold, K. B. (2001). From brand loyalty to e-loyalty: A conceptual framework. *Journal of Economic and Social Research, 3*(1), 43-58.

Gregorio, D. D., Kassicieh, S. K., & De Gouvea Neto, R. (2005). Drivers of E-business activity in developed and emerging markets. *IEEE Transactions on Engineering Management, 52*(2), 155-166.

Gummesson, E. (2002). Relationship Marketing in the new economy. *Journal of Relationship Marketing,* 1(1), 37-57.

Halaris, C., Magoutas, B., Papadomichelaki, X., & Mentzas, G. (2007). Classification and synthesis of quality approaches in E-Government services. *Internet Research, 17*(4), 378-401.

Holland, C., & Naude, P. (2004). The metamorphosis of marketing into an information-handling problem. *Journal of Business and Industrial Marketing, 19*(3), 167-177.

Kaplan, R. S. (1999). *City of Charlotte (B).* Boston, MA: Harvard Business School Press.

Kaplan, R. S., & Norton, D. P. (1992). The balanced scorecard--Measures that drive performance. *Harvard Business Review, 70,* 71-79.

Kaplan, R. S., & Norton, D. P. (1996). Using the balanced scorecard as a strategic management system. *Harvard Business Review, 74,* 75-76.

Kaplan, R. S., & Norton, D. P. (2004). *Strategy maps: Converting intangible assets into tangible outcomes.* Boston, MA: Harvard Business School Press.

Kellogg, C., & Lillquist, R. (1999). *How to market your city.* Retrieved July 16, 2007, from http://www.3cma.org/members/resourcecenter/How%20To%20Market%20Your%20City.pdf.

Kim, C. K. (2007). A cross-national analysis of global e-government. *Public Organization Review, 7*(4), 317-329.

Kim, J. (2008). A model and case for supporting participatory public decision making in E-Democracy. *Group Decis Negot, 17,* 179-193.

Lapierre, J., & Medeiros, R. (2006). Information and community technology usage patterns: A case study. *Journal of Strategic Marketing, 14*(3), 229-244.

Laudon, K., & Laudon, J. P. (2006). *Management information systems: Managing the digital firm.* Upper Saddle River, NJ: Prentice Hall.

Leverick, F., Littler, D., Bruce, M., & Wilson, D. (1998). Using information technology effectively: A study of marketing installations. *Journal of Marketing Management, 14*(8), 927-962.

Matherly, L. L. (2007, August). *Mission: The impact of employee well being on processes, customer satisfaction and financial performance.* Paper presented at the meeting of the National Academy of Management, Philadelphia, PA.

Matherly, L. L., El-Saidi, M. A. & Martin, D. (2008, March). *Using project management to implement strategic planning and a strategic scorecard in a university setting.* Paper presented at the meeting of the Southwest Academy of Management, Houston, TX.

McAfee, A. (2006). Mastering the three worlds of information technology. *Harvard Business Review*, *84*(11), 141-149.

Mecklenburg County, NC. (2007). 2007 Performance Report. Retrieved May 15, 2008, from http://www.charmeck.org/NR/rdonlyres/ea3bzxlsasj7kumjwyegasrdgjf-hxl6m67gl2smozkijpki7wocaqqfnpnfl7mph-ydec7t6rqa5c4spt2bf3gpo7x4g/2007AnnualPerfRpt-final.pdf.

Mecklenburg County, NC. (2007). 2007 Community Survey. Retrieved May 15, 2008 from http://www.charmeck.org/NR/rdonlyres/eknrcrfdromsgtehy7hrcr34rvj2m-dtqswbdnojx4mb7zktgglff7wsmusbwplng-u3xdkhlrzpmxibrzdcc6bmojpmc/2007CommunitySurvey091707.pdf.

Mellor, N. (2006). E-Citizen: Developing research-based marketing communications to increase awareness and take-up of local authority E-Channels. *Aslib Proceedings*, *58*(5), 437-446.

National League of Cities (NLC). (2007). City practices. Retrieved August 22, 2007, from http://www2.nlc.org/dbtw-wpd/exec/dbtwpub.dll.

National League of Cities (NLC). (2008). City practices. Retrieved June 1, 2008, from http://www2.nlc.org/dbtw-wpd/exec/dbtwpub.dll?AC=NEXT_BLOCK&XC=/dbtw-wpd/exec/dbtwpub.dll&BU=http%3A%2F%2Fwww2.nlc.org%2Fexamples%2Fcknsearchtest.htm&TN=EXAMPLES&SN=AUTO19263&SE=164&RN=40&MR=20&TR=0&TX=1000&ES=0&CS=0&XP=&RF=New+Web+Results&EF=&DF=New+Web+Details&RL=0&EL=0&DL=0&NP=3&ID=&MF=&MQ=&TI=0&DT=&ST=0&IR=3509&NR=0&NB=2&SV=0&SS=0&BG=&FG=&QS=&OEX=ISO-8859-1&OEH=ISO-8859-1.

Nedovic-Budic, Z. & Godschalk, D.R. (1996). Human factors in adoption of geographic information systems: A local government case study. *Public Administration Review, 56*(6), 554-567.

OECD E-Government Studies The E-Government Imperative. (2003). *Organisation for Economic Co-operation and Development, 2003*(15), 1-199.

Perryman, M.R. (2006). The market for prosperity: Understanding the economic development process. Retrieved July 22, 2007, from http://www.texasedc.org/files/File/ED%20Publications/Perryman%20report_color.pdf.

Schuler, M. (2004). Management of the organizational image: A method for organizational image configuration. *Corporate Reputation Review, 7*(1), 37-53.

Shark, A. (2007, August). *Revitalizing Rural America.* Paper presented at the Texas Municipal League Technology Summit, Lewisville, TX.

Shih, E., Kraemer, K. L., & Dedrick, J. (2008). IT diffusion in developing countries. *Communications of the ACM*, *51*(2), 43-48.

Struse, D. (2000). Marketing research's top 25 influences. *Marketing Research, 11*(4), 4-10.

Suh, T., & Amine, L. S. (2007). Defining and managing reputational capital in global markets. *Journal of Marketing Theory and Practice, 15*(3), 205-217.

Swope, C. (2007, May). Working without wires. Municipal WiFi is coming. Government employees, not citizens, may be the biggest users. *Governing,* 29-34.

Syfert, P., & Elliott, N. (1998). Charlotte adapts the balanced scorecard. *American City & County, 113*, 32.

Tschirhart, M. (2008). Evaluation of brand use on city government websites: A search for brand type, repetition and consistency. *Journal of Nonprofit & Public Sector Marketing, 19*(1), 35-53.

Webster, F. E. (2005). A perspective on the evolution of marketing management. *Journal of Public Policy and Marketing, 24*(1), 121-126.

Webster, F. E. (1998). Interactivity and marketing paradigm shifts. *Journal of Interactive Marketing, 12*(1), 54-55.

Webster, F. E. (1996). The future is interactive marketing. *Harvard Business Review, 74*(6), 156-157.

West, J., & Berman, E. M. (2001). The impact of revitalized management practices on the adoption of information technology: A national survey of local governments. *Public Performance and Management Review, 24*(3), 233-253.

Section II
Unlocking the Power of City Marketing to City Development

Chapter VII
City Boosterism through Internet Marketing: An Institutional Perspective

María Isabel Huerta-Carvajal
Universidad de las Americas-Puebla, Mexico

Luis Felipe Luna-Reyes
Universidad de las Americas-Puebla, Mexico

ABSTRACT

Local governments around the world are becoming aware of the importance of identifying and marketing their local assets to promote economic competitiveness. Information and Communication Technologies (ICT) have proven useful in supporting marketing activities in the private sector, but there is still little exploration on their use in the public sector. However, ICT effectiveness is constrained by institutional arrangements and the coordination of the marketing efforts with other government processes such as urban planning and strategy development. The purpose of this chapter is to discuss the strategic scaffolding for ICT as a key component of a city's marketing strategy using as an example the city of Puebla in Mexico. Although city marketing efforts and ICT use are still at its initial stages in the city, lessons from current efforts in Puebla are related to the key role of stakeholder networks, ICT interoperability, Geographic Information Systems, and government program continuity.

INTRODUCTION

One of the main objectives of governments around the world is to promote sustainable social and economic development. However, current politico-administrative regimes work on a model in which government provides security, basic infrastructures and services, but economic development relies on private investments that are regulated to different extents. In this way, governments compete to attract investment by providing robust physical infrastructures, promoting the formation of a critical mass of highly qualified human resources, reducing taxes, creating efficient procedures to encourage business and lower investment barriers.

Globalization has intensified the competition for investments among countries and regions (Erickson & Roberts, 1997; Doel & Hubbard, 2002), and similarly to other global phenomena, local manifestations are becoming increasingly important (Frederickson, 2004). Moreover, local governments are playing a key role in global sustainability (Keen & Mahanty, 2006).

In this way, municipal, city and town governments are becoming aware of the importance of identifying and marketing their local assets. However, local government activities, policies and results are both enabled and constrained by institutions, organizations and contextual factors at the local, state and national levels (Lowndes, 2005).

However, Information and Communication Technologies (ICT) not only provide alternative channels to market products and services in the private sector, but also are key enablers of some forms of marketing (Gillenson, Sherrell & Chen, 1999; Mohammed, Fisher, Jaworski & Cahill, 2002; Roberts, 2003). Furthermore, the Internet as a manifestation of ICT is changing traditional marketing paradigms to become a way to aggregate communities that can be engaged in conversations for branding or marketing goods and services (D. M. Scott, 2007; L. Weber, 2007; S. Weber, 2007). However, there is still little exploration of the ways that ICT can support local and city marketing.

The purpose of this chapter is to discuss the strategic scaffolding for ICT as a key component of a city's marketing strategy. Although the interactive and networked characteristics of modern ICT make them an important medium to connect different stakeholders in the marketing process, the main source of value resides in the strategic use of ICT and not in ICT itself. Moreover, both ICT use and marketing activities are shaped by existing and emerging interorganizational networks (enterprises, service personnel, government officials, city planners, consumers, citizens), government structures and institutions (Fountain, 2001; Gil-Garcia, 2005; Hassan & Gil-Garcia, 2007; Luna-Reyes, Gil-Garcia & Cruz, 2007; Luna-Reyes, Gil-Garcia & Estrada-Marroquín, 2008). In this way, the chapter will analyze key factors in city marketing in addition to their interactions with other city management processes, strategic management, and government institutional arrangements.

We will use as an example the city of Puebla, Mexico. City marketing efforts and ICT use is still incipient there, mainly because of institutional constraints and limited commitment among key stakeholders. However, we believe that some developments in city planning and marketing could constitute the basis for a marketing strategy that takes advantage of current ICT.

The rest of the chapter is organized as follows. The second section describes the concepts of strategy, city marketing, ICT for public management and institutions. The third section presents the case of the Mexican city of Puebla to illustrate the city's current strategy and marketing efforts, its use of ICT, the institutional, contextual and organizational factors involved in the planning process, and the lessons learned. The fourth section explores some future trends in the use of ICT in city marketing, and the chapter ends with a conclusion.

BACKGROUND

City marketing is, like many other government activities, a problem that requires collaboration between private and non-profit organizations. The department of urban development needs to create and coordinate the city's infrastructure, to make the city attractive; the department of finance needs to coordinate with legislatures and other government offices to develop the budget; the department of tourism needs to find ways to attract visitors in coordination with hotels, restaurants and the entertainment industry. Citizens are important participants in the creation of a good environment in which to live and to visit. In order to be effective, a coordinated strategy needs to involve all of the key stakeholders (Andersen, Belardo & Dawes, 1994; Eden & Ackermann, 2000; Bryson, 2004a). Moreover, the need to collaborate challenges the traditional hierarchical structures in government, and calls attention to the need for institutional innovation (Gascó, 2004).

Thus, in this section of the chapter we introduce some central concepts and ideas associated with the use of ICT in city marketing. In order to do that, the section starts by using institutional theory as a framework for the concepts presented. After these initial comments, we introduce important concepts related to city marketing and ICT use in government. We end the section of the chapter by describing the importance of government strategy, city planning, and its relationship to ICT use and city marketing.

Institutional Theory

Institutional theory has described phenomena from the organizational, economic and sociological points of view, and it has been used successfully to explain local government actions (Davies, 2005; Moon & Norris, 2005; Norris & Moon, 2005). Institutions are the rules of the game in a society (North, 1999), and in this sense, they shape the ways in which individuals and organizations perform their activities, including strategic development, planning, ICT use or marketing (DiMaggio & Powell, 1983). Institutions are not only reflected in legal structures, but also in norms, cultural and cognitive structures, which influence our day-to-day activities (W. R. Scott, 2001). However, many authors recognize a complex and recursive relation between individual action and institutions (Giddens, 1984; DeSanctis & Poole, 1994; Orlikowski, 1992, 2000). In this way, institutions modify individual action, but individuals and organizations have the potential to generate pressures that change the institutional structure. However, and although institutions suggest permanence, they are subject to change and discontinuity (Hassan & Gil-Garcia, 2007).

Jane Fountain proposes the Technology Enactment Framework, which explains ICT enactment and performance as a result of the functionalities of technology (or objective technology), institutional arrangements, and organizational forms (Fountain, 1995, 2001). An ICT enactment consists of people's perceptions and uses of ICT. Institutional arrangements, as mentioned before, represent the regulatory, normative, and cultural-cognitive structures that constrain government actions. Finally, organizational forms involve roles, routines and relations, particularly inter-organizational networks.

Gil-Garcia (2005) expanded Fountain's enactment framework, including environmental and contextual variables. In his research, environment was found to have both a direct impact on organizational forms and institutional arrangements, and an indirect impact on technology and organizational performance. In addition, environmental and contextual factors help to understand the antecedents of institutions and organizational characteristics.

As Figure 1 shows, the enactment framework can be extended to government activities beyond ICT use. That is to say, the selection of a marketing strategy, or planning process, can be understood as an enactment of government activity, which will result in different organizational outcomes. Such government enacted activities are both facilitated and constrained by organizational structures and institutional arrangements. However, government activities can also modify organizational forms and institutions. Contextual factors (environment) have an impact on institutional arrangements, organizational structures and also government

Figure 1. An institutional view of strategic processes in government (©2005 Jose Ramon Gil-Garcia. Used with permission.)

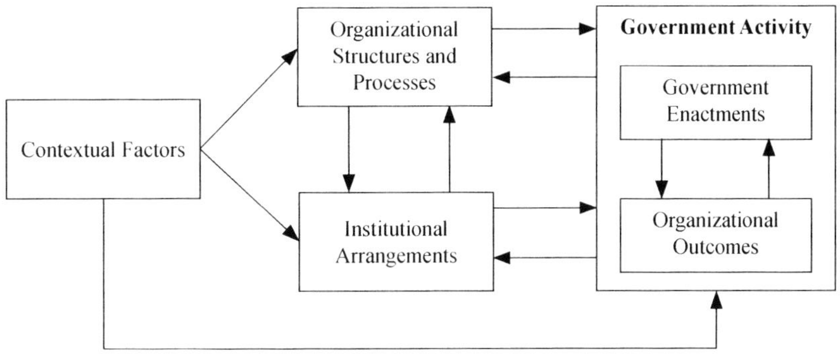

activity. In this way, the sophistication of local institutions and organizations is the result of educational, economic and other contextual factors in the city. Moreover, these contextual factors also condition advances (or lack of advances) in government enactments of ICT, strategic planning, and city marketing.

City Marketing

According to Kavaratzis (2007) "city marketing is a long-term process that cannot be implemented in parts or only to a certain extent. Second, promotional activities constitute only a fraction of the whole process and should not be considered an alternative for strategic marketing" (p. 696).

City marketing strategy implies active participation of key members of the local community, such as leaders of the local public and private sector, citizens and professional planners. The objectives of the strategy are to improve the city image and to set the springboard for urban development, investment and living environments. Hubbard (1996) based on Short et al. (1993) argues that, "place marketing is inevitably accompanied by the fabrication of a new urban landscape, which can therefore be seen as both an expression and a consequence of attempts to re-image the city, playing a crucial role in the entrepreneurial 'selling' of cities" (p. 1444). In this way, concentrations of entertainment and leisure facilities, shopping malls, tourist attractions, or business centers constitute urban flagships developments for many cities (Kolb, 2006).

City marketers implement broad strategies to appeal to stakeholders. Such strategies include marketing the city's image, attractions, infrastructure and people, all of which draw visitors, residents, employees, businesses, and promote exports (Kotler, Gertner, Rein & Haider, 2007).

In terms of the image of a place, two aspects are involved in its creation (Kotler et al., 2007): the individual perception of the image from a personal or general concept (stereotype), and the role of place marketers to design a strategy that enables customers and stakeholders to discover the attractiveness of the place and to drive their buying decision. For instance, the city of Shanghai developed a strategy to host the 2007 Special Olympic Summer Games. To publicize the facilities that could accommodate people with disabilities, organizers made: infrastructure adaptations to facilitate the athletes' performance, designed and promoted a new national sports star, launched a competition for the Olympics logo design and training of volunteers to provide assistance to the athletes (Yuankai, 2007).

However, city image does not always help to host international events, as was the case of Mexico's "Universal Forum of Cultures Monterrey 2007." From the point of view of the media and public opinion, Forum organizers could not attract the attention of the public, in spite of the city's cosmopolitan image.

One way to attain a positive and attractive city image is its strategic management, which consists of tracing image changes in several target audiences to understand their dynamics and how to influence such changes (Kotler et al., 2007). Some target markets or audiences are visitors (business and non-business), residents and employees, business and industries (investors), and export markets (expanding city exports). Business and industries is the most competitive market. Given its perceived potential to promote economic development, one industry that is attracting the attention of governments around the world is the "clean industry," which includes high-tech and software companies.

Associating specific images with the name of a city is a key goal of city branding. A brand is a multidimensional construct that deals with what a city does and have and what the consumers (stakeholders) perceive about it and, it becomes an interface between city and consumers (De Chernatony & Dall´Olmo, 1998; Kavaratzis, 2005; Kavaratzis & Ashworth, 2005). Brand management then consists of controlling the interaction between the city and the consumer (Morgan, Pritchard, & Pride, 2004). The communication of a brand takes place based on the effects of what a city does (primary communication) and the intentional communication through marketing practices (secondary communication) (Karavatzis, 2004; Kavaratzis, 2005).

In this way, home town or home country can have brand equities that build values, qualifications and emotional linkages in the consumer's mind (Anholt, 2004). For instance, a successful global brand, by offering high quality products or services can evoke its place of origin (Anholt, 2000); for example, Japan is associated with high-tech products and France is associated with quality food and wines. Stereotypes can erode or enhance the town, city or country image depending on the effectiveness of brand management (Kotler & Gertner, 2004). In this way, a high tech product from Japan will have advantages over a similar product from China or Taiwan.

Some key factors that can affect the work of city branding are popular perceptions, their level of exposition (previous knowledge) and nationality, media, environment and the strength of a brand (representativeness and functionality) (De Chernatony & McWilliam, 1990; Anholt, 2004; Caldwell & Freire, 2004;). Country brand can be helpful, but also has the potential to work against a particular region or city. For example, the Mexico travel brand is strongly associated with beach destinations, which makes it difficult for colonial cities to position themselves in tourists' and visitors' minds. Building global city brands and shaping and promoting brand equities entails government support and continuous investment (Anholt, 2004).

Finally as Kavaratzis (2007) argues, "city marketing can now be viewed as a fundamental part of urban development" (p. 710) and city branding "as a new episode in the application of city marketing" (p. 704). City marketing entails a series of steps, from a thoughtful analysis of city assets, opportunities and audiences to the evaluation of all activities. We should not forget that a key factor in the attractiveness of any place is its people, who are an integral part of the perceived image of a place (Kotler et al., 2007).

ICT Strategy and City Marketing

The interactive characteristics of ICT such as the Internet have proven effective in the marketing process in the private sector, either to conduct research, sell goods and services or build brands (Mohammed et al., 2002; Roberts, 2003). Moreover, developments on databases and datawarehouses allow the creation of marketing applications that focus on market segments of one, and facilitate the evaluation and assessment of marketing efforts (Gillenson, Sherrell & Chen, 1999). However, most of the front-end applications were limited by the lack of proper back-office processes. As a result, the term "Internet Marketing" has been replaced by the term "e-Business," following a more holistic approach that not only emphasizes the front-end, but extends to the review and integration of back-office applications.

In this way, referring to the relationships between city marketing and ICT strategies, we want to show the need for coordination of front-end and back-office applications. In a recent review of the literature, Gil-Garcia and Luna-Reyes (2006) identified four categories of ICT use in government: providing services to citizens and other stakeholders (e-Services), promoting public participation (e-Democracy), managing back-office processes (e-Management), and developing public policies to promote the information society (e-Policy). City marketing activities can take advantage of ICT in all of these areas. On the one hand, E-Services are an effective way of accelerating both new business creation and alternative channels to provide services to citizens, private organizations and visitors. On the other hand, the coordination of all the actors and stakeholders involved in increasing the attractiveness of the city can be coordinated using back-office interoperable systems, which relies on e-Management. E-Democracy encourages stakeholder participation in city planning, city security and maintenance, and taking care of e-Policy enhances the possibilities of taking advantage of ICT by government and other members of society.

The Internet is, without a doubt, the ICT that has had the greatest impact on marketing practices. Weber (2007) identifies four stages in the technological development of the Internet. During its first years, the predominant Internet technology was Hyper Text Mark-up Language (HTML), which raises the possibility of creating and publishing static content and messages. Internet marketing was born, and marketers around the world started creating content to their prospective markets, using Web pages as an alternative messaging channel. During this stage, the main way of attracting Web site traffic was to place banner ads in paid directories and catalogs.

The second stage involved the use of technologies that allowed for more interactivity between target audiences and businesses and other organizations. Programming languages such as Perl, Java, or JavaScript, as well as database connectivity gave birth to e-Commerce, allowing marketers to use the Internet not only to send and receive messages, but also to sell products and services. Moreover, the use of the Internet for marketing research provided the opportunity to extend reach at a lower cost through online surveys.

The third stage consists of what Weber (2007) and others have called the social Web. From our point of view, the social Web stage is composed of two components. The dotcom bubble attracted attention to back-office processes, promoting the integration with suppliers and other partners by using Internet connectivity to deliver e-business. EXtensible Mark-up Language (XML) and other Electronic Data Interchange (EDI) technologies and standards have been key to promoting integration and interoperability. Marketers started using the Internet to create advantages over their competitors, requiring collaboration with other functional areas inside the enterprise and across organizational boundaries with their business partners. More recently, social Internet applications such as blogs, wikis, networking and social tagging sites have given the Internet a whole different character. XML and hybrid programming approaches like AJAX (combining XML and JavaScript) are some of the technologies behind these social applications. Marketers are still figuring out how to exploit these new tools, but apparently, the Internet is no longer being considered as an alternative channel, but as a way to aggregate relevant contents to target audiences and communities, which become involved in conversations intended to influence buying decisions.

The fourth stage is not yet here, but with the wide adoption of broadband connections to the Internet, it will expand the potential use of multimedia, video, sound and touch for marketing efforts.

City marketers can take advantage of all these technologies to support their marketing strategy. Blogs, wikis, podcasts, and other Web 2.0 tools can be embedded in city Web sites and applications to increase their effectiveness, and to engage in social conversations with citizens and other stakeholders, creating virtual spaces to enhance the city experience (Ishida, 2002).

However, realization of ICT benefits does not come from ICT itself, but from its strategic use (Andersen, Belardo & Dawes, 1994; Janssen & Cresswell, 2005). In other words, ICT use requires for a long term vision and the integration of the specific (sometimes contradictory) visions of main government areas in addition to inputs from other stakeholders in the civil community. In this way, having a web site to market the city adds little or no value if there are no clear objectives for the site, and if those objectives are not aligned with the city's development strategy.

To support this integrated strategy, local governments need to develop a business architecture which has three components: a governance body, ICT policies and standards, a technical and a semantic architecture (Janssen & Cresswell, 2005; Papazoglou & Ribbers, 2006). The governance body constitutes a steering committee for developing e-Government strategy, and it should invite representatives from all government departments or ministries, and representatives from other key stakeholders. ICT policies and standards guide the selection of new technologies and support the prioritization of ICT projects. Technical and semantic architectures are the basis for the creation of interoperable systems that facilitate collaboration among main actors in the city marketing process.

City Strategy and Planning

Developing an integrated strategy is a key means of capitalizing on the benefits of ICT use in city marketing. Strategic urban development plans are the scaffolding upon which local governments pave the way for economic competitiveness. Such plans establish the guidelines, goals and policies for resource allocation to achieve critical goals. Making strategy in the public sector, however, is not a straightforward process. One useful conceptualization for the public sector is that "strategic management is a way of regenerating an organization, through continuous attention to a vision of what the people who make up an organization wish to do. It is a pro-active process of seeking to change the organization, its stakeholders (in as much as they are different from the organization), and the context, or environment, in within it seeks to attain its aspirations" (Eden & Ackermann, 2000, p. 3).

An interesting characteristic of such a conceptualization is that it recognizes strategy making as an emergent process of change. In fact, from Eden and Ackermann's point of view (2000), strategy making is a continuous process of sense-making and negotiation among all stakeholders, who consciously explore and reveal emergent and planned strategies to continue (re-)designing the intended strategy (see Figure 2). Key stakeholders in the strategic process are leaders in the public and private sectors, citizens and professional planners. Given that actors and stakeholders may have different or even conflicting views of the mission and critical objectives, strategy making needs to be understood as a process of continuously developing shared understandings about key problems and objectives (Eden & Ackermann, 2000; Bryson, 2004b). Anticipating future scenarios, public managers can manage environmental risks, build coalitions, and monitor and influence stakeholders. The reflection on power relationships and the political environment are important components of strategy making in the

Figure 2. The strategic process (Source: Adapted from Eden and Ackermann, 2000)

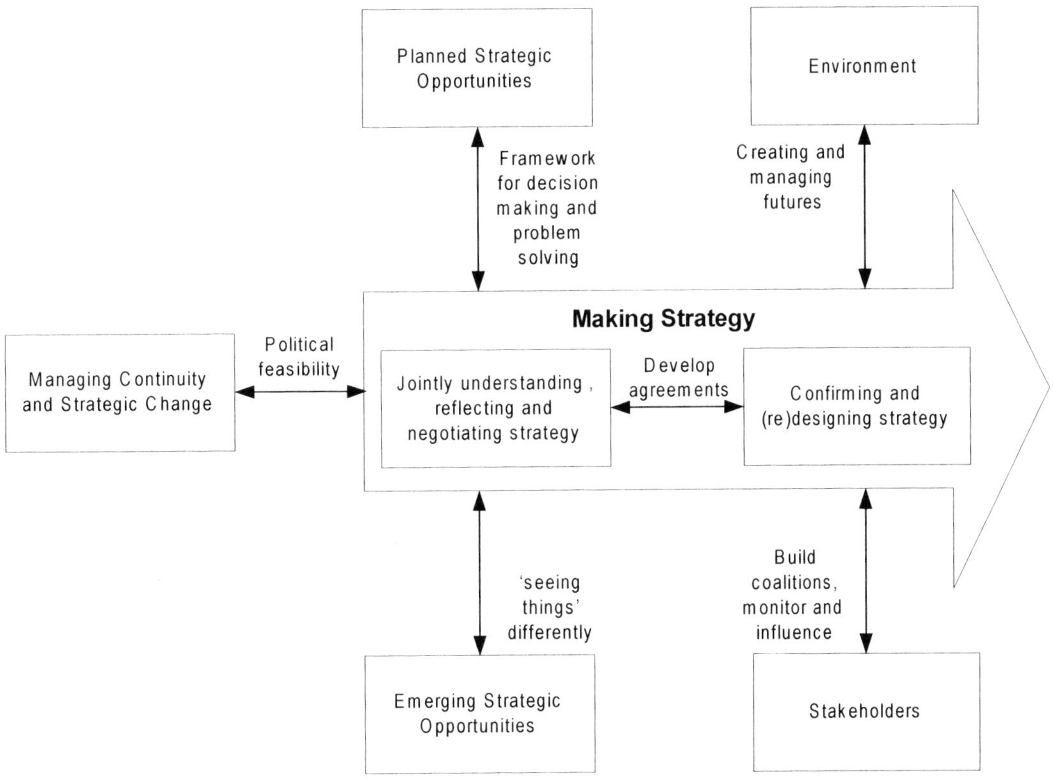

public sector because it helps to balance planned with unplanned opportunities. Moreover, working together in these reflections ensures commitment among participants, allowing them to see things differently and to create alternative frameworks for decision making and problem solving.

The strategic process becomes more important in an environment in which cities become entrepreneurial actors (Jessop & Sum, 2000). In this chapter, the purposes of designing a strategic plan are to take advantage of modern ICT to promote urban development and economic growth, as well as to shape it in terms of strategic formulation, spatial planning and city marketing. There are four interrelated elements of competitive city strategies: urban design (specific character of a place), infrastructure (transportation networks, energy and water, etc), services (street lighting, garbage collection, etc.), and attractions (natural and cultural attractions, historical sites or monuments, etc).

Key technologies for supporting strategic planning for development and city marketing involve Geographic Information Systems (GIS). "GIS is a tool that allows data that can be referenced spatially to be organized and analyzed" (Doyle, 2001, p. 85). GIS can be used in land and natural resource management (Bateman, Lovett & Brainard, 2005), and urban planning (Erbek, Ulubay, Maktav & Yagiz, 2005). Private organizations are already taking advantage of GIS for marketing purposes to identify target audiences, locate business branches, and engage in customer profiling or sales forecasting (Doyle, 2001). GIS will have also an impact on city marketing (see for example Raymond & Brown, 2007). GIS and collaboration platforms also support public authorities and their stakeholders in the execution of the strategy, and in the continuous evaluation and assessment of progress (Rocheleau, 2006). Government Web pages have been adopted to communicate efforts and results in local development (Kotler et al., 2007).

Institutions and organizations have an impact on the strategic process of a city given that, creating and approving strategic plans is constrained by laws and regulations. Normative and cognitive institutions also play a role in the strategic process. For example, "a change in local leadership may often mean the end of a strategic plan because the new leader wants to demonstrate his/her vision for the city future" (Wu & Zhang, 2007, p. 732). Actors in the process take for granted (and naturally) that each time local government changes, its strategic planning does too.

Main challenges for local governments are to accelerate the political and social approval processes of strategic planning, as well as adopting a global vision and a long-term business strategy (Wu & Zhang, 2007). In sum, city governments need to adopt an entrepreneurial vision in their discourse, action and promotion of the city and its regions to attain economic competitiveness, to promote innovative production methods, and to locate sources of raw materials for industry (Schumpeter, 1934; Hubbard, 1996; Jessop & Sum, 2000).

CURRENT DEVELOPMENT OF CITY MARKETING IN PUEBLA, MEXICO

In this section we will describe some current developments in strategic planning, city marketing and the use of ICT in the city of Puebla, Mexico. Although city marketing in Puebla is just beginning, it is an illustration of the initial steps towards an integrated strategy, and the institutional and contextual enablers and limitations in a Mexican city, which can be applied to similar cities in developing countries. General city strategy and planning are conceptually related to city marketing in Puebla although they have not yet been fully connected in practice. One of the staff members made the connection by comparing a city with a person's home, saying that "you do not design your house for your visitors or friends, you design your house for yourself, and you also use the same house to receive them… some of them will like your house and the way you live, and others maybe not, but at the end, your visitors will be assessing your home and the way you live." In this way, a city has to be planned and designed for its inhabitants, which in the words of the informant "will create a social collective, a culture, and a way of living." This way of living and collective culture, with the physical infrastructure and amenities of the city is what visitors will experience. Moreover, the same collective will attract (or drive away) investors.

Recognizing the importance of context and institutions in the enactment of strategies, marketing and ICT use, the section provides some background information on Puebla, and on some Mexican institutions that are important in local government activity. We continue the section with a description of the efforts in Puebla related to city strategy and planning. Although there is not yet an integrated strategy to market the city, we will examine the efforts in city marketing and related ICT use. We will finish the section by exploring the challenges facing the City.

Institutional Context of the City of Puebla

Puebla is the capital of the state of Puebla. Founded in 1531, Puebla was a natural stop for travelers between the seaport of Veracruz and Mexico City. The city is connected by highways with Mexico City, the Gulf of Mexico and the Pacific Coast, and it is the fourth largest city in Mexico. The 1,485,941 inhabitants of the city of Puebla represent about 28% of the total state population.[1] The city is home to about half of the state's hospitality industry (6,641 out of 13,203 available rooms, and

558 out of 984 restaurants),[2] and about 40% of the State Gross Product (80.9 billion pesos out of 209.8).[3] The main economic activity is commerce, particularly with the Mexico's south-eastern states and cities. Some of the city's main attractions are the colonial district (downtown Puebla), cultural attractions, traditional festivals, historical sites and monuments, entertainment, shopping, and living heritage. Puebla City has become an important center of higher education, boasting more than 100 colleges and universities.

Although the autonomy of state and municipal governments is established by the Mexican constitution, a single political party controlled federal, state and local governments for more than 60 years, in a *de facto* centralization of power. A decentralization process started in 2000, but local governments still need to work more towards a more autonomous administration.

The free municipality is the basis for territorial division, and politico-administrative regime in Mexico. The state of Puebla comprises 217 municipalities. At the local level, municipalities have a council chaired by the municipal president, elected by direct voting. This council –called "Cabildo"—has both executive and legislative functions. The "Cabildo" is elected every three years. Although it is possible for municipal presidents to be re-elected, they cannot be re-elected to consecutive terms.

At the municipal level in Mexico, the concept of professional public service is nonexistent, and non-elected positions are appointed. In this way, there are potential knowledge losses every three years as part of the succession process. The Cabildo works in coordination with Auxiliary Councils that are elected through a plebiscite in smaller cities and towns. The Cabildo and the Auxiliary Councils are responsible for the provision of city services such as street lighting, water and sewage, garbage collection, public markets, streets, parks and public security. They also grant building and business permits. The main income sources for municipalities such as Puebla are the property tax, fees for services, and a federal contribution. Municipal income plans, however, need to be approved by the state legislature every year, thus municipalities' capability to invest in local infrastructure depends on state priorities. The Cabildo has the capacity to decide the expense budget. All municipalities in Mexico have the faculty to establish three-year development plans and urban development strategic plans. These plans, however, need to be aligned with national and state level development plans.

Strategic Planning Process

As we pointed out in the background section, a marketing strategy can be successful only when it is guided by urban or regional strategic plans that are oriented to developing a landscape and an infrastructure for the city. Thus in this section we will describe current developments in strategic planning at the state and local levels in Puebla.

Strategic planning in Mexican public administration is relatively new. At the state level, a program promoted in the late 1970s pushed for the first time the creation of long-term urban and regional development plans, and at least in the case of the state of Puebla, the plan was not updated until recently. In general, there is a lack of strategic planning culture at the municipal level, and only 20 out of 217 municipalities have developed long-term urban development programs.

In 2004, the state government decided to review and update the plan developed 27 years before to respond to current needs and opportunities. The plan promotes structural economic competitiveness, and the establishment of conditions to improve the quality of life of population, distributed in nine regions (sustainable environmental urban systems) and 22 homogeneous sub-regions (sustainable environmental urban subsystems).

The plan identifies key population centers as the main sources of regional development in networks of interconnected cities and towns. Puebla City is the most important strategic population center in the state. It offers a broad range of services, and according to state planners, the city has a "great capacity to cover local needs and to exploit its potential in national and international markets." Being a key site for state economic development offers opportunities for city strategy and marketing at Puebla.

In accordance with the state's revitalization plan, local managers joined a strategic planning process for urban development in Puebla City. The plan to create the Municipal Sustainable Urban Development Program of Puebla entailed the involvement of many strategic groups and actors, in addition to an integrated diagnosis of the city's infrastructure and resources. A project manager in the planning process commented that "the process involved more than 150 meetings with specialists, academics, professional planners, public officers and the public." The main objective in the plan is "the accomplishment of a sustainable urban development in a long-term scope." The plan was aligned to the national development plan, the state development plan, and current municipal problems identified for the municipal three-year plan.

The program identifies seven "Activity Polygons" in the city, which are urban areas with special needs or assets, key elements to develop both an urban plan and a city marketing strategy. For example, two of these urban areas are proposed for preservation as ecologic areas with special eco-tourism projects. An important polygon in the area is Puebla downtown, which has been declared World Heritage by the United Nations Educational, Scientific and Cultural Organization (UNESCO).

Although GIS is not yet fully developed in the city, it constitutes an important ICT to support planning at the state and city levels. Figure 3 shows one of the maps in city diagnosis and planning. One of the participants in the planning process commented that "high resolution satellite photographs allow to combine many layers of information associated to them, and linking this information with logical interpretation from government offices and research centers it is possible to develop plans and applications hard to imagine some years ago."

Figure 3. Key zones for urban development in the city of Puebla

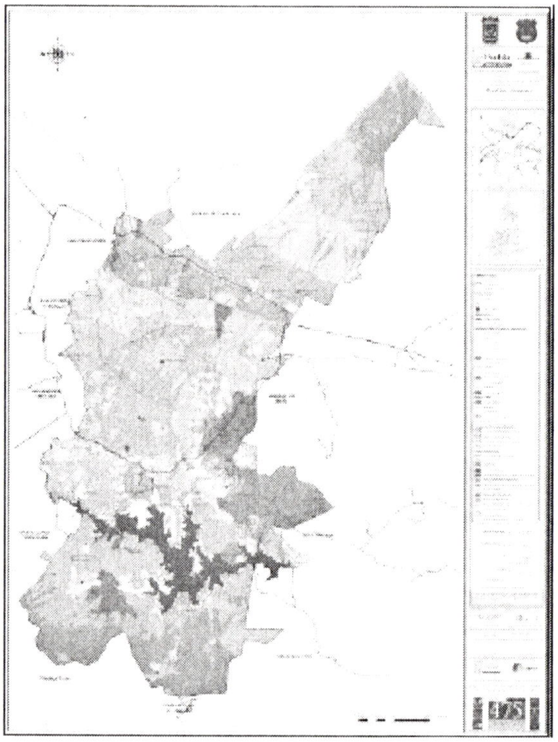

According to our previous discussion, institutions influence government activities such as planning. We have also mentioned the lack of continuity in government activity. In order to promote a long-term vision and to reduce the impact of government transitions, municipal government proposed a Municipal Institute for Strategic Planning to ensure the continuity of the strategic plan. The Institute includes citizens as well as professionals and academics in urban planning; it allows for spatial administration and planning on an ongoing basis. The Institute will be able to maintain the development of strategic projects on behalf of the city despite changes in local administration. The Institute is now in the process of authorization and conformation.

City Marketing Efforts in Puebla

City marketing and branding in the city of Puebla is still a work in progress. One of the government officials interviewed commented, "Puebla is a complex city that can be 'packaged' and marketed to many different target audiences. It is a city with lots of potential, not yet fully exploited." However, planning efforts at the state and municipal level offer an opportunity to start a coordinated strategy among state and local governments and other key stakeholders from the private sector. In this section we will describe the developments in government offices, and some public-private partnerships.

There is no specific effort in terms of city branding, and local authorities use the images that were designed at the state level (see Figure 4). However, state image is influenced by the image of the city of Puebla, which is the state capital. Puebla City is one of the cities in central Mexico that is famous for its gastronomy, and especially for its "mole." Branding efforts at the state level are taking advantage of the cuisine and especially of "mole." As shown in Figure 4, two slogans are used in to attract visitors. The first one, "¿Antojo de Puebla?" literally asks "Craving for Puebla?" The second phrase, "Puebla, mi mero mole," is taking advantage of the famous dish used in the idiom "mi mero mole," which is describes "my favorite thing." Both pictures were taken in the Puebla metropolitan area. Thus, the state image has an important influence on the city of Puebla. As a part of this attempt to promote Puebla, the Talavera (unique pottery) is used as a component of regional identity, and it is recognized internationally as a product with certified origin denotation.

Figure 4. State branding and image efforts

Figure 5. Web presence of the direction of tourism of Puebla city

The city's Web site uses Talavera to reinforce the city brand, using a map of the state covered with adornments similar to the ones used in this handcrafts (see Figure 5). The angel on the Web site is another image associated with the Puebla City brand, based on a legend that attributes the layout of the city's streets to angels.

National brand positioning presents a challenge to attracting visitors. Although there are many sub-brands in the national branding strategy, Mexico is better known nationally and internationally for its "Sun and Beach" destinations, and according to one of the interviewees, sometimes "Sun and Beach" just means Cancun. Puebla belongs to the sub-brand "Colonial Cities," but the strength of the "Sun and Beach" brand makes it difficult to convince visitors to see Mexico's colonial cities. Given that UNESCO has declared Puebla City a World Heritage site, and that Mexico is one of the countries with the most World Heritage Sites in the world, the city hopes to create a sub-brand of "World Heritage Cities." This initiative is fostered by the Director of Tourism.

In terms of marketing efforts, and although there is not yet a coordinated strategy, the city of Puebla has been developing its infrastructure and assets for three target audiences: investors, visitors, and business travelers. Private investors and leisure visitors are the main interest of the Ministry for Economic Development and Touristic Promotion, and business travelers attending conventions and conferences are the main target for the Bureau of Conventions and Conferences.

The Ministry for Economic Development and Touristic Promotion relies on the Economic Promotion Director and the Tourism Director to draw investors and visitors to the city (see Figure 6). The Economic Promotion Director works mainly to attract investors, and the Tourism Director works to attract vacationers.

One strategic objective has been to sell Puebla to transnational firms as an alternative to Mexico City. The economic directors have been consulting with state and municipal urban planners to create infrastructures for industrial parks and for zoning for buildings that will become the headquarters of transnational firms. In addition, the city has been expediting the processes of granting business and construction permits. In fact, they promote the city on their Web site as a place where someone can start a business in two days. ICT has increased the efficiency and transparency of obtaining permits. Potential investors can obtain all permits from city kiosks. One of the objectives of the city is to connect the permits system with the GIS in order to ensure that the distribution of investment according to the needs of each of the city's zones.

Puebla city has many institutions of higher education, and is therefore an important source of highly qualified workers. State and local authorities are using this fact to attract investment. Although a strong infrastructure has been created, the city has not yet capitalized on these efforts, and has not been a significant increase in investment.

Figure 6. Organizational chart for the local ministry for economic and tourism promotion

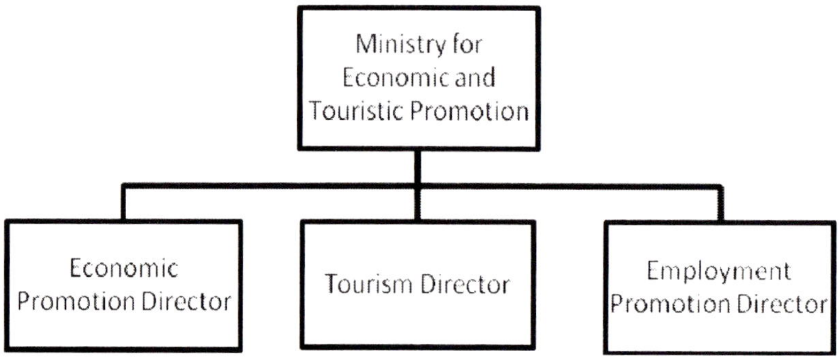

The Tourism Direction has launched several initiatives to attract visitors since 2005. The initiative was undertaken to identify trends in demand. In this way, Puebla is the first Latin American city with a Touristic Observatory registered with the ATLAS Cultural Tourism Research Group. The observatory measures variables such as level of service, reasons for the visit, attractions, and citizen attitudes, and it has been used for seven seasons. The findings are published on the tourism Web page. The observatory has highlighted the lack of coordination among service providers. For example, hotels lack complete information about the city's transportation, restaurants or entertainment and the tourism director has started to meet with local hospitality associations. Creating synergies among service providers has been a difficult task.

The second important activity is the attempt to appreciate city assets by creating a database of its local attractions, buildings and historic sites. Citizen participation has been recognized as a key input in this process, and determining which attractions need to be included in the inventory has been also a challenging task for citizens and visitors. The inventory is available through the tourism Web page, along with a description, photograph, and map of each attraction.

Finally, the target market of business visitors for conventions and conferences has been identified by the Bureau for Conventions and Conferences (BCC). The Bureau is a liaison among professional, government or industry associations and service providers in Puebla City. BCC started as a local association of hotels, restaurants and transportation companies called Destination Puebla, state and local tourism offices, and the Council for the Historical District. BCC is a marketing and sales office (see Figure 7), and its main authority is a Management Board with government representatives, private service providers and citizen councils. The Bureau establishes the city's presence in Meeting Planner Showcases and makes direct sales to professional and industry associations. BCC tries to capitalize not only in the city's cultural and historic assets, but also on its capital investments in a Convention Center in Puebla World Heritage District.

ICT Development and Use

As with city marketing efforts in Puebla, the use of ICT to support the marketing process is also in its earlier stages. However, it is possible to identify efforts in the front-end, and back-office efforts that can give support to the front-end applications.

Figure 7. Organizational chart of the Bureau for Conventions and Conferences

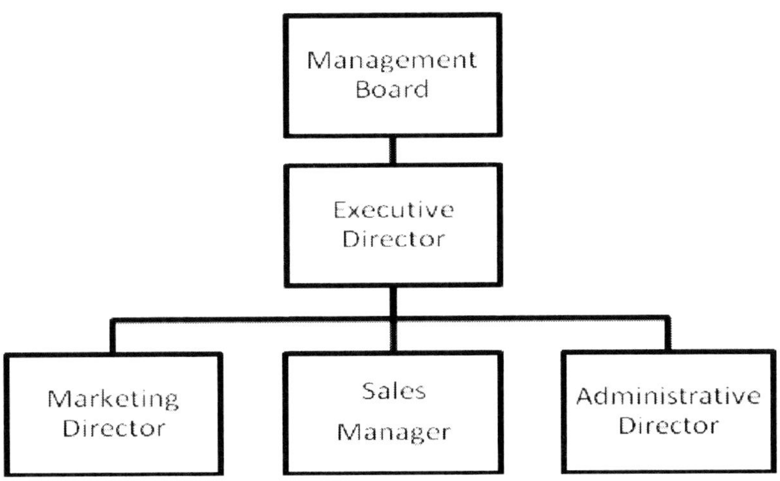

In terms of front-end applications, both the municipal government and the Bureau for Conventions and Conferences are using the Internet to provide information to potential visitors or conference organizers (Figures 8 and 9). Both Web sites are designed to provide information. The municipal Web site (Figures 5 and 8) offers information to potential investors and visitors. Investors can find information about how to open a business in Puebla City, as well as federal, state and municipal programs to promote investment and business creation. Visitors and citizens can find information about places to visit in Puebla. The Bureau for Conventions and Conferences' Web site provides general information, testimonials and a contact address. BCC Web site includes an electronic form to acquire contact information from conference organizers.

In terms of back-office applications, a Geographic Information System (GIS) is being developed for urban planning. However, municipal government officials envision the GIS supporting many other information-intensive applications, such as granting construction permits, analyzing

Figure 8. Google™ maps in the Puebla city portal

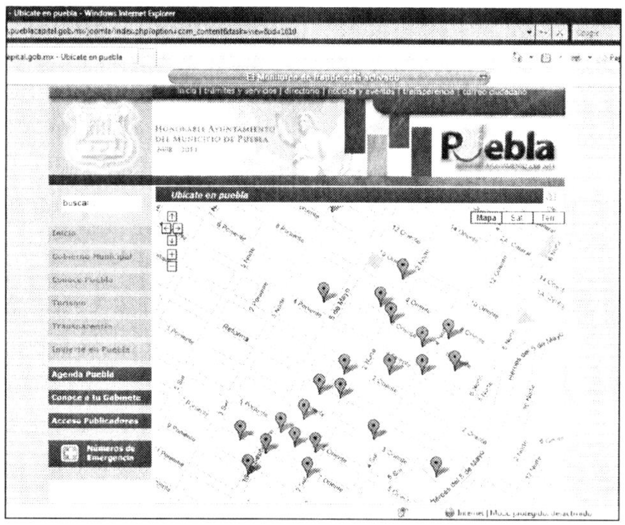

Figure 9. The Bureau for Conventions and Conferences web page

urban growth, infrastructure maintenance, and of course, city marketing. As shown in Figure 8, the Web site takes advantage of Google™ Maps to guide visitors to the city's attractions. The database of local attractions does not include geo-spatial coordinates for all attractions. Completing this information would generate an interactive map of the city. Such an application would be accessed either from the Internet or from kiosks stationed throughout the city.

CURRENT CHALLENGES AND LESSONS LEARNED FOR CITY MARKETING IN PUEBLA

Although the developments described in the previous section are promising and city managers have accumulated expertise and learning, they must face important challenges developing an integrated city marketing strategy in Puebla. In this section, we include some of those challenges and lessons learned.

The first challenge is associated with the succession process in municipal government. A new municipal administration was elected in November 2007, and the new Cabildo took office in February, 2008. The change in elected officials also implies changes in the appointed government officials. Because of the succession process, the activities of the previous municipal government need to be continued by the new officials. Although the proposed Municipal Institute for Strategic Planning could help with program continuity, the Institute itself has not yet been created and its future is at risk.

Puebla has no integrated ICT strategy that allows individual dependencies to choose their own technology and standards; this makes information integration a difficult task. Inside each government office, there is a variety of hardware and software, and a diversity of data models. Nor there is a clear policy or formal procedure to manage the transition from one administration to the next. The administration faces the challenge of developing such a strategy along with institutional elements like laws and regulations.

Another important challenge pertains to the budget for city marketing. Many efforts are being made to ensure coordination and participation of local universities, but there is no specific budget for city marketing. The managers plan to use the results of these efforts to show value, and to obtain public and private resources for city marketing. The new Cabildo needs to include in its expense plan the funds to develop the strategy.

In terms of ICT infrastructure, one of the interviewees commented that even though the GIS could be used by many ministries in local administration, most of the ICT infrastructure is not interoperable. In this way, the next municipal government must also create an ICT strategy that takes into account the development of an interoperable business architecture. In this way, the GIS can be better exploited for the development of front-end and back-office applications to support planning and marketing activities.

Moreover, involving investors and service providers to create synergies has been a difficult task. One of the interviewees commented that in spite of the infrastructure investments and the creation of several industrial parks, both the city and the state have not been able to attract investors to these new developments. Global competition is apparently increasing the bargaining power of investors, who, in the words of one of the interviewees, "do not want to pay taxes, services or utilities to establish in your place or otherwise they establish in a place where they get the best offer..."

Moreover, service providers in the hospitality industry see each other as competitors, and there is little interaction or coordination among them. One of the interviewees commented that "one of the main challenges to improve service to our visitors is to convince private providers that the visitor is an integrated person who needs hotel, food, transport and entertainment." Ideally, hotels should provide information about dining and entertainment, restaurants should provide information about hotels and entertaining. At the present time, it is hard for visitors to find integrated information that will enable them to enjoy the city.

Finally, and given the three-year turnover at the municipal level, an important challenge for the city will be to manage knowledge and to build human capabilities for city marketing. Such knowledge systems will allow the strategic management of city assets. This three-year turnover poses other challenges for continuity of city image. The city portal, for example, changes its design and domain name every three years.

The lessons learned through this case emphasize the role of strategic planning, information systems and ICT tools, and the interaction ability of their users to invigorate city marketing strategies and long-term global strategies with entrepreneurial vision. Likewise, the creation and development of social, organizational and business networks should be considered.

First, systems compatibility and information management help to promote business and economic transactions and to construe them in service and products demands and social and commercial networks. All main leaders are aware of that, and are pursuing this goal. However, they recognize that integration will take several years.

Second, the exploitation and use of ICT tools is crucial to develop and to maintain communication networks among stakeholders, allowing them to meet their expectations as place-users and to facilitate their involvement in decision-making. This kind of ICT use is conducive to the development of social, organizational and business networks.

Third, people involved in the strategic planning process experienced and valued the richness of involving main stakeholders and encouraging public participation. It demonstrates the importance of working with stakeholders in the creation and design of strategic urban development programs.

Fourth, the assets of a city are an integral part of the city's marketing strategies. In this way, urban landscape and design, infrastructure and attractions and people become part of the perceived image of a place. Indeed, city residents need to see themselves as entrepreneurs and marketers of their city in order to support city marketing strategies.

Fifth, creating the infrastructure and developing attractions is not enough. Maintenance of an inventory of such attractions is an important component of planning and marketing. ICT plays a key role in this process.

Another lesson results from the creation and implementation of long-term strategies. It means that, leaders in the public and private sector, citizenship and professional planners have to take into account a broad entrepreneurial vision to fight for market opportunities, investment, place-users and economic competitiveness. The Municipal Institute of Strategy and Planning will play an important role in this matter.

FUTURE TRENDS

This section of the chapter describes some trends in ICT use to support strategic city planning and marketing. We intend to include the developments that are starting to have an impact or that we expect to have an impact on any of these activities.

The case of Puebla shows the potential of open GIS infrastructures for city marketing. One of the free applications that allow users to build their own applications is Google™ Maps (http://www.google.com/maps). Google Maps' API (http://code.google.com/apis/maps/index.html) allows users to embed maps in their own Web pages with routes and locations. A city can use this application to include routes to historic destinations.

The concept of digital cities can also have an impact on city marketing. Digital cities are virtual spaces in which citizens and visitors share knowledge, experiences and interests (Ishida, 2002). A digital city can be as simple as the portals promoted by America Online (AOL), but it also has the potential for interactive three-dimensional applications such as the virtual worlds in Second Life® (http://secondlife.com/). In some cases, ICT can define the image of a city (Dobers & Hallin, 2006).

These two examples (maps and digital cities) are instances of a broader phenomenon on the Internet, which allows marketers to engage in conversations with prospective customers. It is possible now to identify government portals in Mexico that use blogs, wikis and virtual communities. These tools are already influencing marketing products and services in the private sector, and offer a platform for collaboration (D. M. Scott, 2007; L. Weber, 2007).

Mobile applications are an interesting trend that will shape citizens' and visitors' experience of a city. Mobile devices can be linked to GIS to become interactive city guides (Bessa, Coelho, Cruz & Chalmers, 2006). Moreover, devices equipped with Geographical Positioning Systems (GPS) can help citizens and visitors to find historical and tourist sites, in addition to hotels, restaurants and entertainments.

Finally, knowledge-intensive applications will be important for city planning and marketing. Private companies already use Customer Relationship Management (CRM) applications in their marketing strategies, and they are starting to be used as a part of e-Government strategies (Pan, Tan & Lim, 2006). We believe that these applications can be especially relevant in building networks to promote private investments in the city and for convention and conference attendees. Although the city of Puebla is not yet using many of these tools, they merit consideration by city marketers.

CONCLUSION

The promotion of a city is not only about media and supporting tools (Kavaratzis, 2007). Rather, it depends on urban development, spatial planning and economic competitiveness (Jessop & Sum, 2000; Wu & Zhang, 2007), and on the role of the local community in building the city image and imprinting the city brand in the minds of visitors and investors (Anholt, 2004). People enable cities to add city value to the global business, entertainment and leisure arena (Kotler et al., 2007). ICT and the city's ability to exploit Information Technology with a strategic vision become a springboard for the marketing of a city from both the front end and the back office. There has been some progress towards city marketing in Puebla, but local government is still in the process of creating the scaffolding to develop better applications and ICT tools for city marketing.

Although the city of Puebla has directed its promotion strategy through brand management which exploits the representational dimension (emotional links) and the functional dimension (utilitarian aspects), it is crucial to reinforce the emotional link of the brand to personalize the brand by target segment (De Chernatony & McWilliam, 1990; Caldwell & Freire, 2004). Although public and private entities are trying to attract their respective target audiences, a common entrepreneurial vision at a global level and the willingness to cooperate in the common interest is still a work in progress. Current strategic planning efforts at the state and local levels offer an interesting opportunity to advance this collective vision. It must therefore be a priority to build and to cultivate networks of stakeholders and governmental authorities (Eden & Ackermann, 2000; Bryson, 2004b).

Contextual factors in the city of Puebla, such as its economic development and demand for national and international visibility have the potential to stimulate collaboration. However, globalization has at least two faces for local managers. It is an opportunity to compete for global investors, but it also increases investors' bargaining power.

Mexico's institutions limit local capabilities and entrepreneurship. Local income is highly dependent on state priorities, and determined by decisions in the state legislature. On the one hand, local development is constrained by state plans and policies. On the other hand, short three-year government terms without the possibility of consecutive re-election are a serious challenge for strategic continuity. The city of Puebla is trying to meet this challenge through the Municipal Institute for Strategic Planning.

Finally, ICT use in Puebla needs to be consolidated through the creation of a business infrastructure that facilitates interoperability and information sharing among local actors, thus offering better services to citizens, visitors and investors.

REFERENCES

Andersen, D. F., Belardo, S., & Dawes, S. S. (1994). Strategic information management: Conceptual frameworks for the public sector. *Public Productivity and Management Review, 17*(4), 335-353.

Anholt, S. (2000, November/December). The nation as brand. *Across the Board*, 22-27.

Anholt, S. (2004). Nation-brands and the value of provenance. In N. Morgan, A. Pritchard, & R. Pride (Eds.), *Destination branding: Creating the unique destination proposition*. Oxford: Elsevier Butterworth-Heinemann.

Bateman, I., Lovett, A., & Brainard, J. (2005). *Applied environmental economics: A GIS approach to cost-benefit analysis*. New York, NY: Cambridge University Press.

Bessa, M., Coelho, A., Cruz, J. B., & Chalmers, A. (2006). Selective presentation of perceptually important information to aid orientation and navigation in an urban environment. *International Journal of Pattern Recognition, 20*(4), 467-482.

Bryson, J. M. (2004a). *Strategic planning for public and nonprofit organizations: A guide to strengthening and sustaining organizational achievement* (3rd ed.). San Francisco, CA: Jossey-Bass.

Bryson, J. M. (2004b). What to do when stakeholders matter: Stakeholder identification and analysis techniques. *Public Management Review, 6*(1), 21-53.

Caldwell, N., & Freire, J. (2004). The differences between branding a country, a region, and a city: Applying the brand box model. *Brand Management, 12*(1), 50-61.

Davies, J. S. (2005). Local governance and the dialectics of hierarchy, market and network. *Policy Studies, 26*(3/4), 311-335.

De Chernatony, L., & Dall´Olmo, R. F. (1998). Definig a brand: Beyond the literature with experts´ interpretations. *Journal of Marketing Management, 1*(1), 94-105.

De Chernatony, L., & McWilliam, G. (1990). Appreciating brands as assets through using a two-dimensional model. *International Journal of Advertising, 9*(2), 111-119.

DeSanctis, G., & Poole, M. S. (1994). Capturing the complexity in advanced technology use: Adaptive structuration theory. *Organization Science, 5*(2), 121-147.

DiMaggio, P. J., & Powell, W. W. (1983). The Iron Cage Revisited: Institutional isomorphism and collective rationality in organizational fields. *American Sociological Review, 48*, 147-160.

Dobers, P., & Hallin, A. (2006). Slipping into darkness: A Study of the role of ICTs in the making of Stockholm's image. *Journal of Urban Technology, 13*(3), 119-127.

Doel, M., & Hubbard, P. (2002). Taking world cities literally: Marketing the city in a global space of flows. *City, 6*(3), 351-368.

Doyle, S. (2001). Software review: How is geography supporting marketing in today's commercial organizations? *Journal of Database Marketing, 9*(1), 85-89.

Eden, C., & Ackermann, F. (2000). *Making strategy: The journey of strategic management.* London: Sage Publications.

Erbek, F. S., Ulubay, A., Maktav, D., & Yagiz, E. (2005). The use of satellite image maps for urban planning in Turkey. *International Journal of Remote Sensing, 26*(4), 775-784.

Erickson, B., & Roberts, M. (1997). Marketing local identity. *Journal of Urban Design, 2*(1), 35-59.

Fountain, J. E. (1995). *Enacting technology: An institutional perspective.* Cambridge, MA: John F. Kennedy School of Government, Harvard University.

Fountain, J. E. (2001). *Building the virtual state. Information technology and institutional change.* Washington, D.C.: Brookings Institution Press.

Frederickson, H. G. (2004). All public administration is local. *PA Times, 27*(11), 11-12.

Gascó Hernández, M. (2004). E-gobierno en Bolivia y Paraguay. In R. Araya Dujisin & M. A. Porrúa Vigon (Eds.), *América latina puntogob: Casos y tendencias en gobierno electrónico* (pp. 125-150). Santiago: FLACSO-Chile/ AICD-OEA.

Giddens, A. (1984). *The constitution of society.* Berkeley and Los Angeles, CA: University of California Press.

Gil-Garcia, J. R. (2005). *Enacting state websites: A mixed method study exploring e-government success in multi-organizational settings.* Unpublished Doctoral Dissertation, University at Albany, State University of New York, Albany, NY.

Gil-Garcia, J. R., & Luna-Reyes, L. F. (2006). Integrating conceptual approaches to e-government. In M. Khosrow-Pour (Ed.), *Encyclopedia of e-commerce, e-government and mobile commerce* (pp. 636-643). Hershey, PA: Idea Group Inc.

Gillenson, M. L., Sherrell, D. L., & Chen, L. (1999). Information technology as the enabler of one-to-one marketing. *Communications of the AIS, 2*(18), 43.

Hassan, S., & Gil-Garcia, J. R. (2007). Institutional theory and e-government research. In G. D. Garson & M. Khosrow-Pour (Eds.), *Handbook of research on public information technology.* Hershey, PA: IGI Global.

Hubbard, P. (1996). Urban design and city regeneration: Social representations of entrepreneurial landscapes. *Urban Studies, 33*(8), 1441-1461.

Ishida, T. (2002). Digital city Kyoto. *Communications of the ACM, 45*(7), 76-81.

Janssen, M., & Cresswell, A. M. (2005, January). *The development of a reference architecture for local government.* Paper presented at the 38th Hawaii International Conference on System Sciences, Hawaii.

Jessop, B., & Sum, N.-L. (2000). An entrepreneurial city in action: Hong Kong's emerging strategies in and for (inter)urban competition. *Urban Studies, 37*(12), 2287-2313.

Karavatzis, M. (2004). From city marketing to city branding: Towards a theorical framework for developing city brands. *Journal of Place Branding, 1*(1), 58-73.

Kavaratzis, M. (2005). Place branding: A review of trends and conceptual models. *The Marketing Review, 5*(4), 329-342.

Kavaratzis, M. (2007). City marketing: The past, the present and some unresolved issues. *Geography Compass, 1*(3), 695-712.

Kavaratzis, M., & Ashworth, G. J. (2005). City branding: An effective assertion of identity or transitory marketing trick? *Tijdschrift voor Econo-*

mische en Sociale Geografie (Journal of Economic & Social Geography), 96(5), 506-514.

Keen, M., & Mahanty, S. (2006). Sustainability assessment and local government: Achieving innovation through practitioner networks. *Local Environment, 11*(2), 201-216.

Kolb, B. M. (2006). *Tourism marketing for cities and towns: Using branding and events to attract tourists*. Burlington, MA: Elsevier Butterworth-Heinemann.

Kotler, P., & Gertner, D. (2004). Country as brand, product and beyond: A place marketing and brand management perspective. In N. Morgan, A. Pritchard & R. Pride (Eds.), *Destination branding: Creating the unique destination proposition* (pp. 40-56). Oxford: Elsevier Butterworth-Heinemann.

Kotler, P., Gertner, D., Rein, I., & Haider, D. (2007). *Marketing internacional de lugares y destinos: Estrategias para la atracción de clientes y negocios en Latinoamérica*. Mexico City: Pearson Prentice Hall.

Lowndes, V. (2005). Something old, something new, something borrowed [illegible] How institutions change (and stay the same) in local governance. *Policy Studies, 26*(3/4), 291-309.

Luna-Reyes, L. F., Gil-Garcia, J. R., & Cruz, C. B. (2007). Collaborative digital government in Mexico: Some lessons from federal web-based interorganizational information integration initiatives. *Government Information Quarterly, 24*(4), 808-826.

Luna-Reyes, L. F., Gil-Garcia, J. R., & Estrada-Marroquín, M. (2008). The impact of institutions on interorganizational IT projects in the Mexican federal government. *International Journal for Electronic Government Research, 4*(2), 27-42.

Mohammed, R. A., Fisher, R. J., Jaworski, B. J., & Cahill, A. M. (2002). *Internet marketing: Building advantage in a networked economy*. New York, NY: McGraw-Hill.

Moon, M. J., & Norris, D. F. (2005). Does managerial orientation matter? The adoption of reinventing government and e-government at the municipal level. *Information Systems Journal, 15*(1), 43-60.

Morgan, N., Pritchard, A., & Pride, R. (Eds.) (2004). *Destination branding: Creating the unique destination proposition* (2nd ed.). Oxford: Elsevier Butterworth-Heinemann.

Norris, D. F., & Moon, M. J. (2005). Advancing e-government at the grassroots: Tortoise or Hare? *Public Administration Review, 65*(1), 64-75.

North, D. C. (1999). *Institutions, institutional change, and economic performance*. New York, NY: Cambridge University Press.

Orlikowski, W. (1992). The duality of technology: Rethinking the concept of technology in organizations. *Organization Science, 3*(3), 398-427.

Orlikowski, W. (2000). Using technology and constituting structures: A practice lens for studying technology in organizations. *Organization Science, 11*(4), 404-428.

Pan, S.-L., Tan, C.-W., & Lim, E. T. K. (2006). Customer relationship management (CRM) in e-government: A relational perspective. *Decision Support Systems, 42*(1), 237-250.

Papazoglou, M. P., & Ribbers, P. M. A. (2006). *e-Business: Organizational and technical foundations*. Chichester: Wiley.

Raymond, C., & Brown, G. (2007). A spatial method for assessing resident and visitor attitudes towards tourism growth and development. *Journal of Sustainable Tourism, 15*(5), 520-540.

Roberts, M. L. (2003). *Internet marketing: Integrating online and offline strategies*. New York, NY: McGraw-Hill Irwin.

Rocheleau, B. (2006). *Public management information systems*. Hershey, PA: Idea Group Publishing.

Scott, D. M. (2007). *The new rules of marketing & PR*. Hoboken, NJ: John Wiley & Sons.

Scott, W. R. (2001). *Institutions and organizations* (2nd ed.). Thousand Oaks, CA: Sage.

Schumpeter, J. A. (1934). *Theory of economic development: An inquiry into profits, capital, credit, interest and the business cycle*. Cambridge, MA: Harvard University Press.

Short, J. R., & Benton, L. M. (1993). Reconstructing the image of an industrial city. *Annals of the Association of American Geographers, 83*(2), 207-224.

Weber, L. (2007). *Marketing to the social web*. Hoboken, NJ: John Wiley & Sons.

Weber, S. (2007). *Plug your business*. Falls Church, VA: Weber Books.

Wu, F., & Zhang, J. (2007). Planning the competitive city-region: The emergence of strategic development plan in China. *Urban Affairs Review, 42*(5), 714-740.

Yuankai, T. (2007). Shanghai shows its heart. *Bejing Review, 50*(42), 27.

ENDNOTES

1 National Institute of Geography and Statistics (2005). Information retrieved from the Puebla city Web Portal on January, 2008.

2 National Institute of Geography and Statistics (2005). Information retrieved from the Puebla city Web Portal on January, 2008.

3 National Institute of Geography and Statistics (2003). Information retrieved from the Puebla city Web Portal on January, 2008.

Chapter VIII
Official Tourism Web Sites and City Marketing

Pablo Díaz-Luque
Universidad Pablo de Olavide, Spain

ABSTRACT

Large cities are one of the most popular tourism destinations throughout the world. Business and leisure tourists visit these areas every year and before they travel there, they look for useful information on the Internet. This chapter analyses the tourism Web sites developed by Convention and Visitor Bureaus. These Web sites represent the official image of the city on the Internet and trough them tourism organizations can organize the marketing and mix strategy. The chapter studies the concept of a city as a tourism destination, the organizations that manage tourist activities, and the right marketing strategies to be developed on these official Web sites. The strategy begins with the market research to select the right marketing segments and it continues with the right actions from a marketing mix perspective. It means different options in terms of product-destination exhibition, price policies, commercialization, and communication actions.

INTRODUCTION

Big cities worldwide are major international tourism destinations and millions visit them for business or pleasure. From historical cities with architectural heritage to newly created locations with huge skyscrapers, they attract visitors interested in varying aspects of each individual city, directly influencing the target location's economy.

These cities can be seen as products competing at a global level, whose image on the Internet is reflected on official tourism Websites managed by the Convention and Visitor Bureaus (CVBs). Consequently, in this paper, the importance of having an appropriate official tourist Website from a marketing perspective will be considered.

It is generally well known that Tourism is an intensive information sector. Evidently, due to distance, this sector has particular needs and ICT may help. Computer Reservation Systems (CRS) and Global Distribution Systems (GDS) have been good examples of ICT especially designed for (and by) the Tourist Industry. They provide information about transport and hotel vacancies. Nowadays, the Internet represents the most advanced technology and one of the best instruments for cities to promote themselves as attractive tourism destinations all over the world.

The inevitable relationship between Tourism and the Internet is becoming increasingly closer which means that more effort and research should be made in this respect. Similarly, more and more consumers use the Internet as their main tool to plan, book and pay for their holidays and trips.

For some time now, the Internet has identified a clear beneficial distribution platform for the tourist sector. For instance, data from a survey on ICT usage in households and by individuals confirm that 47% of the internet users in the European Union say that they have used services related to travel and accommodation in the three months prior to the survey in 2005 (Knauth, 2006). Many of these international tourists choose big cities as the main destinations for their holidays and researches have demonstrated that the Internet has become an important information source (Tjostheim & Tronvoll, 2002). In the light of this, naturally these cities have been prepared to host many tourists, mainly by creating Official Tourism Websites. The relevance of these Websites lies in the fact that a specific image and information on the city is given. Similarly, the most relevant Websites have also been provided with the necessary tools to include e-commerce and appropriate resources to communicate with tourists.

This paper will help CVBs to realize the importance of having the right image on the Internet, i.e. an Official Tourism Website of each city which is linked to their marketing strategy. The Internet and the *World Wide Web* are considered to be the right technologies to present the city to the target market of tourists and to develop a complete marketing mix strategy.

Over the next pages, the concept of Tourism destination and CVBs will be explained. In the first place, big cities are recognized as a main type of destination for leisure and business trips. Second, the possibilities that the Internet and the official websites offer from a marketing perspective are analyzed. In particular, the chapter refers to a) the opportunities to use the websites and the Internet to develop a complete marketing mix strategy starting with a Website for market research, b) the product and price strategy, and c) the repercussions of the Internet in the channels of tourism products and services. Now CVBs can commercialize local tourism services at an international level via their websites. Finally, the new promotion and communication framework linked to the Internet and *World Wide Web* are presented.

CITIES AS TOURISM DESTINATIONS AND THE ORGANIZATIONS THAT MANAGE THEM

A tourism destination is a complex concept to define. Traditionally, it has been understood as a specific geographic area where tourism resources and products are located and to which tourists travel. However, some writers consider that the actual perception is more relevant whereby the concept is related to the consumer's subjective perception, depending on travel itinerary, cultural background, purpose of visit, educational level and past experience (Buhalis, 2000). For instance, Lisbon can be a considered a destination in itself for a business traveler whereas for someone traveling for leisure, Lisbon could be purely part of a trip encompassing Sintra, Cascais and Estoril.

Authors defending the first type of definition are Bull (1991) and Hall (2000). For them, a tourism destination is a country, a region, an island or a city as the main objectives for the consumers that visit them. On the other hand, Cooper et al. (1998) define tourism destinations as a group of facilities and services designed to meet the needs of tourists. Finally, Buhalis (2000) views tourism destinations as an amalgam of tourism products, offering an integrated experience to consumers.

Whatever the definition we use, it is clear that big cities can be a tourism destination *per se*. Buhalis (2000) mentions cities as a first kind of tourism destination in a typology that includes seaside, alpine, rural, authentic third world and unique-exotic-exclusive destinations. Cities are historically the oldest existing tourism destinations. People have been traveling to urban destinations for politics, religion, business or even sports reasons (e.g. the Olympic Games of Ancient Greece) since the early years of civilization.

Nowadays, large cities concentrate all the tourist resources, services and products to attract visitors. A big city has the necessary transport infrastructure to get there (at least one international airport, large railways and bus connections and stations, motorways and maybe even a port available for transatlantic cruises where possible) and to move around within the city (underground, bus routes and taxis). There are also services and places for relaxation and business (e.g. hotels, restaurants, discotheques and exhibition centers). In addition, most big cities have some tangible attractions such as museums, parks, theatres and events. Additionally, they count on non-tangible aspects such as the environment, culture and friendliness of the locals.

All these aspects make big cities one of the popular tourism destinations all over the world for business or leisure travelers. These two types of travelers are an excellent combination for cities. The former usually travel to cities during working days to attend meetings, conferences and exhibitions, whereas the latter prefer periods of low business activity or holiday periods, such as weekends and school holidays. It is crucial to cater for both.

This is one of the target pursuits by CVBs. These organizations are usually classified as independent and non-profit making, but also can be part of a city government or the local Chamber of Commerce, while others are special legal authorities. CVBs are defined as organizations which represent and market their communities in a very competitive market place that includes meetings and conventions, group tours, and visitors (Gartrell, 1988). The primary mission of a CVB of a city is to promote the tourism business and attract visitors to its city with the ultimate target of enhancing the economy and the city's image through the coordination of all the components of the tourist industry (Yuan, Gretzel & Fesenmaier, 2003).

A similar concept is the Destination Marketing Organization (DMOs); and these are preferred by some authors. Collins and Buhalis (2003) define them as providers of information products and services for customers, intermediaries and suppliers in the tourism value chain. They tend to be part of the local government and have political, legislative as well as financial means to manage resources rationally and to ensure that all stakeholders can benefit long term (Buhalis, 2000).

Tourism is becoming more and more important for cities' economies and these organizations have to develop the right strategy in order to satisfy stakeholders and tourists. Some of the most important cities in Europe have considerably increased their number of tourists in the last few years. For example[1], Rome hosted more than 11 million visitors in 2005 whereas it hardly reached 6 million in the year 2000. Barcelona welcomed 7.5 million tourists in 2006, which is twice the number it reached in 2000. Berlin was visited by more than 7 million tourists in 2006, two million more than in 2000.

INTEGRATING THE INTERNET IN THE MARKETING STRATEGY OF THE CITY: FROM MARKET RESEARCH TO MARKETING MIX

If tourism destinations, including big cities, are seen as a product a marketing strategy must be developed to attract enough customers. But the target should not be only attracting more visitors to the city. Instead, marketing should operate as a mechanism to facilitate city development and to rationalize the tourism activity in order to achieve the strategic objectives. Buhalis (2000) mentions four different marketing objectives for destinations such as cities (Table 1).

Consequently, developing a tourist marketing strategy for a city is a complex process because it affects all these factors. Most cities are an amalgam of independent Small and Medium Enterprises (SMEs) with their own marketing strategies. However, the international tourist industry is becoming an increasingly competitive market where only the best managed cities are likely to succeed. For the city to become a successful destination, CVBs should bring all individual tourist partners together to cooperate rather than compete and to pool resources towards developing an integrated marketing mix and delivery system (Buhalis & Cooper, 1998).

Table 1. Strategic management and marketing objectives for destinations (Source: Buhalis, 2000, p. 100)

• Enhance the long-term prosperity of local people. • Delight visitors by maximizing their satisfaction. • Maximize profitability of local enterprise and maximize multiplier effects. • Optimize tourism impacts by ensuring a sustainable balance between economic benefits and socio-cultural and environment costs.

Integrating Internet Marketing in City Tourism Organization

According to many authors, successful destination marketing strategies require substantial integration and coordination of Internet marketing effort as well as the development of a favorable organizational environment that supports innovation (Buhalis, 2000; Fesenmaier, 2007). In this way, one assumes technology would help to manage this situation. Nevertheless most tourism destination managers are only just beginning to understand and appreciate how they can use emerging web technologies to promote themselves (Gretzel, Fesenmaier, Formica & O'Leary, 2006).

Yuan, Gretzel and Fesenmaier (2003) classified destination in line with their use of technology based on two dimensions: the number of applications and mean length of applications use. Using a hierarchical clustering application, they found five different groups of destinations:

- **Laggards,** these bureaus have used Internet technology for the least amount of time and have not been capable, or not interested, of expanding the use of the technology beyond simple applications such as email and the *world wide web.*
- **Late light adopter,** they have used the Internet for some time but have limited its use to very few applications.
- **Sophisticated followers,** unlike late light adopters, they have been able to adopt the technology in more applications inside the organization.
- **Early light adopters,** those who have used the technology longer than all other CVBs but have not integrated the use of technology into the organization (due to the relatively low number of applications in use)
- **Knowledge adopters,** they have used the Internet longer than many other CVBs and have integrated the technology into many of their business operations.

Based on the two dimensions, these authors describe that two different development paths exist (Figure 1). Path A describes what one may consider an "evolvement" strategy for the implementation of IT within the bureau. i.e. bureaus starting to implement Internet technology might first introduce a few basic applications (initial point, 0) and use these applications until they are fully embedded within the organization. These authors also agree that as staff members gain training and experience over 1 to 2 years, the bureau expands its tool set to include a range of other applications and may be characterized as *Laggards*. This development process continues along the path to *Sophisticated Followers* and *Knowledge Adopters*, where the technology has been fully integrated into the bureau's business processes.

An alternative development path suggested by these researchers appears to follow a less integrative or developmental strategy (Path B, Figure 1). Here, bureaus appear to be innovators, whereby they have adopted Internet technology early on as a substitute for previous technology or as a means to expand current capabilities. However, they appear to limit its use to very basic functions, thereby minimizing its impact on the bureau (and therefore, remaining within the substitution or enlargement phases of technology implementation).

The same authors offer another analysis to explain IT implementation by the bureaus as an evolutionary process (Yuan, Gretzel & Fesenmaier, 2006). According to them advancements in IT implementations did not spread uniformly and appear to occur in six sequential stages, namely: point of adoption, substitution, enlargement, gestation, reconfiguration, and setback.

Differentiation as a Key Strategy

Whatever the stage that a CVB is, they should integrate Internet Technologies in their processes and marketing strategies as soon and as much as

Figure 1. The typology of adoption/diffusions by American CVBs (Source: Yuan, Gretzel and Fesenmaier (2003, p. 247)

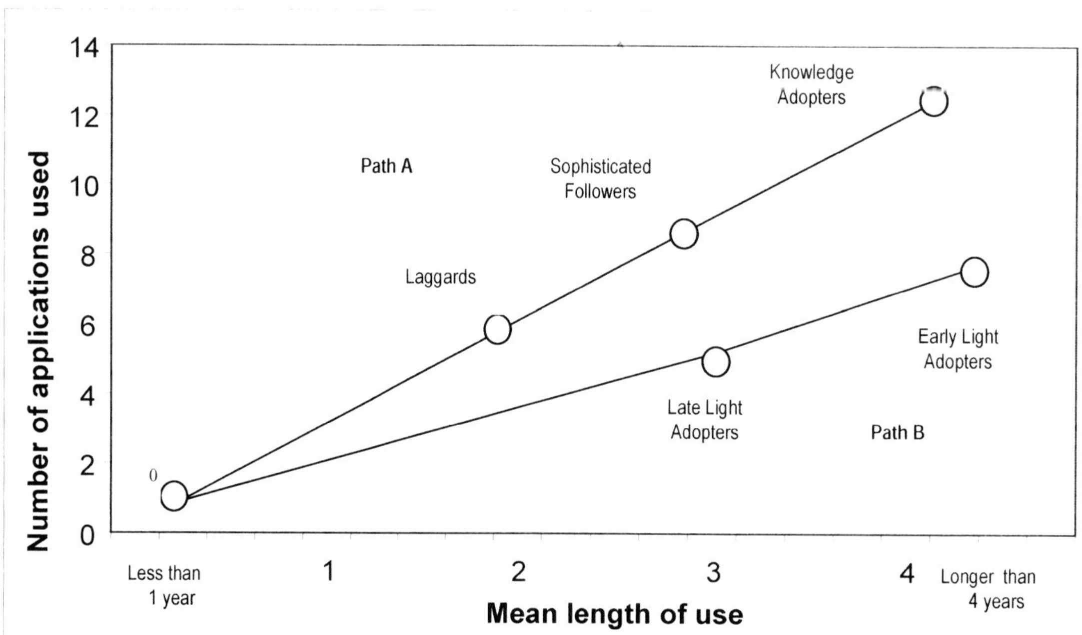

possible. Most analysts agree that destinations must focus on differentiation strategies instead of cost advantage strategies. Buhalis (2000) studied three kind of strategic models (Porter's *Generic Models*, Gilbert's *Strategic Framework* and Poon's *Flexible Specialization*), concluding that tourism destinations should avoid cost advantage strategies as they are based on mass production capacity and resources (Porter, 1980; Gilbert, 1990; Poon, 1993).

As we will we see, the Internet is a technology that allows constant communication with visitors. Just like companies, city authorities must offer competitive advantages, mainly by examining tourists' needs and defining and designing specific products that best fulfill their wishes. The Internet, the official website and cooperation with the stakeholder will enable each city to know tourists' needs and to create the right product for them.

In this sense, each city must create a comprehensive marketing plan with strategies and action points with clear objectives. It is crucial to choose a clear type of consumer as a goal and then reflect what may attract them. Moreover, one should bear in mind the position of the city in comparison to others.

The strategic function is developed through three main actions: market segmentation, selection of the target tourist and the market position. The *market segmentation* is the process by which the number of consumers of a product is divided into identifiable groups depending on their common necessities, wishes and attributes (Kotler, 1997). There are various segmentation criteria, although the most frequently used are the ones considering geographic, socio-demographic, psychographic and behavior variables. By applying these criteria, we have a variety of market segments, although not all would be attractive for the city. In order to know whether a segment is interesting or not, it has to be (Kotler, 1997): 1) Measurable: in order to know how many individuals are grouped in each segment. 2) Accessible: the segments must be realistic so that they can be tackled with adequate marketing strategies and 3) Substantial: the segments have to be big enough to shape their own characteristics, to justify the marketing strategy and its related actions.

Once the existing segments have been evaluated, the city authorities must select which ones they should undertake. In this sense, an *objective market* will be made up of a group of tourists who share the same characteristics or needs and to whom the organization intends to offer their products (Kotler, 1997).There may be three different strategies used in parallel fashion here:

1. **Undifferentiated Strategy:** In this case, the organization does not take into account or develop market segmentation; it addresses consumers in general, focusing on their *general* characteristics and needs,

Figure 2. Kids love London (Source: Retrieved February 21, 2008, from http://www.visitlondon.com)

A special focus on children traveling with their parents to London originated from London tourist authorities. Family tourism is a target segment here. On this website, all the information is closely linked to children.

as opposed to their differences. As we have commented, this is not an adequate strategy according to relevant authors but some cities are unable to focus on particularly unique characteristics.

2. **Differentiated Strategy:** This strategy implies that the organization focuses on certain market segments and designs different offers for each of them (Figure 2, for instance). Through this strategy, a higher profit margin than with the undifferentiated strategy may be obtained.
3. **Concentrated Strategy:** Here the organization is specialized in fulfilling the needs of a *specific* group of consumers, focusing therefore on a good reputation and competitive advantages in services and products (Kotler, 1997). This strategy implies higher risks than the ones mentioned above, due to the dependence created between the niche market segment and potential consumers, who may disappear as potential tourists or whose needs may well change.

The *market position* of the city is simply how tourists themselves define the destination, taking into account the most significant qualities of the place itself. Tourists can be overwhelmed with too much information about products and services and this situation could get even worse if they use the endless possibilities of the Internet without knowing exactly what they want. Unable to evaluate each destination every time they want to choose a different one to visit, they arrange the destinations in different categories; i.e. they "position" the cities together with their distinct branding in order to make a choice.

Once the target markets and the positioning they represent have been selected, one must find a way to target them. Operative marketing is needed to develop the city as an attractive tourism destination, with an adequate price policy and effective promotion. But before this, tourists' needs and desires have to be known and CVBs must develop market research to get this information.

Market Research Through the City's Tourism Website

The Internet, with its constant incoming quantities of information in millions of existing web pages, is a very valuable data source for the tourist authorities in important cities. There are international, national, regional and even local organizations devoted regularly to updating tourist information on the Net. The Internet, in this case, is seen as a secondary source of information.

But the Internet may also be considered a primary source of information for city marketing, as any user may obtain all the useful contents about the city at the same time as information about Web visitors behaviors is saved and centralized in servers. Moreover, the website can be used to develop surveys for visitors to the Website.

Server Activity

Servers where website files are hosted can record all the activity produced by the tourist web site. This is recorded in log files that include information on errors, processing time, bandwidth used, visitor IP address, where visitors come from as well as information regarding their operating system and Internet browser used. Therefore, we can find out where each user is surfing, which are the most visited pages, how long it takes for each user to glean this information, the origin of each visit and even the websites that have been visited immediately before this one.

By means of specialized software, it is relatively easy to manage and present all this information in a comprehensive way. This can be invaluable information for CVBs managers, as they will know the exact number of visitors or web files downloaded, the most frequently visited itineraries and the products acquired. There are many solutions (some of which are free) in the software market to analyze Website statistics (e.g., OpenWebScope™, FastStats™, SurfStats™, WebTrends™, ClickTracks™, AWStats™, MetaTraffic™).

Table 2. Advantages and disadvantages of server activity analysis for marketing purposes (Source: Adapted from Strauss and Frost, 1999)

Advantages	Disadvantages
• Obtain information from all website visits. • On time results and low cost. • Easy management of data obtained. • Useful information about most demanded contents, services and results of campaigns.	• Difficulties to identifying socio-demographic characteristics of website visitors.

Table 3. Advantages and disadvantages of questionnaires on the Internet (Source: Strauss and Frost, 1999, p. 98; Wright, 2005)

Advantages	Disadvantages
• Obtain information from a global and varied community of users. • High speed and low cost. • Answers to personal or embarrassing questions can be answered honestly. • Easy management of data obtained.	• Difficulty to evaluate whether the sample is representative in the population under study. • Possible unreliability as answers may well be incorrect. • Difficulty to guarantee the identity of the participants in the questionnaires.

Online Surveys

A quantitative method useful for obtaining information from potential customers or from tourists who have visited the city is the use of questionnaires on the Website. There are intrinsically both advantages and disadvantages of this, as shown in Table 3 below.

As can be seen, the main advantage is the low cost and speed, as compared to traditional methods of delivering and collecting questionnaires face to face. Also, some of the disadvantages mentioned above can be mitigated if correct systems are used, e.g. through appropriate questioning or prizes for completion of the questionnaire.

There are many ways to develop web surveys. Wright (2005)[2] offers a list of 20 Web survey software packages and online survey-related services available on the Internet. The author examined each of the websites for these 20 online survey product and service companies in order to assess current features, pricing, and limitations, as well as to identify current trends in the online survey product and services market.

Before making an online survey some points have to be taken into consideration. These questionnaires may have particular features adapted to the computed-mediated world in which they are provided (Couper et al., 1998). In order to describe them, a number of criteria related to market research will be considered, mainly in the way they are distributed, the time perspective and their scope. In other words, the tourism web site will serve as the basis to administer questionnaires, either at a specific time, or on a regular or permanent basis, and with a national or international scope:

1. **Self-administered:** This kind of questionnaires is characterized by the absence of the interviewer. The goal is that the interviewees read, understand the instructions and, eventually complete the questionnaire in their own time. However, there is no guarantee that interviewees do the test alone or are not tempted to contact Marketing Organization of the City. To prevent this, the questionnaire should be a CAPI type (Computer Assisted

Personal Interviewing), rather than a CATI (Computer Assisted Telephone Interviewing) questionnaire.

2. **Isolated, periodical or permanent:** The questionnaires outlined in a website of this kind may have all or any of these three features. Due to the simplicity and rapidity to be inserted within the contents of the Web page, this medium can be very useful to obtain data in a short span of time. They can give permanent information which can then be investigated to learn about temporal ideas or changes in some of the studied variables. Periodical questionnaires may be useful to obtain data in specific dates; for instance, after summer holidays or during the Christmas holiday period.
3. **National or International Scope:** this feature is a direct implication of the World Wide Technology. The Net offers the possibility to distribute questionnaires to domestic and international tourists.

In order to place a questionnaire on the Web, a series of issues should be considered i.e. what is going to be asked and to whom. The size of the sample is also crucial. Evidently results must be significant and meaningful. Secondly, a suitable layout must be considered. One must carefully evaluate the exact nature and number of questions and even the order in which these appear[3].

Some tourist and city sites use online surveys. Paris web site, for instance, asked visitors in the summer of 2007 about their geographic and socio-demographic characteristics, website opinions and suggestions, and travel plans[4]. Alaska's Department of Commerce, Community and Economic Development placed a survey online for the first time in 2006 (DCCED, 2007). Those working on the Alaska Visitor Statistics Program (AVSP) distributed "invitation cards" to out-of-state visitors who were leaving Alaska. The card directed respondents to a web address and each card having its own unique password. As an incentive, respondents who completed the survey online were entered into a draw to win one of several prizes. The primary purpose of the online survey was to increase sample sizes, allowing for greater sub-sample analysis. This goal was achieved: in addition to 2,703 intercept surveys, the sample of summer 2006 includes 2,956 online surveys. The online survey introduced a new alternative to the traditional AVSP intercept.

There was however a risk of bias in this online survey sample: it had a response rate of 17.5%, in contrast to the intercept response rate of 85.6%. To address this issue, the study team compared a wide range of demographic variables between the two samples, including gender, origin, age, income, and education. Only origin presented a potential bias: international visitors were less likely to complete the online survey and visitors from certain regions of the USA were slightly more likely to participate. To adjust for this bias, the online sample was weighted by origin so that it reflected the intercept sample. Another bias was apparent in trip planning sources. Online respondents were more thorough trip planners, using more sources at a higher rate than intercept respondents. For trip planning sources, only intercept data was presented in the report.

City Marketing Mix and the Official Website

Once CVBs know target tourist needs, they can develop a marketing mix strategy. Marketing mix is a set of marketing tools used in any company to obtain the expected answer in the targeted audience. It refers to all the actions that a company may develop to influence the demand and direct it towards its product (Kotler, 1997). This definition is oriented towards companies, although the idea is easily extended to other organizations, namely the CVBs.

Once the city organization has launched a general strategy (which includes setting its target market and its development prospects), the moment has come to launch a marketing plan, taking into account its operating budget and a selection of the tools that are necessary to implement it. This is no mean feat. However careful selection of these tools and identification of what can be done by the organization to develop its position in the market (or indeed to attract its target audience) are all crucial factors.

Bearing this in mind, the organization will have to arrange all the factors that are relevant or can be controlled in order to attract the tourists to the particular destination. The combined set of variables that the organization may use is what it can be called *the marketing mix of the city*. If some of the activities are computer-mediated, this is called the *electronic marketing mix of the city*.

Marketing Mix Components

The most relevant components or variables in relation to the marketing mix are known as the four Ps: Product, Price, Place and Promotion. The four Ps summarize the marketing activities that can be developed by organizations as follows (Kotler, 1997):

1. Development of the city product to satisfy the needs of those target segments revealed as central in previous market research and marketing strategies i.e. development of a host of attractive features, infrastructures, services and products which make the city enticing to potential clients.
2. Establishment of prices for services and tourism products located in the city.
3. The selection of the place where a product can be purchased.
4. This involves choosing a set of communication channels and therefore developing the right promotion of the product for the target market. Through this, one can whet the customer's appetite and initiate a process of sale and purchase of the services and tourism products of the city.

PRODUCT AND PRICE STRATEGY IN THE OFFICIAL TOURISM WEB SITE.

Showing Off the City in the Best Light Possible

It is clearly acknowledged that the product is the first component of the marketing mix and, in the vast majority of the cases, the most important one. Through in-depth market research, we can learn about the needs and desires of the consumers and subsequently a satisfying product can be designed and offered via the destination's official Website.

This paper shows how a tourism destination –in this case, the city- can be understood as a product in itself even with its own life cycle as Butler described in 1980 (Figure 3). On the other hand, it is not just the destination which is the product; tourism itself is a product, as stated by Reime and Hawkins (1979). Related to this, Taylor (1980) explained that the real product that tourism generates is a satisfying experience (albeit, difficult to measure), but reaching this goal must be the main objective in the tourist industry.

However, this is not as simple as it seems: to do this we must know about even the first contact that the tourist has with the city. There is very little which can be done when friends and non official Websites have given information but official websites are evidently fundamental and can show all the relevant information and establish communication with tourists in line with the marketing strategies

Needless to say, large cities usually promote the best of the city to satisfy a varied range of tourists' needs and wishes. In the light of this, the most suitable strategy then would be selecting and launching the resources that best fulfill the target market's needs. i.e. simply provide all the desirable services and features that the city can sustain.

Figure 3. Life Cycle of a Tourism destination Source: Butler, 1980, p. 8)

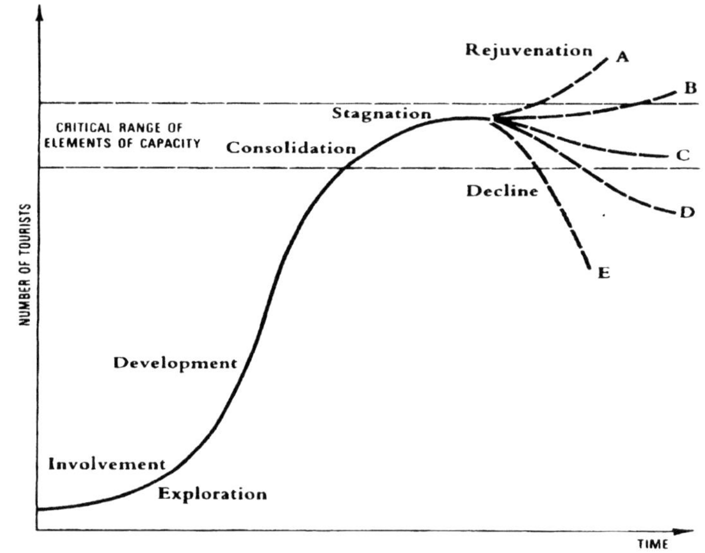

According to Butler a tourism destination has the same cycle of a product in terms of numbers of visitors instead of purchases. Big cities are usually located in higher stages of development. Most of them observe a cyclical movement, alternating between highs and lows.

A tourist product line can be defined as a group of closely related tourism products bound to satisfy the needs of tourists with similar needs or which the segments of the target market demand (Heath & Wall, 1992). Take cities with a mainly cultural offer, based on historical buildings, museums and monuments (Barcelona, Figure 5) or cities such as Las Vegas, where entertainment products such as casinos, discos and theme parks prevails (Atlantic City, Figure 5). Other cities even veer towards business events, with facilities for the celebration of conventions and fairs, together with large hotels and top of the range facilities for meetings and conferences. (Denver, Figure 5).

City authorities will rarely make any effort to develop another tourism product line such as those of sunshine or the beach, even when they are located near the coast or mountains.

Heath and Wall (1992) describe this strategy in two dimensions (Figure 4): each city will be made up of a width and length of tourism product lines. The width shows the number of product lines that the city may hold, whereas the length indicates the number of resources associated to that same line. These must be clearly exhibited on the city website. Indeed, most cities' websites do just this.

This strategy is compatible with the presentation of key attractions which make each city well-known and even unique. This represents the image of the city from outside. In this sense, the webpage is fundamental as it is the medium where this icon is shown, where the city is branded and where symbols are visual and dynamic. This is the case of unique icons like the Venetian Canals, the Eiffel Tower in Paris, the Statue of Liberty in New York and the Little Mermaid in Copenhagen (logo of the official website of Copenhagen).

Cities not only exhibit those tourism product lines and the most interesting icons; it is also necessary to show all the necessary information in an appropriate way so that potential tourists are able to plan their stay and are able to know all the possibilities that the city offers in detail.

Figure 4. City-product mix characteristics(Source: Adapted from Heath and Wall, 1992, p. 131)

-------------- PRODUCT MIX WIDTH ----------------

LENGTH	*Entertainment tourism*	*Historical–Cultural tourism*	*Business tourism*
	Fairgrounds	Museums	Trade fairs
	Discos	Architecture	Accommodation
	Casinos	Opera House	Restaurants
	Concerts	Theatre	.
	.	Art Festivals	.
	.	.	.
	.	.	.

Figure 5. Three home pages from three tourism city websites (Retrieved January 20, 2008)

Atlantic City: www.atlanticcitynj.com
Atlantic city is "always turned on" for those who want entertainment. A picture of the illuminated city by night is seen on the colorful homepage of this website

Barcelona: www.barcelonaturisme.com
Barcelona is another example of a website that wants to portray an image of cultural richness. The modern design is one of its main attractions—a reflection of the city.

continued on following page

Figure 5. continued

Denver: www.denver.org
Denver clearly is focusing on the business market. The skyscrapers and numerous links and suggestions of how to ease the stay of executives are significant

The list of items offered below is not an exhaustive guide of what should be included on a tourist city website; it is as a tool to know which contents would be beneficial on a web page of these characteristics (Tables 3 and 4). These contents will be divided into two categories: tourist information and information about tourism companies. The first category will correspond to all the variables that may help know the city better. These may be divided into three groups: localization and how to get there, basic features of the city and information about attractions of the city. The second category includes companies related to the tourist activity.

The City Created by the Tourist: Interactive Leaflets

As has already been stated, each tourist has a different idea of what a tourism destination is

Table 3. Basic tourist information for city tourism web sites

Location and how to get there	Tourist information and sights	Basic information
Location	Museums	Climate
Explanation of where the destination is located	Typical gastronomy of the city	The weather forecast over the next few days
Situation map of the destination. Graphic explanation.	Architecture	Landscape
How to get there	Cultural agenda (events), leisure and night life.	Means of transport within the city underground, buses, cycle paths
City airports	Craftsmanship and commerce.	History, culture and customs
Port infrastructures	Parks and gardens	Maps, street maps and car parks.
Railways	Suggested visits and tourist routes (advice on places to visit and routes)	Tourist offices and addresses of interest (authorities, medical assistance, police)
Road transport	Specific tourism	Tips and Helpful Advice (Currency, language, visas, local laws and traditions etc.). Accessible tourism for disabled travelers.
	Neighboring destinations (visits to places of interest with departures and arrival to the same city)	

Table 4. Tourist industry information for city tourist web sites

FACTORS LINKED TO THE TOURIST INDUSTRY
Accommodation industry: Hotels (sorted in categories), bed and breakfasts, apartments and youth hostels.
Restaurant industry: bars and restaurants (sorted in categories)
Transport: airline companies, shipping, railway industry and buses with departure and arrival to the city. Rent a car and taxis.
Tourist buses.
Attraction and theme parks.
Theatres, events, shows, casinos and nightclubs.
Exhibition centers
Language Schools.
Sports tourism. Companies offering sport activities for tourists.
Tourist Guides

and in particular, of what kind of services and products they would like to have at hand. For this, the Website will serve as a tool for the tourists to create the features of the destination that better suit their needs.

No longer new or particularly original, there are indeed Web pages where consumers can create a concrete product depending on their preferences. To illustrate this, let us take the example of automobile websites, where one can design the car of one's dreams. By selecting a model, one can select the engine power, color and main optional extras, thus interactively "designing" one' car.

Some city websites allow you to create interactive leaflets[5]. Through selecting various elements such as where to go or stay, what to eat and what to see and do, consumers may build up their own collection of information and tips. Contents are stored so that users have them available whenever they desire. This information can then be printed as a travel guide. In this way, users design their own leaflet, their preferred tourist experiences and destination. On the whole, there is no need to pay a registration fee to use this service although sites like those for Los Angeles and Dublin for instance charge to use this service.

Recommender System

Another useful application for tourism destination websites is a *recommender system*. This tool attempts to present information items that are likely to be of interest to the user. In tourism, travel recommender systems are aimed at supporting the critical travel planning decision that the traveler will face before traveling or while on-the move (Fesenmaier, Werthner & Woeber, 2006). The main objective of a travel recommender system is to ease the information search process and even to persuade the traveler of the appropriateness of the proposed services (Mahmood, Ricci, Venturini & Höpken, 2008).

Traditional recommender systems support simple processes, i.e. they suggest products and tourist services, points of interest, events or activities through analyzing direct questions or the user online activity logs (Ricci & Werthner, 2002). Nevertheless, some systems develop *recommendation strategies*. These systems support a dialogue where one of a set of available actions can be selected in each stage. For instance, Tripadvisor™ (Ricci, Cavada, Mirzadeh & Venturini, 2006) has used a recommendation system that could be described as the following: When dealing with a hotel traveler's preferences, it could employ two strategies: a) ask the user in detail about his or her preferences, and use this information to

Figure 6. Miami's interactive leaflets (Source: Retrieved February 20, 2008 from http://www.gmcvb.com)

Miami: www.gmcvb.com
Miami Itinerary saves all visitors' options selected and one can delete these at any stage. The entire itinerary can later be printed or sent to email addresses.

extract a small product subset or b) propose a set of products to the user, and exploit the user feed to refine future recommendations.

Nevertheless, researchers are developing complex recommender systems such the *Adaptive Recommender Systems* (Mahmood et al., 2008). These systems adapt its suggestions according to the measurable answers of the user that is, in fact, part of the recommendation agents in continuous communications with the environment (the user and the information system). The recommendation agents use optimal policies (rather than optimal strategies) in the context of Reinforced Learning to learn from the answers and give the best recommendations by the result of applying iteration algorithms.

Dynamic Packaging

According to the European Travel Commission, "*the winners of the future will be those destinations and suppliers that develop user-friendly websites allowing for direct bookings and dynamic packaging*" (ETC, 2004, p. 37) . Nevertheless, Dynamic Packaging (DP) has not yet been extended into city tourism websites. DP is a user centered, cheaper and more flexible way of assembling and booking a personalized holiday, using the web and associated application of technology (Markus & Lassing, 2008). A cited definition is given by Kabbaj (2003), according to him DP means Dynamically (i.e., in real-time) Putting together—and pricing—a package of several major travel components, e.g., flight details and stopovers, hotel nights, car rental etc., from heterogeneous suppliers and information sources or back-end reservation services.

To build a DP system travel components must be bundled at the single contact point of sale and in dynamic response of the customer. The management of the dynamic packaging and pricing process is enabled by the appliance of business rules which determine the price (discounts) and combination options of assembled travel components. Finally, the service provider assumes *the legal responsibilities*, (for instance, the guarantees for the package).

As a real-time or instant packaging service, DP should not be confused with non-instant and non-automated packaging services that are assembled by traditional tour operators in a response to the customers' requirements. Also, DP should not be confused with pre-packaged travel arrangements, which are not dynamically interrelated and therefore allow only the selection of different fixed packages. Figure 8 illustrates the distinction between traditional component selling, pre-packaged travel arrangements and DP.

Nowadays some internet tourism portals offer packaging services. In these cases, marketers have adopted more commercial descriptions for this technology: "Tailor-make and save!" (Expedia™), "Tailor Made Holidays", "Flight + Hotel" (LastMinute™ and ebookers™). From a destination point of view DP could be beneficial for marketing and customer service. A whole array of regional tourism services might be integrated and prepared for packaging and booking. Nevertheless, in order to exploit DP potential enormous organizational and technological challenges must be taken. Only the biggest and best-managed cities will be able to afford this challenge (Markus & Lassing, 2008).

Price in City Tourist Web Sites

Price is the second component of the marketing mix and one of the most difficult to handle as regards strategy to be followed by the city marketing authorities. The vast majority of companies are able to establish a clear price strategy (market prices, cost prices plus a profit margin or promotion prices, for instance). Even though the city tourist managers do not have the power to set prices for the services and products in this sector of its influence areas, city authorities must establish clear goals and price orientation of the main tourist products and services cities offer (Buhalis, 2000).

Nevertheless, this task may involve serious difficulties:

1. Authorities may select an adequate price strategy, but this does not necessarily match

Figure 7. Dynamic Packaging in relation to traditional pre-packaged travel services (Source: EC, (2006, p. 110)

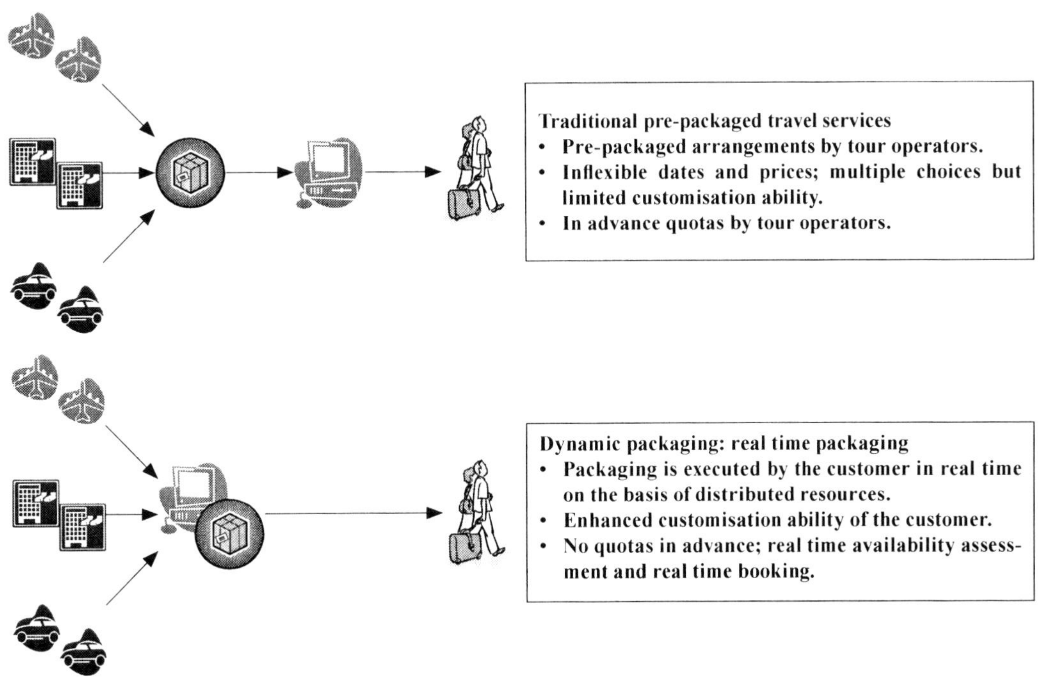

the particular interests of the influential companies in this sector.

2. Prices in tourism are very much influenced by national politics and international conditions and these in turn may hinder any proper price strategy. Particularly, any emerging city wanting to develop a low cost price policy may face currency devaluation by neighboring countries, which may cause weakened price competitiveness. Economic, political, climatic and international crises have historically influenced the tourism sector in particular, in the same way that they have had an effect on the failure of tourism price strategies.
3. Seasonal variation of demand directly influences the price of products and services. Even though this feature is more closely linked to other destination types (sun and beach), it also has a clear impact on big cities. Naturally, prices have to be lowered in those seasons when there is less inflow.
4. Finally, it is a well known fact that tour operators play a role in determining the price consumers pay for products locally. Tour operators in Europe exercise a bargaining and coercive power due to the large volume of tourists they represent and reduce the price.

As we can see, pricing the destination and all the individual elements of the local tourism product is very complex. CVBs or DMOs can control pricing through regulation, advice and partnerships with the private sector. However, these organizations often provide guidelines by suggesting minimum prices to protect small suppliers from competition and maximum prices to protect consumers from overcharging (Buhalis, 2000). For example, several cities set maximum prices for transportation and taxi services and this information is usually on their websites[6].

Although developing pricing policies is extremely difficult, city managers should formulate partnerships with major layers and teach them the importance of maintaining a fairly standardized pricing structure and policy. In this way, city websites could help to control the prices charged by the private sector, at least via the official site- and offer it as a reliable way of acquiring products and services.

Cities can only charge premium prices if they offer their visitors a unique experience. Venice for example charges substantially higher prices due to its exclusive characteristics. But, according to Buhalis (2000) consumers are increasingly unimpressed by tourism facilities and products, as they have traveled extensively and have acquired a wealth of experiences. Global competition may well affect cities which disregards their customer's perception of prices. Tourists now have more information and facilities to travel than ever before and can choose other destinations if they consider prices too high or feel they are not getting value for money.

Although cities can not control prices of the private sector, it is true that they offer several offers through their websites. Most of the websites have specially designed spaces for these purposes and are a key attraction for visitors of these sites. These offers usually concern accommodation but extend to other services. Figure 8 for example shows the search section for offers on the website for Las Vegas. Selling special tourist cards for discounts is now also highly popular. (See Figure 9). These are more linked to promotional strategies than price strategies but attract visitors who are interested in getting good prices and discounts.

PLACE. CVBS AS INTERMEDIARIES

The third factor of the Marketing mix, place, refers to the location where the product is distributed and commercialized. This is a key factor in the tourism sector when compared to other products, due to the existing distance between consumers and tourist products. (P. Kotler, Bowen & Makens,

Figure 8. Las Vegas offers search (Source: Retrieved February 26, 2008, from http://www.visitlasvegas.com)

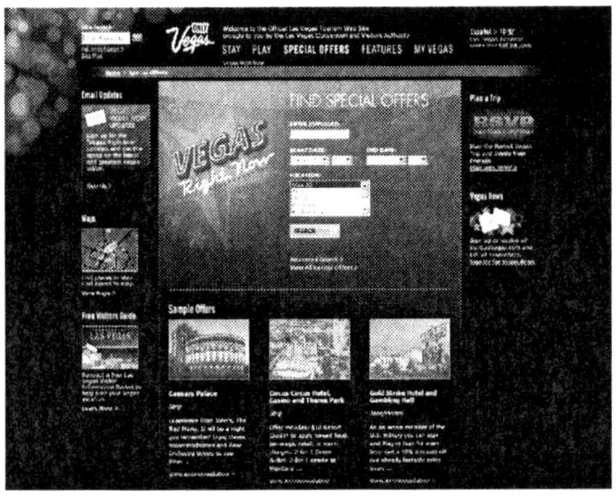

Las Vegas: www.visitlasvegas.com
This Las Vegas website includes a search engine for all the offers located on its Website. Most of them are related to accommodation but it is also used for finding tourist packages or discounts for golf courses.

Figure 9. Two city cards discounts included (Source: Retrieved February 28, 2008)

Paris: www.parisinfo.com
Paris City Passport offers discounts for 33 sites selected by the Paris Convention and Visitors Bureau (discounts from -10% to 50% in general. This includes museums, attractions, excursions, shopping and evenings out.

Barcelona: www.barcelonaturisme.com
City card featuring free travel on public transport, discounts and free offers at museums, cultural venues, theatres, leisure facilities, night-clubs, shops, restaurants and entertainment, other services and a variety of means of transport.

1996) Because tourist products are created and are expected by customers far away from points of acquisition, an adequate distribution and commercialization channel is fundamental. As can be imagined, what is mainly distributed is information about destinations, its services, availability and booking of their services and products (Poon, 1993).

Traditionally, the tourism sector has been very much linked to intermediaries such as travel agencies and tour operators in order to distribute and commercialize products, formally giving relevant information about the destination to wherever consumers were. These traditional intermediaries, having studied different formulae and economic structures, create the channels to promote commercialization and economic exchange. This exchange, however, is not immediate, requiring a constant flow and exchange of information in the form of fulfilling needs, information about the features of the destinations, availability of services, booking, payment and confirmation possibilities (Poon, 1993) (Figure 10).

Distribution or marketing channels are defined as a set of interdependent organizations involved in the process of making a product or service available for use or consumption (Buhalis, 2000). The traditional distribution scheme of the sector is shown in Figure 10, where information flows are detailed. It also shows the exchange of information and data that exists between companies offering products and services (flight companies and hotels) and their intermediaries. Information about flights, availability, tariffs, schedule variations, and routes, together with purchase confirmations and sales commissions go from the companies themselves to the travel agencies. In the opposite direction, reservations and payments go from the travel agencies to tour operators and from tour operators to flight companies. Similarly, hotels send information about vacancies, localization and tariffs, together with commissions and booking confirmations. In turn, travel agencies and tour operators send reservations and incomes to the hotels.

Within this structure, it is clear that both effective communication and an appropriate transmission of information are essential in the tourism industry. The World Tourist Organization (WTO, 1999) states that the communication and information systems are vital because the tourism products do not exist at the time they are acquired. It means that, when a trip or hotel room is booked, they are still mere information, i.e. not tangible products. Travel agencies and tour operators first, and computer-mediated distribution systems later (CRS and GDS), have been traditional intermediaries in charge of processing information and making it available for consumers.

Since CRS and later GDS have arrived to the tourist distribution scenario, the possibilities of integrating this technology in Destination Marketing Organization activities have been commented on. In the early 90s, Archdale (1993) noted that the majority of what he called Public Tourist Offices neither operate on a commercial basis nor carry out reservation activities; but where this happened it was almost always at local level and functions principally as an after-sales service. Computerized systems were mostly developed as a support for their product database, holding information on the tourism product such as hotels, event and tourist attractions in the area. These databases were regarded as an internal management function and were not considered as a marketing resource (Archdale, 1993).

Figure 10. Traditional information flows in the tourism sector (Source: Poon, 1993, p. 155)

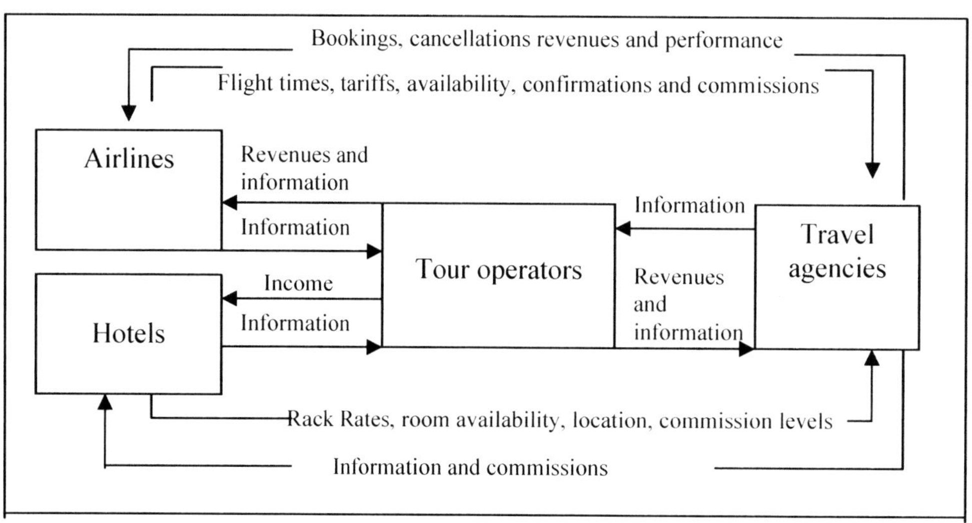

The development of destination databases formed by "product databases" and "customer databases" would have huge potential for marketing purposes and to finally develop a reservation system. Moreover, according to Archdale (1993) it would have others positive targets such as it:

- Allows the DMOs concerned to develop a range of database marketing and other promotional activities either alone or with other commercial partners;
- Allows the DMOs to demonstrate that it is actively playing a role in what is still a relatively new field and to quantify the effectiveness of its work.
- Allows the DMOs to carry out the role of intermediary on behalf of its local tourism business, a particularly important issue in relation to what are know as small or medium sized enterprises.

Nevertheless, the huge investment needed to build an e-commerce platform in the early years of CRS meant that DMOs were highly reticent. Also, the argument that DMOs should restrict themselves to so-called traditional marketing activities has been always present against the development of this service. But ICT prices have decreased constantly whilst improvement and connectivity have increased rapidly; and as Archdale (1993) mentioned, Technology can help to quality control the complexity of a destination amalgam. Moreover, if the public sector abstained from developing initiatives, then parallel networks would be likely to develop. These could then threaten the current rationale and existence of the DMO.

The Internet means a significant change in the traditional structure that we have been examining. Now we count on Communication and Information Technology, accessible for all organizations, where all the information related to the tourism sector flows together. Under these conditions, the city marketing organizations have made use of this technology to launch themselves as a relevant commercial agent.

Not so long ago, various authors started to forecast that destination organizations could be one of the agents promoting a new distribution framework thanks to ICT. In particular, Buhalis (1998) claimed that destination organizations would be the main developers of regional systems, whose presence, image and direct reservations would be promoted with the help of new Information and Communication Technologies.

Moreover, a great deal of arguments for and against disintermediation were given at this time, when researchers realized the opportunity for suppliers to avoid intermediaries (Buhalis & Licata, 2002; Worthier & Ricci, 2004). Eventually, a re-intermediation with some new agents occurred. Then, the arrival of the Internet and e-commerce was followed by the beginning of a new tourism e-mediaries. New Web based travel agencies (Expedia), off-line travel agencies that invested in their new web sites (Thomas Cook), internet (Yahoo®) and vertical portals (www.ski.com) were good examples of this activity (Buhalis & Licata, 2002). The tourist suppliers firms also developed e-commerce on their web pages. For instance flight companies (www.brithisairways.com), international hotel brands (www.marriott.com) and rent a cars (www.avis.com) started to offer this service (Buhalis & Licata, 2002).

The new distribution framework can be schemed as in Figure 12. Now tourists could choose between looking directly for tourist suppliers, asking traditional or new intermediaries, or visiting DMO websites. According to Werther and Ricci (2004) it is the needs of consumers are changing, they are increasingly less loyal, take more frequent vacations of shorter duration and take less time between choosing and consuming a tourist product.

It is true DMOs were intermediating before through call centers and tourist offices; But now they have the opportunity to do it to a far larger extent whilst having more benefits such as controlling its position in the marketplace; and increasing its market by getting closer to the customer and providing greater customer satisfaction (Hussmann & Baker, 1996).

The most developed CVBs have built profitable e-commerce platforms on their websites. Two different systems could be identified. The first one is completely managed by city authorities. In those cases, the CVB employ their own resources or use outsourcing (which they control). In the second case, city websites offer booking possibilities but the final purchase is made on the supplier platform. Barcelona[7] official tourist website can be cited as an example of the first group while Paris[8] is part of the second group.

City websites, in general terms, were started by commercializing hotel vacancies. However, with a later development of websites, new products and services have been included. To illustrate this, it is worth mentioning the website of the city of Barcelona, which offered (Summer 2007), the possibility to book and pay for a great variety of products and services. In particular, in section "BCN SHOP", there was the possibility to make reservations of hotel rooms, restaurants, tickets for shows and sports events, maps and tourist guides, tours to areas of tourist interest nearby, tickets for funfairs and even guided scooter trips.

PROMOTION AND COMMUNICATION IN A NEW MEDIA

Promoting a destination implies the development of communication channels with clientele and other stakeholders to increase awareness and persuade them to purchase products (Buhalis, 2000). Designing an effective promotional mix for a city is difficult because of the diversity of tourism suppliers and the spread of consumers throughout the world. Traditionally, CVBs lead promotional campaigns, whilst suppliers participate and contribute. To develop these campaigns, a wide range of techniques are used. Some of them are considered *"above the line"* (TV, radio and press advertisements as well as poster campaigns), while others are called *"below the line"* (i.e. participating in Tourism and travel fairs and producing and distributing brochures) (Buhalis, 2000).

Internet is a communication channel which has its own communication and promotional features, making it different to other techniques. Traditionally, there has been a difference between personal media (such as e-mail, phone or face

Figure 11. Internet basic tourist intermediation framework (Source: Adapted from Worthier & Klein, 1999; and Buhalis, 2003)

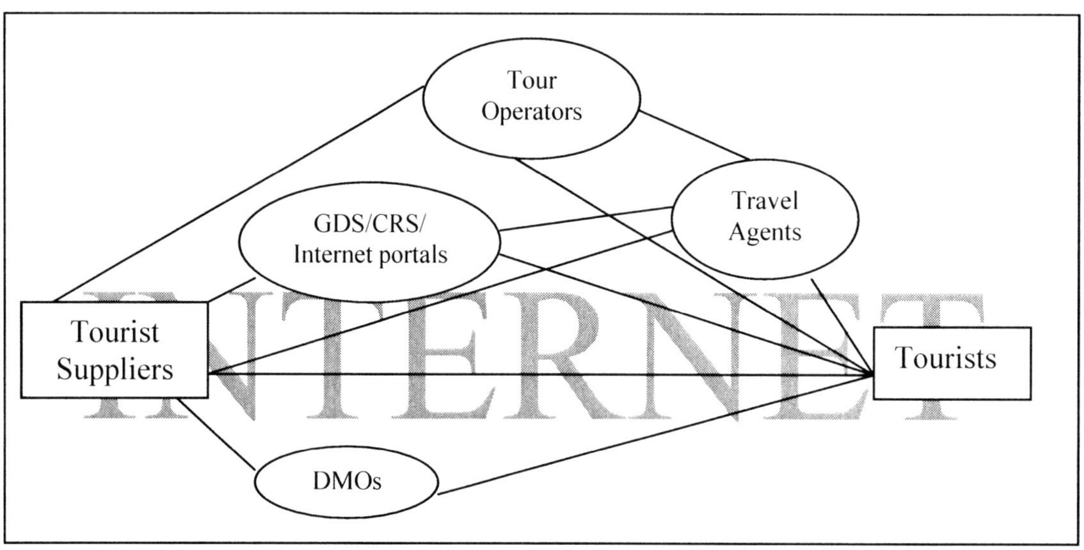

–to-face meetings) as opposed to mass media. The former focus more on one being able to adapt to one's interlocutor but the number of participants in this type of communication is limited.

On the contrary, traditional mass media such as the press, radio, television and cinema, allows the transmission of information to larger groups. Nevertheless, given the large size of the target audience, this leads inevitably to a communication problem: messages cannot be personalized. Moreover, it can be very complicated to measure the repercussions of promotional campaigns in the mass media, so that specific research studies on the after effects of sales or changes in consumer attitudes are vital.

In this sense, there is a twofold situation. As far as traditional means of communication are concerned, there is a high interactivity (with fewer participants) and mass media, (greater numbers but far more impersonal). On the other hand there is the Internet, a communication channel allowing interpersonal communication mainly via e-mail or online conversation, such as chat[9]. However, its possibilities go way beyond these frontiers, as this allows communication between more than two people thanks to mailing lists and the well-known *World Wide Web*. We return to the example of the official tourist websites.

Internet and Hypermedia Environments (Hoffman & Novak, 1996) allow communication between organizations and individuals. Organizations expose their own attributes and products, if they are companies, or the city as a product, if they are city marketing organizations. These spaces can then be open to interactivity with users, tourists and suppliers (Figure 12).

Figure 12. Communication processes on the Internet and tourism agents (Source: Adapted from Hoffman and Novak, 1996, p. 53)

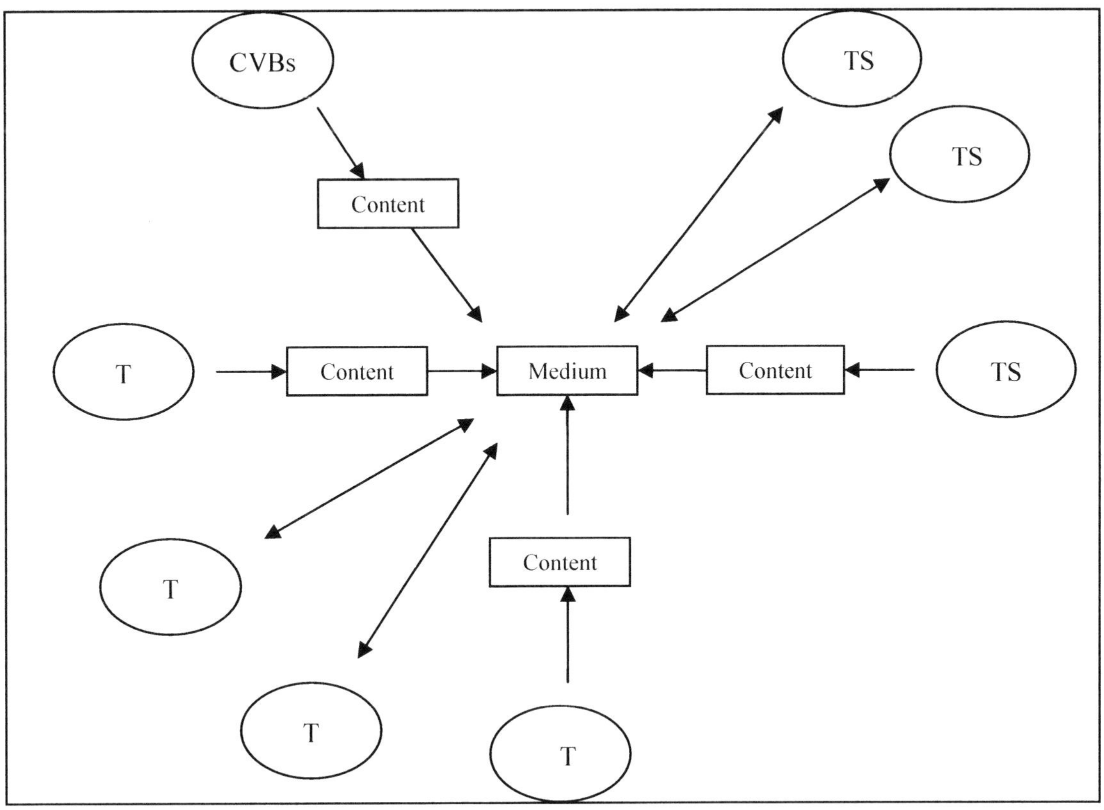

T: Tourist. TS: Tourist Supplier.

Moreover, organizations can make use of various tools to gain a greater insight into just how successful (or not) their Web pages are, to find out how long the site is visited and the origin of the query. Thanks to these tools, organizations are able to alter or update the contents to subsequently make them more interesting and user friendly.

Hoffman and Novak (1996) describe the Internet form of communication as a multi communication model. They concretely define hypermedia computer-mediated environments as a dynamic distributed network, potentially global in scope, together with associated hardware and software for accessing the network, which allows consumers and firms to 1) provide an interactively hypermedia content, and 2) communicate via this medium. The World Wide Web is the most relevant development of being able to reach far and wide on the Net. The Web allows extensive access to multimedia contents, as well as the possibility of sending and receiving messages, providing each user has the necessary hardware, software and Internet connection. The hypertext structure means users have freedom of movement.

Gretzel, Yuan and Fesenmaier (2000) agree that as compared with traditional media, the Internet combines and integrates the following functional properties as a form of advertising media:

- Information representation,
- Collaboration
- Communication
- Interactivity
- Transactions

They claim that this flexibility makes the Web rich and appealing but also very complex and difficult to deal with. Installing communication tools is easy; handling the increased volume of information is not. It means that collaborative efforts must be made and a possible consequence would be the building of virtual communities or virtual organizations. It will require flexible and open structures, a change in the organizational mindset and new business processes (Grenier & Metes, 1995). Therefore, the real opportunity of the Web is to rethink the business model in terms of delivering value to the customer and in building relationships with clients, suppliers and other business partners (Hagel, 1999). Hagel (1999) calls this new business model "*collaboration marketing*".

The Web offers powerful resources that will lead to deeper relationships and greater personalization of contents, products and services. Using these capabilities will enhance relationships with customers and reinforce the participation of suppliers. The goal is to use *Interactive Mark*eting using information from the customer rather than about the customer (Day, 1998).

Moreover, five factors must be taken in account to develop successful interactive marketing on the Web (Gretzel et al., 2000) (Figure 13, Table 5):

1. Attracting users,
2. Engaging users' interest and participation,
3. Retaining users and ensuring they return,
4. Learning about user preferences and
5. Relating back to user to provide customized interactions.

Some studies have analyzed the usefulness and success factors of CVBs websites. For instance, Wang and Fesenmaier (2006) agree that that motivating visuals and trip information functionality are significant predictors of Web site usefulness. Moreover, the usefulness of Web site is a significant predictor of intent to travel to the destination, whereas previous visits to the destination or its Website have significant but small negative influence on intentions to travel.

Finally, Kaplanidou and Vogt (2006) confirm that a successful Web marketing strategy requires the integration and coordination of Website features, Website promotional techniques, and Customer Relationship Management (CRM) programs. However, the American CVBs are still

Figure 13. Interactive marketing framework (Source: Adapted from Gretzel, Yuan and Fesenmaier, 2000, p. 148)

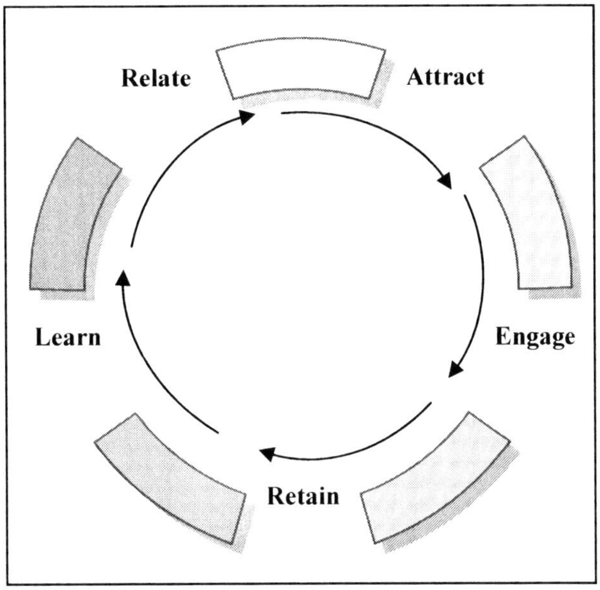

Table 5. Explanation of Figure 14 (Source: Gretzel, Yuan and Fesenmaier, 2000, p. 148; and Parsons, Zeisser and Waitman, 1998)

Activity	What	How
Attract	Attract consumer to the application	Audience creation. Mnemonic Branding. "Piggy-back" advertising.
Engage	Generate interest and participation	Intuitive interface or navigation. Interactive content. User-generated content.
Retain	Make sure customers come back	Dynamic content. Transaction capabilities. Online communities.
Learn	Learn about consumers' preferences	Information capture. Continuous preference learning.
Relate	Customize interaction and value delivery	Personalized/customized communication and products/services. Real time interactions. Link to core business.

in the preliminary stages of technology adoption, incorporating into their Websites relatively simple brochure materials and other information-oriented features related to activities, attractions, and accommodation but offering limited e-commerce and dialogue activities through their Web sites. This lack of development appears to be a manifestation of the CVBs' inability to effectively adopt and manage IT necessary to support more sophisticated operations and business processes.

EMERGING WEB 2.0 RESOURCES IN CITY WEBSITES: TOURISTS GENERATED CONTENT AND BLOGS

Most city Web sites are still unaware of the unique possibilities of interactive media and continue to broadcast information rather than letting the consumer interact with the web site and create content themselves. But the Internet provides new ways for individuals to learn about tourism destinations and their products and services directly from other consumers. This has created a new method of communication that is similar to word-of-mouth and empowers consumers. Travelers can e-mail one another, post comments and feedbacks, publish online blogs, and form communities on the Internet.

It is believed that community will gain more importance as the Internet becomes even more pervasive in the new global economy, and it will become the dominant organizing metaphor in the next decade (Y. Wang, Yu & Fesemaier, 2002). The most new notable developments in the Internet applications have been in the area of User Generated Content (UGC) and peer-to-peer applications, collectively known as Web 2.0. More frequently visited websites of this new trend include MySpace.com™, YouTube.com™ and Flickr.com™. What is different about these types of sites is that the content is relatively unstructured and content development is not managed by the host organization. Instead, content is loaded directly onto the websites by users, with varying (usually minimal) levels of moderation.

As one type of digital word-of-mouth, User Generated Content (UGC) poses new possibilities and challenges for tourism marketers from CVBs. They are widely considered to be more persuasive than advertising, yet many destinations have been slow to recognize the benefits, preferring instead to maintain a tight control over their marketing message. But with the growing number of peer-to-peer tourist sites such as RealTravel™, Travelblog™, Travelpod™, or Tripadvisor™, it is fast becoming impossible to ignore UGC.

This kind of sites allows consumers to exchange recommendations and opinions about various destinations and tourism products. In some cases, UGC takes the form of diaries or 'blogs' about travel experiences, while others are relatively short entries rating a particular hotel or tour.

Some destinations have incorporated these resources into their websites. In this way, they have put their Websites in the forefront of the new trends of the Net. For instance, Visit London[10] has teamed up with TripAdvisor™ to offer a wide range of user-generated reviews. According to this website, Visit London is the first tourist board in the world to provide user-generated reviews on its website.

There are more examples of UGC from bigger destinations such as actual countries and regions. On these sites, tourists not only offer reviews of tourist services and products, they also post their experiences and photos (see the Websites below for examples in Sweden[11], Switzerland[12], England[13] and Arizona[14]).

Blogs are gaining more and more popularity. Blog is a shortened word originating from "web log" (Chow, 2005). The format consists of familiar, frequently updated, reverse-chronological entries on a single Web page (Blood, 2004). Audio and video blogging from mobile devices are also available (Baker & Green, 2005). In March 2007, according to Technorati[15], there were more than 70 million weblogs, and approximately 120,000 new weblogs were being created worldwide each day, making about 1.4 blogs created every second of every day.

Kumar et al. (2004) demonstrated that the blogspace consists of at least three layers: the individual bloggers who are defined by their demographic characteristics, a middle layer of pairs of bloggers based on friendship, and a third higher layer of interest groups and virtual communities with geographic or demographic correlations. Blogspace can also create virtual relationships and communities, and its influence moves far beyond the readers of the blogs, creating a new type of reality through search engines online.

Travel blogs express the experiences of the visitor at a specific destination, and given the ever-increasing number of Internet users, digital word-of-mouth communication will inevitably become more preferable as travel information sources (Crotts, 1999). For this reason, tourism marketers need to understand blogs as a new technological phenomenon with implications for marketing and promotion of a destination.

Again, some cities have incorporated blogs as a promotional tool. For instance, the official site for the Netherlands's Tourist Board and Convention develop eleven personal blogs to promote capital and major cities (and one region) of this country (Figure 14). Most city websites have special links to these blogs from their homepage. In these cases, the positive message seems to be controlled by the authorities but a personal experience is still shown.

There is a great deal of new researches about blogs and tourism destination promotion. Schmallegger and Carson (2008) noted that the most immediately useful application of travel blogs for destination marketing organizations and tourism businesses is for monitoring the attitudes of travelers. This was also the intention of Pan, MacLaurin and Crotts (2007). They used semantic network analysis and content analysis methods to find a total of 177 positive and negative sentences articulated about Charleston (USA) in the three travel blogs analyzed (www.travelblog.org , www.travelpod.com , and www.travelpost.com), 134 being positive and 43 negative. Thus, three out of four comments made about the city of Charleston were positive (75.5%). They also conclude that "*travel blogs are an inexpensive means to gather rich, authentic, and unsolicited customer feedback. Information technology advances and increasingly large numbers of travel blogs facilitate travel blog monitoring as a cost-effective method for destination marketers to assess their service quality and improve travelers' overall experiences*" (Pan et al., 2007, p. 99).

CONCLUSION

According to Buhalis, innovative marketing led by research and using new Technology, will be the only way to manage and market competitive destination in the future for the benefits of stakeholders (Buhalis, 2000). As we have seen, the Internet and the Web could be used as a clear

Figure 14. City blogs developed by the Netherlands's Tourist Board and Convention (Source: Retrieved February 21, 2008, from http://citybreaks.holland.com; ©2007 Nederlands Bureau voor Toerisme. Used with permission.)

resource to establish and develop a marketing and mix strategy.

A sector such as tourism, intensive in information, and ICT such as the Internet were bound to be linked from the very beginning. We could metaphorically say that both have always been destined to understand each other, even when their origins and rationale are very much apart.

CVBs from big cities could not allow themselves to fall behind in this respect, so they have incorporated the Internet into their marketing strategies. However, for a competitive destination it is not enough to have a website only with information. The high possibilities of the applications and the increased participation of users in an interactive scenario give destinations the opportunity to develop rich and successful Websites.

Thanks to this interactivity we can get to know our visitors (tourists and website visitors) better. The way visitors use the Website, online surveys or User Generated Content and blog analysis will allow CVB to understand better what the tourists and visitors are like and what they want. This knowledge will be very useful to organize the products and services offered via the Website as satisfying the target market is naturally the main objective.

Moreover, destinations have the opportunity to establish themselves and their websites as a new channel of commercialization for the stakeholder of their areas. Some CVBs appear now as new travel agents with an extended offer not only of accommodation but also for restaurant bookings, visits and events.

The content of the Web pages is extremely important for websites of tourism destinations because it directly influences the perceived image of the destination. Nevertheless, the most important challenge for CVBs is to develop a beneficial relationship with their tourist and website visitors. Internet users are not only looking for information but also seeking a different kind of experience and expect interaction. Successful CVBs will be those who embrace change and build effective bridges to this end.

REFERENCES

Archdale, G. (1993). Computer reservation system and public tourist offices. *Tourism Management, 14*, 3-14.

Baker, S., & Green, H. (2005). Blogs will change your business. *Business Week, 3931,* 56-67.

Blood, R. (2004). How blogging software reshapes the online community. *Communications of the ACM, 47(12)*, 53-55.

Buhalis, D. (1998). Strategic use of information technologies in the tourism industry. *Tourism Management, 19,* 409-421.

Buhalis, D. (2000). Marketing the competitive destination of the future. *Tourism Management, 21*, 97-116.

Buhalis, D. (2003). *E-Tourism, information technology for strategic tourism management.* Essex, UK: Prentice Hall.

Buhalis, D., & Cooper, C. (1998). Competition or cooperation? Small and medium sized tourism enterprise at the destination. In E. Laws, B. Faulkner & G. Moscardo (Eds.), *Embracing and managing change in tourism* (pp. 307-323). London: Routledge.

Buhalis, D., & Licata, M. C. (2002). The future E-Tourism intermediaries. *Tourism Management, 23*, 207-220.

Bull, A. (1991). *The Economic of travel and tourism.* Melbourne: Pitman.

Butler, R. W. (1980). The concept of a tourist area cycle of evolution. *Canadian Geographer, 24*, 5-12.

Chow, J. (2005). The new frontier. *National Post Business Magazine,* 40.

Collins, C., & Buhalis, D. (2003). Implementation of a new strategic framework for survival of destination management systems In A. Frew,

P. O'Connor & M. Hitz (Eds.), *Information and communications technologies in tourism* (pp. 202-211). Wien: Springer-Verlag.

Cooper, C., Fletcher, J., Gilbert, D., Shepherd, R., & Wanhill, S. (1998). *Tourism: Principles and practices.* England: Addison-Wesley, Longman.

Couper, M. P., Baker, R. P., Clark, C. Z. F., Martin, J., Nicholls, W. L., & O'Reilly, J. M. (1998). *Computer assisted questionnaire information collection*: John Wiley & Sons Inc.

Crotts, J. (1999). Consumer decision making and prepurchase information search. In Y. Mansfield & A. Pizam (Eds.), *Consumer behavior in travel and tourism* (pp. 149-168). Binghamton, NY: Haworth Press.

Day, G. S. (1998). Organizing for interactivity. *Journal of Interactive Marketing, 12(1)*, 47-53.

DCCED. (2007). Alaska visitor statistics program (Fall/Winter 2006-07): Department of Commerce Community and Economic Development.

European Commission (EC) (2006). *ICT and E-Business in the tourism industry. ICT adoption and e-business activity in 2006. European Commission.* Retrieved February 20, 2008, from http://www.ebusiness-watch.org/studies/sectors/tourism/documents/Tourism_2006.pdf

European Travel Commission (ETC) (2004). *European tourism insight 2004. European Travel Commission.* Retrieved February 20, 2008, from http://www.etc-corporate.org/resources/uploads/ETC_EuropeanTourismInsights_2004.pdf

Fesenmaier, D. R. (2007). Introduction: Challenging destination promotion. *Journal of Travel Research, 46*, 3-4.

Fesenmaier, D. R., Werthner, H., & Woeber, K. (2006). *Destination recommendation system: Behavioral foundation and applications.* Oxford: CABI Publishing.

Gartrell, R. B. (Ed.). (1988). *Destination marketing for convention and visitor bureaus.* Dubuque: Kendall/Hunt.

Gilbert, D. (1990). Strategic marketing planning for national tourism. *The Tourist Review, 1*, 18-27.

Grenier, R., & Metes, G. (1995). *Going virtual.* Upper Saddle River, NJ: Prentice Hall.

Gretzel, U., Fesenmaier, D. R., Formica, S., & O'Leary, J. T. (2006). Searching for the future: Challenges faced by destination marketing organizations. *Journal of Travel Research, 45(2)*, 116-126.

Gretzel, U., Yuan, Y., & Fesenmaier, D. R. (2000). Preparing for the new economy: Advertising strategies and change in destination marketing organizations. *Journal of Travel Research, 39*, 146-156.

Hagel, J. (1999). Net gain: Expanding markets through virtual communities. *Journal of Interactive Marketing, 13(1)*, 55-65.

Hall, C. M. (2000). *Tourism planning: Policies, processes and relationship.* Harlow: Prentice Hall.

Heath, E., & Wall, G. (1992). *Marketing tourism destinations, a strategic planning approach.* New York, NY: John Wiley and Sons.

Hoffman, D. L., & Novak, T. (1996). Marketing in hypermedia computer-mediated environments: Conceptual foundations. *Journal of Marketing, 60*, 50-68.

Kabbaj, M. Y. (2003). *Strategic and policy prospects for semantic web services adoption in US online travel industry.* Doctoral dissertation. Retrieved December 20, 2008, from http://ebusiness.mit.edu/bgrosof/paps/kabbajmasters-thesis-travel+sws.pdf

Kaplanidou, K., & Vogt, C. (2006). A structural analysis of destination travel intentions as a function of web site features, *Journal of Travel Research, 45(2)*, 204-216.

Knauth, B. (2006). *Tourism and the Internet in the European Union*. Luxembourg: European Communities.

Kotler. (1997). *Marketing management. Analysis, planning, implementation and control.* New Jersey: Prentice Hall.

Kotler, P., Bowen, J., & Makens, J. (1996). *Marketing for hospitality and tourism.* New Jersey: Prentice Hall

Kumar, R., Novak, J., Raghavan, P., & Tomkins, A. (2004). Structure and evolution of blogspace. *Communications of the ACM, 47(12)*, 35-39.

Mahmood, T., Ricci, F., Venturini, A., & Höpken, W. (2008). Adaptative recommender systems for travel planning. In P. O'Connors, W. Höpken & U. Gretzel (Eds.), *Information and communication technologies in tourism* (pp. 1-11). Innsbruck, Austria: Springer Verlag.

Markus, M., & Lassing, M. (2008). Some critical remarks on dynamic packaging from the perspective of SMEs and small tourism destination. . In P. O'Connors, W. Höpken & U. Gretzel (Eds.), *Information and communication technologies in tourism* Innsbruck, Austria: Springer Wien.

Pan, B., MacLaurin, T., & Crotts, J. C. (2007). Travel blogs and the implications for destination marketing. *Journal of Travel Research, 46*, 35-45.

Parsons, A., Zeisser, M., & Waitman, R. (1998). Organizing today for the digital marketing of tomorrow, *Journal of Interactive Marketing, 12(1)*, 31-46.

Poon, A. (1993). *Tourism, technology and competitive strategies*. Oxon, United Kingdom: Cab International.

Porter, M. (1980). *Competitive strategy: Techniques for analyzing industries and competitors.* New York, NY: Free Press.

Porter, M. (1985a). *Competitive advantage*. New York, NY: Free Press.

Porter, M. (1985b, Winter). Technology and competitive advantage. *The Journal of Business Strategy,* 60-70.

Reime, M., & Hawkins, C. (1979). Tourism development: A model for growth. *Hotel and Restaurant Administration Quarterly, 20*, 67-74.

Ricci, F., Cavada, D., Mirzadeh, N., & Venturini, A. (2006). *Case based travel recommendations.* Paper presented at the Destination Recommendation Systems: Behavioral Foundations and Applications, Oxford.

Ricci, F., & Werthner, H. (2002). Cased-based querying for travel planning recommendation. *Information Technology and Tourism, 4(3/4)*, 215-226.

Schmallegger, D., & Carson, D. (2008). Blogs in tourism: Changing approaches to information exchange. *Journal of Vacation Marketing 14(2)*, 99-110.

Strauss, J., & Frost, R. (1999). *Marketing on the Internet. Principles of online marketing.* New Jersey: Prentice Hall.

Sussmann, S., & Baker, M. (1996). Responding to the electronic marketplace: Lessons from destination management systems. *International Journal of Hospitality Management, 15*, 99-112.

Taylor, G. D. (1980). How to match plan with demand: A matrix for marketing. *International Journal of Tourism Management, 1*, 56-60.

Tjostheim, I., & Tronvoll, B. (2002). *The Internet and city tourist: A study of preferences for information sources in travel planning.* Paper presented at the Cities Tourism 2002, Vienna, Austria.

Wang, Y., & Fesenmaier, D. R. (2004). Modeling participation in an online travel community. *Journal of Travel Research, 42*, 261-270.

Wang, Y., & Fesenmaier, D. R. (2006). Identifying the success factors of web-based marketing strategy: An investigation of convention and visitors bureaus in the United States. *Journal of Travel Research, 27*, 326-341.

Wang, Y., Yu, Q., & Fesemaier, D. R. (2002). Defining the virtual tourist community: Implications for tourism marketing. *Tourism Management, 23(4)*, 407-417.

Werthner, H., & Klein, S. (1999). *Information technology and tourism, a challenging relationship.* Vienna: Springer Verlag.

Werthner, H., & Ricci, F. (2004). E-Commerce and tourism. *Communications of the ACM, 47(12)*, 101-115.

Wöber, K. W. (Ed.). (2002). *City tourism 2002.* Vienna: Springer Wien.

Wright, K. B. (2005). Researching Internet-based populations: Advantages and disadvantages of online survey research, online questionnaire authoring software packages, and web survey services. [Electronic Version]. *Journal of Computer-Mediated Communication, 10(3).* Retrieved from http://jcmc.indiana.edu/vol10/issue3/wright.html

WTO. (1999). *Marketing tourism destinations online: Strategies for the information age.* Madrid: World Tourism Organization. .

WTO. (2001). *eBusiness for tourism: Practical guidelines for destinations and business.* Madrid: World Tourism Organization

Yuan, Y., Gretzel, U., & Fesenmaier, D. R. (2003). Internet technology use by American Convention and Visitors Bureaus. *Journal of Travel Research, 41(3)*, 240-255.

Yuan, Y., Gretzel, U., & Fesenmaier, D. R. (2006). The role of information technology use in American Convention and Visitor Bureaus. *Tourism Management, 27*, 326-341.

ENDNOTES

1 Arrivals in all paid forms of accommodation establishments in city area only. Market: Total foreign and domestic. Source: Retrieved August 08, 2007, from http://www.tourmis.info

2 Retrieved February 28, 2008, from http://jcmc.indiana.edu/vol10/issue3/wright.html

3 For more detailed information see Couper et al. (1998)

4 Retrieved August 19, 2007, from http://www.crmmetrix.fr/projects/etour/otp/fr/tscreen.asp

5 "My saved trip" in http://www.onlyinsanfrancisco.com (San Francisco), "Miami itinerary" in http://www.gmcvb.com (Miami), "My travel planner" in http://www.greaterlosangeles.com (Los Angeles) or "My Denver suitcase" in http://www.denver.org (Denver), and "My Dublin" (http://www.visitdublin.com) . Retrieved February 20, 2008.

6 Taxi Services information in London website: http://www.visitlondon.com/travel/getting_around/taxis

7 Retrieved February 20, 2008, from http://www.barcelonaturisme.com

8 Retrieved February 20, 2008, from http://www.parisifo.com. Paris website permits bookings in fastbooking.com platform.

9 Some Destination Websites offer a form of informative chat line with local operators from official tourist offices, e.g. Valencia Regional website: www.comunitatvalenciana.com/contacto/contacto.htm . Retrieved February 20, 2008.

10 Retrieved February 22, 2008, from http://www.visitlondon.com/people/tripadvisor

11 Retrieved February 22, 2008, from http://www.communityofsweden.com

12 Retrieved February 22, 2008, from http://www.myswitzerland.com

13 Retrieved February 22, 2008, http://www.enjoyengland.com

14 Retrieved February 22, 2008, http:// www.goseearizona.com

15 Retrieved February 21, 2008, http://www.sifry.com/alerts/archives/000493.html

APPENDIX: CITY TOURISM WEBSITES

Amsterdam	http://www.iamsterdam.com/
Atlantic City	http://www.atlanticcitynj.com
Barcelona	http://www.barcelonaturisme.com/
Berlin	http://www.berlin-tourist-information.de/
Budapest	http://www.budapestinfo.hu/
Buenos Aires	http://www.ohbuenosaires.com/
Chicago	http://www.choosechicago.com/
Copenhagen	http://www.visitcopenhagen.dk/
Denver	http://www.denver.org/
Detroit	http://visitdetroit.com/
Dublin	http://www.visitdublin.com/
Edinburgh	http://www.edinburgh.org/
Florence	http://www.firenzeturismo.it
Hamburg	http://www.hamburg-tourism.de
Helsinki	http://www.hel2.fi/tourism
Houston	http://www.visithoustontexas.com/
Istanbul	http://www.istanbul.com/
Las Vegas	http://www.visitlasvegas.com/
London	www.visitlondon.com/
Los Angeles	http://www.greaterlosangeles.com/
Lyon	http://www.lyon-france.com/
Manchester	http://www.visitmanchester.com/
Mexico D.F.	http://www.mexicocity.gob.mx/
Miami	http://www.gmcvb.com/
Munich	http://www.muenchen.de
New York	http://www.nycvisit.com/
Ottawa	http://www.ottawatourism.ca/
Paris	http://www.parisinfo.com/
Rome	http://www.romaturismo.it/
San Francisco	http://www.onlyinsanfrancisco.com/
Stockholm	http://www.stockholmtown.com/
Tokyo	http://www.tourism.metro.tokyo.jp/
Valencia	http://www.turisvalencia.es/
Zurich	http://www.zuerich.com

Chapter IX
E-Tourism Image: The Relevance of Networking for Web Sites Destination Marketing

Lluís Prats-Planagumà
Universitat de Girona, Spain

Raquel Camprubí
Universitat de Girona, Spain

ABSTRACT

The competitiveness of tourism destinations is a relevant issue for tourism studies, moreso, is a key element on the daily basis of tourism destinations. In this sense, the management of tourism destinations is essential to maintain competitive advantages. In this chapter tourism destination is considered as a relational network, where interaction and cooperation is needed among tourism agents, to achieve major levels of competitive advantage and a more effective destination management system. In addition, the perceptions of tourists are obtained from two main sources. The first one is the social construction of a tourism destination previous to the visit and the second one is obtained from the interaction between tourists and tourism destination agents during the visit. In this sense, the management of tourism destination to emit a homogenous and collective image is a factor that can reduce the gap if dissatisfaction from the previous and real tourist perception. The authors specifically discuss the importance of a common agreement of tourism agents on virtual tourism images projected through official Web sites, considering that the literature focused mainly in how to promote and sell destinations trough Internet but not in terms of exploiting a destination joint image. Finally, in order to analyze the integration of a tourism product and determine their consequences in tourism promotion an empirical research has been done, using the case of Girona's province. The main findings determine that, although interactions among tourism agents can improve destination competitiveness, little cooperation in tourism promotion on Web sites is achieved, as well as a few uses of technological resources in the Web sites to facilitate to tourists a better understanding of tourism resources in the area.

INTRODUCTION

Each tourism destination can be considered a market in itself. At these destinations tourism suppliers (i.e., accommodations, restaurants, museums, and tourism offices, among others) interact simultaneously with the tourists who consume these products or services. For that reason a market approach is more appropriate than a supply or a demand one.

A tourism destination is the geographical area where a set of tourism agents interact and intervene in tourism activities. These interactions, from a supply point of view, help develop a relational network at the destination. A relational network is the set of economic and personal relationships established among a number of agents who share goals, cooperation systems, knowledge, reputation, and image, among other elements, in common. These elements help the destination network generate collective learning and knowledge, and consequently, achieve greater levels of competitiveness than individual agents would obtain.

In addition, from the demand point of view, these interactions within the destination help minimize the existing gap between perceived and real images. All tourists have a socially constructed image of a destination (Urry, 1990; Galí & Donaire, 2005; Larsen & George, 2006), which conditions their decision-making, and it is important for the tourism agents involved in the network to control the image of a destination.

This control has two simultaneous benefits. The first one, related to the tourism demand, is the potential to influence tourist decision-making. The second one is related to the tourism supply chain: the competitive advantage brought to tourism destinations by projecting a real image.

Internet is a very important channel that helps tourism agents to achieve these two benefits derived from this control in three aspects. First, a number of authors assume the relevance of Internet as a tourism destination image generator (Baloglu & Pekan, 2006; Choi, Lehto & Morrison, 2007; Hashim, Murphy & Muhammad Hashim, 2007); although "research on the Internet as an image formation agent is still at an infancy stage" (Choi, Lehto & Morrison, 2007, p. 118). Second, Internet brings a great number of opportunities to tourism image formation, contributing to destination imagery formation to consumers (Hashim, Murphy & Muhamd Hashim, 2007) and giving to tourism destination an opportunity to improve destination marketing through the use of "Internet's unique features, such as geographical interactivity with audience, low-cost accessibility, world-wide, hyperlinks with other travel suppliers and design flexibility, to attract more tourists and better position their state in the intense competition for visitors" (Lee, Cai & O'Leary, 2006, p. 816). Third, Internet and destination websites, in particular, act as an information tool for tourists, being an influencing element in their decision-making. This article will discuss the attainment of these two benefits and the relevance of Internet in them using the tourism image and social network theories to clarify how supply and demand interact in a tourism destination. A conceptual model will be proposed as part of a theoretical market approach to tourism destinations, which integrates supply and demand, explains interactions between them and highlights the relevance of this scope of analysis to better understand the dynamics of a tourism destination and the possibility of improving its competitive advantage. In addition, the article demonstrates the necessity of using this integrated approach for planning and managing a tourism destination to improve its competitiveness and highlight this theoretical view.

One of the elements that can be planned with an integrated approach, mainly because it helps to establish scale economies in terms of promotion, it is the promotional website content of a destination. Usually different agents take part in this promotion, Destination Marketing Organization's (DMO) local governments, private companies or associations creating different sources of information. Apart from these possible scale economies this will also help in terms of unification of the destination image. If this image is homogeneous gives also an extra value of competitive advantage preventing incoordination.

Finally, a case study is conducted in order to analyze network configuration trough promotional websites in Internet and determine if the tourism product of a destination is integrated and promoted globally or, on the contrary, each tourism agent acts independently. Results show the existence of a reduced network, that means a little integration of the tourism products in the province of Girona.

This article is organized in five main sections. The first explains the process of tourism image formation and how the tourism agents that intervene in this process affect the image that tourists have of tourism destinations, from the perspective of the social construction of tourism images. The second focuses on the network configuration of the destination, taking into account the tourism agents who take part in the tourism system and how networking can generate competitive advantages. The third part presents a theoretical model of an integrated market approach to tourism destinations. The forth section presents a case study which analyze the network configuration of tourism products in the province of Girona. Finally, the conclusions based on the theoretical model and the case study is drawn, the model's limitations are considered and proposals for future research are made.

HOW TOURISTS PERCEIVE DESTINATIONS

Social Construction of Tourism Destination Images

Images have been used in a number of contexts and disciplines: psychology perceives the image as a visual representation; thought behavior geography emphasizes the association of impressions, knowledge, emotions, values and beliefs; and marketing focus on the relationship between image and behavior of consumers (Jenkins, 1999). The majority of academics from the 1970s to the present day agree that tourism image is "the sum of beliefs, ideas, and impressions that a person has of a destination" (Crompton, 1979).

Gunn (1988), in her main academic study mentions "all of us have images of destinations, whether or not we have traveled to them. These images may be sharp or vague, factual or whimsical, but in all cases they are indicative of likes and dislikes." (p. 23) This means that all places have an image, which has not appeared out of nowhere, but they have consciously or unconsciously been created by "somebody". In this sense, one needs to think about how a tourism destination image is constructed.

From the realization of a task and its reiterated repetition by people in a society, this task ends up being institutionalized by this society. The acceptance of this task as habitual makes it "settle" in this society and form part of its traditions, so in that sense, the reality of this society has been constructed collectively (Berger & Luckmann, 1968). Using this approach to tourism image, it could be contemplated that a tourism image is constructed socially in the same way as a task is accepted as a normal way to do something in a society.

Tourism images are full of visual elements and signs that evoke socially constructed images (Urry, 1990), for example a couple of lovers in Paris suggest romantic Paris. In this sense, the image construct implies some overriding impression or stereotype (Mazanec & Schweiger, 1981). However, this tourism image does not always reflect the reality, because "the tourism image is, at the same time, a subjective construction (that varies from person to person) and a social construction, based on the idea of collective imagination" (Galí & Donaire, 2005, p. 778).

Variations in tourism image are complex if one considers how these images are formed, a little bit at a time. As Gallarza, Gil and Calderón (2002) have exposed "image is not static, but changes depending essentially in two variables: time and space". (p. 72) The influence of time on image is demonstrated in a number of studies on tourism image (Gartner, 1986; Gartner & Hunt, 1987; Chon, 1991; Selby & Morgan, 1996), especially if one considers its formation as a process (Gunn, 1972). At the same time, the space variable also influences the image of a tourism destination. Some studies in this field show that the distance between potential tourists and the tourism destination affects the perceived image of the place (Miossec, 1977; Talisman-Kosuta, 1989; Gallarza, Gil & Calderón, 2002). Considering the dynamic nature of tourism, image is useful if the effect of marketing actions on time and space variables (Gallarza Gil & Calderón, 2002; Talisman-Kosuta, 1989) is taken into account. In this manner the periodic evaluation of tourism image is relevant (Talisman-Kosuta, 1989).

How is the Image of a Tourism Destination Formed?

Accepting as valid the fact that tourism image is socially constructed (Urry, 1990; Larsen & George, 2004; Galí & Donaire, 2005), some studies point to the existence of factors or components that form part of every tourism image and influence its formation process (Baloglu & McCleary, 1999; Gallarza, Gil & Calderón, 2002; Beerli & Martín, 2004). In this sense, Gartner (1993) mentions that some authors have systematized the elements that influence the process of tourism image formation in different conceptual models. At the same time, it is possible to find a number of authors who focus on the existence of a formation process of the tourism image, which is made up of different stages that contribute to how a tourism image is formed (i.e., Gunn, 1972; Govers & Go, 2004).

One of the most important models that show how a tourism image is formed is the seven-stage process of tourism experience, which has been

Figure 1. Seven-stage model of a tourism experience (Source: Gunn, 1972)

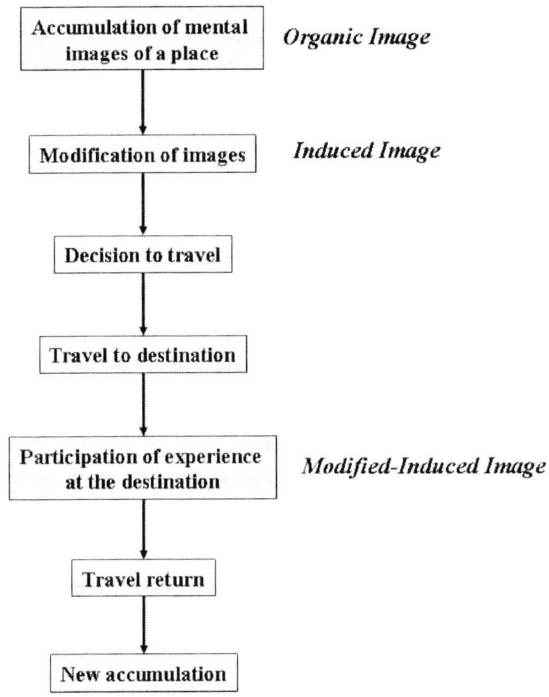

developed by Gunn (1972). This model shows that images held by potential visitors, nonvisitors, and returned visitors differ (Gunn, 1972).

At the first stage, potential tourists assimilate general information, such as, newspapers, television documentaries, books, and school lessons. This process generates an organic image of the destination; this is because the mere mention of these places evokes images, which are not necessarily tourism images.

The second stage implies a modification of perceived images based on consulting tourism information (i.e., tourism posters, guides, articles in specialized reviews, etc.). These changes in perceived images are influenced by induced images, which are the result of a conscious effort to develop, promote, and advertise a destination.

When the potential visitor has a perceived image based on the organic and induced images of the place, then they are prepared to make a decision. Other factors such as previous experience or the money available are also taken into consideration.

Travel to the destination may condition the image that a visitor has, but the key factor of a new change in a visitor's perceived image is their personal experience at the destination, as well as their participation in different activities, such as, visiting museums or the use of tourism services such as accommodation. At this stage, visitors have a modified-induced image, which is the result of the balance between the perceived image before visiting the destination and the perceived image after the visit.

Returning home after traveling, visitors evaluate and make reflections about their experience and discuss it with other travelers. At the final stage, tourists accumulate new information if one considers that this is a circular process. In this sense, it is widely recognized in academic literature that experienced tourists will become a "source of information" for other potential visitors, which will be based on their experience at the destination (Balogru & McCleary, 1999; Beerli & Martin, 2004).

As is noted in this model, the creation and modification of tourism images are constant and demonstrate the dynamism of the tourism image. The space variable shows these phenomena through the contact of visitors with tourism destinations.

Image Management as a Competitive Advantage

Academic literature recognizes the need to manage tourism image, as it is one of the most important factors that influences the decision-making process of tourists that choose a destination to spend their holidays (Gartner, 1993; Govers & Go, 2004). Gunn (1972), in her model explains that tourism images are conditioned by the actions of a number of agents that influence the creation of tourism images. Although it is agreed that the tourism image is socially constructed, agents intervene in this process emitting images, which end up being consolidated and accepted as valid in a specific society.

According to Gartner's (1993) agents' classification, there are four types of agents. The first, Overt Induced is a kind of agent who promotes the creation of a specific tourism image of the tourism destination in a conscientious way, to influence a tourist's process of decision-making. Gartner (1993), makes a distinction between these agents, who are of two types. On one hand, Overt Induced I are "the promoters of the destination [that] construct an image of the salient attributes of the destination in the minds of the targeted audience" (Gartner, 1993, p. 197) with the traditional forms of advertising (i.e., television, radio, brochures, etc.). In this case, one could also include tourism businesses of the destinations, such as, accommodation, restaurants, activities, and so on. On the other hand, Overt Induced II are usually "tour-operators, wholesalers or organizations who have a vested interest in the travel decision process, but which are not directly associated with a particular destination area" (Gartner, 1993, p. 199). As Gartner (1993), mentions "destination area promoters do have some control over the images projected through tour operator" (p. 199) because if the tourism image does not conform to the reality of the destination it could create dissatisfactions to both locals and visitors (Govers & Go, 2004).

The second, Covert Induced are agents that apparently emit a tourism image that is not induced. In this case the author also defines two types of Covert Induced agents. The first is called Covert Induced I, who is related to a recognizable spokesperson who recommends a destination to support a higher level of credibility of tourism destination advertising. The second is Covert Induced II, this category corresponds to people or organizations who write articles, reports or stories about a particular place. Often this published information is a result of a familiarization tour for travel writers or special interest media groups. These actions increase credibility and allow destination promoters to project a specific image.

The third kind of agents is called Autonomous. These agents are people or organizations who produce reports, documentaries, movies, and news articles independently without the specific aim of creating a tourism image of a place.

The last group of agents that Gartner (1993), identifies is called Organic, and is related to information and opinions about a place that a person receives from other people, from their previous experience in this place, Unsolicited Organic corresponds to people who give information about a destination where they have been, without having been specifically asked by the other interlocutor, for example, when this is a topic of conversation with friends in colleges. The existence of Solicited Organic agents implies that individuals actively search for information about a destination and somebody informs them using their own experience. Friends or relatives usually constitute these kind of agents, who have a high level of credibility and are an extremely important part of the destination selection process.

When people visit a destination they become an Organic (pure) agent, having the capability to give information in a solicited or unsolicited way.

This model shows that the task of Overt Induced agents, in this context, is undeniable, especially if one considers the sustainable competitive advantage of the destination. Sustainable competitive advantage is generally based on either core competences or unique resources that are superior to those possessed by competitors and are difficult to imitate (Johnson & Scholes, 1999; Aaker, 2001). Govers and Go (2004, p. 169), established that superior resources for a tourism destination "are generally to be found in either its unique and natural environment (climate, wildlife or landscape) or its cultural heritage" and also mention that "competitive advantage might be created through core competences, such as, the host community's existing unique capabilities in attracting visitors from outside." (i.e., destination's ability to stage world class events, festivals or exploit its folklore and prevailing traditions).

Following these considerations, the management of tourism image is viewed as a management tool (Ritchie, 1993). Govers and Go (2004), propose that it is necessary "to formulate a plan for projecting the 'right' image" (p. 170) as one of the essential parts of tourism development strategy. Gartner (1993), mentions the importance of considering the "image mix", as a continuum of factors that have to be taken into account to decide which agents will intervene in the formation of tourism image, as well as, the amount of money budgeted for image development, characteristics of target market, and demographic characteristics or timing. This task is obviously attributed to promoters of the destination who can select the right mix of image formation agents to maximize their scarce resources (Gartner, 1993).

THE NETWORK CONFIGURATION AND A TOURISM DESTINATION

Tourism Destination Agents

Tourism agents are an essential part of the system and of any destination, therefore they have been identified. Some definitions that are more applicable to industrial destinations, like innovation systems, clusters, milieux innovateurs, or industrial districts, always consider three main types of agents. Moreover, it is considered that tourism specificity needs another main agent who helps to define the situation of the system and therefore needs to complete it.

Two other types of agents support these main agents at all times and are also necessary to maintain the main set of agents stable. Figure 2 illustrates these agents in detail.

Following this scheme, those organizations that take part directly in generating the tourism

Figure 2. Tourism destination agents

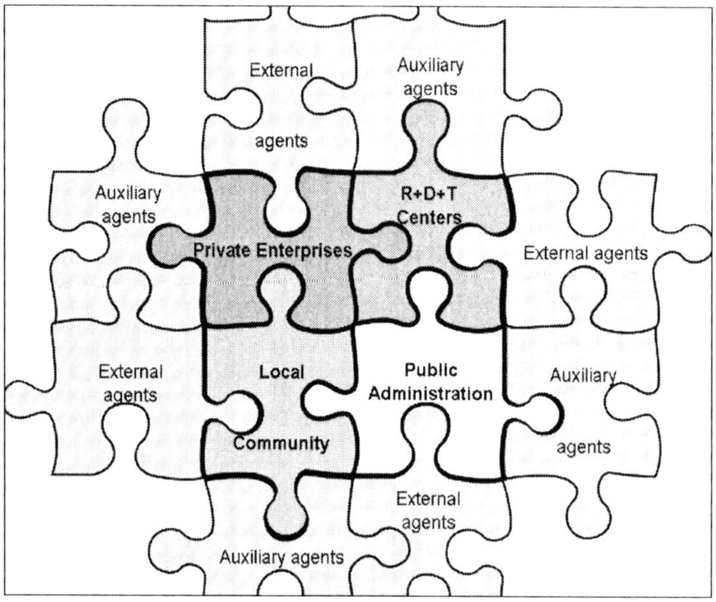

experience are private companies. These include basic tourism companies, and also others whose main activity is related to tourism.

Public administrations are those organisms of governmental function that take part in the tourism processes and whose intervention can generate new legislations, give incentives for research, planning, and others.

Research, development, and training centers (R+D+T) are the essential elements capable of generating specialized training and/or research in the scope of tourism, such as, universities, research institutes, or consultants.

These three main agents appear in the academic literature on innovation systems (Lundvall, 1992; Nelson, 1993), as well as, in clusters and industrial districts. In addition, the tourism scheme proposed by Gunn (1997), and later adapted by the OMT (1999) is taken, and the tourism industry as a functional tourism system (Prats & Balagué, 2005) is conceptualized. They demonstrate that the local community also has an essential role in the development of tourism activity, and consequently of the system. The local community is defined as the inhabitants of a territory. These people are individuals or organizations without economic aims, such as, NGO's, civic organizations, or others. The relevance of local community in tourism is emphasized, seeing that civic movements have been able to modify important decisions in city-planning, ecological subjects, or others, restraining or impelling tourism.

After describing the basic elements, the tourism auxiliary agents can be defined as those agents who do not have activities directly related to the tourism industry, but who support the main agents. Looking at the economic theories, the auxiliary agents are some of the receivers of the multiplying effect (McIntosh, Goeldner & Ritchie, 2000). And the external agents are those tourism agents who are part of other destinations, but who interact with one or more internal agents.

The set of agents in a tourism destination is basically located in the same geographical territory. However, a territory by itself does not have enough conditions for their collective coordination, and also the proximity does not generate synergies by itself, but it can contribute to their effectiveness with other dimensions shared between the agents (Zimmermann, 2001). A good example of an agent who belongs to a distant destination could be a specialized tour operator who commercializes destinations, which are geographically distant but relationally close.

Relational Networks

The use of relational networks in the analysis of a company's competitive advantage can be related to several approaches in the fields of economics or sociology, among others (Sorensen, 2004). Therefore, in the most static frame, this analysis has appeared within the network of individual companies who have useful and important connections with other companies, becoming more than just a unit inside an atomized market (Håkansson & Snehota, 1995). In this sense, these companies must be analyzed considering their relationships with other companies outside the network and also the existing relationships among other companies within the network (Holmen, Pedersen & Torvatn, 2005).

In relation to the most dynamic frame, it is observed that it was contemplated not to see the innovation process as a linear and consecutive process, meaning that the result of the initial stage brings up the following one and so on. Innovation is considered an intensive activity in both knowledge (Sundbo, 1998; Roberts, 2001) and learning. It is also totally accepted as a key element in the innovation process. Thus, innovation also arises and takes place through the interactions between companies (Sorensen, 2004), and between these companies and other relevant actors who are important for their activity (Prats & Guia, 2005). These ties must be understood as intense flows of knowledge and, therefore, essential for innovation, and also for competitiveness.

However, Sorensen (2004) presents a definition that considers networks as the set of conscious and accepted business relationships, whether formal or informal, with transmission of resources, immaterial or material, within the company's scope. In any case, it is useful to adopt the perspective of social network analysis, which studies specific relationships between a defined series of elements, like people, groups, organizations, countries or events, among others (Molina, 2001). It is necessary to consider that social network analysis is based on relationships and not on the attributes of elements. Then, a social network can be defined as the group of people, organizations or other social entities connected by a set of significant relations (Wellman, 1997).

Granovetter (1985) and Hite (2003), affirm that the existing relationships within social networks influence economic actions, and Hite (2003), distinguishes seven different types of ties that can take place inside a social network: the main three are business ties, personal ties, and hollow ties, and the other four types are formed as a result of the relationship between the main types.

Porter (1990), with his five forces model and his later approach to clusters, universalized the necessity to maintain the business or commercial ties that had been previously valued by Becattini (1979), and other authors. Other theories such as the industrial districts theory show that personal relationships have to be considered as a value that contributes to empower the agents' ties making them more efficient and trustworthy (Becattini, 1979). Hollow ties appeared only recently in network theories and have become very common, because they represent all those ties that you accept with the mediation of a third person, so your trust in the relationship is not with the agent to whom you are related, but with the agent who did the mediation (Prats, Camprubí & Comas, 2005).

It seems evident then, that a tourism destination can be defined as a relational network. In every single destination exist relationships among its agents (Prats, Guia & Molina, 2007), considering the specificities of the tourism product, and the existence of the different types of ties able to generate an active and beneficial set of agents and relationships. As an example, a tourism package is integrated by different items that are provided by various companies, which are linked though some kind of relationship. However, tourists do not distinguish that these items belong to different companies, although they need a perfect integration of them, in order for the tourism products to be successful. Another example can be the usual sectorial associations, where hotels, travel agencies or other tourism agents are associated in order to gain power in front of the suppliers or in front of the administration.

Networking as a Generator of Competitive Advantage

If different agents interact among themselves, it can be argued that these interactions often allow the agents to have joint benefits from infrastructures, common engineering, and transfer of tacit knowledge. It also makes productive combinations and interactions more difficult to carry out in atomization or individual isolation.

Even with the continuous growth of the on-line travel expenditure, the academic literature focused mainly in how to effectively build and evaluate hospitality and tourism websites (Han & Mills, 2006). Moreover the analysis is focused on individual perspectives comparing different websites as example, but there are not published results related to collective image construction trough websites. "Because destination images influence tourist behavior, a destination must be favorably differentiated from its competition and positively positioned in the minds of consumers" (Hudson & Ritchie, 2006, p. 388), and this can be achieved also by the common internet promotion.

Belonging to a destination or relational network involves interacting with other members, which is usually transformed into routines of the organization. This is what Rallet and Torre (2004) call the belonging logic. This logic and interaction will be easier a priori if there is a common knowledge; this is called logic of similarity.

The interaction of these agents generates a number of factors that determine if a destination or local innovation system is successful or not in all scopes. A first and fundamental factor is the internal and external relationships that take place in the system. These relationships can be very different and they have been summarized into two characteristic groups.

On the one hand, depending on the relational structure that is adopted in a system, the degree of success will vary. In this factor the key element is the degree of connectivity that is obtained, understanding that the better the connectivity between the agents is, the closer it will be to "the ideal" system. It is understood that good internal connectivity will contribute to a more fluid circulation of knowledge between the agents, and this will increase the trust among them. But at the same time, an excess of internal connectivity can make the trust on external agents decrease to such an extent that they are considered intruders (Zimmerman, 2001). The lack of trust between external agents could have serious consequences in the new knowledge generation, because the closure of relational networks in itself could limit information flows that come from outside, blocking the possibility of generating new knowledge and collective learning (Lazerson & Lorenzoni, 1999).

On the other hand, however, it must also be observed that the quality of relationships within a system such as this, affects its success. The key element in this factor is trust, as a greater trust between the elements of a system will transmit more relevant information, and greater benefits for the whole destination will increase.

Another determining factor is the macro-environment, which is divided into five elements: (1) political, such as, decisions or political elements that affect the system; (2) economic, for instance economic situations that affect the system; (3) technological, which has two levels: (a) the hard level such as the automation level, and (b) the soft level such as the training level of the population; (4) social, this contributes to the system culture, for example the degree of associationism or the cultural level; and finally, (5) historical macro-environment, which gives perspective and historical experience, such as, political periods or natural disasters.

Using the agents' interactions and macro-environment variables, tourism destinations should be able to generate essential collective knowledge and learning for the evolution of the system. The main purpose of this collective knowledge and learning is being able to generate a constant innovation capacity that will bring dynamism to the system as shown in Figure 3.

This innovation capacity allows the system to obtain four successive outputs, which can be observed in Figure 4. Each stage must be achieved to obtain the desired results. If an "ideal" configuration of the system is obtained, the four outputs will also be obtained, and this will revert again to the tourism destination.

The first unquestionable output of the innovation capacity is innovations in any of their modalities. In the opinions of Prats and Guia (2005) the innovation must allow the system to generate a competitive advantage, allowing the destination to satisfy the needs of the tourists better than the competitor's destinations.

The competitive advantage, consequently, must contribute to the system's collective wealth, which in turn increases the wellbeing of all the agents who join it. Wellbeing is understood to be an improvement of the quality of life of all

Figure 3. Tourism local innovation system model (Source: Prats & Guia, 2005; Prats, Guia and Molina, 2007)

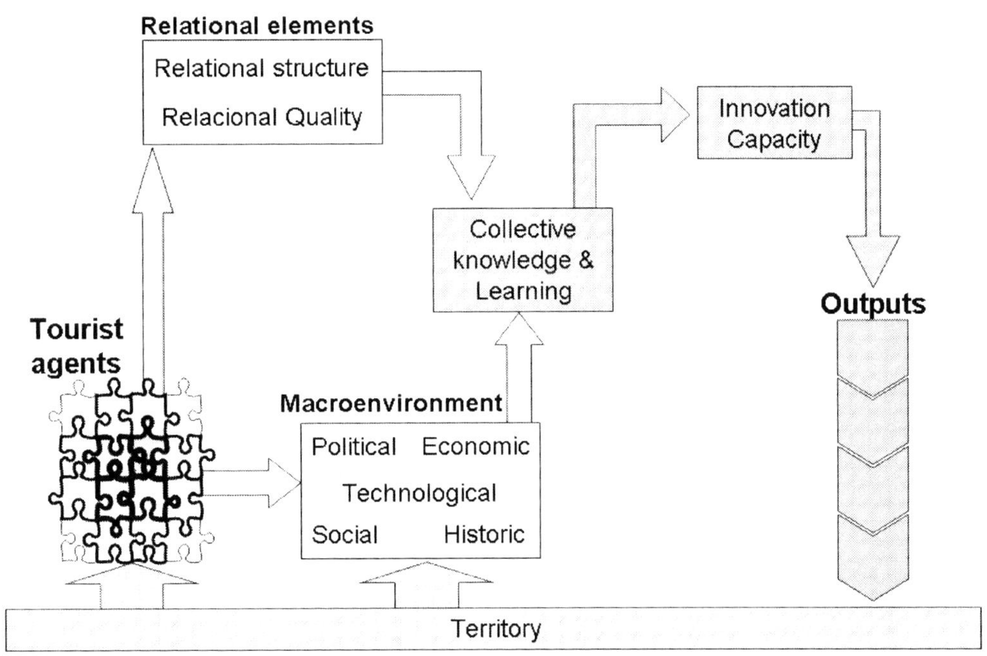

Figure 4 Tourism local innovation system outputs (Source: Prats and Guia, 2005)

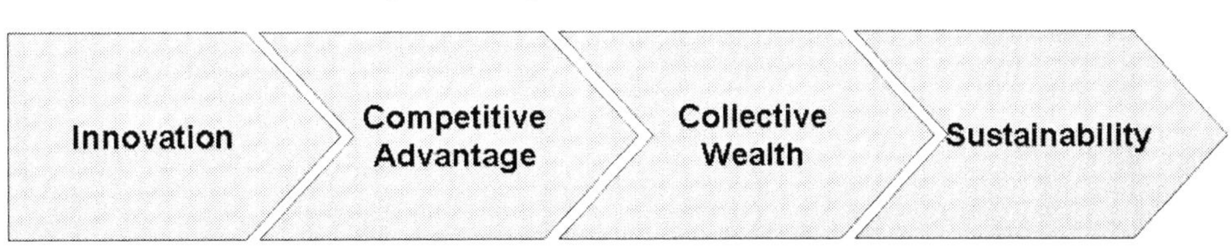

the elements, which is not based solely on the economic, environmental, or social benefits at an individual level, but is a perfect balance between all of them at a collective level.

This balance allows the system to become sustainable and generates a new and better situation that is a territory improvement, and which also feeds the agents and the macro-environment, varying the behavior of the system constantly, forcing it to reframe itself, and be constantly dynamic.

DESTINATION TOURIST PERCEPTION & NETWORK CONFIGURATION: A THEORETICAL PROPOSAL

The tourism image perceived by tourists and represented in Gunn's model (1972) has a close connection with tourism destinations, and in particular, with agents that interact in the promotion of the destination.

The seven-stages of tourism experience (Gunn, 1972), show interactions between tourists and tourism agents. This materializes, initially, in the process of searching for information, which is done voluntarily by tourists; and later, if tourists travel to the destination the tourism image is again modified by direct contact with tourism agents.

In this context, direct contact and coordination among internal and external agents of the tourism destination are also necessary, so that tourists can perceive a real tourism image of the destination both before and after traveling to the destination. If this situation occurs, it will be easier for tourism destinations to maintain their competitive advantage in a sustainable way.

Therefore, tourism destinations, which are established as a network and based on trust among its members, can better guarantee a unique tourism image, which is more coherent with the reality of the tourism destination. This configuration has a close relationship with the structure of network,

Figure 5 Market approach of tourism destination: Conceptual model

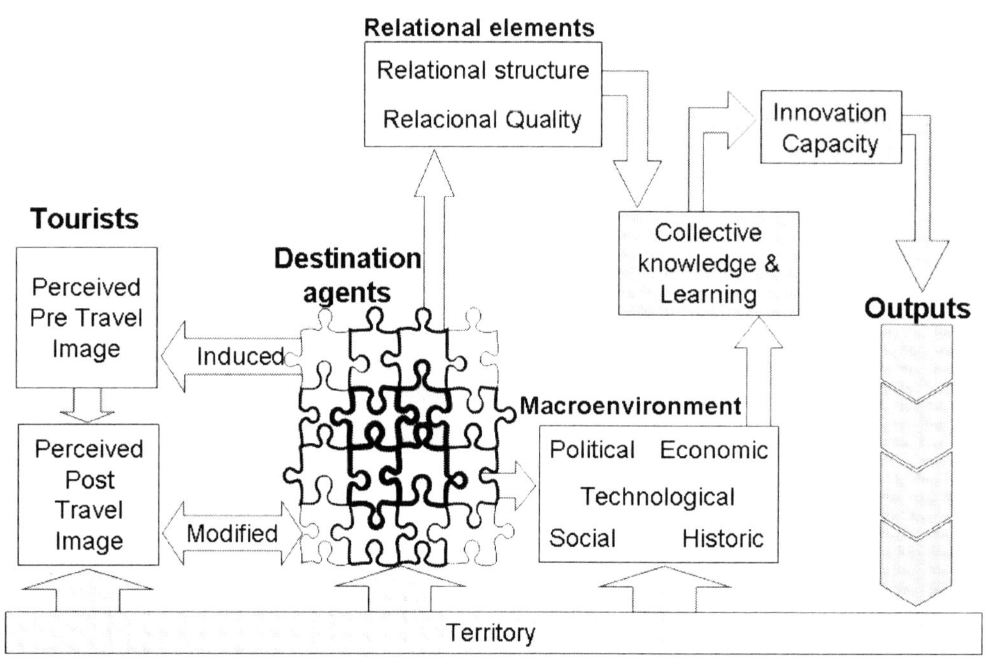

the quality of ties, and its macro-environment. Moreover, the innovation capacity generated has to be useful to adapt the induced image to the tourism product.

Figure 5 shows a market approach model, which focuses on demand and supply simultaneously and explains the relationship between them, taking into account the multiple factors that affect the behavior of tourism agents. As a consequence the competitive advantage of tourism destinations is explained.

In this model the interaction between tourism agents and tourists is highlighted, giving an overall picture of what happens in a destination. In this context, those tourism agents who interact frequently with tourists tend to be public administrations, private companies, and the local community.

First, public administrations have a direct relationship with tourists through the promotional actions of tourism destinations. These agents act as induced agents of tourism images by acting as promoters and developers of destinations. In this case, following Gartner's classification (1993), public administrations act as Overt induced agents I, who emit an induced tourism image, which influences both the tourists' perceived image of a destination and the decision-making process at the moment of choosing a destination (Gartner, 1993).

Second, tour operators, as they form part of the private companies of a destination, also act as Overt induced agents II, because they have a clear interest in influencing the decision-making process of tourists at the time of selecting a destination. Public administrations, as well as tour operators, influence the "perceived pre-travel image" of tourists.

When the potential tourist travels to a destination and becomes a real tourist a direct interaction between tourists and tourism companies takes place, and this influences the perceived image of the place that tourists had before going there, creating a new image of the place (Gunn, 1972).

Finally, the local community has a strong relationship with the tourist and usually the tourist's real image is strongly modified by this kind of contact. In the author's opinion a key factor at this level is the perception the local community has of the tourism activity and the benefits that the inhabitants receive from it, because if a local community thinks that the benefits and damages that tourism causes are in perfect balance or in a more beneficial situation for the local community, these inhabitants will contribute to the tourists well-being, otherwise they will behave to the contrary.

Tourism research and training centers play a secondary, but fundamental role, especially as they might condition the induced tourism image through research projects, and simultaneously, they could also influence perceived tourism images indirectly by training the tourism workers who help tourists during their stay.

Therefore, the influence on "perceived post-travel image" comes from tourism companies, as well as the local community and research and training centers.

This model shows that, the relationships between tourists and tourism agents are systematic and necessary throughout the whole process. This means that this interaction is essential both before a tourist travels to a destination as well as during his/her stay.

If tourists do not go to a destination, this might mean that tourism agents cannot control the factors that generate the appropriate knowledge and transmit the right image to convince tourists. However, other uncontrollable factors exist, such as, the travel time needed, the distance to the destination, the money available or to what extent a tourism product fulfils tourist needs.

When tourists are dissatisfied with their visit because the "perceived post-travel image" is extremely different from the "perceived pre-travel image", there is another scenario where the agents have not transmitted the reality of the destination. In this situation tourism agents had the innovation capacity, but they had not used it in the correct way to obtain the desired image outputs. This context shows the importance of communication and coordination among all the tourism agents of a destination, to induce a real and homogenous image.

CASE STUDY

As we mentioned before, tourism products and even more evident tourists doesn't understand political boundaries, but regional and local governments use it to divide the territory. This situation causes management and commercialization problems that don't help to emit a coherent image of the whole destination to the possible tourists. These scenarios can be avoided developing networks for product commercialization that includes all the Overt induced I agents. Is one of the easiest way to start a network, because in that sense they can share marketing costs entering to scale economies.

We analyzed the public institutions' websites of a tourism destination in a local and regional level in order to know if tourism product is configured globally through collaboration between destination's tourism agents; or in a contrary way the tourism product is fragmented. This can help us to understand the level of tourism image coherence that is transmitted to tourist through Internet.

Methodology

In order to analyze how a tourism product is promoted in a local and regional level a website analysis of public institutions from the Girona's province, which has been conducted during the first trimester of the year.

The Girona's province is situated in the north-eastern of Catalonia and their capital is the city of Girona. It is a very rich region in terms of natural resources and heritage; and it has a very privileged placement between the sea and the oriental Pyrenees. Tourism activity is mainly developed in the coastal area, although in Pyrenees tourism is also a relevant activity for their economy. In general terms, Girona's province is integrated by two tourism brands: (a) Costa Brava for the coastal

Figure 6. Relational map of direct web links

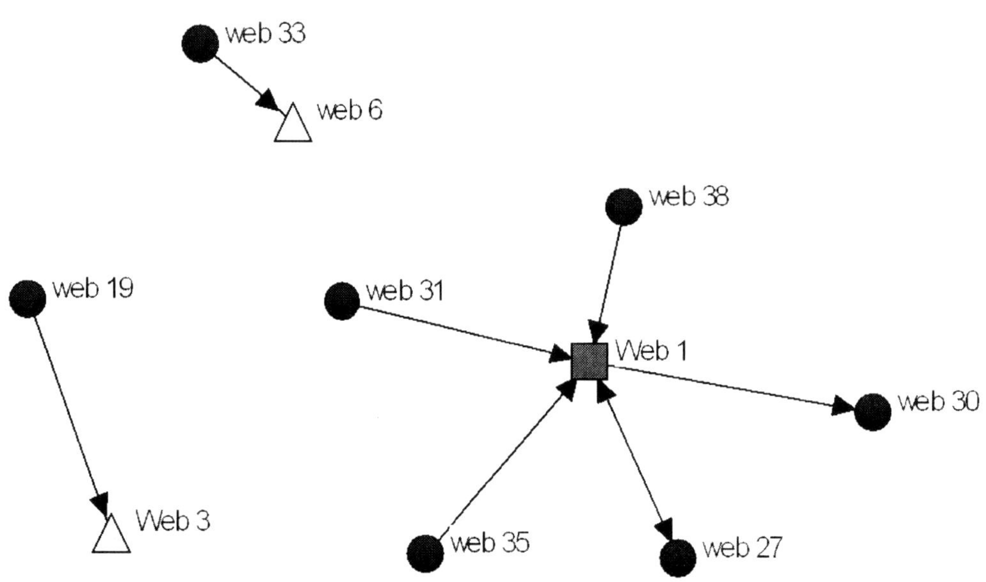

Table 1. Use of technological and interactive web resources

Type	N
Advanced Compelling	18
Video	2
Sound	1
Photographic galleries	11
Interactive maps	4
Total	36

area, and (b) Girona's Pyrenees for the mountain area. These two brands divide their territory and facilitate their promotion.

Websites of 41 public institutions have been analyzed, taking into consideration three labels of public institutions: (a) DMO of the province, (b) regional institutions, and (c) local institutions.

The existence of a section in their website related to tourism activities has been the criteria to select websites. In this context, we have analyzed the role of public agents that act as Overt induced I, following the Gartner's nomenclature (1993) as we mentioned above.

Data collection has been done taken into consideration the level of public institution searching the existing relationship among them. Three kinds of data have been collected. Firstly, it has been detected direct and indirect links that are placed into de website and which is the linked agent. Secondly, it has been observed the quality of detected links and it has determined their typology (advertising links, friends' links or partners' links). Finally, it has been also analyzed the e-marketing resources that have been used in the websites.

In order to analyze data, it has been used UCI-NET 6.0 software (Borgatti, Everett & Freeman, 2002), which is a software specialized in social network analysis in a qualitative and quantitative way. In our case it has been used the application to represent graphically the social network.

Figure 7. Percentage of technological and interactive web resources

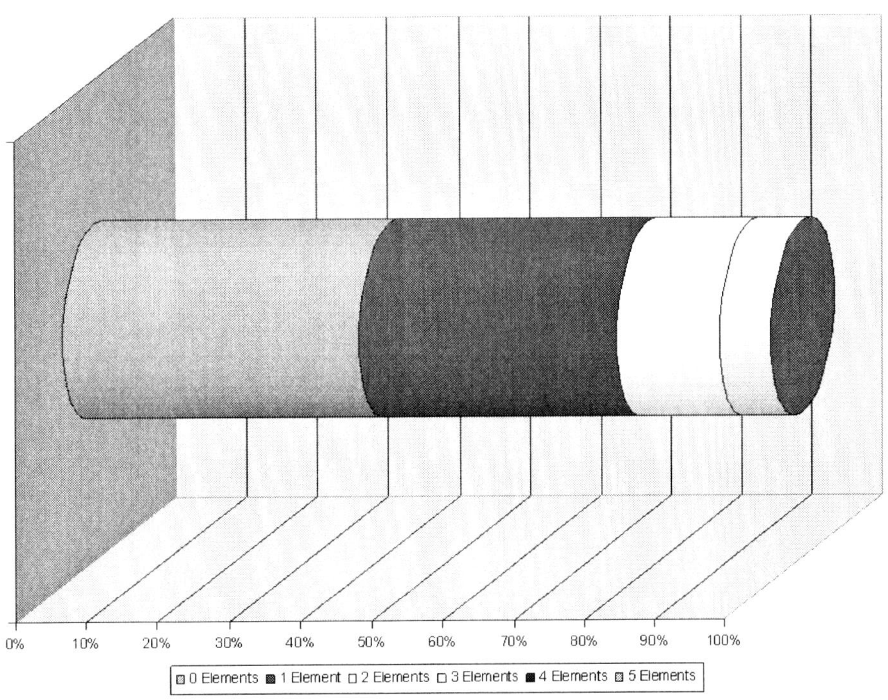

Data Analysis

Observing the network that can be drowned after the analysis, is possible to assure that don't exist a common commercialization network in the Costa Brava destination. From the 41 analyzed webs only 10 have direct links with other promotion agent. The main problem is that only one of these links is bidirectional. This means that the rest maybe are not well developed links.

Another element to extract from this relational map is the fact that only 2 of 8 regional institutions appeared on it, and always as link receiver and not as a link creator, which means that they promote the tourism elements independently from the municipalities that they represent.

Jumping to the technological elements included into the webs that can help to emit a better image five different types of them were analyzed, the type of compelling web system, the multimedia systems divided into sound, video and photography galleries, and finally the interaction with the tourists represented by interactive maps (Table 1).

What is really shocking is that the main Catalan destination, only have little technological and interactive elements to attract tourists. This is done by the historical tourism tradition of the destination when an intensive promotion to attract tourists never was needed. Most of the mature tourism destinations have a similar problem.

One third of the tourism Web sites of the destination don't have any of the mentioned elements and nearly half of them only have one. This situation sorts out that the emitted image trough websites and technological elements are really poor. At least one of the webs that have three of these elements is the one of the main DMO maybe the most visited site. The problem is that in this site there is not all the information related with destination as we can imagine for the link system showed before.

CONCLUSION

Discussion

Initially, it has to be mentioned that it is important to observe demand and supply in an integrated way. This integrated vision gives a greater innovation capacity, in particular, regarding the destinations' tourism agents; who have a broader view of the possibilities of maintaining and improving the sustainable competitive advantage of the destination; allowing to consider the tourists' key role and how the tourism agents' interact with them, as well as when this interaction takes place, and what the basic tools that maintain this relationship are.

In this context it is assumed, as justified earlier, that a tourism destination's image is constructed socially during a complex process in a seven-stage tourism experience. Moreover, the necessity to manage this image is accepted as a method that influences the tourists' process of decision-making and for that reason it is necessary to pay special attention to the agents who take part in this.

Nowadays, Internet is positioned as a relevant tool to contribute to destination imagery formation as well as and information source for tourists that can influence their decision-making. At the same time, it is commonly accepted that city marketing differs in many ways of destination marketing, but if it is analyzed the common projected image trough websites the division line disappears. So we can assume that a big city like Barcelona can have the same number of official emitting image websites than a regional destination like Costa Brava.

In the authors' opinion, social network analysis is a perfect approach for studying tourism destinations. In this sense, they have highlighted the interaction among tourism agents to create a tourism product or service adequate to tourists' needs, as one of the most important factors. One of these interactions can be done in terms of public agents' promotion for the whole destination; this can be materialized though official websites and links those appear in it.

As can be seen, in the case study, the promotion through websites in Girona's province does not help to create a coherent and jointly destination image. This situation can be avoided with a better and wide interaction among the public agents, materializing it through their websites. In addition, probably, the low use of technological and interactive resources in websites is done by the hard sedimentation of tourism industry in this region, and in our opinion these elements will need to be improved if tourism destination goes into a declining process of their life cycle.

Limitations and Future Research

The work that has been presented in this article opens a wide field of future research that takes into consideration all the agents mentioned in the presented model. This empirical analysis will contemplate if having a whole picture of the market will contribute to really improving the planning and management of tourism destinations, as discussed.

In this first conceptualization the autonomous and organic agents from Gartner's model (1993), who generate an uncontrolled tourism image by induced agents, have not been considered. In future research, it would be necessary to revise the conceptual model and include autonomous and organic agents, to have a better proxy, taking into account that tourists also interact with autonomous and organic agents before traveling to the destination.

In the authors' opinion, it could also be interesting to consider the difference between real and potential tourists in future revisions of the model. This reflection could have relevant implications, especially for observing the induced tourism image and searching for explanations for a tourist's reasons to travel to a specific destination and not to others.

REFERENCES

Aaker, D. (2001). *Strategic market management.* (6th ed.). New York, NY: John Wiley and Sons.

Baloglu, S., & McCleary, K. W. (1999). A model of destination image formation. *Annals of Tourism Research, 26*(4), 808-889.

Baloglu, S., & Pekan, Y. A. (2006). The website design and Internet site marketing practices of upscale and luxury hotels in Turkey. *Tourism Management, 27,* 171-176.

Becattini, G. (1979). Dal sttore industriale al distretto industrialle. Alcune considerazione sull'unità dell'economia industriale. *Revista di Economia e Politica Industriale, 1.*

Beerli, A., & Martín, J. D. (2004). Factors influencing destination image. *Annals of Tourism Research, 31*(3), 657-681.

Berger, P. L., & Luckmann, T. (1968). *Social construction of reality.* Amorrortu: Buenos Aires.

Borgatti, S. P., Everett, M. G., & Freeman, L. C. (2002). *Ucinet 6 for Windows: Software for social network analysis.* Harvard: Analytic Technologies.

Choi, S., Lehto, X. Y., & Morrison, A. M. (2007). Destination image representation on the web: Content analysis of Macau travel related Web sites. *Tourism Management, 28,*118-129.

Chon, K. S. (1991). Tourism destination image modification process: Marketing implications. *Tourism Management, 12*(1), 68-72.

Crompton, J. L. (1979). An assessment of the image of Mexico as a vacation destination and the influence of geographical location upon the image. *Journal of Travel Research, 18*(4), 18-23.

Galí, N., & Donaire, J. A. (2005). The social construction of the image of Girona: A methodological approach. *Tourism Management, 26,* 777-785.

Gallarza, M. G., Gil, I., & Calderón, H. (2002). Destination Image – Towards a conceptual framework. *Annals of Tourism Research, 29*(1), 56-78.

Gartner, W. C. (1986). Temporal influence on image change. *Annals of Tourism Research, 13*, 635-644.

Gartner, W.C. (1993). Image formation process. *Journal of Travel and Tourism Marketing, 2*(2), 191-215.

Gartner,W., & J. Hunt (1987). An analysis of state image change over a twelve year period (1971-1983). *Journal of Travel Research, 16*(2), 15-19.

Govers, R., & Go, F. M. (2004). Cultural identities constructed, imagined and experienced: A 3-gap tourism destination image model. *Tourism, 52*(2), 165-182.

Granovetter, M. S. (1985). Economic action and social structure: The problem of embeddedness. *American Journal of Sociology, 91*, 481-510

Guia, J., Prats, L., & Comas, J. (2006). The destination as a local system of innovation: The role of relational networks. In L. Lazzeretti, & C. Petrillo (Eds.), *Tourism local system and networking.* Elseivier: Amsterdam.

Gunn, C. A. (1972). *Vacationscape: Designing tourist regions.* Washington DC: Taylor and Francis/University of Texas.

Gunn, C. A. (1988). *Vacationscape: Designing tourist regions.* New York: Van Nostraud Reinhold.

Gunn, C. A. (1997). *Vacationscape: Developing tourist areas.* Washington DC: Taylor and Francis.

Håkansson, H., & Snehota, I. J. (Eds.) (1995). *Developing relationships in business networks.* London: Routledge.

Han, J. H., & Mills, J. E. (2006). Zero acquaintance benchmarking at travel destination Web sites: What is the first impression that national tourism organizations try to make? *International journal of Tourism Research, 8*, 405-430.

Hashim, H. H., Murphy, J., & Muhamad Hashim, N. (2007). Islam and online imagery on Malaysian tourist destination Web sites. *Journal of Computer-Mediated Communication, 12*(3), article 16.

Hite, J. M. (2003). Patterns of multidimensionality among embedded network ties: A typology of relational embeddedness in emerging entrepreneurial firms. *Strategic Organization, 1*(1), 9-49.

Hjalager, A. M. (2000). Tourism destinations and the concept of industrial districts. *Tourism and Hospitality Research, 2*(3), 199-213.

Holmen, E., Pedersen, A. C., & Torvatn, T. (2005). Building relationship for technological innovation. *Journal of Business Research, 58*(9), 1240-1250.

Hudson, S., & Ritchie, J. R. B. (2006). Promoting destinations via film tourism: An empirical identification of supporting marketing initiatives, *Journal of travel research, 44*, 387-396.

Jackson, J., & Murphy, P. (2002). Tourism destinations as clusters: Analytical experiences from the World. *Tourism and Hospitality, 4*(1), 36-52.

Jenkins, O. H. (1999). Understanding and measuring tourist destination images. *International Journal of Tourism Research, 1*, 1-15.

Johnson, G., & Scholes, K. (1999). *Exploring corporate strategy.* Harlow, UK: Prentice Hall.

Larsen, G., & George, V. (2004, February). The social construction of destination image – A New Zeland firm example. *Working Papers*, 4/01.

Lazerson, M., & Lorenzoni, G. (1999). The firms that feed industrial districts: A return to the Italian source. *Industrial and Corporate Change, 8*(2), 235-266.

Lee, G. M., Cai, L. A., & O'Leary, J. T. (2006). WWW.Branding.States.US: An analysis of brand-building elements in the U. S. state tourism Web sites. *Tourism Management, 27*, 815-828.

Lundvall, B. A. (Ed.) (1992). *National systems of innovation. Towards a theory of innovation and interactive learning.* London: Pinter Publishers.

Mazanec J., & Schweiger, G. (1981) Improved marketing efficiency through multi-product brand names? An empirical investigation of image transfer. *European Research, 9*(1), 32-44.

McIntosh, R. W., Goeldner, C. R., & Ritchie J. R. B. (2000). *Turismo: Planeación, administración y perspectivas.* Mexico D.F.: Limusa Willey

Miossec, J. M. (1977). L'image touristique comme introduccion à la géographie du tourisme. *Annales de Géographie,* 55-70.

Molina, F. X. (2001) European industrial districts: Influence of geographic concentration on performance of the firm. *Journal of International Management, 7*(4), 277-294.

Nelson, R. (1993). *National innovation systems.* New York, NY: Oxford University Press.

OMT (1999). *Desarrollo turístico sostenible: Guia para administradores locales.* Madrid: OMT.

Porter, M. E. (1990). *The competitive advantage of nations.* London: The McMillan Press.

Prats, L., & Balagué, J. (2005). Cohesión y sostenibilidad, elementos clave en la competitividad del territorio turístico: El caso de la Costa Brava. *Retos Turísticos, 3*(3), 8-15.

Prats, L., Camprubí, R., & Comas, J. (2005). Network ties relevance on the destination business relationships. In Universidade do Algarve (Eds.), *Recent developments in tourism research.* Faro: University of Algarve.

Prats, L., & Guia, J. (2005). The destination as a local system of innovation. In C. Petrillo & J. Swarbrooke (Eds.), *Networking and partnership in destinations and development management* (pp. 121-136). Arnhem: ATLAS.

Prats, L., & Guia, J., Molina, F. X. (2007). Tourism local innovation systems or how tourism destinations evolve. In M. Smith & L. Onderwater (Eds.), *Destinations revisited. Perspectives on developing and managing tourist areas* (pp. 35-53). *Arnhem: ATLAS*

Rallet, A., & Torre, A. (2004). Proximité et localisation. *Economie Rurale, 280.*

Ritchie, J. R. (1993). Crafting a destination vision: Putting de concept of residence responsive tourism into practice. *Tourism Management, 14*, 279-289.

Roberts, J. (2001). Challenges facing service enterprises in a global knowledge-based economy: Lessons from the business services sector. *International Journal of Services Technology and Management, 2*(3), 402-433.

Selby, M., & Morgan, N. G. (1996). Reconstructing place image: A case study of its role in destination market research. *Tourism Management, 17*, 287-294.

Sorensen, F. (2004). *Tourism experience innovation networks.* Doctoral dissertation. Roskilde University, Denmark.

Sundbo, J. (1998). *The theory of innovation: Entrepreneurs, technology and strategy.* Cheltenham, UK: Edward Elgar.

Telisman-Kosuta, N. (1989). Tourism destination image. In S. F. Witt & L. Moutinho (Eds.), *Tourism marketing and management handbook* (pp. 557-561). Cambridge: Prentice Hall.

Urry, J. (1990). *The tourist gaze.* London: Sage:

Wellman, B. (1997). An electronic group of virtually a social network. In S. Kiesler (Ed.), *Culture of the internet*. Mahwah: Lawrence Erlbaum.

Zimmermann, J. B. (2002). Des «clusters» aux «small-worlds» une approche en termes de proximité. *Géographie, Économie, Société*, *4*, 3-17.

Chapter X
Marketing Your City's Industries to the World: Building and Retaining Export Oriented Clusters through Strategic ICT Investments

Nicholas P. Robinson
McGill University, Canada

Prescott C. Ensign
University of Ottawa, Canada

ABSTRACT

This chapter discusses the importance of making strategic investments in information communication technologies (ICTs) in order to benefit from globalization and the benefits created by robust export-oriented business clusters. Examples of investments made by local governments in India, Jamaica, and Hong Kong will illustrate how an adept ICT strategy can position a city to grow local industries and encourage exports. The authors argue that a coherent and comprehensive city marketing plan can attract and retain investments and play an instrumental role in the city's future prosperity. Intelligent investments in ICT infrastructure that address a city's unique attributes, objectives, needs, and competitive advantages can open up new export markets, especially in the service industry where ICTs can make local labour globally accessible.

INTRODUCTION

Wealthy trading city-states like Hong Kong have embraced them as a vehicle to spur growth, developing economies have claimed that they are an effective way to connect local merchants to international markets, and the United Nations believes that they are an important tool for municipalities in third world states in the fight against poverty (Business World, 2004). Indeed, information communications technologies are rapidly becoming an important element of many cities' marketing strategies to grow local industries and encourage exports—especially given that ICTs are an important factor driving the institutional and organizational transformations occurring in public administration (Gascó, 2003). Developing an export oriented business community is a vital part of ensuring long-term survival and prosperity (Stough, Haynes & Salazar, 2005, p. 23). Regardless of whether a city aims to export services or manufactured goods, making strategic investments in relevant information communications technologies is a prerequisite.

Intelligent investments in ICTs are a critical element of a successful marketing plan, especially at the product design phase and product promotion phase of the marketing mix. Cities, as the product being marketed, can be moulded to attract and retain investment and export-oriented businesses. This entails designing and promoting the city to achieve the goals of policy makers and export intensive businesses. In other words, utilizing ICTs in a fashion that both works within local constraints and gives exporters a competitive advantage to locating in a certain municipality will spur the growth of an export-oriented business community.

Cities, in collaboration with national governments, in developed and developing nations can re-design themselves using ICTs to create an atmosphere that is conducive to attracting and retaining export-oriented businesses. Examples of success stories in the service sector will be used to illustrate how making critical investments in ICTs and other areas can spur economic growth and diversification. By investigating the best practices being used by cities, such as Hong Kong, Montego Bay, and Bangalore, this chapter will expose politicians and city planners to innovative city design and promotion projects that have been proven to be effective in attracting and retaining export-oriented businesses.

Arguably, making strategic investments in ICTs is an effective arsenal for any city that wishes to look beyond its borders and offer its wares to the world. Improved ICT infrastructures will open up service industry markets in a manner similar to how free trade has opened markets for tangible goods. Improvements in ICT systems and infrastructure will enable municipal economies to compete on the world stage as barriers to exporting services are removed (Graham & Marvin, 2001, p. 352), and cities will need to correctly position themselves to capitalize on the opportunities presented by ICTs in a globalized world.

This chapter will serve to elaborate a methodology on how cities can identify city design barriers and remedy them with ICT investments to encourage the development of export-oriented business communities. Examples of projects in less developed country (or "LDC") cities such as Montego Bay, Jamaica and Bangalore, India and developed economies, such as Hong Kong, will illustrate how ICT investments can bridge gaps and help cities to meet the needs of potential investors (Black, 2002, p. 267). Further, examples of how these cities have used Internet based promotional tools to market these industries will be given.

BACKGROUND

For those marketers interested in developing a healthy export-oriented business community, the purpose of city marketing is to ensure that the city is equipped to meet the needs of export-oriented businesses. Developing a thriving export-oriented business community is an essential element in ensuring long-term prosperity as cities can increase their productivity, income, and employment levels by maintaining a positive trade balance. In fact, it can be said that "cities accumulate and retain wealth, control and power because of what flows through them, rather than what they statically contain" (Graham & Marvin, 2001, Acknowledgements). Exports have "a dynamic effect on overall economic growth" (Stough, Haynes & Salazar, 2005, p. 23) and cities that adopt "an outward orientation should be able to benefit from technological spillovers generated by trade and better participate in the international flow of goods and technologies" (Stough, Haynes & Salazar, 2005, p. 23). One critical move that cities can make to adopt this outward orientation is to "embrace ICTs" as an integral element of their city design plans (Ahmad, 2005). Making strategic ICT investments is necessary to "fully participate in ...new opportunities" (Ahmad, 2005).

ICTs can play a particularly integral role in a city's *marketing mix* (consisting of *product*, *promotion*, *price* and *place*) at both the city design (otherwise know as the *product design* phase) phase and the *promotion* phase (Berkowitz, Crane, Kerin, Hartley & Rudelius, 2003, p. 15). The marketing mix, being the "marketing manager's controllable factors", represents the range of influence that city marketers can exert over their cities (Berkowitz, Crane, Kerin, Hartley & Rudelius, 2003, p. 15). As the focus of this chapter will be on the product and promotion aspects of the marketing mix, price and place, the last two elements of the mix, will not be discussed.

For the city marketer who has an interest in developing his/her city's capacity to attract and retain export oriented industries, product can be defined as "all the attributes of the area (land, communications, skilled labour, training facilities, quality of life) and the service that they (and their partners) can offer to potential investors" (Fretter, 1993, p. 167). Similarly, promotion can be defined as all communication "which is undertaken with the clear objective of influencing the behaviour of those at which it is targeted" (Ashworth & Voogd, 1990, pp. 98-99).

Secondary elements of the product, such as websites and certain city services that aren't essential to an export-oriented business' operations may be referred to as *accessory* product features. These features constitute an important part of the product offering but most often are not essential to the consumer. Collectively, the 4 Ps represent the range of elements that the city marketer can influence and manipulate to achieve the goal of encouraging trade and export-oriented business. At the product phase, ICTs are used to design and mould the city's core attributes, characteristics and accessory product features. This could mean making strategic investments in ICT infrastructure, such as telecommunications networks, or simply improving access to important information and services through a city-run website, which is an accessory benefit. At the promotion phase, ICTs can be applied to communicate with an international audience, especially via innovative Internet applications, and promote the city globally. The contemporary city marketer therefore should not presume that "the urban product can be assumed and the task of the city marketer is only to seek out and capture a suitable market for a pre-existing product" (Ashworth & Voogd, 1990, p. 65). The contemporary city marketer can use tools, such as ICTs, to design and promote the city in order to "satisfy the needs of the consumer while also trying to achieve the organization's goals" (Berkowitz, Crane, Kerin, Hartley & Rudelius, 2003, p. 21). This means using ICTs to satisfy the needs of export oriented companies while achieving internal organizational goals (i.e. budgetary constraints, economic growth targets). Export oriented companies, as the *target market*, or the "group of potential consumers toward which" the city directs its marketing program" (Berkowitz et al., 2003, p.15), require ICT infrastructure that allows them to sell internationally and compete globally. The phenomenon of *geographical detachment*, which occurs when companies can offer services and products to consumer markets remotely, means that ICT investment is more important than ever. Alternatively, ICTs also enable firms to market and provide services customized to local consumers on a global basis—known as the *glocal* phenomenon (Graham & Marvin, 2001, pp. 376-377). Websites like eBay allow local communities to exchange ideas and products through a medium that is otherwise global. Similarly, some scholars have used the word *disintermediation* to describe how "customers and providers of a service no longer need to be located in the same place" (Graham & Marvin, 2001, p. 352). Further, it can be argued that ICTs have allowed for the disintermediation of both governments and markets. Le Galès and other scholars have remarked that states themselves have lost part of their "central role" allowing for greater "horizontal restructuring" (Le Galès, 1998, p. 501).

These phenomena have led to the rise of a new breed of corporation—one that thinks internationally and acts locally. *Export-Oriented Companies*, or "EOCs", which are part of this new breed of corporations, are essentially firms that manufacture goods and services principally for consumption in foreign markets. Subcontractors and subsidiaries that participate in *outsourcing* activities, which involve moving manufacturing or service operations abroad, and generally away from the company's primary customer markets, are perhaps the best examples of EOCs (Black, 2002, p. 337). These sorts of firms are sought after by cities given their ability to generate jobs and earn income locally. Competition for EOCs is intense and has produced somewhat of a race to the bottom, with cities sparring "to provide investors with ever speedier development approvals, ever larger tax concessions, ever weaker environmental regulations, and ever friendlier business environments" (Graham & Marvin, 2001, p. 342). This being said, there is substantial evidence that attracting and retaining EOCs can be an important element of building a healthy economy (Stough, Haynes & Salazar, 2005, p. 23). Cities, and more generally nations, "succeed in particular industries because their home environment is the most forward looking, dynamic and challenging" (Kotler, Haider & Rein, 1993, p. 282). Re-designing the city-product through ICT investments and promoting the city using certain ICT communications vehicles, like the Internet, can therefore present an exciting opportunity for economic development.

In this regard, cities have utilized novel structures, such as *public-private partnerships* or "PPPs", where EOCs partner with governments (through joint ventures or simply coordinated development projects) to create environments that are suitable for export-intensive business operations. Some of the most notable examples of PPPs have come from *Less-Developed Countries* that lack the infrastructure necessary to support many industries but have ample low cost labour. In fact, government intervention in the economic development process is generally very high in many LDCs, countries for which a prominent feature is low per capita incomes (Ramachandran & Sougata, 2005, p.162). This fact makes sense given that "one of the key means of pushing rapid development is by the creation of physical infrastructure" (Ramachandran & Sougata, 2005, p.162).

The importance of government interventions in applying ICTs to re-design and promote the city cannot be underestimated. When determining if a certain ICT investment or promotional strategy is suitable, city planners must assess whether the investment would create consistency amongst internal and external factors. According to Fretter (1993, p. 166), all place marketing, city marketing included, involves a number of key elements. These elements must be consistent with one another in order for the city marketer to achieve organizational goals.

These include having a clear *(1) vision*, or an understanding of what you wish to achieve that is shared by stakeholders such as EOCs (the customer), municipal officials and local business leaders; having an understanding of the municipality's *(2) capabilities*, including budgetary constraints, environmental issues and even social, economic and cultural issues; knowing one's *(3) customers*, or knowing what EOCs need, their limitations, and objectives; *(4) Designing the product* , or in other words adapting and changing the city's attributes to suit the needs of the targeted customers (i.e. the EOCs) and the objectives of city planners; Understanding how a city is situated relative to its *(5) competitors*, or other municipalities that could compete viably for the same customers and markets; Communicating the city's offering to customers and stakeholders with *(6) one voice*, meaning creating a coordinated and clear campaign that has a unified message; and, finally, *(7) Differentiating* the city from others by offering a unique competitive advantage to the target customers and communicating that advantage (Fretter, 1993, pp. 165-172). This model provides a useful framework for analyzing place-marketing issues but can be adapted to analyze the opportunities that cities face when trying to attract and retain EOCs. City marketers concerned with attracting and retaining EOCs must answer questions such as:

- **What are the city's organizational objectives?** For instance, this could entail job creation or diversification of the local economy
- **What are the current attributes of the city-product?** This could include labour force characteristics and the state of local infrastructure
- **What are the needs of the target EOCs?** This could include cost-savings, or access to specialized labour or office facilities
- **What competitive advantages does the city have over competing cities?** Superior ICT infrastructure, better promotional tools, and low cost labour are examples
- **What deficiencies need to be remedied for the city to succeed?** The cost of doing business in a locality or the absence of certain ICT infrastructure needed to attract EOCs could be deficiencies
- **What actions must the city take to achieve its objectives (i.e. considering the city's attributes, deficiencies, competitive advantages, and opportunities)?** This might mean investing in certain types of telecommunications infrastructure or other ICTs

If the city marketer can craft a plan that creates consistency amongst these elements it is then possible to achieve organizational goals. By matching the needs of target EOCs to the objectives, attributes, and competitive advantages of the city, and remedying deficiencies, cities can determine which ICT investment actions might have the highest probability of attracting target EOCs. Cities such as Bangalore, India, Montego Bay, Jamaica and Hong Kong have all undertaken promotional and city design projects that have achieved consistency amongst these elements. Both Bangalore and Montego Bay have achieved notable growth in certain export-oriented service industries while Hong Kong has managed to differentiate itself by deploying a website that achieves consistency amongst these elements.

MAIN THRUST

Issues, Controversies, Problems

Contemporary city marketers must contend with a number of issues when designing the city-product and promoting the city-product. Economic liberalization, technological advancements, and the removal of barriers to international trade have meant that city marketers today must plan to compete in a world that is *geographically detached*. In other words, city marketers must design cities to compete internationally and must promote their cities globally. This fact has meant that city marketers must contend with several challenges. Firstly, city marketers are often unable to implement ICT solutions without the collaboration of private sector partners and need to employ novel solutions in order to achieve city design goals. Secondly, city marketers must find ever-more innovative ways of differentiating their city's offerings against global competition. Lastly, city marketers must choose ICT investments that will attract and retain export-oriented industries despite geographical detachment by ascertaining the industries in which they can build and keep a competitive advantage.

This being said, geographic detachment has led many observers to believe that improved telecommunications networks and other ICT will make the physical location of many export-oriented industries irrelevant. Being able to connect to international markets from remote locations therefore poses a great opportunity and challenge to cities that wish to remain internationally competitive and expand their export-oriented industries (Esbjornsson & Vesterlind, 2003). Additionally, cities that wish to retain existing industries must ensure they remain competitive or risk losing these industries to more competitive markets. City marketers that wish to benefit from the age of geographical detachment must utilize promotional tools, like the Internet, to communicate opportunities with export-oriented companies, and invest in ICTs that will position their cities to compete effectively.

Many scholars, such as Sassen (2001, p. 257), have stated that "the dispersal capacities emerging with globalization and telematics led many observers to assert that cities would become obsolete". Indeed, obsolescence is a very real threat for those municipalities that are not competitive on a cost-basis and cannot offer some other competitive advantage. Geographic detachment means that companies can operate in both service and non-service (such as manufacturing) industries without being hindered by their physical distance to primary consumer markets.

Software design firms in Bangalore, India offering customized services at competitive prices to US customers, client service representatives operating out of Montego Bay, Jamaica addressing the needs of British citizens filing insurance claims, and physical goods being manufactured in Hong Kong to be sold in Australia are all concrete examples of the new reality of a globalized world where geographical detachment is a norm. For the city marketer that is interested in capitalizing on these opportunities making the right investments in ICTs is important. As city marketers in Bangalore, Hong Kong and Montego Bay know, working with private sector partners to ensure the development and deployment of ICT systems is a critical first step in competing globally and encouraging exports.

Similarly, Hong Kong and many other cities in developed economies that are reliant on trade are competing globally to create greater value for customers, specifically exporters and international buyers. Through the creation of Internet based ICT promotional tools these cities provide users with accessory benefits that allow them to differentiate themselves and develop a sustainable competitive advantage. Through the shared efforts of industry leaders and public sector officials at the city and national level, creative ICT tools can be developed to combat geographical detachment's threat to developed economies.

Geographical detachment poses a challenge to cities with developed economies that wish to retain their positions and cities with underdeveloped economies, which wish to increase their exports and diversify their economies. In fact, the effect of geographical detachment has been so great that many media pundits have publicly condemned its effects. CNN's Wolf Blitzer, a US-based journalist and commentator, has decried geographical detachment's effect on the US economy in TV segments discussing the "outsourcing of America" (Blitzer, 2006). Similarly, Le Galès notes "certain territories are organizing politically to resist the often devastating effects of the market on local societies" (Le Galès, 1998, p. 503). In other words, local governments are fighting back by restructuring themselves to be more competitive in high-growth industries. Designing a competitive city despite geographical detachment requires that city marketers determine which areas they can succeed in and then select intelligent investments in ICTs to attract EOCs and retain existing EOC industries. ICTs, in this respect, can be considered a trade promotion tool that can give a city the competitive advantage needed to attract (or in some cases just simply access) international business markets. Choosing which investments are best, thereby re-designing the city-product, is perhaps the most difficult element of the product design process for city marketers interested in encouraging export-oriented trade. Reviewing projects undertaken in cities in developed and developing economies will provide insights.

Examples of What Has Been Done

Example 1: *Hong Kong retains and attracts export-oriented trade by providing value added online services to international buyers and local exporters.*

Despite increasing competition from ports in mainland China, Hong Kong has continued to see increased activity in its shipping and re-export business (Kelly, 2005, p. 375). "Exports are the lifeblood of city-states such as... Hong Kong, whose natural resources are too limited to produce everything they need, nor can their limited populations absorb all the goods they can produce" (Kotler, Haider & Rein, 1993, p. 32). The city, which is now part of China's Special Administrative Region (or "SAR"), is renowned for its position as a manufacturing centre (especially in clothing and textiles) and a centre for international trade, financial services, shipping and other "invisible exports" (Kelly, 2005, p. 375). Other municipalities in the region, such as Shanghai pose a serious challenge to Hong Kong's position as a leader in international trade and shipping, thus making it imperative for the port to distinguish itself from its competitors in mainland China to retain its share of export-oriented businesses.

In this light, Hong Kong's Trade and Development Council (the "HKTDC") has implemented a number of unique web based applications that provide buyers and sellers with additional value. The HKTDC, itself a public sector organization whose directors come from both the public sector and private industry, helps the city compete against other emerging Chinese ports through its online forum for Hong Kong based businesses and international buyers (www.tdctrade.com). The HKTDC acts as the official marketing arm for Hong Kong's manufacturing and export industries, and is the official public sector contact point for importers and exporters in the Special Administrative Region (www.tdctrade.com).

The website is just one of several ways in which Hong Kong has re-designed its city product to retain export-oriented business. Hong Kong, as the product being marketed, includes the HKTDC website and all other accessory benefits that an EOC might derive from having operations in the city (Fretter, 1993, p. 167). In other words, similar to how a warranty on a computer constitutes an "accessory" element of the product, Hong Kong's HKTDC website acts as an accessory element of the city product and adds value for consumers, namely EOCs, and differentiates the city from the competition.

The website helps Hong Kong's export industries benefit from geographical detachment by making Hong Kong based companies accessible to the world. The sourcing section of the website offers buyers access to hundreds of thousands of products from Hong Kong, Mainland China, and Taiwan, ranging from toys to cars and auto parts (www.tdctrade.com). Detailed information regarding suppliers, manufacturers, product characteristics, sales and shipping terms are offered in addition to contact information (www.tdctrade.com). In fact, the HKTDC website makes it possible for overseas buyers to quickly and efficiently assess potential purchase opportunities. Buyers can connect with sellers by submitting an enquiry, but cannot make a purchase directly through the online forum.

The HKTDC website provides equally important information regarding the local economy, trade shows and events, and international opportunities (www.tdctrade.com). The forum effectively has a second function—that of a promotional tool that showcases local events and business opportunities. Lastly, in addition to the online services offered, the HKTDC promotes local manufacturing industries, the re-export business, and Hong Kong's seaport through trade magazines, trade fairs (both overseas and at home) and consulting services.

The Hong Kong Trade and Development Council's website is an excellent example of how ICTs can be mobilized to encourage the retention and growth of local export oriented companies, and promote a city's offerings to the world. The sophistication of the portal and the unique benefits it offers is incomparable to other such projects in Asia and thus gives Hong Kong a notable competitive advantage over emerging Chinese port cities that do not have the marketing savvy and organizational finesse to provide a similar service. The website differentiates Hong Kong in a way that may help it retain its leadership position while promoting the city's offerings to the world. The website positions Hong Kong to benefit from geographical detachment and compete effectively with other regional port cities.

Example 2: *Public-private collaboration and strategic telecom investments spur the development of Montego Bay's export-oriented service industry*

Hidden away along the coastline of Montego Bay, Jamaica's second largest city and most popular tourist destination, lies a facility developed through the collaboration of public and private sector partners (Graham & Marvin, 2001, p. 357). Inside the Jamaica Free Trade Zone (or "FTZ") in Montego Bay, one of two such projects undertaken by the Jamaican government with the cooperation of municipal and corporate partners, the operations of Jamaica Digiport ("JDI") can be found. The FTZ is classified as being a special economic zone where foreign export-oriented companies can operate without the hindrances of certain government regulations and high-taxes. Further, the zone offers foreign companies exceptional ICT infrastructure and connects Jamaica to international markets, especially the United States and Canada. The FTZ includes the Jamaica DigiPort facilities, a privately operated for-profit company, created through the collaboration of public and private sector players, that caters to foreign firms looking to capitalize on low cost Jamaican labour while benefiting from high-end telecommunications infrastructure, similar to what could be found in Europe or North America. The company removes one of the major barriers that prevent foreign companies from operating in Jamaica by providing the ICT infrastructure necessary to do business in Jamaica without the difficulties companies might ordinarily face.

JDI is the result of considerable collaboration between local authorities, the private sector and the Jamaican government and incorporates these stakeholders into its mission, which is "to facilitate the development of industry in Jamaica, by providing the highest quality telecommunication services at a cost that will allow ...customers to provide services at rates that will be internationally competitive." (www.jadigiport.com)

The company offers international clients looking to establish call centres, data processing centres, and other labour intensive service based operations, state of the art telecommunications systems allowing multinational corporations ("MNCs") to benefit from the economic climate in Jamaica while maintaining close contact with home offices. JDI provides dedicated Internet and telephone lines for data and voice transmission, access to a "high speed IP based network" which utilizes "Multi-Protocol Label Switching (MPLS) technology to provide a fast, secure, and scalable connection" for internet users running on JDI's systems (www.jadigiport.com), international direct dial service for call centres which wish to make outbound calls to Canada and the United States at rates that are competitive with those which would be charged in the continental United States (www.jadigiport.com), and a plethora of different toll free service arrangements that allow MNCs to offer customer service direct from Montego Bay (www.jadigiport.com). JDI can offer these services at rates comparable with "tier 1" US telecom carriers given a "Nortel DMS 100/200 switch equipped to provide ACD & Centrex services", its partnership with local telephone company Cable and Wireless, and a direct connection to the "CJFS/Maya 1 submarine cable system" (www.jadigiport.com). Satellite to earth stations and a stable internal backup power source further reinforce JDI's ability to offer international clients services that are comparable with what they would receive locally. The company plays host to numerous Fortune 500 firms who employ roughly 5000 locals, thereby making a considerable contribution to the local economy and increasing the city's international income through service oriented exports (Graham & Marvin, 2001, p. 357).

JDI itself is a product of government intervention, as the company was originally founded as a public-private partnership between the Jamaican government, Cable and Wireless, and US-based telecom giant AT&T (www.jamaicaobserver.com). The government's intervention, through investment in telecommunication infrastructure and the creation of the Free Trade Zone in Montego Bay, has allowed call centres based in Montego Bay to offer phone service direct to the United States at six cents per minute, which is several multiples lower than rates a decade earlier (www.jamaicaobserver.com). Despite now being privately owned, the Jamaican government continues to promote the establishment of service-based industries that will use the telecommunications networks developed by JDI and other new telecom companies.

The Jamaican Trade Promotion Board, known as Jampro locally, assists in marketing Digiport's solutions to international clients (www.jamaicatradeandinvest.org). JDI's success is an example of how forming the right product can attract EOCs. The country's base of over 3 million English speaking people who reside in close geographical proximity to the United States and Canada, and low wages give the nation a clear competitive advantage in service based industries that are labour intensive.

Jampro promotes Digiport and other local EOCs through a team of trade professionals and certain web based ICT tools. The Jampro website boasts that Jamaica's telecommunications system is "among the best in the world" and connects website users to trade representatives who are ready to assist them in finding the information necessary to make export-oriented investments in Montego Bay and other cities in Jamaica. Access to a supplier's database that lists the names of service industry suppliers, like Jamaica Digiport, and other tools further reinforce the strategy to promote Montego Bay as a viable destination for service based industry (www.jamaicatradeandinvest.org).

The promotional strategy employed by Jampro and the City of Montego Bay responds to the issues that need to be addressed by place marketers. According to Kotler, Haider and Rein (1993, p. 162) every place marketer attempting to promote a city's image must ask certain questions, including (1) Who the target audience is, (2) What broad influence tools are available, (3) What major advertising channels are available and what are their characteristics, (4) What criteria should be used in choosing specific advertising media vehicles, and finally (5) How the advertised messages should be timed.

The approach taken in marketing Digiport responds to both the needs of Jamaican cities to attract export-oriented industries while working within the budgetary limitations of local and national governments. Jampro's utilization of a website allows it to cost effectively communicate with foreign EOCs and put them in direct contact with the authorities that can facilitate their investment in export-oriented business in Jamaica. The target audience, decision makers working in EOCs, is often best influenced through personal marketing techniques and can easily be contacted via the Internet. Criteria such as whether the advertising medium can serve to communicate detailed information to a decision maker, whether the medium can cost-effectively provide this information to foreign EOCs, and whether the medium can do so in a time-efficient manner are all critical. The fact that the Internet trade portal created by Jampro can offer access to detailed, time-sensitive information at an affordable cost makes the medium preferred for countries that have limited resources.

Further, interested EOCs can develop personal relationships with JamPro by contacting a trade representative and personal marketing techniques can then be used to court the potential investor. Often, large-scale investments that involve considerable risk, as is the case when establishing a subsidiary or branch operation in a foreign country, require more intimate marketing approaches. Personalized promotional approaches play a critical role in developing an EOC cluster. Indeed, personal promotional approaches were used to initiate investments in the telecommunication sector in Jamaica. The island nation's Minister of Commerce himself courted potential investors and has negotiated several new telecommunications licenses. By doing so, the Jamaican government has worked with local authorities to help to cut telecommunications costs by granting new licenses to other companies that compete with Jamaica Digiport. Recently, licenses were granted to companies that have agreed to offer data and voice transmission services at prices that are 70% less than current prices. Interventions of this kind by governments push operating costs down and make cities more competitive internationally (www.jamaicaobserver.com). The beneficiaries of the new licenses, namely Fibralink Jamaica and the Trans-Caribbean Cable Company ("TCCC"), have worked with the local government to further enhance the attractiveness of Montego Bay as a business destination by offering lower cost telecommunications options to EOCs (www.jamaica-gleaner.com). This is consistent with the argument put forth by Le Galès that central governments are yielding power to local governments in order to better facilitate economic development (Le Galès, 2006, p. 718).

Example 3: *Bangalore attracts export-oriented companies in the information technology industry by re-designing itself through ICT investments.*

Like Montego Bay, the Indian city of Bangalore has had success in developing its service-based export sector. Bangalore hosts numerous software development and IT help desk operations which employ skilled, often computer savvy and university educated, locals to complete work that would cost several multiples more in Europe or North America. Like Montego Bay, enabling Bangalore to compete internationally in service industries was a matter of removing certain barriers and building modern ICT infrastructures that could put the city on par with locales in Europe, Canada, and the US. Bangalore also has a large English speaking population and its residents have benefited from technical programs offered at Indian universities in areas such as computer programming and software design. Cooperation between the Indian government and local authorities has allowed Bangalore to repackage itself into a highly marketable commercial centre for EOCs in the IT industry. Bangalore's transformation is another example of how re-designing the city-product using ICTs can lead to the successful development of EOC clusters. India, on the whole, has over 8,000 software firms alone, and "the software industry in India grossed an annual revenue of US$ 12 billion during 2002-03, up from US$10 billion in 2001-02, registering an overall growth of 18.8% in dollar terms" (Ramachandran & Sougata, 2005, p. 149). These companies cater primarily to foreign customers (Ramachandran & Sougata, 2005, p. 149) and offer a broad spectrum of software development services.

The growth of the software industry in India is a product of efforts by private and public authorities. In order to "promote the growth of the IT industry, ...many state governments have created software technology parks in which both economic and social infrastructure are readily available...the parks have played an important role in enabling clusters to develop" (Ramachandran & Sougata, 2005, p. 151) and have been a key element in Bangalore's rise to become the "Silicon Valley of India" (Ramachandran & Sougata, 2005, p. 152). In fact, the city has blossomed from a "mere 13 software firms in 1991-1992" to "over 1100 software firms working in areas such as chip design, systems software and communication software" and now employs more than 80,000 people in IT (Ramachandran & Sougata, 2005, p. 152). Bangalore is the result of an intensive city design effort by local, state and federal governments in India to create an environment where a software industry could flourish.

One of the best "means of pushing rapid development is by the creation of physical infrastructure" (Ramachandran & Sougata, 2005, p. 162) thereby changing the nature of the product offering being made to foreign EOCs. Governments play "a crucial role in shaping the cluster by developing and promoting educational and research institutions, attracting investment in high technology areas by providing better factor conditions and creating local demand" (Ramachandran & Sougata, 2005, p. 162). In both Montego Bay and Bangalore, government interventions have been an important element of city marketing plans and have yielded astounding results. In the software industry's infancy, companies such as Infosys had to compensate for the shortcomings of local governments by investing in systems to provide a stable power supply, proper office facilities, and even water supply (Ramachandran & Sougata, 2005, p. 157). Both Bangalore and Montego Bay are a testament to the idea that smart investments geared towards shaping the city are often key in developing EOC clusters. India's investment in telecommunications networks, and other ICT facilities and Montego Bay's investment in high-tech facilities both yielded considerable returns for local and national governments.

In order to facilitate the development of India's export oriented software and customer services clusters, local and national governments in India made considerable ICT and infrastructure investments. Bangalore STPIB program, an acronym for Software Parks of India Bangalore, is an example of this type of investment. STPIB is a government owned and operated program that assists in the development of technologically advanced office space for hi-tech companies interested in establishing operations in Bangalore. These companies cater almost exclusively to foreign markets, especially the United States, where labour is much more expensive (www.blr.stpi.in). The first software policies developed in the 1980s in India "emphasized the concept of software development and export through data communication links" and kick-started investment in the basic ICT infrastructure necessary for the sector's development (www.blr.stpi.in).

The STPIB concept focuses exclusively on one sector, information technology, and associated areas such as customer service for software users (www.blr.stpi.in). The scheme is "100 percent export oriented" and focuses on the "development and export of computer software, including export of professional services using communication links or physical media" (www.blr.stpi.in/). Like in Montego Bay, the Indian government established zones dedicated to export intensive industries, such as the software industry. Conventional technology parks are located within these "Export Processing Zones" further increasing the attractiveness of these centres to foreign investors wishing to establish operation in cities like Bangalore. The technology parks created by STP, such as the Bangalore facilities described above which house that city's IT services and software development industries, offer services such as private leased lines for Internet and telephony to connect with offices and customers overseas. For instance, Softpoint, one of the most popular services offered uses IPLC transmission lines and "Intelsat, New Skies satellite, APSTAR" technologies to connect offices in Bangalore with any point in Europe or North America (www.blr.stpi.in). The parks are also equipped with facilities to ensure the integrity and security of data stored and information transferred from the technology parks (www.blr.stpi.in).

The role of governments in developing these ICT infrastructures was critical and has put Bangalore's software industry at the forefront. Indian software companies, like Infosys, are now amongst the world's largest and have grown from small start-ups to global powerhouses thanks, in part, to a marketing plan that emphasized the creation of export-oriented zones equipped with the technologies needed for the local software industry to thrive. More interestingly, the Indian government has placed the offices that administer the STPI program within the technology parks themselves, keeping government officials in close contact with the industry (www.blr.stpi.in). This type of close collaboration between city marketers working for the Indian government and business has spurred on the innovation process.

SOLUTIONS AND RECOMMENDATIONS

Solution to Problem 1: Employing Novel Solutions to Implement ICT Investments

As has been demonstrated by the examples of both Montego Bay and Bangalore, novel solutions must often be employed to effectively implement ICT solutions. In both cities, the creation of parks and zones dedicated to specialized export oriented business activity was a key element of their city design plans. In Bangalore, government developments, designed specifically to house export oriented software and IT businesses, equipped with the most advanced telecommunications equipment was the product of public-private co-operation. Investments in office space and ICT infrastructure coupled with industry consultation have made Bangalore an internationally renowned centre for IT activity. Foreign investment has spawned the creation of new facilities and led to the development of an Indian software industry, with software giants like IBM, Infosys and Wipro housing themselves in Bangalore (Graham & Marvin, 2001, p. 338). Infosys, in return, has collaborated with local governments and has made private investments in things like roads, electricity and water systems (Ramachandran & Sougata, 2005, p. 157).

Similarly, in Montego Bay, governments have partnered with the telecommunications industry, becoming a shareholder in companies like Jamaica Digiport at the formative stages, and have helped to fund the development of improved ICT networks. Submarine cable networks and satellite uplinks now make Jamaica one of the most wired nations in the Caribbean and the presence of a business community that makes use of these networks for the purpose of exporting services is an extraordinary bonus. Both Jamaica Digiport, based in Montego Bay, and its competitor, the Trans-Caribbean Cable Company, have been instrumental in spurring on the growth of service exports from the island. Public-Private partnerships and strategic government investments that satisfy industry needs, such as technology parks, are novel solutions that can help countries effectively implement ICT investments in a geographically detached world.

Solution to Problem 2: Differentiating the City Using Unique ICT Tools

ICTs have also enabled cities like Hong Kong to differentiate themselves and develop a sustainable competitive advantage in the face of rising regional competition. Hong Kong's development of Internet-based applications that provide accessory benefits to business communities by connecting local industry to international markets, while promoting the city is highly innovative. The HKTDC website is both a product design exercise, as it adds value to doing business in Hong Kong, and a promotional tool, as it communicates the city's capabilities internationally. The website differentiates the city from other ports and trading centres in the region, most of which do not offer a comparable service, and is a potent tool in fighting the threat of geographical detachment on Hong Kong based industries. By creating value for EOCs through unique ICT tools cities can remain competitive and differentiate themselves from other locales that compete purely on the basis of their core product offering. Accessories, like the HKTDC website, are an effective way of achieving this.

Solution to Problem 3: Choosing the Right ICT Investments to Attract and Retain EOCs in a Geographically Detached World

Selecting ICT investments that will assist the city marketer in attracting and retaining export-oriented businesses while achieving municipal objectives is difficult. However, considering ICT developments in Bangalore and Montego Bay, one can see that certain factors must be considered before an investment is undertaken. By viewing ICT investment opportunities in terms of the city's objectives, attributes deficiencies, and competitive advantages, and the needs of the target EOCs, one can better understand the probability of an investment's success.

Using the example of the evolution of Bangalore's software development industry, the ICT investment rubric discussed in the background section can be applied. In the case of Bangalore, the city's attributes included considerations such as relatively low cost labour, highly educated software engineers with technical knowledge and a culture that is conducive to the development of a software industry. Infrastructure deficiencies, in both areas such as basic utilities and information communications technologies were barriers to connecting Indian software developers to international markets. The objective of developing an Indian software industry required the city to take certain actions, such as the creation of office space and investment in infrastructure. EOCs based in the United States, Canada, and Europe required low cost skilled labour—an important input for most EOCs, and something which Bangalore has in abundance. Ultimately, investments in office space, telecommunications infrastructure and basic utilities have increased employment and diversified the local economy in Bangalore (Ramachandran, 2005, p. 162). The effect of the investments made in Bangalore has meant that the city has been effectively re-designed so as to attract the target market, which includes international IT and software companies such as Microsoft, IBM, Nortel and Infosys. City marketers, therefore, play an important role in identifying a niche and shaping the city to suit the needs of a specific industry. In the case of Bangalore, investments in telecommunications infrastructure, office space and even upgrades in the local power system were necessary to attract international investment and lead to the development of the IT and software cluster. In other words, action by empowered local governments has yielded increased involvement by the city's corporate citizens who have in turn made investments in Bangalore. However, individual citizens have been shown to be unmoved by the decentralization of governance functions and the empowerment of local governments (Hoffmann-Martinot, 1998, p. 198).

According to one Indian software guru, "India has all the prerequisites to emerge as a software superhouse" (www.dewangmehta.com). Putting the country in this position was a matter of removing a few critical roadblocks and offering incentives to foreign companies that wish to do business in cities like Bangalore. Deficiencies, such as the lack of competitive connectivity, problems with the power supply in many regions of India, and regulatory barriers were removed by creating the specialized STPIB parks equipped with facilities that are competitive with those in North America and Europe. A "window clearance" scheme that allows foreign investors to receive clearance for their investment by all levels of government for all necessary issues using one single application cuts regulatory red tape and new start-ups are given a tax-free grace period (www.hyd.stpi.in). The State Technology Parks (or "STP") office developments, coupled with other government incentives, have spurred on the growth of the software sector in India. Marketers who collaborated with other levels of government in the development of these parks have been successful in using ICT upgrades to help attract EOCs to Bangalore and compete in the face of geographic detachment. Bangalore, like Montego Bay and Hong Kong, has benefited from geographical detachment because it has re-designed itself to achieve internal objectives and help EOCs satisfy their needs.

Similarly, Montego Bay can be viewed through the rubric of this model. The attributes of the city, the objectives of city marketers, opportunities to satisfy EOC needs, the city's deficiencies and possible competitive advantages must all be considered under this model. Montego Bay, like India, has a competitive advantage given that it can almost uniquely offer American companies low labour costs, an English-speaking population, and close proximity to two important consumer markets (i.e. Canada and the United States). The city's competitive cost of labour makes locating labour intensive industries, like call centres and data processing centres in Montego Bay appealing to EOCs based in the United States (the target market) and satisfies their need to locate labour intensive business functions in cost effective regions. The city, in seeking to diversify its economy away from its dependence on tourism, had to re-design Montego Bay as a competitive international destination for EOCs by addressing some of the barriers to setting up a labour intensive service industry in the city. This meant addressing deficiencies such as the island nation's inferior telecommunications system, and regulatory issues. By working with local private sector partners and investing in ICTs, Montego Bay has positioned itself as a destination for call centre outsourcing.

Likewise, Hong Kong has re-positioned itself by changing the nature of the accessory benefits it offers in order to compete with other regional ports that provide the same core benefits. Whereas Shanghai and other ports can compete on the basis of cost and can offer similar value, in terms of strategic needs, Hong Kong has the advantage of access to an incredibly skilled and business savvy labour force that is capable of providing unique accessory services to re-export and shipping clients. In order to achieve the port's objective of maintaining its leadership position, despite the increased competition, the port needed to find a basis for differentiation. The Hong Kong

Table 1. Bangalore software industry summary

Element	Application to Bangalore Call-Centre Industry
OBJECTIVE (City):	To increase hi-tech service exports and diversify the local economy
ATTRIBUTES (City):	English proficiency, exceptional IT training facilities and skilled cheap labour, poor basic infrastructure and international telecommunications infrastructure
NEEDS (EOC):	IT and software industry is under pressure to cut costs and improve investment returns, requires technical/ skilled labour
ADVANTAGE (City):	Low wages, stable country, skilled workforce are the main competitive advantages
DEFICIENCY (City):	Need superior ICT infrastructure, stable power supplies, labour security
ACTION (City)	Investment in software technology parks equipped with specialized telecom switches, localized power supply stabilizers, etc.

Table 2. Montego Bay call centre industry summary

Element	Application to Montego Bay Client-Care Industry
Objective (City):	To stimulate investment by EOCs in Jamaica's service industry in order to diversify the local economy and create job growth
Attributes (City):	Over 3 million English speaking people, weak telecom networks, close proximity to massive US and Canadian markets
Needs (EOC):	Low cost, English-speaking labour within proximity of consumer markets.
Advantage (City):	Cost-competitive labour, geographical proximity to the markets being served, some cultural similarities between local and foreign market, native English speakers, relatively safe and stable country, attractive climate and locale
Deficiency (City):	Lack of suitable office space, unstable telecommunications networks that are costly for MNCs to use
Action (City):	Investment by local and national city planners in affordable high-quality office space, telecommunications networks, tax-incentives and some basic language training

Trade Development Council offers companies that source goods from Hong Kong access to a database of potential suppliers and gives those who export from Hong Kong more opportunities to promote their wares than other regional ports. The port's Internet site, and the other accessory benefits offered in conjunction with the website (such as trade shows and support services), create a clear point of differentiation and deliver better value for companies that choose Hong Kong instead of a competing port.

Montego Bay, Bangalore, and Hong Kong have all managed to benefit from geographical detachment, attracting international EOCs and diversifying their economies. By choosing ICT investments that will match the objectives, attributes, and deficiencies of the city with the needs of the target EOCs, city marketers can be more effective in successfully attracting and retaining EOCs. An inconsistency between these elements may expose an investment opportunity or an irresolvable barrier. In the case of both Bangalore and Montego Bay, deficiencies were remedied easily through ICT investments and other actions. Similarly, in Hong Kong, the city's need to differentiate itself merited an investment in an accessory service. This rubric therefore provides a good methodology for initially reviewing ICT investment opportunities.

FUTURE TRENDS

In the future, investments by countries like Jamaica and India will lead to increased competition and a greater need for cities in developed nations

Table 3. Hong Kong re-export and shipping industry summary

Element	Application to Hong Kong Re-export & Shipping Industry
Objective (City):	Objective of maintaining and possibly even growing the shipping industry despite increased competition from other regional ports
Attributes (City):	Exceptional port facilities, strategic locale at the cross roads of major trade routes, access to very skilled labour and business expertise, close proximity to other major ports and major manufacturing centres such as India and mainland China
Needs (EOC):	EOCs require better value than what is available from similar regional ports, including better marketing and promotional benefits and support from local organizations.
Advantage (City):	Ability to develop accessory services, trade promotion services, access to superior internet based databases and technology that make it easier for Hong Kong exporters to market their products internationally
Deficiency (City):	Lack of a point of differentiation, potentially not cost competitive with other ports in the region
Action (City):	Investment in a website that offers accessory benefits to companies that re-export or ship from Hong Kong

to differentiate themselves in a way that cannot be readily replicated. Improved connectivity in third world nations will mean that glocalization will accelerate and expose first world states to increased competition in service industries. Increased competition from cities in LDCs will make it necessary for cities in North America and Europe to invest more heavily in accessory ICTs, like websites, to provide added value and differentiate themselves from international competition. As in Hong Kong, mature urban economies may come under threat and need to find new ICT tools to distinguish themselves from the competition.

This being said, some scholars have suggested that economic and technological development, moving in lock step, follow a cycle. Kindleberger (Shaw, 2001, p. 293) has suggested that places achieve "economic primacy" because of "a series of innovations" and then "successfully consolidate those methods into institutional structures that in time became resistant to further change, only to be overtaken in turn by another nation more open and receptive to further innovation". Considering Kindleberger's thesis, it is therefore imperative for cities to compete to make themselves most receptive to innovation, or risk falling behind. Choosing to make strategic ICT investments is one way of meeting this challenge.

CONCLUSION

A city's ICT infrastructure is one of its defining characteristics and can often be influential in its success or failure in the economic arena. Cities with superior ICT infrastructures are better connected and, overall, better positioned to benefit from the globalization of trade in services and goods (Ahmad, 2005). Industry leaders, city officials, and academics have all suggested that ICTs can play a pivotal role in a city's future competitiveness— acting as a city's future natural resource after contemporary industries have evolved or even disappeared. Re-designing the city by making strategic investments in ICTs is critical to attracting and retaining export-oriented industries, and is therefore a vital element of long-term economic success.

In fact, some have suggested that by embracing ICTs cities are in a better position to spawn "new entrepreneurial forms" (Carrier, Raymond & Eltaief, 2004, p. 349) and lead new industries. This is because "markets ultimately create products rather than the reverse" (Ashworth & Voogd, 1990, p. 65), and cities need to be responsive to markets and understand the needs of both industry and consumers in order to ensure long-term economic success. Export-oriented industries require superior infrastructure, skilled labour, and competitive forward-looking conditions. Cities can position themselves to benefit from the presence of EOCs by incorporating strategic ICT investments into their city design plans. This may include improving the core benefits offered by the city through investments in sophisticated ICT facilities in business parks or could mean simply differentiating the city through accessory benefits such as a website. Irrespective of whether accessory or core benefits are modified to meet the needs of EOCs and correct deficiencies, re-designing the city, as the product being marketed, using ICTs can stimulate the growth of export-oriented industries. Given that IT is "fast becoming a vital engine of growth for the world economy", cities can position themselves for success through ICT investments (Kenny, 2006, p. 67) that meet the needs of EOCs and achieve municipal organizational objectives. The decentralization of governments has allowed local governments to undertake new initiatives using ICTs and work to "institutionalize patterns of cooperation between" actors in industry and government (Le Galès, 2001, p. 181). As in Hong Kong, Montego Bay, and Bangalore, ICT investments can be deployed to promote and design the city so as to be better prepared to meet the needs of export oriented business sectors.

REFERENCES

Ahmad, A. (2005, March 28). Intel chief puts spotlight on digital era: Barrett wants GCC states to invest more in information communication technology. *Gulf News*, A7.

Ashworth, G. J., & Voogd, H. (1990). *Selling the city: Marketing approaches in public sector urban planning*. London: Belhaven Press.

Berkowitz, E., Crane, F., Kerin, R., Hartley, S., & Rudelius, W. (2003). *Marketing* (5th Canadian ed.). Toronto: McGraw-Hill Ryerson.

Black, J. (2002). *Oxford Dictionary of Economics* (2nd ed.). New York, NY: Oxford University Press.

Blitzer, W. (2006, March 14). The situation room transcripts. *CNN.com*. Retrieved August 4, 2007, from http://transcripts.cnn.com/TRANSCRIPTS/0603/14/sitroom.02.html

Business World (2004, January 20). *ICTs can help fight poverty*. Manila: Author.

Carrier, C., Raymond, L., & Eltaief, A. (2004). Cyberentrepreneurship: A multiple case study. *International Journal of Entrepreneurial Behaviour and Research*, *10*(5), 349-363.

Editorial. (2005, January 6). Fibre optic licenses awarded. *The Jamaica Observer*. Retrieved August 5, 2007, from http://www.jamaicaobserver.com/news/html/20050105T220000-0500_72709_OBS_FIBRE_OPTIC_LICENCES_AWARDED.asp

Esbjornsson, M., & Vesterlind, D. (2003). Mobility and social spatiality. In M. Hard, A. Losch, & D. Verdicchio (Eds.), *Proceedings from transforming spaces: The Topological Turn in Technology Studies Conference*. Darmstadt, Germany. Unpublished.

Fretter, A. (1993). Place marketing: A local authority perspective. In G. Kearns & C. Philo (Eds.), *Selling places: The city as cultural capital, past and present* (pp. 163-174). New York, NY: Pergamon Press.

Gascó Hernández, M. (2003). New technologies and institutional change in public administration. *Social Science Computer Review*, *21*(1), 6-14.

Gordon, S. (2006, April 12). Flow is neither new nor different say competitors. *The Jamaica Gleaner*. Retrieved August 5, 2007, from the http://www.jamaica-gleaner.com/gleaner/20060412/business/business2.html

Graham, S., & Marvin, S. (2001). *Splintering urbanism: Networked infrastructures, technological mobilities, and the urban condition*. London: Routledge.

Hoffmann-Martinot, V. (1998). Urban Political Parties: Role and Transformation. In V. Hoffmann-Martinot, T. N. Clark (Eds) & M. Gromala, *The new political culture* (pp. 195-218). Oxford: Westview Press.

Hong Kong Trade and Investment Council. Retrieved August 4, 2007, from http://www.tdctrade.com/.

Jamaica Trade and Invest. Retrieved August 4, 2007, from http://www.jamaicatradeandinvest.org/.

Jamaica Digiport. Retrieved August 4, 2007, from http://www.jadigiport.com

Kelly, P. (ed.) (2005). *Economies of the world*. London: Routledge.

Kenny, C. (2006). The Internet and economic growth in LDCs: A case of managing expectations. In A. D'Costa (Ed.), *The new economy in development: ICT challenges and opportunities* (pp. 67-88). New York, NY: Palgrave Macmillan.

Kotler, P., Haider, D., & Rein, I. (1993). Marketing places: Attracting investment, industry and tourism to cities, states, and nations. New York, NY: The Free Press.

Kotler, P., & McDougall, G. (1983). Principles of marketing (Canadian ed.) Scarborough: Prentice-Hall Canada.

Le Galès, P. (1998). Regulations and governance in European cities. *International Journal of Urban and Regional Research, 22*(3), 482-506.

Le Galès, P. (2006). New state space in Western Europe? *International Journal of Urban and Regional Research, 30*(3), 717-721.

Le Galès, P. (2001). Urban governance and policy networks: On the urban political boundedness of policy networks- A French case study. *Public Administration, 79*(1), 167-184.

Mehta, D. Dewang Mehta. Retrieved August 4, 2007, from http://www.dewangmehta.com/main.htm.

Ramachandran, K., & Sougata, R.(2005). Creating information technology industrial clusters: Learning from strategies of the early and late movers. In T. Thatchenkery & R. Stough (Eds.), *Information communication technology and economic development: Learning from the Indian experience* (pp. 149-166). Northampton: Edward Elgar Publishing.

Sassen, S. (2001). Cities in the global economy. In R. Paddison (Ed.), *Handbook of urban studies* (pp. 256-272). London: Sage Publications.

Shaw, D. (2001). The post-industrial city. In R. Paddison (Ed.), *Handbook of urban studies* (pp. 284-295). London: Sage Publications.

Stough, R. R., Haynes, K. E., & Salazar, M. E. (2005). Economic Development Theory and practice: the Indian development experience. In T. Thatchenkery & R. Stough (Eds.), *Information communication technology and economic development: Learning from the Indian experience* (pp. 11-28). Northampton: Edward Elgar Publishing.

Technology Parks of India Bangalore. Retrieved August 4, 2007, from http://www.blr.stpi.in

Technology Parks of India Hyderabad. Retrieved August 4, 2007, from http://www.hyd.stpi.in

Williams, P. (2002, December 13). C&W Lowers Rates at MoBay Digiport. *The Jamaica Observer*. Retrieved August 5, 2007, from http://www.jamaicaobserver.com/news/html/20021213T000000 0500_36551_OBS_C_W_LOWERS_RATES_AT_MOBAY_DIGIPORT.asp.

Chapter XI
WEB 2.0, Social Marketing Strategies and Distribution Channels for City Destinations:
Enhancing the Participatory Role of Travelers and Exploiting their Collective Intelligence

Marianna Sigala
University of the Aegean, Greece

ABSTRACT

During the last decades, the use of Web 2.0 applications for the generation, dissemination, and sharing of user-generated content (UGC) and the creation of new value added services are enormous. Web 2.0 tools have tremendously changed the way people search, find, read, gather, share, develop, and consume information, as well as on the way people communicate with each other and collaboratively create new knowledge. UGC and Web 2.0 are also having a tremendous impact not only on the behaviour and decision-making of Internet users, but also on the e-business model that organizations need to develop and/or adapt in order to conduct business on the Internet. Organizations responsible to market and promote cities on the Internet are not an exception from these developments. This chapter aims to inform city tourism organizations responsible for the development of city portals about (a) the use of the major Web 2.0 tools in tourism and their impact on the tourism demand and supply; and (b) the ways and practices for integrating the use of Web 2.0 into their e-business model and e-marketing practices.

INTRODUCTION

During the last years, the number and use of numerous Web 2.0 tools, whereby Internet users produce, read and share multimedia content (User Generated Content, UGC), is mushrooming (eMarketer, 2007a). It is estimated (eMarketer, 2007b) that 75.2 million USA Internet users currently use UGC, and this is expected to increase to 101 million by 2011. eMarketer (2007c) also found that over 25 million USA adults regularly share advice on products or services online.

The Web 2.0 technologies and applications (e.g. tags, RSS, blogs, wikis, podcasts, etc.) are considered as the *tools of mass collaboration*, since they empower Internet users to collaboratively produce, consume and distribute information and knowledge. In other words, Web 2.0 tools do nothing more than realizing and exploiting the full potential of the genuine concept and role of the Internet (i.e. the network of the networks that is created and exists for its users). This has tremendously changed the way people search, find, read, gather, share, develop and consume information, as well as on the way people communicate with each other and collaboratively create new knowledge (Sigala, 2008). UGC and Web 2.0 technologies are also having a tremendous impact not only on the behavior and decision-making of Internet users, but also on the e-business model that organizations need to develop and/or adapt in order to conduct business on the Internet (Bughin, 2007).

The tourism industry is not an exception from such developments. On the contrary, as information is the lifeblood of tourism, the use and diffusion of Web 2.0 technologies have a substantial impact of both tourism demand and supply. Indeed, more than ¼ of Internet users have used a weblog to review information about a destination or travel supplier in the last 12 months (Harteveldt, Johnson, Epps & Tesch, 2006), many new Web 2.0 enabled tourism cyber-intermediaries have risen challenging the e-business model of existing online tourism suppliers and intermediaries who in turn need to transform their e-business model and e-marketing practices in order to survive (Adam, Cobos & Liu, 2007). As the Internet plays an important role for the e-marketing of city destinations (Sigala, 2003; Yuan, Gretzel, & Fesenmaier, 2006), Web 2.0 tools and applications also create both threats and opportunities for organizations developing and maintaining destination management systems and portals. In this vein, this chapter aims to inform city tourism organizations responsible for the development of city portals about: a) the use of the major Web 2.0 tools in tourism and their impact on tourism demand and supply; and b) the ways and practices for integrating the use of Web 2.0 into their e-business model and e-marketing practices.

WEB 2.0 TOOLS IN TOURISM: USE, IMPACT AND APPLICATIONS IN CITY MARKETING

RSS (Really Simple Syndication)

Definition, Features and Use

RSS allow users to subscribe to a webpage for receiving new content, e.g. subscribe to receive online distributions of news, blogs, podcasts etc, and so, RSS allows the creation of links and interactive communication amongst other Web 2.0 applications and users. This is done either through a news aggregator (similar to an email inbox) or a news reader (a web-based environment) (Winer, 2005). By doing so, one does not have to visit each individual website that he/she is interesting to read any new information, but rather the RSS feeds all new updated information to the users' RSS reader. RSS readers enable Internet users to gather and read all new information that is customized to the user's profile within one consolidated message. Many free RSS exist on the Internet,

Figure 1. Communication types

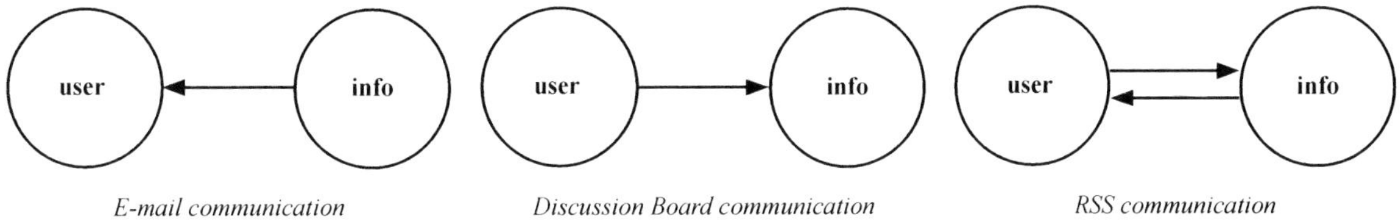

e.g. FeedDemon, NewsGator, Rojo, software on the website of Google™, MyYahoo®, etc.

RSS allows new communication and interaction modes with information (Figure 1) (Farmer, 2004). In e-mail, the control of the communication channel is held entirely by the instigator of the communication. Consequently, e-mail communication is characterized at times by flame wars, antisocial behavior and feelings of intrusion by the participants, while the information artifact is transitory, unfixed and not archived except in individual instances. In discussion boards, information artifacts are fixed, frequently archived and can be interacted with through threading and comments, but for accessing information, the user must deliberately visit a dedicated online area. In contrast, in RSS, both the communicator and the reader of information have control of the communication process, i.e. the former sends information only to those that users have selected to aggregate the RSS feed, while the later select from where and how (e.g. summaries, titles, or full entries) to receive communication. Further, as most RSS aggregators are either integrated with or standalone desktop/web applications, readers need only to check the aggregator for new items.

Impact on Tourism Demand

RSS feeders have a tremendous impact on the way consumers search and read information nowadays. RSS has the following benefits for users:

- saves a lot of time spent on information searching;
- provides users with consolidated personalized information;
- is less obstructive and more personalized to users' interests than other Internet based communication, and so, RSS entice subscribers to visit the related websites, thus helping in building website traffic and visitation;
- RSS boosts viral marketing and online word-of-mouth as users tend to forward items in RSS feeds to their friends, family and co-workers, much like the 'forward this message' feature in eNewsletters.

Business Applications for City Marketing

As RSS is an information distribution technology that is characterized as a demand pull rather than a supply push model, many tourism destination organizations have adopted and incorporated RSS feeds in their websites in order to communicate with their potential and current travelers in a less disrupted and personal way. Some examples of RSS include:

- Keep a communication with their travelers such as sending them Newsletters and/or updates of the programme of cultural events organized in the city.
- RSS helps organizations to enhance their Search Engine Optimization by creating inbound links to a company's website and by informing search engines whenever new content is uploaded on a website, so that they can index it.

- RSS is used for syndicating content to other Websites expanding the original website's readership and reach.

For example, the destination marketing organization of Las Vegas has included an RSS on its website (visitlasvegas.com), whereby users can subscribe to feeds that automatically notify them of current travel specials. Almost all of the information (e.g. events' news, weather updates, special offers, etc.) on the official city website of Dublin created by the city tourism board (visitdublin.com) is available to any traveler and/or other website through RSS. RSS are offered for free for anyone for reading and/or enriching his/her own web site, provided that the latter follows the proper format, terms and conditions and attribution, e.g. attribution such as *"Content provided by Dublin Tourism"*. In such a way, visitdublin.com aims to enhance readership of its content, continuous personal communication with its customers, viral marketing, and search engine optimization through content syndication and incoming links.

Blogs (or Weblogs)

Definition, Features and Use

Weblogs began as personal writing spaces that store and update regularly multimedia content (in reverse chronological order) and links of interest to the author. Thus, blogs are used for recording its author's journey and sharing it with others by using links, RSS, trackbacks, comments, taglines, archives, permanent links, blogrolls, etc. (Blood, 2000). Weblogs are defined as a "… site consists of dated entries" (Blood, 2000), whereby entries are episodic or conversational in a diary or "story telling" format. Motivated by different reasons (Forrester Research, 2006), such as documenting one's life, providing commentary and opinions, expressing deeply felt emotions, articulating ideas through writing, and forming and maintaining community forum, weblogs (or blogs) are a *"new form of mainstream personal communication"* (Rosenbloom, 2004, p. 31) for millions of people to publish and exchange knowledge/information, and to establish networks or build relationships in the world of all blogs. Indeed, blogging tools enabling between-blog interactivity are building up the "blogosphere" whereby social networks among bloggers are created. Du and Wagner (2006) identified the following characteristics of blogs:

- **Personalized:** blogs are designed for individual use and their style is personal and informal. Blogger.com offers a "team blog" collaborative feature enabling also multi-person weblog.
- **Web-based:** blogs are easy to access and frequently maintain by simply using a web browser.
- **Community-supported:** Weblogs can link to other weblogs and websites (e.g. photos, videos, web-texts), enabling the synthesis and linkage of ideas from different users, and so, stimulating meta-knowledge, i.e. the generation of new knowledge through sharing.
- **Automated:** Blogging tools are easy to create and maintain without the need of HTML programming skills and knowledge; so, bloggers can solely concentrate on the content.

Du & Wagner (2006) also identified 3 types of blogging tools and features: type I features of blogging (such as text, diary, hyperlinks, user friendly editor) provide easy-to-use and learn tools for editing, presentation, publishing and interlinking of content. Such blogs are heavily used by those that solely seek a channel of self-expression. Type II blogging tools (such as Indexed archive, search, "permalink": a permanent URL for each weblog entry enabling referencing of specific past entries like other web-source. "Trackback": a reverse hyperlink tracking the referrer weblogs "making these formally invisible connections visible", categorisation & syndication) are used by bloggers who wish to easily share rich media (e.g. videos, pictures etc), to have a sophisticated content management system and to enable between-blog commenting or hyper linking, e.g. through "permalink" or trackback (Blood, 2000). Nowadays, the emerging Type III of blogging tools provide improved content distribution and between-blog connectivity (e.g. "pingback", alert of other bloggers' comments or new posts), as well as integrated applications for further enhancing social networking and community building such as the following examples:

- Workflow or project management (e.g. Lycos Circles workflow for a party, from invitation to management of responses and to travel directions),
- Polls, Intrasite messaging (e.g. ModBlog allows users to track friends' newest entries, or to know who are the most "recent visitors"),
- Web invitation, picture/music sharing (e.g. MSN Space picture/music sharing, and remote posting of updates via email or mobile devices).

Impact on Tourism Demand

Numerous examples of general and-or (content or user) specific blogs have been created in the tourism industry, such as tripadvisor.com, hotelchatter.com, bugbitten.com, placeblogger.com, realtravel.com, travelpod.com, igougo.com, gazetters.com (a B2B blog for travel agents). Many travelers and tourists also develop and maintain their own blogs for sharing their travel experiences with others and distributing their feedback (reviews) of travel suppliers for achieving fun, social recognition, prestige and-or self-expression. Due to the unbiased information shared in blogs based on first-hand authentic travel experiences, many travelers tend to use and trust blogs' information for searching for travel information, tips and selecting travel suppliers and destinations. Blogs have the power of the impartial information and the electronic word-of-mouth that is diffusing online like a virus. Hence, blogs are becoming a very important information source for international travelers for getting travel advice and suggestions. Moreover, when reading others' travel experiences through weblogs, this also creates to the reader the willingness to travel and visit the same destination or suppliers. Indeed, research has shown that UGC at blogs has a similar AIDA effect to users as paid advertisements have. The latter is because blog's content can (Lin & Huang, 2006): 1) **A**ttract the attention, eyeballs of other Internet users and increase traffic on a website, 2) create **I**nterest to users who can now seek more and additional information, 3) develop someone's **D**esire to also visit a destination and/or buy the product and 4) foster an **A**ction (e.g. book a hotel or organize a trip to a destination). Of course, it should be noted that the power of blogs can also be negative, i.e. spread a bad experience of a tourist to million of online Internet users. Therefore, it is very important that tourism companies authorize a public relations staff as the responsible representative of the company to first scan and read blogs' content and then respond to formally any positive and negative users' comments. Guidelines and corporate policies for responding to UGC should also be established. Nowadays, many blogs take the form and are presented in a video format (Vlog, video blog). The first travel website to implement vlogs exclusively is endlesseurope.com. Due to the multimedia features of video content and the intangible nature of the tourism product, it is argued that vlogs are going to have a much greater impact and influence on travelers' decision making and evaluation of alternative tourism products and suppliers.

Business Applications for City Marketing

Blogs generate and distribute a plethora of UGC related to travelers' experiences, suppliers' and destination reviews, travel tips and advices. City destination management organizations can exploit and use such content for:

- monitoring and influencing electronic word-of-mouth;
- conducting an easy, free, timely and reliable market research about travelers' preferences, feedback and profile;
- communicating with current and prospective travelers in a very personal and informal way;
- gathering travelers' feedback and responding to customers' complaints;
- enhancing search engine optimization: blogs are becoming very important tools affecting information search since their links, content and popularity can dictate the position of a company on a search engine search.

There are many search engines that one can use for identifying and locating weblogs. The most popular one is technorati.com, which also provides statistics about the online activity of weblogs, e.g. about the popularity of a blog and its potential influence on search engines results.

Apart from exploiting others' blogs, many tourism suppliers and destination management organizations have also adopted a pro-active strategy by creating and incorporating blogs on their own websites. For example, Marriott has created its own blog on its website (www.blogs.marriott.com), while Starwood has created a blog to communicate with its Preferred Guests and enhance their loyalty through the website www.thelobby.com. Company initiated and moderated blogs can offer the following benefits: solicit and gather feedback from customers; conduct free online market research; become recognized as an expert on a specific topic; communicate and update your customers with latest news; and use others' customers' suggestions for helping customers select and evaluate products such as what amazon.com is doing by allowing users to upload books' reviews on its website. For example, Eurostar has initiated a blog (www.voiceofacity.com) whereby it has commissioned local Parisians to post blogs for creating a travelers' guidebook with a truly ground-roots feel. The destination organization of the city of Los Angeles has created a blog supporting the sharing of bloggers' experiences and insights on their adventures of the diversity of Los Angeles' arts and culture (http://blog.experiencela.com/). The fully Web 2.0 enabled official portal of Holland features a blog capability (i.e. the triplog) enabling Dutch travelers and locals to share and post the experiences they lived in Holland (http://us.holland.com/blog.php?sf=e.pagerank&so=DESC&sel=popular). In this vein, blogs are becoming a useful tool for enabling local communities to get more involved in destination marketing, communicate and blur with tourists. As a result, blogs and web 2.0 tools can support and foster community participation in city tourism development and marketing practices. Community participation can ensure a better blend between locals and tourists reducing any inter-cultural conflicts, creating social relations and respect and understanding amongst different cultures as well as enabling multi-stakeholder understanding and communication in tourism decision making and activities.

Social Networking–Collaborative Networking

Definition, Features and Use

Social networking websites enable users to create their profile and invite others with similar profile to take part in their online community. The most popular websites such as myspace.com and bebo.

com reflect the willingness of Internet users to transform websites as a gathering place of people with similar profiles.

Impact on Tourism Demand

Many social networking websites have been created in the tourism industry allowing travelers and prospective travelers to network with one another based on shared interests or attributes, such as tripmates.com, gusto.com, triporama.com, triphub.com, traveltogether.com and wayn.com. Travelers log into websites and create a personal profile detailing their travel experience and interests, then network with others o share travel advice and stories, and even plan trips together. Hence, social networking websites have a tremendous impact on how tourists nowadays create, organize and consume tourism experiences. Many tourists nowadays prefer to have the reassurance of other users sharing similar profile with them that the trip, the travel company, destination and/or the itinerary that they have scheduled is a good one and it matches their preferences and tastes. Many tourists also wish to use the Internet for collaboratively organizing a group trip with their friends. Websites such as tripup.com, traveltogether.com and travelpost.com enable tourists to create an itinerary, e-mail and share it online with others, who in turn can edit, modify and enhance it, post it back to others for further comments and / or invite and read other travelers' comments and advices on the trip they organized in order to finally achieve a consensus and proceed to a group booking.

Business Applications for City Marketing

Since sharing travel experiences in a social website can significantly inspire travel and boosts one's willingness to visit a destination or supplier, several tourism websites are incorporating social networking tools in their e-business models. For example, existing cyberintermediaries, such as Yahoo!® Trip Planner has adopted a collaborative trip organizing and booking tools. The official website of the city of Philadelphia has also features a collaborative trip planning tool (www.planit.pcvb.org), that potential travelers can use for organizing their itinerary in Philadelphia with friends as well as soliciting feedback and comments from other travelers and locals.

Lufthansa created and operates its own social networking website, named as Jetfriends, for enabling its young flyers to share flight experiences and indoctrinate them into the Lufthansa brand and frequent flyers' club (http://www.jetfriends.com/jetfriends/kids/). Sheraton also re-organized and re-designed its website (which is nowadays titled as the Sheraton Belong Neighborhood), whereby Sheraton's guests can subscribe to the website, upload their experiences, stories, pictures, videos and comments for sharing them with other website visitors and users. The social networking of Sheraton's website enables potential travelers to organize and book their holiday and hotel experience at any Sheraton property that matches their profile and preferences by reading and reflecting on the comments and first-hand experiences written by previous Sheraton customers. The impact of social networking features for persuading potential travelers to select a particular hotel and/or destination is very powerful, because through social networking websites, travelers can search website content based on keywords and stories contributed by other travelers that may be more relevant and make more sense to them than keywords and experiences being created and pushed by the website developers themselves. For example, on Sheraton's website one can search an hotel experience based on the comments and tags contributed by a previous guests referring to *"nice walking in beaches nearby Sheraton hotels"* or *"relaxing family holidays in Sheraton properties"* rather than using the Sheraton's search engine to find hotel based on its location, facilities etc.

Nowadays, many city destination organizations have also incorporated social networking features into their e-business model and strategy in order to further enhance their communication with customers and take benefit of the electronic word-of-mouth that they can create. For example, the official website of Los Angeles invites any cultural and art organization-institution to become a partner with experiencela.com in order to share and distribute related content on their website. Experiencela.com has also created a cooperation e-business strategy with the social networking websites clickr.com and myspace.com. Specifically, the destination management organization of LA created a special webpage for LA on flickr.com (http://www.flickr.com/groups/21164279@N00/pool/) and myspace.com (http://www.myspace.com/experiencela) in order to enable the users of the formers that are also funs and travelers of LA to share personal photographs, comments and stories about LA. The city of LA has realized that such UGC can crucially drive traffic to their website, boost their search engine optimization strategy, instil travelers' desire to visit LA and use customer intelligence for providing reliable and timely advice and suggestions for trips to potential travelers to LA.

Tagging (Social Search and Tag clouds)

Definition, Features and Use

Tagging represents a new way for categorizing information. Users tag a piece of content (e.g. an audio, a picture, a word) with a meaning (a word or phrase) and then this information is categorized in categories based on this meaning. Community tagging is a bottom-up, grass-roots phenomenon, in which users classify resources with searchable keywords. The tags are free-form labels chosen by the user, not selected from a controlled vocabulary. Tagging is also known as consumer-generated taxonomy. Forrester (2006) defines tagging as 'the act of categorizing and retrieving Web content using open-ended labels called tags'. Tagging provides customer value, because it allows them to assign their own word or words to mark products and content online in order to categorize content that they find relevant, i.e. such as what bookmarking allows users to do. Words that users choose for categorizing website and content then become a navigation shortcut that a person can use to browse and search content throughout that site.

Tagging is used not only for saving and sorting a user's content but also for sharing content with others. Websites with tagging capabilities can also allow users to share their personal tags and navigation ways with other users. Moreover, some tag enabled websites enable users not only to share their tag navigations, but also their profile. In this way, users can see who has tagged something, and try to search and find information based on the search behavior of users with similar profiles and mental maps with them (personalized social searching). In this vein, tagging has a great effect on how search engines identify and present information results in keyword searches to users.

Flickr.com represents the first wide-spread of tags, whereby users can add their own tags to any photo they wish to share, aggregate pictures into photosets, create public or private groups, search photos by tags and easily add flickr-stored photos to a blog. Nowadays, there are numerous websites enabling tagging and searching based on tags, e.g. del.icio.us, a bookmarking service, Technorati, a blog cataloging site, and digg, a gathering place for tech fans. These sites create clickable "tag clouds" for resources, groupings of tags arranged alphabetically, with the most used or popular keywords shown in a larger font. In this way, these websites present other websites that users think are important or relevant to them. Many such sites make use of RSS to notify interested users of changes and new developments, e.g. in flickr.com, RSS feeds can be attached to individual tags, or to photos and discussions. In addition to RSS, flickr.com and other social networking sites typically offer functions such as search (for users and tags), comments (and comment trails), and APIs (application program interfaces) for posting to or from the tools, that can be used in combination with blogs. An interesting use of RSS that is combined with tagging is the Flashcard exchange, where, one can view or subscribe to all flashcards posted for learning Spanish (or other languages).

Although the tagging process is by no means simply technical—a way of categorizing resources—it has a strong social dimension as users of the website find common interests and create on-line communities. It represents another example of the fuzziness separating consumers and creators on the Web. A contribution to a tagging site, seen by other users, may cause additional tags or comments to be added, automatically building and updating and thus ultimately defining a resource. Instead of one person making a judgment about a blog entry, photo, or other resource, a consensual classification is created. In effect, a text or object identifies itself over time. This creation of "folksonomies" (as the user defines how to sort information which in turn defines how others search and find information) can be seen as a democratic implementation of the Semantic Web. Thus, for some, tagging helps and boosts the creation of the semantic web (Web 3.0), whereby web content and search is directly related to its meaning for the users.

Impact on Tourism Demand

Several websites offer the capability for users to sort, share and categories travel related content based on tagging, e.g. flickr.com (for pictures sorting and sharing), travbuddy.com (for travel experiences sharing), travelistic.com (for tagging video content). However, although more and more users are using collaborative tools for identifying and sorting travel content, tagging is still an emerging technology: only 5% of USA online leisure travelers—slightly more than 5 million of the 114 million USA adults who travel for leisure and go online regularly—tag Web pages or other content on sites like del.icio.us or Flickr™ (Epps, 2007). Moreover, the social capabilities of tagging for community building and social collaboratively construction of concepts' meaning and of travel experiences could have numerous innovative applications in tourism as well. For examples, travelers may be enabled and offered the opportunity to build structural tags in a text using XML for creating word groups or simply finding appropriate keywords to describe a travel experience. This would offer additional options to Internet users to collaboratively develop travel itineraries and search of travel information with others sharing a similar profile with them.

Business Applications for City Marketing

Because of the power of the folsconomy to provide enhanced user-value and influence the results/page ranking of search engines' search, many tourism firms nowadays include and consider tagging when designing their websites and e-business strategy. For example, Thomson's website provides an affiliate link to deli.ci.ous.com (http://www.thomson.co.uk/), so that its users can tag and sort its website content through this technology.

However, tourism firms may adopt different strategies regarding the way they use and incorporate tagging into their websites (Epps et al., 2007). For example, Triporama.com has launched its tagging system, titled "Triporama Bookmarks", which allows its website users to download a free software in their PC for tagging content in their own words from anywhere on the Web to their Triporama group trip plan, which they can then share with their travel companions. Such a solution provides differentiation customer value for Triporama, because as travelers visit many different Websites while planning a trip, Triporama's bookmarking tool lets travelers collect, label, and share the content they have found on the Web with other members of their group. Users also have the option to make their tags publicly available, while Triporama also aims to edit and curate these public tags into features like "top 10" lists to give other users ideas for planning their own group trips. When redesigning its website, Sheraton introduced its "Vacation Ideas" feature whereby guests are invited to write stories about their hotel stay, users give Sheraton their consent for publishing their stories online and the entire story becomes a tag. A tag cloud is created, titled the "Buzz Barometer", whereby word occurring most frequently in stories appear in bigger fonts, while based on the number of stories shared containing different words a "Vacation Ideas page" is created (http://www.starwoodhotels.com/sheraton/index.html). For example, by clicking on the "Beach" guide brings up the five hotels whose stories mention "beach" most frequently (weighted by the number of stories relative to the size of the property). By making storytelling the method by which the tags are created, Sheraton has made tagging so friendly and easy for its guests that they do not even know that they are tagging content. Sheraton benefits from this tagging strategy because:

- It helps first time website users: Vacation Ideas gives travelers, not knowing where they want to go and/or not familiar with the Sheraton brand, a more creative, user friendly and understandable way to search and book hotels than the customary destination-based and company pushed search.
- It helps Sheraton to build long lasting relations with its guests by maintaining a close relationship with the travelers before, during and after the trip. This is because the website provides guests having stayed at a Sheraton's hotel to return to the Sheraton's website in order to contribute, solicit or read other travelers' stories.
- It improves organizational learning, since Sheraton gets insights into Sheraton's hotel properties and customers' experiences. Instead of conducting expensive and time consuming research, Sheraton uses tagging as a simple and reliable way for gathering customer feedback and intelligence about its products and services. Based on the customer knowledge that is gained, operation managers can improve organizational processes, while marketers get to know what and how customers think and talk about the brand, in order to better position the Sheraton brand in the market and enhance the guest experience at the hotels in ways that reinforce guests' perspective of the Sheraton brand.

Yahoo!® Travel introduced new tagging features into its Yahoo!® TripPlanner that enable users to tag their own or others' Trip Plans with suggested or custom tags, which are later analyzed and used by Yahoo! for identifying and feeding recommendations on the Yahoo! Travel home. Users can set their preferred level of privacy at the level of the Trip Plan (private, shared with invited friends, or visible to any user). Users are provided around 30 tags (e.g. ("budget," "luxury", "weekend", "honeymoon") to choose from for labeling trip plans from the style of the trip, however, Yahoo! monitors the tags used most frequently by its users for augmenting its list of proposed tags. By using taxonomy-directed tagging, Yahoo!® eliminates many of the inherent problems of folksonomies created when users label similar things differently using synonyms or different forms of the same word (e.g. lodging, accommodation, hotel etc.). Tagging has helped Yahoo!® to: a) make its content (750,000 Trips Maps, photos, users' comments etc.) more useful, accessible, searchable and understandable to its users; and b) to gather, analyze and use customers' intelligence (where they live, where they have traveled, and what content they have viewed) and further refines that knowledge through the lens of the tags they use to search in order to create targeted, personalized recommendations for destinations and deals sold through its vertical search website

Yahoo!® FareChase. In other words, tagging helps to further refine the collaborative filtering process that Yahoo! uses in order to provide personalized recommendations and suggestions to its users. Personalized suggestions for cross and up sales can significantly drive and enhance booking and sales levels as well as provide additional functional and emotional value to website users that in turn enhances customer loyalty.

In the same vein, the official destination portal of Holland uses tagging technology in order to provide travelers an easy way to search the website content and its multimedia information (video, photos etc.) (http://us.holland.com/). Actually, tags are used as an user constructed and defined search engine rather than providing a search engine designed by the website developer that reflects a top down business defined search process. Tag clouds appear on the left with different font sizes to reflect words used more or less frequently, while "Top 10" suggestions for each tag (e.g. restaurant) are also constructed and updated continuously when new content and UGC is shared on the website. The portal also provides users the possibility to comment each others' contributions and comments as well as to tag the Holland's webpage content by using different social booking technology such as Digg and Furl.

Overall, it becomes evident that city tourism organizations should consider including tagging into their websites, as tagging can help them overcome the following issues (Epps, Harteverldt & McGowan, 2007):

- Very frequently websites do not speak the same language and they do not use the same terminology as their users. City destination organizations should consider using tagging in order to make their website content more accessible, understandable and appealing to its users. For example, the marketer of a city might promote as the major value of the destination its easy accessibility by air transportation, however, travelers may perceive as the most valuable feature of the destination the fact that it is "a safe city to walk around".
- Tagging can help and further enhance keyword search by supporting nuanced, adjective-based searches. Tagging also enables social search whereby users can see who has tagged something, how credible or relevant its suggestions are based on his/her profile and his/her evaluation by other users.
- Tagging helps organize and display user-generated content uploaded on websites. As more and more city tourism organizations invite their users to upload and share their UGC (e.g. reviews, itineraries, photos, videos, and podcasts), they later struggle to make this UGC relevant and accessible to their users and tags can help in addressing the latter.
- City tourism organizations can gather reliable and timely customer intelligence and feedback regarding the image of their destination, the mental maps of their travelers and how they view and perceive their destination etc. Such customer knowledge can be later used for marketing campaigns as well as for improving the products and services of the city as a destination.
- Customer information gathered through social tagging can also be used for improving search engine optimization campaigns. For example, words used frequently as tags by travelers can be used as metatags-metadata for building the portal's website as well as for spending money on keyword sensitive search engines such as Google™ AdWords.

However, when deciding whether and how to use tagging, city tourism organizations responsible for the development of the city portal should also decide the process, the policy and the way for developing their tagging system regarding the following issues: a) does the company edit the tags incorporated by users? This is important specifically if tags are uploaded with spelling mistakes or they include anti-social and embarrassing words. In other words, editing and a clear policy may be required in order to protect the consistency, the ethos and the good image of the website; b) are the tags and taggers' profiles made publicly available for everyone? What consent and agreement are required to take from the users and how the privacy policy of the website should be amended to incorporate the former?; c) is a software going to be purchased to manage the tagging process or is this going to be done manually? Are the required and skilled labor sources available? and d) how tags are going to be created? Are the tags going to be provided by the users or are the tags restricted by the website owner? Forrester (2006) recommends that companies use *taxonomy-directed* tagging, as it makes the tagging process more efficient and easier to use and it promotes consistency among tags. This is because when users create a tag, they can choose from existing suggested tags, or they can add their own.

Wikis

Definition, Features and Use

Wiki is a piece of server software permitting users to freely create and edit (hyperlinked) content via any browser and without the need to have access to and know to use any programming language. A wiki is a collaborative website whose content can be edited by anyone who has access to it. Wiki features include easy editing, versioning capabilities, and article discussions. So, wiki technologies enable users to add, delete, and in general edit the content of a website. Wiki users-creators are notified about new content, and they review only new content. As a result, such websites are developed collaboratively through their users, and a wiki becomes a collaboratively expandable collection of interlinked webpages, a hypertext system for storing and modifying information—a database, where each page is easily edited by any user with a forms-capable Web browser client. Neus and Scherf (2005) defined wiki as web content management systems allowing collaborative creation, connection and edition of contents, while Pereira and Soares (2007) defined wiki as a shared information work space that facilitates access to information content, organizational communications, and group collaboration. In other words, wikis represent another way of content publishing and communication as well as for group collaboration. In this vein, wikis and blogs have some similarities but they differ regarding the notification of new content, editing format, and structure. In other words, *'a wiki can be a blog, but a blog does not have to be a wiki'*.

Impact on Tourism Demand

The most popular wiki is the famous online encyclopedia, titled wikipedia.com, that is created and continually updated by its users. In tourism, wikitravel.org represents a wiki based effort of Internet users to collaboratively create and continuously update an online global travel guide including world-wide destinations. The number of readers, creators and content at wikitravel.org are continuously mushrooming. At wikitravel.org, one can find guides for any destination irrespective of its size and/or geographic location, as well as create a guide for any destination that he/she wishes. Wikitravel.com is further enriched with other web 2.0 tools and technologies such as maps, tags, podcasts etc.

Business Applications for City Marketing

Many tourism organizations take the opportunity to promote and create links to their websites through wikitravel.com in order to create and drive traffic to their own websites (http://wikitravel.org/en/London). Many other destination management organizations exploit and incorporate the wiki technology in their website portals in order to enable its users (travelers and locals) to collaboratively create and share their perceptions and mental images and opinions about their destination. For example, the National Library of Australia has included a wiki on its portal (http://wiki.nla.gov.au) inviting users to share their understanding and knowledge of local Australian dances as well as negotiate their meaning and create metaknowledge by synthesizing different views and perspectives. The National Library of Australia has also developed a wiki and social network website (www.pictureaustralia.org) whereby users can share their pictures about Australia and tell their story. In this way, the Library aims to help democratize history and establish a collective memory of places and events around the country. Ancient Times website (http://ancient.arts.ubc.ca/community.html) includes several collaborative tools, such as a wiki, blog and an arts metaverse enabling any user and history student to collaboratively develop and negotiate the meaning and construction of old cities and destinations, such as Giza and Athens. These cultural guides can significantly enhance the appeal and the interpretation of the cultural artifacts of historical cities and destination by providing several edutainment services and benefits to their users/visitors (Sigala, 2005a). Other wiki applications can also be provided on the city portals in order to boost website loyalty, repeat traffic, and travelers' desire to visit the destination. For example, a destination organization can design and incorporate a wiki on its portal for enabling potential travelers and locals to exchange and collaboratively develop recipes of local dishes and food.

Podcasting and Online Video

Definition, Features and Use

Podcasting refers to the uploading of audio and video files by users on websites. The most popular website for sharing such content with others is youtobe.com. Podcasting represents repositories of audio (podcasts) and video (vodcasts) or "video podcasting" materials that can be "pushed" to subscribers, even without user intervention, through RSS aggregate feeds of audio and video content facilitating users to search the latest services. Podcasting-capable aggregators or "podcatchers" are used to download podcasts. These files can also be downloaded to portable media players that can be taken anywhere, providing the potential for "anytime, anywhere" learning experiences (mobile learning). Podcasting's essence is about creating content (audio or video) for an audience that wants to listen when they want, where they want, and how they want. Podcasting differs from webcasting. A podcast has a persistent site, capable of synchronizing with a portable multimedia device, e.g. an MP3 player or iPod, whereas webcasting is streamed from the internet and requires the user to be connected to the internet while playing or viewing the webcast files. Webcasting is closely related to real-time downloading and synchronous broadcasting. Podcasting adds spatial flexibility to the temporal flexibility that webcasting offers and affords itself for creating personally-customizable media environments. Podcasting offers customer value in terms of the flexibility possibilities to hear personalized content whenever and whatever device one wishes, e.g. one can download the "Economist"'s or "CNN"'s personalized news' and press releases to his/her iPod and listen to its favorite news while he/she driving at work. As podcasting does not rely on the visual senses, it allows users to carry out other tasks while listening.

Impact on Tourism Demand

Tourism experiences are intangible. One cannot experience, feel and try a travel experience before he/she buys and before he/she travels to a destination. As a result, the purchase risk of a travel—tourism experience is high and it is difficult to persuade a user for the qualities of a tourism service. Due to its multimedia features, podcasting helps users to better and easier evaluate travel alternatives by experiencing in someway a travel experience before they decide to buy and consume it and/or travel to a destination. This is because audio and video files of hotels, destinations, and other travel products created and uploaded for sharing by other users are considered as more unbiased information and not staged experiences produced by the supplier aiming to promote his/her own product as the best one. Podcasting has also been used as mobile guides for travelers, e.g. Virgin Atlantic provides through its website free podcasts-guides of cities whereby they fly to.

Impact on Tourism Supply

Many tourism suppliers are using Podcasting as a marketing, information and customer communication tool. For example, Jumeirah hotel uploads podcasts on its website for delivering and updating its potential guests about what is happening in its properties at every single day, and-or delivering to website users and potential buyers the experiences of VIPs that have stayed at their property. Tate Gallery enables their visitors that have experienced their paintings and exhibition to record themselves, upload their audio-video on the Tate Gallery website, and which others can later download and use them as a mobile interpretation guide while visiting the gallery. Orbitz.com provides podcasting of many destinations that travelers can download to their MP3 players and use them as guides while visiting the destination. In a similar way, MGM Grand Hotel Las Vegas has lauded online video on its website under the title "Maximum Vegas" in order to better illustrate to its potential guests the experience and services of its hotel and gaming resort. Similarly, city destination organizations should consider enhancing the content and marketing appeal of their website portals by enabling podcasting opportunities, i.e. either allow users to share content or push their own created podcasting content (e.g. http://www.visitlondon.com/maps/podcasts/, Podcasts at the official portal of London). For developing podcasts, city tourism organizations can outsource this function to companies such as soundwalk.com, podtrip.com and heartbeatguides.com that specialize in the development and dissemination of destination podcasts.

METAVERSES: MASSIVELY MULTIPLAYER ONLINE ROLE PLAYING GAME (MMORPG)

Definition, Features and Use

Metaverses are three dimensional virtual worlds whereby Internet users collaboratively play "online MMORPG games" with others. However, these platforms are wrongly perceived as "simple games" and "virtual" worlds, since they frequently represent an extension to our physical day-to-day world to which users add new socio-economic and political situations. MMORPG are games that are played by numerous players (e.g. millions of users) and they could be considered as an intermediate step from 'computer' to 'ambient' era. Some of these games (e.g. World of Warcraft) develop around a theme defining the goals of the game, while other games, such as SecondLife.com, there.com, cokemusic.com, habbohotel.com and http://play.toontown.com/about.php, encourage a free-style of playing, allowing the users to make what they want out of it. Metaverse environments are internet-based 3-D virtual world whereby their users, called residents, can interact with each other through motional avatars (an internet user's representation of him/herself) providing an advanced level of a social network service. Although it is difficult to measure the size and growth of such games, it is estimated that the market for massively multiplayer online games is now worth more than $1bn in the West world (Book, 2003). For example, one can simply consider the size of and growth of Second Life®® itself. Second Life® has more than 5 million users, while about half a billion US$ are being transacted every year on Second Life's® website (as reported on SecondLife.com on April 2007).

Impact on Tourism Demand

Tourists and travelers participate in such games either for fun and-or for ways of expression of oneself and for achieving satisfaction through task—accomplishment, self-actualization and creation—design of something new. For example, many people dream and try to become and excel on a profession that they could not achieve in their real life, others try to design a new product and service hoping that other players will adapt and pay for it and so they can gain money and/or head hunters would spot their talents and recruit them in their real or virtual companies.

Business Applications for City Marketing

Many tourism and travel related companies have already created their representative offices and headquarters in metaverse environments such as SecondLife.com. Embassies (e.g. that of Sweden), Tourism Authorities (e.g. that of Maldives) of many countries and many tourism companies (e.g. TUI, Burj Al Arab Hotel, Marriott, Costa Cruises) have created their offices and companies on islands of SecondLife.com for boosting their marketing practices such as enhancing customer communication and education about their products/services, building brand reputation and user communities, and achieving word of mouth (WOM) and advertising. Hyatt used residents of Second Life® and exploited their intelligence and knowledge for designing a new hotel concept, named as Aloft; architects and guests were involved in designing the hotel providing their feedback, preferences and specialist knowledge (read the related blog at http://www.virtualaloft.com/). As a result of the popularity of the new hotel, the first Aloft hotel will open and operate in real life in New York in January 2008. Apart from collaborative new product development, a firm can further exploit the social intelligence gathered and generated at SecondLife.com and other metaverse environments in order to conduct market research and to test new product ideas and new advertising campaigns, e.g. Toyota first tested the campaign of its new brand Scion on SecondLife® and then, widely broadcasted the new campaign in real life. Other companies, use SecondLife® for recruiting and identifying new talents e.g. CNN does head hunting of new journalists online.

Many destinations are also moving into the futuristic world of virtual reality and metaverse, as many city and country destination organizations create their virtual destinations. Netherlands Tourism Board recently opened a national tourism board in SecondLife® (http://us.holland.com/secondlife.php), the city of Galveston launched a virtual replica of itself in SecondLife® (http://www.galveston.com/secondlife/), providing their visitors with the chance to become part of an interactive community (Figure 10 and 11). The aim is to provide digital travelers the chance to take guided virtual tours, learn about the history, culture and daily life of the destination, and interact with new virtual friends from around the world. Tourism Ireland has also launched the world's first tourism marketing campaign in SecondLife® (http://dublinsl.com/index.php) including the sponsorship of a range of events and activities, including concerts, fashion shows, and photographic exhibitions, in Second Life's® replica city of Dublin. Dublin's representation in SecondLife.com is the first place-location in Ireland that the Tourism Board created its representation in Second Life®. Similiar to the Dublin creation, Amsterdam in Second Life® comes complete with Dutch signs, canals, trams and a lot of attention to detail. Overall, when investigating the impact of SecondLife.com on its residents' behavior, it becomes evident that historical landmarks and buildings such as Tour

Eiffel, London Bridge, Ajax Football stadium etc., have a great effect in building virtual communities of people spending a lot of time on dwelling them. Since it is apparent that real world modern-day cities and their landmark attraction are probably the most effective at driving and retaining visitor traffic, city tourism organizations should exploit this inherent advantage and exploit their cultural and heritage assets in metaverse environments for boosting their city brand name, recognition and promotion.

Mash-Ups

Definition, Features and Use

Mash-ups describe the seamlessly combination of two or more different sources of content and-or software for creating a new value added service to users. Many mash-ups enrich their services with some geographical content such as Google™ Maps; e.g. *The New York Times Travel Section's "36 Hours In..."* mash-up, which allows users to search the *"36 Hours in ..."* story archive from a Google™ Map. For example, when visiting the website traintimes.org.uk, one can see on real time where trains are located and when they will arrive at destinations, since the website combines information from Google™ maps, and information from the British rail website about train time tables, delays etc.

There are several mash-up applications in tourism such as new cyberintermediaries including mapping services (e.g. earthbooker.com, tripmojo.com, reservemy.com) and meta-search engines such as farecompare.com. Other examples include: www.43places.com that combines Flickr photos, RSS feeds and Google™ Maps with tagging and user-generated content, allowing users to share their favorite destinations; www.randomdayout.co.uk combines a number of data sources to create a mapped itinerary, using Virtual Earth (Microsoft's equivalent of Google™ Maps). An innovate example related to destination marketing management is the case of the city of Pennsylvania (http://www.visitpa.com): based on a project amongst Google™ Earth, Carnegie Mellon University, NASA, the Pennsylvania Tourism Office and the National Civil War Museum, virtual tourists would have the chance to view Pennsylvania's Civil War trails online. More sophisticated examples of mash-ups are the "Marco Polo" function on triptie.com and the "Trip Planner" function on Yahoo!® Travel, which allow users to integrate content from other websites into the user's own itinerary planning toolkit on the host website.

Impact on Tourism Demand

Travel decisions are complex and involve the searching, comparison and combination of several information located in many different websites. For example, a decision to travel to a destination requires various and a plethora of information about weather conditions, exchange rates, travel and accommodation alternatives and prices, attractions etc. As a result, tourists increasingly demand and expect to combine and cross-check information from different sources, so that they can better and easier make a holistic decision. For example, tourists may not be able to clearly understand where a hotel may be located when the description of the hotel websites states that the hotel is located on the beach, near the beach etc. Tourists easily get confused from different descriptions found in different websites. On the contrary, mash-up websites empowered with maps (e.g. earthbooker.com) enable users to see where exactly a hotel or other attraction is located (sometimes even locate the exact orientation and view of a hotel room and then decide whether to book this room at this hotel).

Business Implications for City Marketing

Mash-up applications have empowered the rise of new cyber- and info- intermediaries offering new sophisticated information services (e.g. flightcompare.com search and compare all flight information from different websites in order to provide comparable flight information within one webpage to its users). Moreover, many tourism suppliers and organizations also enrich their website content with maps in order to make it more user friendly and useful to their visitors, e.g. the official website of London and Dublin use Google™ Maps with geotags for enabling tourists to identify points of interest, hotels etc. Moreover, many companies leave their software as an Open Application Programme Interface (API), so that users can create limitless combinations of their services. For example, backstage.bbc.co.uk represents BCC's services and opportunities offered to its users, who are enabled to take content from the BBC, re-structure it and present it the way they prefer. Enabling user innovation is another way that companies aim to exploit on users' creativity and intelligence instead of investigating solely on company's R&D efforts.

FUTURE TRENDS

It has become evident from the above mentioned analysis that the two major impacts of web 2.0 and its UGC on consumer behavior and marketing practices are: 1) the electronic word-of-mouth that is created; and b) the opportunities to build and maintain customer communities for enhancing the practices of Customer relationship Management and social marketing. Exploiting web 2.0 for city marketing can have a tremendous effect on the marketing effectiveness, since, as the following analysis and discussion illustrates, both previous issues significantly affect consumer loyalty and purchasing behavior.

Web 2.0 and Electronic Word-of-Mouth (WOM)

Word-of-mouth (WOM) is very important in tourism and in services in general, since objective information about a service experience cannot be easily provided before one buys and consumes the services themselves. Services are intangible and so they are difficult to be tested, tried and evaluated before buying them. Consumers also tend to rely more on consumer reviews when purchasing high involvement products (Park, Lee & Han, 2007), such as several travel products e.g. a honeymoon trip, an adventure travel etc. In this vein, tourism decisions are very complex and risky. Indeed, the literature about information search in the tourism field has recognized the important role of WOM in travel planning and decision making (Hwang, Gretzell, Xiang & Fesenmaier, 2006; Murphy, Moscardo & Benckendorff, 2007). WOM has been found to be one of the most influential information sources for travel (Morrison, 2002). Research has also shown that those with past experience with a specific travel destination and that engage in digital word-of-mouth communication are most likely to be the most preferred and the most influential source of information in the pre-trip stage of travel decision making (Crotts, 1999).

To make travel decisions easier, travelers need to reduce the inherent information complexity of travel decisions as well as the risk related to the service firm (i.e. is that a good and reliable company), the service risk (i.e. is that a service that fits my preferences and needs?) and the purchasing risk (i.e. is that a trustworthy booking and buying channel to use for buying a travel service?). To achieve that, consumers use recommendation-based heuristics and other users' feedback to reduce uncertainty, eliminate the related risks as well as filter and process the plethora of information that must be processed when making decisions (Olshavsky & Granbois, 1979). WOM-based information is heavily used and trusted by consumers for taking travel decisions, because it is seen as more vivid, easier to use, and more trustworthy as it is based on actual experience and typically provided without direct benefits (Smith, Menon & Sivakumar, 2005).

As demonstrated in the above mentioned analysis of UGC in Web 2.0 websites, electronic WOM can take different names and forms such as virtual opinion platforms, consumer portals, social networking, blogs' comments, tag words, podcasting, virtual communities and online feedback mechanisms (Armstrong & Hagel, 1997; Bellman, 2006; Sigala, 2008). Users of Web 2.0 websites and tools may post their own experiences, videos, share their opinion, give advice, or look for answers to their questions. Consumers also perceive electronic WOM to be a reliable source of information (Gruen, Osmonbekov & Czaplewski, 2006). Dellarocas (2003) identified three different characteristics of online WOM relative to traditional WOM: 1) electronic WOM is larger in scale (both in terms of quantity and people impact) due to the Internet's low-cost and networking features; 2) electronic WOM is a powerful and reliable market research tool giving organizations the ability to monitor on real time their operations; and 3) it is difficult to convey contextual cues (e.g. facial expression) through the Internet and peer review websites for example, and so not knowing or seeing who the information provider is makes, it is harder to interpret the subjective information in online interaction. To address this problem, websites often display demographic or other data about reviewers (for example, the length of membership, their location, etc.) in order to help build credibility and trust. Websites may even provide the possibility to users to upload and share their own feedback and evaluation (by incorporating each review into a rating of the reviewer) about the quality of the reviews written by other members. Moreover, because Web 2.0 enables users to identify and use personalized and contextual information (e.g. look at what others' with similar profiles are saying), electronic WOM is considered as both more relevant and unbiased than traditional WOM, whereby one cannot easily track and relate the content with the profile of its original messenger.

Smith, Menon and Sivakumar (2005) claim that consumers prefer such peer recommendations over other forms of input, while Amis (2007) advocated that social network sites have as much influence on consumers as television and more than newspapers. Statistics actually provide evidence of consumers' reliance on electronic word-of-mouth. More than 80% of web shoppers said they use other consumers' reviews when making purchasing decisions (Forrester, 2006). eMarketer (2007d) reports that nearly 6/10 consumers prefer websites with peer-written reviews, and that websites with reviews experience greater conversion rates.

Overall, Dellarocas (2003) summarized organizations' benefits of electronic WOM in the following: brand building; customer relationship management; customer acquisition; addressing customer complaints; market research; product development; quality control and supply chain quality assurance activities.

Web 2.0, Customer Relationship Management (CRM) and Social Marketing

The major aim of CRM is to personalize business services and products as well as develop a 1:1 communications and long lasting relation with profitable customers (Sigala, 2005b). eCRM also requires the development of customers' communities for providing loyal customers with functional, emotional and social benefits and value (Sigala, 2006). The previous section provided practical examples illustrating the way in which web 2.0 applications and tools enable the formation and development of customer virtual communities. By identifying and reviewing the limited related studies that have been conducted so far, the following analysis further supports the capability of web 2.0 to build virtual communities of users and enhance the community benefits (functional and emotional/social) to its users.

Ying and Davis (2007) and, Lento, Welser, Gu and Smith (2006) illustrated how blogs create and maintain strong online communities through their social ties tools such as blogrolls, permalinks, comments and trackbacks. Indeed, many authors (e.g. Lin, Su & Chien, 2006; Ying & Davis, 2007) have started to apply social network analysis for measuring and illustrating the social bonds, networking and communication structures created within the blogsphere. Li and Stronberg (2007) summarized blogs' benefits for firms as follows: search engine optimization; e-word-of-mouth (eWOW); improved brand perception and visibility; instantaneous client feedback; market research and insight; increased sales efficiency; and reduced impact from negative user-generated content. Damianos et al. (2007) advocated that social bookmarking generates social influence and bonds as well as creates value by: enabling resource management, information sharing and discovery, expert finding, and social networking; providing teams with a place to share resources; forming and supporting social networks around interest areas; and feeding expertise finding and user profiling. Awad and Zhang (2007) discussed the marketing benefits of eWOW generated in online review communities and debated firms' efforts and strategies addressing it. By examining the communication tools and social cues of myspace.com, Dwyer (2007) demonstrated the impact of social networking sites on developing customer interrelations and communities. In their study of videos' tags on Del.ici.ous, Paolillo and Penumarthy (2007) found that social tagging can generate community benefits such as: easy retrieval (as users use words they can remember and have useful meaning to them); contribution and sharing; attract attention; opinion expression; play; and self-presentation. Thus, since tagging can be used for providing functional services, creating social ties, market research on users' opinions and interests, and WOW, social tagging's ability in creating user communities is evident. Forrester (2006) demonstrated web 2.0's ability to generate customer and business value in different processes: customer service (e.g. community self-service savings); sales (e.g. community loyalty and sales reduces commissions and price competitions); marketing (e.g. credibility of eWOW); production (e.g. co-design reduces waste); and R&D (e.g. community input raises success rate).

A significant amount of literature also highlights the business benefits from developing virtual communities particularly in the area of CRM and social marketing. Analytically, Wang and Fesenmaier (2004) illustrated that virtual tourist communities are useful for managing customer relations by: attracting customers through in-depth, focused and member-generated content; engaging customers through social interactions; and retaining customers through relation building with other members. Online communities also build customer value (Wang & Fesenmaier, 2004) by generating users with all types of relational benefits namely functional, social, hedonic and psychological (Gwinner, Gremmler & Bitner, 1998). Kim, Lee and Hiemstra (2004) provided evidence of the impact of virtual communities on travelers' loyalty and product purchase decision making. Andersen (2005) explored the use of online brand communities for developing interactive communication channels and establishing social and structural bonds with devoted users. Jang, Ko and Koh (2007) showed that online brand communities posses and develop features - such as, quality and credibility of information, service quality, member interaction and leadership, brand reputation and (intrinsic and extrinsic) rewards for members' activities- that in turn, contribute to increased users' brand loyalty, commitment and sales. Erat, Desouza, Schafer-Juger and Kurzawa (2006) discussed how different types of communities of practice (e.g. B2C, C2C) can be used for acquiring and sharing customer knowledge in order to improve business processes and performance. Beyond collecting customer knowledge, online communities can also be used for co-operating with customers for New Product Development (NPD) and innovation (Rowley, Teahan & Leeming, 2007). A plethora of cases and research studies (e.g. in Lagrosen, 2005; Pitta & Fowler, 2006) reflects the possibility to use virtual communities for NPD as well.

Proposed Models for Exploiting Web 2.0 in Enhancing Marketing Communication and CRM

The previous analysis and industry examples illustrate that web 2.0 tools and applications have a twofold impact on the way CRM is implemented: 1) web 2.0's networking and connectivity capabilities provide enormous opportunities to communicate and co-operate with customers and industry partners in many different directions (e.g. many-to-one, many-to-many) (Table 1) the social intelligence and knowledge created collaboratively in web 2.0 platforms (i.e. the user-generated content) can be exploited in different ways for identifying, developing, enhancing and maintaining relations with profitable customers (Table 2).

Table 1. Web 2.0 extended CRM implementation

	Low market integration	High market integration
High customer integration	**Many-to-one** Target: clients' networks Active customers' involvement *e.g. Lonelyplanet.com, Sheraton.com*	**Many-to-many** Co-exploitation of customers' profiles with other network partners *e.g. mash-ups, earthbooker.com, flightcompare.com*
Low customer integration	**One-to-one** Target: individual customers	**One-to-many** Ecosystems of partners offering a seamless experience to individual clients (cross-selling, products' bundling) *e.g. travelocity.com*

In other words, CRM cannot anymore be considered as synonymous to one-to-one communication and personalized service at an individual basis. Web 2.0 augment CRM practices and implementation to include various forms of communications with clients and business partners. Following Gibbert, Leibold and Probst (2002), Table 3 reflects a two dimensional matrix, whereby the vertical axis represents how firms integrate customers into their value chains and the horizontal axis represents the integration of business partners into the firm's value chain. Companies can use web 2.0 technologies to communicate and enable dialogues and interactions not only between them and their customers, but also between customers themselves (C2C), between business partners, among all of them etc. When engaged in two directional communication both customers and partners, firms can involve the former in their value chain in order to create customer value and benefits. For example, as explained earlier, when customers communicate with other customers in virtual communities, customers provide social and emotional support to others as well as functional benefits (e.g. free consultancy in trip planning). Also, when co-operating and sharing content and applications with other businesses (e.g. in mash-up websites), firms can collaborate with and integrate other partners in their value chain in order to provide additional services to their clients, e.g. a holistic tourism product-services such as a dynamic packaging.

Moreover, in developing successful relationships with profitable clients, firms need to understand and manage all phases through which relations are evolved, as each phase is characterized by differences in behaviors and orientations and so, it requires different CRM approaches. Theory and practical evidence has shown that customer relations evolve over three major distinct phases related to the customer lifecycle (see Sigala, 2008): initiation, maintenance and retention or termination. Hence, all CRM implementation models reflect practices that col-

Table 2. Exploiting social intelligence for managing and enhancing customer relationships through their lifecycle

Phase	Type of customer information/intelligence	CRM implementation activities
Acquisition	Of the customer information: transaction and personal data	Create brand awareness and recognition amongst customers and virtual communities by building and supporting electronic word-of-mouth
		Develop brand reinforcement and trust by educating and informing customers about the brand, its services, functionalities etc
		Use customer intelligence in order to identify and target new customers, e.g. clone the profile of existing product-service users, use the connections and recommendations of existing customers etc.
		Use customer intelligence to understand how customers use the service, what functionalities they prefer or not
		Use customer intelligence for profiling customers
Retention	For the customer information: relationship and product data	Use customer intelligence for enhancing customer service and transactions
		Use customer intelligence for personalizing services and products
		Build and develop community of customers-users
		Use customer intelligence for innovation & NPD
Expansion	For the customer information: relationship and product data	Use customer intelligence for cross selling, e.g. suggest compatible products based on other users' purchases
		Use customer intelligence for up-selling
		Use customer intelligence for developing affiliation and loyalty programmes
Win back	By the customer information: feedback and monitoring data	Use customer intelligence (feedback, reviews etc) for identifying pitfalls and faults
		Use customer intelligence and communities for handling customers' complaints
		Use customer intelligence and communities for monitoring and managing the firm's reputation, status and prestige

lect and use three forms of customer information / intelligence in order to manage each relational phase. "Of-the-customer" information includes customers' personal and transaction data for understanding and measuring their profile, e.g. sales, profitability, purchasing patterns, preferences. "For-the-customer" information refers to product, service and firm information perceived as useful by clients for making more informed decisions. "By-the-customer" information reflects customer feedback (e.g. customer complaints, suggestions, reviews) used for new product development or business improvement. As illustrated previously, Web 2.0's user-generated content mushrooms these three types of customer information and provides firms with several opportunities not only to collect, but also to get access to such types of customer intelligence. In other words, Web 2.0 platforms can be exploited as a free and real time market research and intelligence tool. Table 2 summarizes how firms can exploit web 2.0 tools and platforms for collecting and analyzing this customer intelligence for augmenting and supporting their CRM practices.

Overall, it becomes evident that web 2.0 enabled CRM reflects a cultural shift from product *'designing for customers'* to *'designing with'* and *'design by'* customers. For firms to achieve such a cultural shift, crucial organizational changes should also take place. Importantly, the role of marketers should be changed from being sales people to becoming community builders and perceiving customers not as targets to identify and sell, but as partners to collaborate with. Firms should also realize that they should use customer intelligence not only for learning about their customers and identifying new target markets (opportunistic behavior), that they should also use customer intelligence for learning and improving processes and products with their customers as well as with different business partners (partnership relation). In other words, firms derive and realize maximum benefits when they exploit web 2.0 tools for establishing and maintaining co-creation and co-learning adaptable and flexible ecosystems with their customers and business partners (Table 3).

CONCLUSION

Internet users and travelers are nowadays empowered to create and synthesize in their own way the travel content that they also wish to distribute and share it with others through users' controlled distribution channels. In this vein, Web 2.0 technologies enable Internet users to become the co-producers, the co-designers, the co-marketers and the co-distributors of tourism experiences and services as well as the co-entrepreneurs of new tourism products and new e-business models. As the diffusion of web 2.0 applications becomes wide and consumers incorporate them within their daily and professional life, travelers expect tourism firms and organizations to provide similar Web 2.0 enabled services. The previous analysis aimed at identifying and illustrating the business implications created for tourism and hospitality enterprises as well as strategies and tactics that they can adopt for eliminating threats while exploiting the arising opportunities. Therefore, as Web 2.0 is here to stay, it is evident that unless a city tourism organization adopts and incorporates Web 2.0 tools into its e-business model and strategies for marketing and managing its destination, the competitiveness of the latter is threaten. Nevertheless, in order to be successful, the adoption and use of any web 2.0 tools should be accompanied with appropriate organizational and cultural changes within the firm regarding the roles, job descriptions and tasks of its staff, users and business partners. Further research is required in order to understand and examine how firms are achieving and trying to implement such organizational changes when incorporating web 2.0 into their e-business model.

REFERENCES

Adam, J., Cobos, X., & Liu, S. (2007). *Travel 2.0: Trends in Industry Awareness and Adoption*. New York University and PhoCusWright Inc.

Amis, R. (2007, May). You can't ignore social media: How to measure Internet efforts to your organization's best advantage. *Tactics*, 10.

Andersen, P. H. (2005). RM and brand involvement of professionals through Web-enhanced brand communities: Coloplast case. *Industrial Marketing Management*, *34*(3), 285-297.

Armstrong, A., & Hagel, J. (1997). *Net gain: Expanding markets through virtual communities*. Boston, MA: Harvard Business School Press.

Awad, N. F., & Zhang, S. (2007, January). *Stay out of my forum! Evaluating firm involvement in online ratings communities*. Paper presented at the 40th HICSS, Waikoloa, Big Island, Hawaii.

Bellman, S., Johnson, E., Lohse, G., & Mandel, N. (2006). Designing marketplaces of the artificial with consumers in mind. *Journal of Interactive Marketing*, *20*(1), 21-33.

Bickart, B., & Schindler, R. M. (2001). Internet forums as influential sources of consumer information. *Journal of Interactive Marketing*, *15*(3), 31-40.

Blood, R. (2000). Weblogs: A history and perspective. Retrieved October 15, 2007, from http://www.rebeccablood.net/essays/weblog_history.html

Book, B. (2003, July). *Traveling through cyberspace: Tourism and photography in virtual worlds*.

Paper presented at the conference Tourism & Photography: Still Visions - Changing Lives in Sheffield, UK.

Bughin, J. R. (2007, August). How companies can make the most of user-generated content. *The McKinsey Quarterly*, Research in Brief, Web Exclusive.

Crotts, J. (1999). Consumer decision making and prepurchase information search. In Y. Mansfield & A. Pizam (Eds.), *Consumer behavior in travel and tourism* (pp. 149-168). Binghamton, NY: Haworth Press.

Damianos, L., Cuomo, D., Griffith, J. Hirst, D., & Smallwood, J. (2007, January). *Adoption, utility, and social influences of social bookmarking.* Paper presented at the 40th HICSS, Waikoloa, Big Island, Hawaii.

Dellarocas, C. (2003). The digitization of word-of-mouth: Promise and challenges of online feedback mechanisms. *Management Science, 49*(10), 1407-1424.

Du, H. S., & Wagner, C. (2006). Weblog success: Exploring the role of technology. *International Journal of Human-Computer Studies, 64*, 789-798.

Dwyer, C. (2007, January). *Digital relationships in the 'MySpace' generation: Results from a qualitative study.* Paper presented at the 40th HICSS, Waikoloa, Big Island, Hawaii.

eMarketer (2007a). *Web 2.0 sites draw more visitors.* Retrieved May 2, 2007, from http://www.eMarketer.com.

eMarketer (2007b). *UGC users outnumber creators.* Retrieved July 2, 2007, from http://www.eMarketer.com.

eMarketer (2007c). *The rising roar of word-of-mouth.* Retrieved June 29, 2007, from http://www.eMarketer.com.

eMarketer (2007d). *Reviews boost e-commerce conversions.* Retrieved May 25, 2007, from http://www.eMarketer.com.

Epps, S. R. (2007). *Demystifying tagging for travel sellers.* Forrester Research Report.

Epps, S. R., Harteveldt, H. H., & McGowan, B. (2007). *Executive Q&A: Social tagging for ebusiness. Answers to E-Business professionals' common questions about social tagging.* Forrester Research.

Erat, P., Desouza, K., Schafer-Jugel, A. & Kurzawa, M. (2006). Business customer communities and knowledge sharing: studying the critical issues. *European Journal of IS, 15*, 511-524.

Farmer, J. (2004). Communication dynamics: Discussion boards, Weblogs and the development of communities of inquiry in online learning environments. In R. Atkinson, C. McBeath, D. Jonas-Dwyer, & R. Phillips (Eds.), *Beyond the comfort zone* (pp. 274-283). *Proceedings of the 21st ASCILITE Conference.*

Forrester Research (2006). *Social computing.* Retrieved September 14, 2007, from http://www.forrester.com.

Gibbert, M., Leibold, M., & Probst, G. (2002). Five styles of CKM, and how smart companies use them to create value. *European Management Journal, 20*(5), 459-469.

Gruen, T. W., Osmonbekov, T., & Czaplewski, A. J. (2006). eWOM: The impact of customer-to-customer online know-how exchange on customer value and loyalty. *Journal of Business Research, 59*, 449-456.

Gwinner, K. P., Gremmler, D. D., & Bitner, M. J. (1998). Relational benefits in services: The customer's perspective. *Journal of the Academy of Marketing Science, 26*(2), 101-114.

Harteveldt, H., Johnson, C. A., Epps, S. R., & Tesch, B. (2006).*Travelers embrace social com-*

puting technologies. Guidelines for travel e-commerce and marketing executives and managers. Cambridge, MA, USA: Forrester Research.

Haven, B. (2007). *Making podcasts work for your brand.* Cambridge, MA, USA: Forrester Research.

Hwang, Y., Gretzel, U., Xiang, Z., & Fesenmaier, D. (2006). Information search for travel decisions. In D. Fesenmaier, H. Werthner & K. Wöber (Eds.), *Destination recommendation systems: Behavioral foundations and applications* (pp. 3-16). Cambridge, MA: CAB International.

Jang, H. Y, Ko, I. S., & Koh, J. (2007, January). *The influence of online brand community characteristics on community commitment and brand loyalty.* Paper presented at the 40th HICSS, Waikoloa, Big Island, Hawaii.

Kim, W. G., Lee, C., & Hiemstra, S. J. (2004). Effects of an online virtual community on customer loyalty and travel product purchases. *Tourism Management, 25*(3), 343-355.

Lagrosen, S. (2005). Customer involvement in NPD: A relationship marketing perspective. *European Journal of Innovation, 8*(4), 424-436.

Lento T., Welser, H. T., Gu, L., & Smith, M. (2006). The ties that blog: Relationship between social ties and continued participation in blogs. *Workshop on Weblogging* Edinburgh.

Li, C., & Stromberg, C. (2007). *The ROI of blogging.* Cambridge, MA: Forrester Research.

Lin, Y., & Huang, J. (2006). Internet blogs as a tourism marketing medium: A case study. *Journal of Business Research, 59,* 1201-1205.

Lin, Y., Su, H. Y., & Chien, S. (2006). Knowledge-enabled procedure for customer relationship management. *Industrial Marketing Management, 35,* 446-456.

Morrison, A. (2002). *Hospitality and tourism marketing* (3rd ed.). Albany, NY: Delmar.

Murphy, L., Moscardo, G., & Benckendorff, P. (2007). Exploring word-of-mouth influences on travel decisions: friends and relatives vs. other travelers. *International Journal of Consumer Studies, 31*(5), 517-527.

Neus, A., & Scherf, P. (2005). Opening minds: Cultural change with the introduction of open-source collaboration methods. *IBM Systems Journal, 44*(2), 215-225.

Olshavsky, R.W., & Granbois, D.H. (1979). Consumer decision making: Fact or Fiction? *Journal of Consumer Research, 6,* 93-100.

Paolillo, J. C., & Penumarthy, S. (2007, January). *The social structure of tagging internet video on del.icio.us.* Paper presented at the 40th HICSS, Waikoloa, Big Island, Hawaii.

Park, D. H., Lee, J., & Han, J. (2007). The effect of online consumer reviews on consumer purchasing intention: The moderating role of involvement. *International Journal of Electronic Commerce, 11*(4), 125-148.

Pereira, C. S., & Soares, A. L. (2007). Improving the quality of collaboration requirements for IM through social networks analysis. *International Journal of Information Management, 27,* 86–103.

Pitta, D., & Fowler, D. (2005). Online consumer communities and their value to new product developers. *Journal of Product & Brand Management, 14*(5), 283-291.

Rosenbloom, A. (2004). The blogosphere. *Communications of the ACM, 47*(12), 31–33.

Rowley, J., Teahan, B., & Leeming, E. (2007). Customer community and co-creation: A case study. *Marketing Intelligence & Planning, 25*(2), 136-146.

Sigala, M. (2003). Developing and benchmarking internet marketing strategies in the hotel sector in Greece. *Journal of Hospitality & Tourism Research, 27*(4), 375–401.

Sigala, M. (2005a). In search of online post-modern authenticity: Assessing the quality of learning experiences at eternalegypt.org. In M. SIGALA & D. Leslie (Eds.), *International Cultural Tourism: Management, implications and cases* (pp. 123–136). Oxford, UK: Butterworth Heinemann, Elsevier.

Sigala, M. (2005b). Integrating customer relationship management in hotel operations: Managerial and operational implications. *International Journal of Hospitality Management, 24*(3), 391–413.

Sigala, M. (2006). e-Customer relationship management in the hotel sector: Guests' perceptions of perceived e-service quality levels. *Tourism: An International Interdisciplinary Journal, 54*(4), 333-344.

Sigala, M. (2008, January). Developing and implementing an eCRM 2.0 strategy: Usage and readiness of Greek tourism firms. *ENTER 2008 conference*, Innsbruck, Austria.

Smith, D., Menon, S., & Sivakumar, K. (2005). Online peer and editorial recommendations, trust, and choice in virtual markets. *Journal of Interactive Marketing, 19*(3), 15-37.

Wang, Y., & Fesenmaier, D. R. (2004). Modeling participation in an online travel community. *Journal of Travel Research, 42*(3), 261-270.

Winer, D. (2005). What is a .River of News. style aggregator? *Really Simple Syndication*. Retrieved June 22, 2005, from http://www.reallysimplesyndication.com/riverOfNews

Ying Z., & Davis, J. (2007, January). *Web communities in blogspace*. Paper presented at the 40th HICSS, Waikoloa, Big Island, Hawaii.

Yuan, Y., Gretzel, U., & Fesenmaier, D. R. (2006). The role of information technology use in American convention and visitors bureaus. *Tourism Management, 27*(2), 326-341.

Chapter XII
Developing Patterns for Thinking About City Marketing Initiatives

José-Rodrigo Córdoba
University of Hull, UK

Nicolas Jullien
Môle Armorcain de la Recherche sur la Société de l'Information et l'Usages d'Internet, France

Jocelyne Trémenbert
Môle Armorcain de la Recherche sur la Société de l'Information et l'Usages d'Internet, France

ABSTRACT

This chapter defines three different patterns to understand how city marketing initiatives are designed and implementing, and how information and communication technologies (ICTs) can support the implementation of such patterns. The word pattern is used to characterize particular ways of thinking which embed values and beliefs about city marketing. The first pattern (idealist) assumes that a visionary type of image of cities can be developed to attract people to cities. The second pattern (strategic) emphasizes dialogue and participation in shaping up marketing initiatives by considering the underlying beliefs and values of people and how these can be "branded". The third pattern (power-based) aims to be inclusive of intended and unintended effects of marketing so as to develop initiatives for the people and by the people. In the chapter, the role(s) of ICTs are discussed. Experience of using these patterns to understand the situation of two cities (Hull, UK, and Brest, France) leads us to suggest the intertwining of these patterns, and hence the importance of people's engagement to facilitate better use of ICTs in the context of city marketing initiatives.

INTRODUCTION

Current thinking in city marketing emphasizes the renewal of cities through imaging, flagship projects and improvement of quality of life. Cities are important part of societies and 'hubs' to sustain economic and cultural growth as suggested by those advocating a network paradigm for society. Despite efforts to increase cities attractiveness using marketing, we still see often how cities continue marginalizing certain groups at the expense of others by using information technologies in a very uni-directional ways (Criado & Ramilo, 2003). This, together with marketing strategies guided by only economically driven goals make little or no impact in important citizens groups.

Take for example the city of Hull in the UK, a medium size city with approximately 300,000 inhabitants and with one of the largest council estates in Europe. Economic development after World War II has seen a mixture of results. Generally speaking, the city is still regarded as a pocket of social deprivation, low educational levels and little in terms of economic opportunities despite being declared as a 'pioneering city' by council officials. The city has offered a number of opportunities for immigrants and asylum seekers, and one can say the city can still get by. What is interested is whilst those in charge of city development aim to create a city image through regeneration of the city centre, the provision of better transport, shopping and leisure facilities, the motivations for people to live and stay in the city are still unknown. These strategies are very similar to those adopted by the city of Brest (Le Cam, Ruellan, & Cabedoche, 2006). Both cities share a history of deprivation, demographic decline in surrounding rural areas and the loss of the fishing industry (Clout, 2006). Both cities thrive to become more attractive places for people to live.

These and other examples prompt us to ask a number of questions: But what are the main drivers about city marketing for policy makers and city officials? How can we better understand the reactions of those groups of people who will get affected by city marketing initiatives? What can be improved to guide use of information technologies in city marketing? From a systemic and socio-technical perspective, this chapter aims to provide some answers to these questions. It will do so by proposing a number of 'thinking patterns' that provide some insights to understand city marketing in the context of a worldwide network society paradigm which is affecting cities (Castells, 1991).

The word 'pattern' refers to a set of ideas, assumptions, beliefs and activities regarding the use of information technologies. These are shared across different groups and organizations, and inform information systems development (Hirschheim & Klein, 1989; Hirschheim, Klein, & Newman, 1991; Mingers & Willcocks, 2004). Through the definition of these patterns, the chapter will also refer to the role(s) of information technologies in improving the marketing of cities.

Our chapter aims to respond to what we see as a lack of theoretical elements to describe city marketing worldwide and in particular the use of information and communication technologies (ICTs). In this regard, the work of Kavaratzis (2005), seems to be an exception in proposing a conceptual framework which could be used by those responsible for ICTs for cities. The lack of theory is compounded by what we see as an uncritical acceptance of the rise of the 'entrepreneurial city' (Hall & Hubbard, 1998) which aims to rescue the economy and life in modern cities through marketing or branding (Paddison, 1993). Taking city marketing as a current point of analysis, we develop a number of ways of thinking about it on the basis that marketing is about "trying to have what the consumer wants" (Holcomb, 1994).

To begin our discussion, we continue where we left with a story about the city of Hull in the UK.

On Official and Unofficial Strategies

Official and unofficial marketing strategies of cities collide. In the afternoon of June 25th, 2007, the United Kingdom suffered one of the biggest floods in the last 25 years. The amount of rain that was expected for the whole of June was deposited in a matter of few hours and this was really felt in Hull. After the initial shock, the city's website showed some changes which were accommodated within the main website's structure. Two additional sections built in the main page: 1) Latest news and environmental health advice and 2) Flood support. In the first, people could access up to date news on schools; get to know what the council was doing (e.g. identifying first the victims of floods). There was also a facility to make donations on line to the Hull Flood Fund, which is an extension of online payments service (e.g. fines, council tax). In a secondary section of the website (linked to the main page), some advice was given on what to do in affected areas and in the case of new floods; a leaflet could be downloaded showing how the council has reacted by setting up customer service centers to provide advice. The information displayed in the city's website was focused on the administrative aspects of the flood.

There was very little which reflected how collective action could be undertaken by citizens and the council working together. Indirectly we happened to know that Hull city council 'delegated' some support to other organizations which legally were supposed to interact with citizens. In parallel to this, electronic forums showed a strong feeling that still is with many today: Being left to their own devices. This was despite the fact that the council were implementing a strategy to "identify those who are most vulnerable, in the aftermath of the floods, and ensure that we provide direct support to them quickly" (Hull City Council, 2007). In the meantime, members of the council successfully attracted the attention of national news by declaring openly that the UK central government regarded Hull as a 'forgotten' city (Hull, 2007). Whilst the city council tried to take away any blame, in other websites people made their voices heard to 'devolve' this blame. Action, reaction and 'being left alone to act' become part of marketing in the digital era of cities.

THE INFORMATIONAL CITY

Castells (1991) has warned us that the technological revolution affects societies and how cities would be transformed. Societies would become, according to Castells, 'nodes' or 'hubs' of economic and cultural development. This would leave smaller cities and rural areas challenged to become connected into a network of economic and cultural flows of information.

Castells (1991; 2001) develops and studies a number of hypotheses about the transformation of societies from their industrial (i.e. Fordist) state towards a stage of development that he calls 'informationalization', in which the roles of information and communication technologies (ICTs) are paramount. ICTs are the vehicles to allow capitalists societies to generate new modes of operation. These modes are based on the exchange of flows of electronic information and the accumulation of knowledge as a new source of wealth and competitiveness. In terms of geographic and demographic changes, production processes become dependent (and integrated into) processes of information generation, making cities 'nodes' or 'sub-networks' of the global economy. As Miles et al (2000) describe it:

Typically this [transformation] has involved an aggressive redefinition of the city identities and images and the pursuit of private sector and tourist investment through the development of post-industrial infrastructures of convention centers, waterfront developments and landscapes of business and leisure. (p. 5)

This aims to connect cities to the new economy of knowledge, to which Castells (1996) sees opportunities but also profound impacts. Under this view, the city is bound to transform itself or be transformed as a network node, leaving out any other possibilities to conceive of it in alternative ways. The city becomes subsumed by technological change, and the most it can do is to try to keep abreast of networks dynamics.

Under a 'network view' of society, knowledge becomes concentrated in certain territories or localities to which people migrate to participate in knowledge networks; these are sources of culture and innovation (Castells, 2001). Interestingly, the purpose(s) of these networks are unknown "even for many of the entities integrated in the network of exchanges...places become *de-structured* in the process of selective reorganization of work and residence" (Castells, 1991, pp. 348-349) (italics added). Large cities become the nodes-metropolises of a vast global network, with smaller places being subsidiaries, of a web of suppliers and customers, so that "...locally-based labor will be able to provide the skills required in the production system at the precise point of its connection in the network of productive exchanges" (Castells, 1991, p. 351). Cities could connect to global networks, but this can have the result that those living inside or outside the city come together for other reasons than for instance, cultural or historical grounds.

The above network perspective for societies and localities shows one side of the story: A predetermined, inevitable and visionary view, mainly driven by technological and economic changes. However, a number of current developments show that the picture of cities is far from becoming a coherent one in relation to this vision (Madon, 1997; R. Mansell & Steinmueller, 2000; Graham, 2002; Beck, Madon, & Sahay, 2004). As with the development of the information society, the possibilities of configuration of interactions between citizens, technology providers, businesses and governments with technology is uncertain, dynamic and changing (R. Mansell, 2002). There is something that gives us the possibility of thinking outside the box in terms of cities. Castells suggests that information technologies are helping groups of people to resist some of the effects of dominant networks. For instance, local governments and localities (including cities) could regain their cultural identity by taking advantage of electronic communication, and engaging their citizens into collective action to enhance their participation in decision-making locally or globally: "...Social movements must think local (relating to their own concerns and identity) and act global —at the level where it really matters today" (Castells, 2001, p. 143).

CITIES TODAY

The concept of marketing has become an extension of marketing for profit organizations (Kotler & Levy, 1969) and has been used to attract investment by promoting the 'unique character' of cities (Short & Kim, 1998), some of which can be also be fostered by the development of cultural industries with the support of information technologies (Wood & Taylor, 2004). During the 1990s, cities have adopted these or other strategies to achieve economic objectives by 'standing out' of the competition of other cities (Paddison, 1993; Holcomb, 1994). As described before, re-vamped city marketing involves cities adopting ways to be seen (internally and externally) as centers of business, culture, creativity and innovation (Harloe, 2001). To achieve this, the advertising of the city involves "targeting specific types of activity which both reflect and bolster the [city's] image" (Paddison, 1993, p. 340), and commonly involves flagging up a number of projects to show that they are positively evolving (Holcomb, 1994). Projects also show that cities participate not only in certain regions of influence but also global spaces (or networks) of flows (Doel & Hubbard, 2002).

Historically, the above view of city marketing coincides with the emergence of strategic thinking in business organizations. City marketing can also be seen as an extension of strategy because marketing is about giving advantage, selling the city as a unique place with unique attributes; in organizations this required developing a vision of what the organization wanted to be in the future (Steiner & Miner, 1982). During the 1980's businesses were making use of information technology to support their strategies. Alignment of technologies with corporate plans and business operations was a key element to generate competitive advantages (Henderson & Venkatraman, 1999), which resulted in businesses offering unique value propositions to customers through products and services. In the realm of information systems development this was seen as the prominence of the functionalist paradigm in which a system was conceived of to meet pre-defined goals and objectives (Hirschheim & Klein, 1989; Mingers, 2001). What we are witnessing now is an extension of strategic planning now for cities, where they are using their strengths to sell themselves, and by doing so addressing their weaknesses, one of which seems to be a need for stronger self-representation and identity (Urban, 2002). In the case of cities though, both the 'new' (strategic) aspects and the 'old' inevitably co-exist in the minds of people (Criado & Ramilo, 2003; Van der Meer & Van Winden, 2003). Following this 'strategic' way of thinking, one could also think that they need to be 'aligned' respond to a view of the city as a future place to live.

Despite the degree of appeal that this type of strategic (and vision oriented) thinking can offer, two important aspects need to be mentioned. First, strategic thinking has become better characterized as a continuous process of learning, in which not only rationally defined goals and objectives are to be implemented; 'emergent' strategies are as important as those formally defined. They are the result of enabling people to include their own views and concerns, and innovate in their understanding of markets, businesses, consumers and technology (Ohmae, 1982; Mintzberg & Waters, 1985). Second, those advocating the role of information systems and technologies in strategy implementation suggest that it is essential to generate appropriate support environments for people to come together and make use of systems to continuously generate new opportunities and strategies (Ciborra, 1994; Ward & Griffiths, 2002; Galliers, 2004). Rather than generating 'one-off' actions to support a particular strategy, it is important to create conditions for people to interact, tinker, try and err continuously so they can learn and improve.

In cities, attempts to provide a more holistic views of marketing so that different aspects (e.g. social welfare) of the city and some values are explicitly addressed and presented have been put forward (Paddison, 1993; Kavaratzis & Ashworth, 2005;). It remains to be seen how this integration effectively enables flexibility to facilitate inclusion of new ideas, concerns and perspectives in marketing plans, and how information and communication technologies (ICTs) can be employed for a variety of purposes other than developing a vision or a unique strategy for the city.

CITY BRANDING

Attempts to develop wider perspectives of the city in marketing have resulted in looking at the notion of city *branding*. Branding is about permanence, working on relationships, construction of the city as a place in the mind of those living or working in it (Kavaratzis & Ashworth, 2005). Branding involves not only planned 'city advertising' interventions, but managing people's perceptions and images, selecting the values that they want and managing such selection.

According to Kavaratzis and Aswhorth (2005), a brand embodies a whole set of physical and socio-physical attributes and beliefs which are associated with a product. Branding involves communications in which organizations provide certain attributes and values and customers select and incorporate them. Kavaratzis and Ashworth suggest using this notion to facilitate the improvement of the ways in which cities represent themselves. Cities can strive to position themselves in the *minds* of people, what they associate a city with. This is not specific effort to sell the city as a product to be bought. As Kavaratzis and Aswhorth (2005) say: "The city is simultaneously a place of residence and a place of work for the people that live in it, a destination for the people that visit it (or plan to do so), a place of opportunity for the people who invest in it" (p. 512).

The above definition enables marketing efforts to consider that "places have more varied 'users', 'owners', and 'governors' than do commercial corporation and thus not only are the products more varied, so also are the goals of the producers and the utilities of the consumers" (Kavaratzis & Ashworth, 2005, p. 511). In this regard for instance, not only administrative products could get the attention of new or old citizens, but a variety of possibilities can follow; cities can for example design and deliver online services to support life events of their citizens (i.e. marriages) or help in the day-to-day ones (i.e. reporting crimes, providing advice) (Van der Meer & Van Winden, 2003). Moreover, in branding 'consumers' (citizens) can be active part of the process of value creation. They can continuously select and live by the attributes that make up for the city as a brand, what they value of it as a whole rather than as a collection of parts. For instance, in the case of the Swedish city of Kista, branding it as a 'wireless valley' could help conveying a number of attributes, as well as fostering different collaborations between interested parties to develop this brand. For citizens, adequate management of relationships that align them to a particular 'brand' like this would need to be ensured and implemented through provision of online services and information.

In addition to enable the integration of various efforts in marketing around a more amenable and wider notion (that of a brand), information technologies can support continuous identification and development of *relationships* between citizens. Technologies like customer relationship management (CRM) are fit for this purpose. These technologies can enable decision makers in the city to segment the city's audiences; to explore behavioral patterns; to respond to customers' requests (or events), and design particular products and services to maintain alive positive perceptions and attributes about the city. In short, they can lead to the creation of citizen relationship management systems (Bailor, 2007).

Two examples can help us to illustrate how a city brand is being implemented (Bailor, 2007; Gentile, 2007). In the city of Minneapolis, (the US), the use of a CRM system via a centralized call centre centralizes attention to citizens whom "by dialing one simple-to-remember number...learn how to apply for a building permit, find out how late the library is open, or report a stray dog..." (p. 33). The system enables local officials see where most commonly reported problems occur, allocate resources accordingly and respond to service requests. In another example (Gentile, 2007), the city of Albuquerque (New Mexico) implemented a CRM system to allow local government to clear the city of abandoned vehicles. After an extensive telephone call campaign, a geographical database was built. Appropriate procedures were put in place and the city was able to challenge the assumption that only vehicle abandonment occurred only in certain areas.

At first sight branding—and as it seems relationship management within it—can be useful to include a number of concerns and variety of views and perspectives about the city, and to represent them in systems which deal with instance of such relationships. This strategy can be directed to plan, develop and renew more stable relationships with current or potential citizens. This intangible nature of cities branding (Kavaratzis & Ashworth, 2005) could be developed by accommodating and responding to a variety of needs from different citizens' audiences. Key messages can be sent and received via multiple channels. Requests can be processed with appropriate information from a centralized geographical database. In addition, attention to these requests can be timely and geographically directed.

However, the above resembles a number of possibilities that just now have started to become more popular (Van der Meer & Van Winden, 2003). With the particularities of each city and a strong tendency to reproduce administrative information above all, it is not clear which or how other social concerns of city stakeholders are to be addressed or included in the branding (in particular, if they are perceived as being negative). Definition of city brands is still very much in the hands of city decision makers and if not, on ICT suppliers or experts (e.g. consultants) who know the intricacies and related functions that information systems are to support (Dunleavy, Margetts, Bastow & Tinkler, 2006), with consequences often detrimental for local administrations and citizens in particular.

Despite this potential shortcoming in branding and ICT support for cities, in the information systems literature we find theoretical developments that not only provide room for more participative ways of designing information systems and services (Mumford, 1983), but which also provide room for the 'unexpected'. As said before, relationships between strategy and ICTs can be better seen as learning processes. Unintended consequences and effects of information systems need to be included into our design / understanding of plans and initiatives (Robey & Boudreau, 1999; Sabherwal, Hirschheim & Goles, 2001) if we are to be comprehensive and consider the dynamics of change and appropriation of technologies. In other words, we need to enable change to happen in different ways (both positively and negatively). The inevitability of technological pervasiveness in different realms of human life also means that organizations, information systems are increasingly seen as instruments to advance certain groups' interests at the expense of others; however people have possibilities to resist (Bloomfield & Coombs, 1992; Doolin, 2004). With these ideas a similar point can be raised for cities and the support for city marketing. Different groups of individuals (government officers, technology suppliers, citizens) can gain expertise and control of systems technologies, and can therefore use them according to their own agendas; policies should be designed to allow people to test, experiment, resist, or make ICTs their own (Ciborra, 1994), so that different possibilities can be implemented.

THREE PATTERNS FOR THINKING ABOUT CITY MARKETING AND INFORMATION TECHNOLOGY USE

To sum up, we have considered a number of challenges that arise when cities aim to integrate themselves into wider and worldwide knowledge economies. One of the key challenges is to develop alternative ways to study city marketing that enable the accommodation of the needs and concerns of different city stakeholders, and which enable better (intended, unintended) interactions among them with the help of information and communication technologies.

Despite shortcomings detected in mainstream thinking about city marketing as presented above, we still think that efforts are to be encouraged when it comes to improve the quality of life in cities in the context of current societal transformations. Our aim is then to provide a more detailed description of what we see as ways of thinking about city marketing, and how these ways could impact the design and implementation of information technologies.

We propose to use the word 'pattern' to encapsulate a set of beliefs, values and activities that could be used to characterize city marketing initiatives. These sets could be shared by groups of people at different levels (governments, citizens, other groups involved and affected by city marketing initiatives). They contain assumptions about the nature of the phenomena to be explored (in this case city marketing), how they will be developed, and how information and communication technologies (ICTs) are to support marketing initiatives.

In a way, with these patterns we respond to a growing need to categorize the use of ICTs in cities, and relate these to the formulation of appropriate policies and plans in terms of content, infrastructure and access (Van der Meer & Van Winden, 2003). In this regard we take a slightly different direction: We focus on identifying ways of thinking about and developing city marketing initiatives and then on drawing a number of implications for managers, ICT providers, users and policy makers of such initiatives. The following table presents an overview of three (3) of such patterns.

Under this pattern, those responsible for city marketing and technologies would need to 'let go' of from visionary and strategic ideas about the city and enable people to promote the city in their own ways. They would rather need to provide assistance and support to use technology. Difficulties could arise due to lack of 'trust' and the dynamics of relations in which agency is allowed to change such relations, and influences between actions become greater.

The nature and complexity of power relations that are assumed as a background could render this pattern dangerous to use for decision makers. But to its credit, a better understanding of how power operates in relation to a city can throw light to the emergence of contradictory, competitive and unintended views of the city, which to some risk representing orthogonal images of the city (In France, Puteau is a well known example of such case3). With a power-based pattern, questions could be asked about the effects of this or any type of action in individuals and groups, with a view of enabling them to re-define their own representations about the city, and their use of information technologies to support their activities.

Table 1. Patterns for city marketing analysis

Pattern	Idealist	Strategic	Power-Based
Characteristics	Portraying a vision and mission of a city within the global economy network	Communication, ideally through 'two way' interactions; building of consensus and 'diversity' around the city, within an explicitly democratic background.	Blurred boundaries between city groups Contradictory, conflicting views about the city.
		Hybrid approaches to develop city initiatives; consideration of local (technological, human) aspects	Tensions, asymmetry. Unintended consequences Opportunities to redefine who we are within power relations
Website features	Administrative information Sections to attract visitors	Contact us facilities; section of citizens own information (my pictures of the floods)	Representing different views about what needs to be done
	News and events Positive images	Sections about current problems and what is being done	Information spread across websites which are linked from the city council one.
Supporting technologies	Virtual reality tours; online tourism	Electronic forums; online and mobile transaction services (e.g. CRM systems)	Those enabling communities, centralization and decentralization, institutional power and individual autonomy;

Weakness

This pattern could easily be interpreted as 'anything goes' in terms of systems, ICT and its uses. There is a dimension that should be included in developing city marketing initiatives under a power-based perspective. It is the *ethical dimension*, by which individuals and groups decide on ways of relating to themselves and others, and decide on which of those ways are appropriate and not.

As Foucault (1984) suggests, in relation to power, awareness should be continuously developed to avoid certain ways in which ways of relating to oneself and another generate situations of normalization, in other words situations in which there are no possibilities for doing otherwise (so there is no power being left to maneuver). For cities, this means that not every possibility of relation through marketing should be accepted. When a possibility becomes standard (for instance by sticking to a particular image or brand), it should be examined and challenged. This means that we need to continuously rethink how we interact, how we sell the city, how we allow people to relate to each other, so that freedom is not being compromised.

THE PATTERNS USED: IMPLICATIONS

In Table 2 below we make an attempt to draw some insights on the existence of these patterns in the city of Hull and the city of Brest. The table also contains a proposal for the technologies that we see more appropriate to support marketing under each of the patterns identified. We base our analysis in the cities' websites (mainly council websites), as well as some first-hand knowledge of activities and supporting technologies being developed to promote them. We have also consulted resources like YouTube® and MySpace® to look for information about cities, and some electronic media (newspapers, forums).

Patterns in Detail: Idealist

The above Table 1 contains a definition of the main features of these patterns and their representation in city websites and electronic systems related. Our definition is by no means exhaustive but can help us to 1) understand the assumptions and rationale behind the development of these elements, and 2) to help us think about how we

Table 2. Patterns analyzed in the cities of Hull and Brest

City	Idealist pattern	Strategic Pattern	Power-Based
Hull features	'Pioneering' city campaign	Online transactions Website information about what to do about flood claims and liabilities (Figure 2).	Tensions between, for instance, local and central government (s) about which vision should be portrayed (pioneering, forgotten).
	'Forgotten' city campaign	Websites messages of "You could send us your own pictures and videos" of the flood in news and council websites	Electronic forums show people talking about lack of support from council; other resources (YouTube) also show positive views about the city. Opportunities to redefine the city among 'pioneering' and 'forgotten'
Brest features	Invest in Brest dedicated Web site (http://www.investinbrest.com/)	Tramway dedicated Web site	No real on-line debate

can improve the provision of information through these resources.

We first define an *idealist* pattern as a set of activities to show an often single and visionary image of a city which aligns the city with global trends and national policies. The vision put forwards the objectives of economic growth and the improvement of the quality of life of citizens through work in different (but separate) action fronts. A city is seen as something that will meet a number of goals, be they economic, technological and social. An idealist pattern of activities would seek to transform the image of a city through portraying an outward look, possibly with some flagship urban projects, and historical information which can also serve to enhance cities' self-representation. Historical information and events are being seen for some time as the preferred sections of city websites' visitors (van Limburg, 1998). Information technologies enhance these and the overall presence of this idealist pattern by enhancing virtually what a city is about and aims to do; uses of technologies include virtual simulation (to navigate the city, and even experience a day in one of the city's squares). Some advanced technologies including virtual tours, videos, podcasts, online tourism and other technologies can get people close to the reality of cities.

The use of new technologies can help keeping the futurist nature of the city and its representation; predefined (mainly governmental) goals to achieve guide the design and implementation of information systems and technologies. These goals are assumed to be shared by most if not all stakeholders involved (Hirschheim & Klein, 1989).

Strength

By following an idealist pattern of marketing, cities can start marketing by creating their first view of themselves to grapple with. Then, cities can engage into providing an alternative vision for people. The aim would be to attract visitors and show how the city is being able to 'catch up' with globalization or 'leapfrog' their current situation, and thus avoid previous pre-conceptions or notions about it which can be deemed as negative for marketing. A vision could be a new beginning for the city. This vision should *align* both activities in the city and technologies to support them.

Weakness

Although adopting idealist patterns can help cities 'leapfrog' their current situation, this adoption could lose their market value due to potential generalization of visions, in other words several cities can be doing the same (Holcomb, 1994). This issue, together with the accelerating pace of technological change and the possibilities of using information technology to have several visions (e.g. represented in several websites) of the city at the same time, can make it difficult to unify a vision. Furthermore, temporary visions, images of events and initiatives are difficult to retain by people (Urban, 2002). As said before, city marketing initiatives and supporting technologies would need to become more flexible to accommodate *different* views about the city, and facilitate their continuous updating. The use of information systems should also consider how people interpret the world around them, how they make sense of it, and therefore, how they can use information technologies to pursue their own goals (Checkland & Holwell, 1998). A possible alternative is to focus city marketing not the visionary content of the city, but instead in the *participation processes and dialogues* that take place.

Patterns in Detail: Strategic

We define a *strategic* pattern for city marketing that reflects the city efforts to build and represent dialogue between city's communities (insiders and outsiders). The nature of 'strategic' comes from the idea that through some degree of interaction, people can then align themselves with the city's visions or plans and participate more actively in their development. Here in this pattern alignment of activities and technology is not so much important as alignment of people to activities. A strategic pattern assumes willingness of people to participate in the context of democratic environments to do so. Information technologies could help in facilitating the development of such environments, as well as in empowering individuals to communicate via different channels (Dunleavy, Margetts, Bastow & Tinkler, 2006). CRM systems, electronic forums, instant messaging, e-mail, feedback and online transaction technologies can facilitate (but not exhaust) the development of this pattern.

The view of an information system for marketing is that of a tool to serve a variety of purposes, including that of enabling different groups and their interests to be accommodated (Checkland & Holwell, 1998). Interaction and participation can be included as elements that facilitate dialogue and accommodation. Technology wise, some 'strategic' and visible elements of this pattern are present in cities websites like for instance 'contact us', 'feedback' and 'follow up' facilities, or, for key city initiatives in dedicated Web sites. Information kiosks can also widen the spectrum of possibilities and enable connection between citizens, tourists, local and private facilities and services.

An example of this pattern is the website dedicated to the construction of the tramway system in the city of Brest[1]. In the future, this pattern can be represented in online transactional systems that will respond to citizens requests, and enable them to form communities so that they can share knowledge, and even exert accountability over their city councils (Córdoba, 2005; Cegarra & Córdoba, 2006).

Strength

To its benefit, a strategic pattern of thinking about city marketing can help people to 'join in' the development of the city via consultation and electronically mediated dialogue. By doing so, a variety of ideas, perspectives and goals from different groups could be received and addressed. Different ways of representing the city can be allowed, developed and connected. Communication channels can be tailored to different audiences (i.e. telephone, internet, mobile phones, etc). The provision of online communities, electronic forums and even CRM systems can also pave the way for more active forms of participation of citizen groups, as they see that the boundaries of local governments can now include them (Hubbard & Hall, 1998) and that they can exert some degree of accountability in the city's governments (Dunleavy, Margetts, Bastow, & Tinkler, 2006).

Weakness

A potential shortcoming in this pattern though, is that electronic participation might not be considered as different from other types of participation, and electronic communication can become embedded in traditional ways of interacting with citizens (e.g. via formal letters after emails have been received) (Criado & Ramilo, 2003; Cegarra & Córdoba, 2006). Electronic communication resources can be used by city councils to automate their existing administrative processes (or meeting their own targets) without much consideration to needs of citizens. In terms of city marketing, this would mean that decision makers can develop an 'efficient' vision of the city (and then follow an idealist pattern) which is not strong in facilitating social inclusion or dialogue, while defining and using electronic participation as a means to their ends. Appropriate conditions would need to be met to enrich interaction once systems technologies are being put in place to allow flexible learning, communication and emergence of communities. A strategic pattern for city marketing would need to look at different issues from the context in which it is being implemented (Heeks, 2005) so that incremental and dialogical approaches can be used. Through dialogue and incrementalism, the implications of inclusion and marginalization of people via marketing can be explored and addressed. The end result of this exploration could include changes to facilitate a more dialogical and participative approach, where flexibility and communication happen in various ways and citizens are given opportunities to form their own communities (with the support of technologies). An unintended consequence of this is that people take marketing according to their own purposes, as the next pattern shows.

Patterns in Detail: Power-Based

Participation, consultation and the existence of a democratic environment to support marketing can be guaranteed in many cities worldwide; marketing supposes willingness to be engaged in sending and receiving information, in other words to recreate a number of images and take part of the attributes of the city (Kotler, 2000; Kotler & Levy, 1969). However, a problem arises when this condition is not met, or if it is met by not considering multiplicity of perspectives, some of which are contradictory to what is official, or most of which is a reinterpretations of daily life events by individuals to meet their own purposes (De Certeau, 1988). The third pattern challenges the existence of harmony, single visions, community, equality and predictable consequences of the use of information technology for the development of marketing initiatives in a city.

A power-based pattern reacts against an intended and linear set of consequences for marketing initiatives. It assumes instead that people can exert power for their own purposes. Castells (2001, p. 282) opens up this possibility of *agency* in the network society with the following imperative:

So, either we enact political change (whatever that means in its various forms), or you and I will have to take care of reconfiguring the networks of our world around the projects of our lives... If you do not care about the networks, the networks will care about you. (original parenthesis)

What Castells implies is that life in the cities is *inevitably* networked and that we should acknowledge our contribution to it by either being active or passive. The type of action that Castells envisages can be inferred as political, given the possibilities that people have in establishing and maintaining networks. But action should not be bounded by (or limited to) the forms that governments or citizen groups use. Choice for action need not to be limited to the 'official' (e.g. democratic) ways of relating, but also to ways of operating within these and to new, imagined ways. These ways can also be supported electronically.

Castells is suggesting to use the power we have as individuals to influence action. Although he opens the possibility, little can be said about how to act. Marketing initiatives do not offer much either. It is necessary then to conceive of action about marketing in relation to power which sees it not only in explicit and 'official' forms. We offer now a view on power based on the ideas of Foucault. According to Foucault (1984), power is an analytical concept which allows us to understand the ways in which we have become 'subjects' of modern society. Power operates in any relation we establish with ourselves and others in different spheres of life; inevitably in these relations any action influences other actions. This clarifies the nature of action that Castells has envisaged, and Foucault also gives further advice (1977):

It seems to me that power must be understood in the first instance as the multiplicity of force relations immanent in the sphere in which they operate and which constitute their own organization; as the process which, through ceaseless struggles and confrontations, transforms, strengthens, or re-

verses them...and lastly, as the strategies in which they [relations] take effect, whose general design or institutional crystallization is embodied in the state apparatus, in the formulation of the law, in the various social hegemonies. (pp. 92-93)

Both Castells and Foucault are highlighting the need to take reflective action about our roles as individuals in society. For the case of marketing, it is possible to say that a notion of power in which action is framed suggests the importance of understanding power as a map, a 'terrain', a battle ground as a complex set of relations to which we cannot be exterior and upon which we need to act. This terrain is asymmetrical, pervasive and dynamic, and constitutes the background in which people involved and affected by city marketing interact.

Strength

A power-based pattern would enable a better degree of inclusion of concerns and interactions of people in relation to city marketing. It would allow us to see how we are part of a grid of relations, and study its potential effects. Moreover, it would also allow us to see how our agency and use of information technologies could reinforce, resist or re-create inclusions, exclusions and marginalizations. Complexity, unintended consequences and contradictory views about information technology use would need to be taken into account (Robey & Boudreau, 1999). Radical uses of technology could also be included to support emancipation of individuals and groups from constraints derived from traditional (hierarchical) ways of managing (Hirschheim & Klein, 1994).

Thus, a power-based pattern would see how marketing is being developed, which relations among people, institutions and organizations can enable it to flourish or fail. It would also allow us to see what power we have available to counteract any potential effect that we see as harmful for citizens. This reflection can start by looking at how they are being influenced by current marketing initiatives and associated technologies, some of which are aimed at 'empowering' them (Foucault, 1980). But more importantly, a power-based pattern would give us as individuals possibilities to act in the light of power relations according to what we think is ethical (Córdoba, 2006). By using technologies in different ways, people can do their own marketing about a city, be its own consumers but also producers, they can create and satisfy needs. This can mean for instance resisting official initiatives or adopting them for their own purposes. People can use their own judgment and decide what they can do. By enabling reflection, this pattern should also help individuals to continuously re-define their relations to the city and use technologies to articulate such redefinition.

Overall, it is worth noting that each of the patterns has some representation in the cities analyzed. We now draw on some more detailed implications for each pattern. First, we see that there could be different 'ideals' (*idealist* pattern) that are developed by cities like Hull, which so far provide a degree of reference to any marketing initiative but also could constrain them. Hull can be following an idealist pattern by declaring itself as a pioneering city, and this can be inferred by looking at the council's website. But then we find another idealist pattern (this time rather negative), where Hull becomes a 'forgotten city', led by council members complaining about the support of central government. The 'forgotten city' ideal gained force by being referred by several citizens' emails and comments in electronic forums. Inevitably people participated and voiced their concerns. The city council's website did only provide advice and guidance to the flood. Currently, this information is no longer available. This prompts us to highlight three more detailed insights of our analysis:

Firstly, and with the dynamics of websites and other ICTs associated, we are witnessing a dynamic accommodation of both ideals, and a continuous emergence of new visions and ideals for the city. In the case of the city of Hull, efforts to position the city as a pioneering one continue, and new visions are being developed. But the floods could come again, and the city should be prepared for this; information and communication technologies (ICTs) could be used to support the management of potential eventualities and contingencies; a clear case of this is the management of disasters and emergencies in the city.

Furthermore, the existence of multiple visions about the city can enhance flexibility in marketing. These multiple visions can also help to manage the cohabitation between different "voices" from different City services or city groups. In the case of Brest, the "Citizenship and New Technologies" City service has opened a space on the Web dedicated to allow inhabitant to speak about their country, making emerging another communication, "the voice of the people"[2]. This website though, is not referenced in the official website. An idealist pattern for marketing would need to allow different visions to emerge and co-exist.

Secondly, the existence of a *strategic* pattern in cities seems to be limited to forms of dialogue subsumed to official versions (idealist patterns). In the case of Hull, advice and information is given, but when it comes to facilitate other types of two-way communications we only find facilities like citizens sending videos or pictures (in this case of the flood situation). The city council's website allows people to do online registrations and payments, and the system enabled to receive donations for the flood victims. But despite being located in a democratic environment (UK), marketing of the city, participation, formation of communities with the support of technologies, inclusion of different perspectives has not taken off as an interactive process, at least officially.

In Brest, a similar situation could be inferred, and the city council's aims to develop social cohesiveness within the city community but more than a way to develop a communication towards the outside, rather than strengthening the links between people and the City council in varied ways. If a strategic pattern is to be adopted for cities, people should be allowed to be part of the design of information systems, and to keep their own space to dialogue with other city stakeholders. Forms of dialogue, communities, forums (not only comments about news), and accountability activities should be designed and fostered. This can contribute to generate attributes of participation, governance and cohesion which could enhance the image of cities to the inside and outside.

Thirdly, the emergence of electronic forums and opinions shows an increasing level of 'unofficial' engagement of people and the city, and this can be interpreted as a sign that *power-based* patterns are emerging, in which people use electronic resources to provide their own representations of their cities. Although it might be difficult to further assess how people go beyond the 'official' and experiment with technology to influence other people actions, we see that available technologies are making it easy for people to contribute through different channels (for instance YouTube, Facebook and MySpace). There is a variety of perceptions, images, views, stories and activities that are being reflected about a city in public electronic resources. We suggest that city decision-makers should be more aware of this situation, and try to recognize their existence by for instance being more inclusive of such efforts in their own website and other resources.

For citizens in general, the availability of these resources can show that there are opportunities to reflect on what type(s) of city are really emerging, and use these resources. Although it cannot be said that the use of electronic resources is available for all, city decision makers can foster education and awareness on their existence, and with this facilitate the development of new (intended and unintended) images of their cities. We see many opportunities to enhance, enrich and continue efforts to make cities better, and we see that technology is also giving power to do so. It is up to us to take our chances forward.

CONCLUSION

In this chapter, we have explored city marketing from a number of perspectives. Our exploration has taken us to suggest a number of patterns that we see useful to think about our cities and to develop ways of improving what we witness as marketing. These patterns should help decision makers and citizens in many ways, and we now suggest the following:

1. They should help city decision makers to become aware of the assumptions that fuel their city marketing initiatives. In our discussion, we have provided strengths and weaknesses of each pattern. An idealist pattern can help cities to kick off, and provide a first common view of a city. It should then give way to the existence of many different visions, some of which should be allowed to flourish via dialogue and participation of citizens (via a strategic of patterns), and some of which emerge unintentionally but can be fostered to facilitate inclusion (power-based pattern).

Moreover, these patterns can help decision makers to initiate or improve their existing efforts. An idealist pattern can help them to ask questions related to the development of a 'leap frog vision' (for instance what sort of vision if any is being developed? What technologies are being used?). A strategic pattern can help them to assess the inclusivity of citizens in initiatives (with questions like how is interaction being encouraged and managed? Are there any forums, communities online or systems that help us accommodate different visions of the city?). A power-based pattern would allow them (and citizens) to understand how marketing can become better deployed (with questions like what relations are we encouraging? How can we support new relations within the city?)

2. Thee patterns should help citizens to become more confident in understanding why and how cities are currently engaged in marketing themselves as places that offer better quality of life and better futures than others. This though needs to be taken cautiously, as our discussion has shown. Not only by using 'make up' or working on a particular brand to embellish a city, and providing technological support, issues of participation and inclusion of variety of concerns are going to be effectively addressed. Citizens should strive to be given possibilities for participation, dialogue, community development and learning. These can become locally adapted attributes that could contribute to improve life in cities.
3. The patterns also give citizens and other people interested in cities the possibilities of contributing to their improvement by using electronic resources, and to reflect their own concerns, purposes and ideas so they can make better use of technologies available and get more involved in developing the city.

Finally, our discussion has raised some possibilities that information technologies and their use can give us to contribute to developing our network (or networked) societies. It is up to us to decide how we engage with these societies in our physical locations. We hope this chapter offers other opportunities to reflect on the type of cities we want, and in the support that information technologies can continue giving us in the future for this purpose.

REFERENCES

Bailor, C. (2007, January). When disaster does not strike. *Customer Relationship Management*, 42-43.

Beck, E., Madon, S., & Sahay, S. (2004). On the margins of the 'Information Society': A comparative study of mediation. *The Information Society, 20*, 279-290.

Bloomfield, B., & Coombs, R. (1992). Information technology, control and power: The centralization and decentralization debate revisited. *Journal of Management Studies, 29*(4), 459-484.

Castells, M. (1991). *The informational city: Information technology, economic restructuring, and the urban-regional process*. Oxford, UK: Basil Blackwell.

Castells, M. (1996). *The rise of the network society*. Cambridge, MA: Blackwell Publishers.

Castells, M. (2001). *The internet galaxy: Reflections on internet, business and society*. Oxford: Oxford University Press.

Cegarra, J. G., & Córdoba, J. R. (2006). Assessing and developing E-Government use by SMEs. *Business School Research Memoranda, 56*, 1-21.

Checkland, P., & Holwell, S. (1998). *Information, systems and information systems: Making sense of the field*. Chichester, UK: John Wiley and Sons.

Ciborra, C. (1994). The grassroots of IT strategy. In C. Ciborra & T. Jelassi (Eds.), *Strategic information systems: A European perspective* (pp. 3-24). Chichester, UK: John Wiley.

Clout, H. (2006). Vive la Bretagne! *Modern and Contemporary France, 14*(1), 79-83.

Córdoba, J. R. (2005). Communities and evaluation of E-Government services. In S. Clarke & E. Coakes (Eds.), *Encyclopaedia of communities of practice in information and knowledge management* (pp. 32-34). Hershey, PA: Idea Group Publishing.

Córdoba, J. R. (2006). Using Foucault to analyze ethics in the practice of problem structuring methods. *Journal of the Operational Research Society, 57*(9), 1027-1034.

Criado, J., & Ramilo, M. C. (2003). E-government in practice: An analysis of web site orientation to the citizens in Spanish municipalities. *International Journal of Public Sector Management, 16*(3), 191-218.

De Certeau, M. (1988). *The practice of everyday life*. Berkeley, CA: University of California Press.

Doel, M., & Hubbard, P. (2002). Taking world cities literally: Marketing the city in a global space of flows. *City, 6*(3), 351-368.

Doolin, B. (2004). Power and resistance in the implementation of a medical management information system. *Information Systems Journal, 14*(4), 343-362.

Dunleavy, P., Margetts, H., Bastow, S., & Tinkler, J. (2006). *Digital era governance: IT corporations, the state, and e-government*. Oxford: Oxford University Press.

Foucault, M. (1977). *The history of sexuality volume one: The will to knowledge* (Vol. 1). London: Penguin.

Foucault, M. (1980). Truth and power. In P. Rabinow (Ed.), *The Foucault reader: An introduction to foucault's thought* (pp. 51-75). London: Penguin.

Foucault, M. (1984). The ethics of the concern of the self as a practice of freedom (R. e. a. Hurley, Trans.). In P. Rabinow (Ed.), *Michel Foucault, ethics subjectivity and truth: Essential works of Foucault 1954-1984* (pp. 281-301). London: Penguin.

Galliers, R. (2004). Reflections on information systems strategizing. In C. Avgerou, C. Ciborra & F. Land (Eds.), *The social study of information and communication technology: Innovation, actors and contexts* (pp. 231-262). Oxford: Oxford University Press.

Gentile, A. (2007, September). Help! 311 and CRM systems generate quick responses to residents' ordinary problems. *American City & County*, 32-36.

Graham, S. (2002). Bridging urban digital divides? Urban polarization and information and communication technologies. *Urban Studies, 39*(1), 33-56.

Hall, T., & Hubbard, P. (Eds.). (1998). *The entrepreneurial city: Geographies of politics, regime and representation*. Chichester, UK: John Wiley.

Harloe, M. (2001). Social justice and the city: The new 'liberal' formulation. *International Journal of Urban and Regional Research, 25*(4), 889-897.

Heeks, R. (2005). E-government as a carrier of context. *Journal of Public Policy, 25*(1), 51-74.

Henderson, J., & Venkatraman, N. (1999). Strategic alignment: Leveraging information technology for transforming organizations. *IBM Systems Journal, 38*(2&3), 472-484.

Hirschheim, R., & Klein, H. (1989). Four paradigms of information systems development. *Communications of the ACM, 32*(10), 1199-1216.

Hirschheim, R., & Klein, H. (1994). Realizing emancipatory principles in information systems development: The case for ETHICS. *MIS Quarterly, 18*(1), 83-105.

Hirschheim, R., Klein, H. K., & Newman, M. (1991). Information systems development as social action: Theoretical perspective and practice. *Omega, 19*(6), 587-608.

Holcomb, B. (1994). City make-overs: Marketing the post-industrial city. In J. Gold & S. Ward (Eds.), *Place promotion: The use of publicity and marketing to sell towns and regions* (pp. 115-131). Chichester, UK: John Wiley.

Hubbard, P., & Hall, T. (1998). The entrepreneurial city and the 'new' urban politics. In T. Hall & P. Hubbard (Eds.), *The entrepreneurial city: Geographies of politics, regime and representation* (pp. 1-23). Chichester, UK: John Wiley and Sons.

Hull. (2007). Flood-hit Hull a 'forgotten city'. *BBC news, available at http://news.bbc.co.uk/1/hi/england/humber/6270236.stm.*

Hull City Council. (2007). *Hull City Council website.*

Kavaratzis, M., & Ashworth, G. J. (2005). City branding: An effective assertion of identity or a transitory marketing trick? *Tijdschrift voor Economische en Sociale Geographie, 96*(5), 506-514.

Kotler, P. (2000). *Marketing Management*. Upper Saddle River, NJ: Prentice-Hall.

Kotler, P., & Levy, S. (1969). Broadening the concept of marketing. *Journal of Marketing, 33*(1), 10-15.

Le Cam, F., Ruellan, D., & Cabedoche, B. (2006). *Public service information. The editorial identity of city councils' websites in Brittany*. Brest: Mole Armoricain de la Recherche sur la Societé de la Information (M@rsouin).

Madon, S. (1997). Information-based global economy and socioeconomic development: The case of Bangalore. *The Information Society, 13*, 227-243.

Mansell, R. (Ed.). (2002). *Inside the communication revolution: Evolving patterns of social and technical interaction* (1st ed.). Oxford: Oxford University Press.

Mansell, R., & Steinmueller, W. (2000). *Mobilizing the information society: strategies for growth and opportunity*. Oxford: Oxford University Press.

Miles, M., Hall, T., & Borden, J. (Eds.). (2000). *The city cultures reader*. London: Routledge.

Mingers, J. (2001). Combining IS research methods: Towards a pluralist methodology. *Information Systems Research, 12*(3), 240-259.

Mingers, J., & Willcocks, L. (Eds.). (2004). *Social theory and philosophy for information systems*. Chichester, UK: John Wiley and Sons.

Mintzberg, H., & Waters, J. (1985). Of strategies, deliberate and emergent. *Strategic Management Journal, 6*(3), 257-272.

Mumford, E. (1983). *Designing human systems for new technology: The ETHICS method*. Manchester: Manchester Business School.

Ohmae, K. (1982). *The mind of the strategist: The art of Japanese business*. New York, NY: McGraw Hill.

Paddison, R. (1993). City marketing, image reconstruction and urban regeneration. *Urban Studies, 30*(2), 339-349.

Robey, D., & Boudreau, M. (1999). Accounting for the contradictory organizational consequences of information technology: Theoretical directions and methodological implications. *Information Systems Research, 10*(2), 167-185.

Sabherwal, R., Hirschheim, R., & Goles, T. (2001). The dynamics of alignment: Insights from a punctuated equilibrium model. *Organization Science, 12*(2), 179-197.

Short, J. R., & Kim, Y. H. (1998). Urban crisis / urban representations: Selling the city in difficult times. In T. Hall & P. Hubbard (Eds.), *The entrepreneurial city: Geographies of politics, regime and representation* (pp. 55-75). Chichester, UK: John Wiley and Sons.

Steiner, G. A., & Miner, J. (1982). *Management policy and strategy*. New York, NY: MacMillan.

Urban, F. (2002). Small town, big Web site? Cities and their representation on the internet. *Cities, 19*(1), 49-59.

Van der Meer, A., & Van Winden, W. (2003). E-governance in cities: A comparison of urban ICT policies. *Regional Studies, 37*, 407-419.

van Limburg, B. (1998). City marketing: A multi-attribute approach. *Tourism Management, 19*(5), 475-477.

Ward, J., & Griffiths, P. (2002). *Strategic planning for information systems*. Chichester, UK: John Wiley and Sons.

Wood, P., & Taylor, C. (2004). Big ideas for a small town: The Huddersfield creative town initiative. *Local Economy, 19*(4), 380-395.

ENDNOTES

1. http://www.letram-brest.fr/
2. The official website: http://www.puteaux.fr/ and the dissident citizen's one: http://www.monputeaux.com/
3. http://www.wiki-brest.net

Section III
Stories from the Battlefield: Finding Out the Power of City Marketing

Chapter XIII
The Use of Internet in Building the Brand of "Stockholm: The Capital of Scandinavia"

Peter Dobers
Mälardalen University, Sweden

Anette Hallin
Royal Institute of Technology, Sweden

ABSTRACT

This chapter discusses a current example of the ongoing efforts of city managers to promote their cities, also known as place marketing or place selling. After introducing the concept of city branding, and a model of how Web site elements communicate brand values and messages, we analyze a recent attempt of city managers in Stockholm to promote the brand of "Stockholm: The Capital of Scandinavia". The authors hope that the empirical illustrations of how city managers of Stockholm have worked to provide a broader understanding to the complex Web of communication and brand building on the Internet, both empirically and conceptually.

INTRODUCTION

Branding, i.e. the building of brands, is something that since the 1980s have become more and more common in everyday language denoting, that which gives a product meaning and identity in time and space. With the growing realization that the brand is important terms like "brand equity" (the financial value of the brand) and "brand management" (the managerial activities aiming at creating, sustaining and developing the brand) have also entered our language (Kapferer, 1997). The ideas of branding have also influenced the managers of cities, who, in a world of hardening competition for investors' money, tourists and the most intelligent and creative workers, have begun to grasp the importance of the building of a strong city brand.

But despite the rapidly growing interest in branding both among practitioners and researchers, there is still a lack of empirical studies that shed light on how the process of branding a city actually takes place. There is also a need for case studies that reveal how cities work with the new possibilities provided by information- and communication technologies (ICTs) in their branding work. Thus this chapter provides experience, concepts and knowledge for practitioners to reflect upon their own branding situation and for scholars to further develop.

Cities, if seen as a complex web of actors of interacting, fighting, organizing and disorganizing, certainly provide challenges to the brand managers. The aim of this chapter is to explore how the branding of a city takes place by presenting and discussing a case study from the Swedish capital of Stockholm. In 2005 Stockholm launched the brand "Stockholm: The Capital of Scandinavia" and behind this cheeky statement lay several years of work, including city managers and brand specialists such as Dowell//Stubbs, a reorganization of the city organization and the building of a partnership involving 43 of the municipalities in the greater Stockholm region. In this the Internet was chosen as one channel to communicate the brand both with the partners geographically spread around Stockholm as well as with the presumed international audience. By telling the story of how Stockholm developed the brand and tried to engage as many actors as possible in this work and by examining closely the web pages of the City of Stockholm our aim is to shed light on the branding of cities and the use of ICTs in this work.

Having a long history of developing and using information- and communication technologies, it is not surprising that the City of Stockholm recognized the importance of IT as a tool in building the new brand. Even though local governments today are increasingly represented on the Internet (Urban, 2002), they still represent a sector where online branding has played a limited role, though, a state which will is expected to change in the future due to the digitalization of the economy (Rowley, 2004a, 2004b). Therefore, a case study of how the City of Stockholm has worked with online branding as a step in building "Stockholm: The Capital of Scandinavia" should provide interesting insights into the role of ICTs, especially information technologies, in the building of city brands today. The questions to be answered here are thus: How do the City of Stockholm and its partners use their web pages when it comes to promoting the brand? How was the brand developed and what was the rationale behind forming the partnership?

To answer these questions we have carried out nine interviews with the people responsible and involved in the work such as City managers and people from business. In addition to this we have studied Internet pages and collected all kinds of printed material about "Stockholm: The Capital of Scandinavia" such as brochures and articles from magazines and newspapers. After giving a brief background to the theory of city branding in general and to brand-building on the Internet in particular, we will based on the material mentioned above tell the story of how the idea of "Stockholm: The Capital of Scandinavia" developed. Then we will take a closer look at how the Internet has been (and is) used in the work before discussing our findings. Finally, we make some concluding remarks.

Why City Branding?

Big cities have the power to absorb the interests of many. But today cities have in the increasingly global and networked world found themselves in competition with other cities (Bennett & Nathanson, 1997; Chevrant-Breton, 1997; Stewart, 1996). Competition between city regions occurs nationally in areas such as governmental funding of sustainable development efforts (Bergström & Dobers, 2000) and internationally in areas of tourism and corporate inward investments (Kotler, Haider & Rein, 1993). As one answer city managers lead the work of forming metropolitan alliances (Bennett & Nathanson, 1997) with the aim of strengthening their competitiveness on the global scene for attracting new businesses, highly educated labour, experts from industry and academia to seminars and conferences, and the everyday tourists. In the wake of globalisation cities are becoming important change agents in regional development formulating progressive strategies and images of their cities as part of place promotion (Ashworth & Voogd, 1990; Kearns & Philo, 1993a, 1993b; Ward, 1998; Ward & Gold, 1994), filling the city and business life with entertainment (Hannigan, 1998; Wolf, 1999), and initiating large urban marketing and propaganda projects (Boyle, 1997; Kotler et al., 1993).

Not seldom do these alliances evolve around new technology, because ever since the underground sewage and water systems of ancient Rom, typical urban visions prevail that new technologies and infrasystems almost immediately will lead to a better quality of life or a more effective management of the city (Brotchie, Batty, Hall & Newton, 1991). In late modernity telecommunication and information technology has been identified as an obvious driver of urban prosperity and new kinds of cities: the wired city (Shostak, 1982; Hepworth, 1989), the information city (Hepworth, 1990), or the tele-mediated city (Graham & Marvin, 1998, 2001). The OECD backed up such visions or prophecies when the organisation showed an interest in urban issues and new technologies enhancing the trend towards the so called information or network society (OECD, 1992). Through new technologies certain cities and their environment have been put on the (world-) map such as Stockholm as the Internet Capital of Europe (Dobers, 2002).

Cities and regions are turned into critical agents of economic development by introducing city images like the *technopoles* or *technoburbs* (Castells, 1989; Castells & Hall, 1994; Fishman, 1987). Thus, constructing and using images of technodreams related to information technology has been one way of managing the city and emphasising its *past*-modernity by which the digitalisation of information is the divide between modernity and *past*-modernity. Modern examples have been Osaka as a *city of intelligence*, Barcelona as a *city of telematics*, Amsterdam as a *city of information*, Manchester as a *wired city* (Mark E Hepworth, 1990). More recent and *past*-modern examples include Stockholm, of course, as an *IT City* or a *Mobile Valley*, Boston as the *Cyber District*, Colorado Springs as the *Silicon Mountains* and the beach area between Santa Barbara and San Diego as the *Digital Coast*. The images of the city and ICTs that prevailed at the turn of the millennium and a few years after seem to have faded away though (Dobers, 2006), perhaps due to them not always having been the result of a more or less conscious branding strategy of the city managers of the city. But the recognition of the importance of "place marketing" or "place branding" seems to have grown.

Promoting places (Ward & Gold, 1994; Wells Jr & Wint, 2000), selling places (Ashworth & Voogd, 1990; Kearns & Philo, 1993b; Chevrant-Breton, 1997; T. Hall, 1998; Ward, 1998) or marketing places (Erickson & Roberts, 1997; Kotler et al., 1993; Sadler, 1993; Smyth, 1994; Kotler, Asplund, Rein & Haider, 1999; Beckerson, 2001; Rainisto, 2003) have become common concepts in how city managers attempt to improve the situations of cities and regions. These concepts have in common that they attempt to promote, sell or market places by being specific and clear about which advantages a place or city has or is claimed to have. This approach resembles an effective branding strategy that is crucial "in order to create the common ground necessary for the whole marketing endeavour" (Kavaratzis & Ashworth, 2006, p. 17). More and more cities seem to be involved in these kinds of activities. The reason is most often a sharpened competition for example visible in different kinds of rankings of cities all over the world. The global real estate service provider Cushman & Wakefield Healey & Baker has for example conducted annual surveys on Europe's major business cities since 1990. Called the "European Cities Monitor" the survey examines issues deemed important for where companies decide to locate, and compare how leading business cities in Europe perform on each issue such as easy access to markets, availability of qualified staff, transport links to other cities, quality of life or freedom from pollution. In this monitor Stockholm ranks quite ordinary and in the middle of 33 ranked cities in Europe in 2006 (Cushman & Wakefield, 2006). London, Paris and Frankfurt have topped the list since when the survey started in 1990. New cities have been added during the years so that all neighbouring capitals to Stockholm (Oslo, Copenhagen and Helsinki) are represented in the monitor in 2006. Although Stockholm is in the middle of the ranked European cities it tops the countries in Northern Europe. The impact of these rankings is probably high, not only for businesses and investors, but for politicians and city managers.

BUILDING CITY-BRANDS THROUGH THE INTERNET

A general assumption is that poor images of a city can have devastating effects on the city's economy as it reduces the likelihood of inward investments, undermines business community activities and has a detrimental effect on the number of visitors to the city (Kotler et al., 1999; Trueman, Klemm & Giroud, 2004). At the same time it has been noted that there is generally a lack within the public sector of a coordinated approach concerning branding activities often due to the difficulty of finding a brand which the very wide community of the public sector can agree upon, being engaged in a vast number of different activities, serving a broad group of citizens. Within branding theory the agreement of a common brand is considered necessary though as it "allows the establishment of strong brands... that can act as umbrella brands for a range of activities and initiatives including those that are both stable and time-limited" (Rowley, 2004a, p. 134). For branding in general it has also been suggested that it is important to develop a robust multi-channel strategy for branding communication, not only using traditional channels, but also new media such as Internet, e-mail, mobile TV etc. This makes it difficult to differentiate, theoretically as well as practically, between online and offline branding (Rowley, 2004b).

At the same time the Internet has become a major tool of developing the city's image and local governments are indeed increasingly represented on the World Wide Web, probably due to an increasing involvement in national and global policies as a result of economic globalization. On these sites local and regional functions (town-hall options) are often juxtaposed with national and global functions (place branding options). In a study of 20 cities in different countries all over the world it was concluded though that there is no relationship between the extension of the real and the virtual city (population and amount of websites) but that there is a clear relation between the wealth level of a country and the size of the city websites (Urban, 2002). Also, the very largest cities, the so called "world cities" (P. Hall, 1977, 1966) did not have tourist information on their web sites probably due to the impossibility of the public organization having the resources to cover and update this kind of fast-changing information (Urban, 2002).

One of the advantages on being present on the Internet is that it is a way of being present in the "customer's" life, the Internet being a virtual storefront where the customer not only can find information about that which is interesting but also engage in communication with the seller who in turn can adjust the message to suit the particular customer. Internet thus functions as a communication channel, a transaction channel as well as a distribution channel (Kiang, Raghu & Shang, 2000).

Online branding includes everything (for instance documents, web pages or products) that carries the same brand and thus can be seen to emanate from the same source. The challenge with the webpage compared to traditional marketing channels is that the webpage has a limited space making it necessary to focus the brand values and the messages. One way of doing this is to keep the structure formulated by Rowley (2004a, b) in mind (see Figure 1). The web elements of logos, graphics, text and copy, currency and news, colour, shapes, layout and combination of images, relationship features as well as wallpapers and screen savers together form a whole and should, in the best of cases, communicate a unified message. Does it in the Stockholm-case?

Before we take a closer look at what these elements look like on the web pages in the Stockholm case we must first describe how the idea behind "Stockholm – The Capital of Scandinavia", offline and online, was born.

THE NEED FOR A BRAND

To me the situation was clear: the biggest problem Stockholm had was that is was saying too many things. Not too few. It was confusing people with what it actually stood for. Our mission was to cut through that and to crystalize it into something, an umbrella to put all these fantastic assets under and still make it work. Then, whom do we really compete with: Copenhagen? The Öresunds region? Why do people come here and not there? (Julian Stubbs, CEO and Senior Brand Planner of Dowell//Stubbs, in an interview on November 23rd, 2006)

During the spring of 2005 the City of Stockholm shocked the other Scandinavian capitals by announcing that Stockholm was the "Capital of Scandinavia". To understand this strategical move it needs to be pointed out that after the turn of the millennium, and with the experience of the downturn within the telecom industry having extended effects on the Stockholm region due to the major influence of the telecom company Ericsson, sev-

Figure 1. Web site elements that communicate brand values and messages (Source: Rowley, 2004a, 2004b)

Graphics and logotype:	indicates the content and nature of the service and may include pictures, logos and other images, of which logotype is the shorthand for everything that the brand stands for
Text and copy:	sets the tone of voice and determines whether a message is intelligent, comprehensible and relevant
Currency and news:	communicates a live and dynamic Web and ensures that users have access to the latest information
Colour:	have different culturally defined messages but communicates different values to the users
Shapes:	say different things depending on how shapes of pictures or graphics, buttons and menu options are formed
Layout and combination of images:	can be used as a metaphor to render similarities with for instance a print newspaper (webpage arranged in columns) or a village (webpage create familiarity)
Relationship features:	the features and functions that take the user beyond the web page and its initial impact
Wallpapers and screen savers:	to be downloaded and work as a reminder of the brand

eral actors in Stockholm expressed the need for a new plan for growth. Not having had a growth strategy since the end of the 1970's, different parts of the Stockholm city organization had worked in different ways to attract international investors and more visitors. As a result several images of the city had evolved. Stockholm was viewed as the city of culture (Porsander, 2000), as an advanced environmental city (Adolfsson, 1999, 2003), or as a city of information technology (IT) (Dobers, 2001; Dobers & Strannegård, 2001). Regarding the latter, city managers often referred to IT-ventures such as the Stockholm Challenge Award, Stokab (the broadband company of Stockholm City) and the Mobile Valley of Kista. Through the Challenge Award Stockholm challenged world cities to become best at IT; with Stokab the city has constructed one of the world's more far-reaching broadband nets; and by exploiting the area of Kista since 1973 Stockholm has created and became a hot spot for high tech ventures in tele- and mobile communication (Dobers, 2002). Now, however, the politicians and city managers agreed to form a coherent strategy to build the Stockholm of the future.

When looking at how other cities had organized this work it became clear that it was necessary to develop a brand strategy for the city and to gather everyone who worked with these issues under the same roof. The result was a reorganization of the city and the launching of Stockholm Business Region (SBR) in January of 2005. Fully owned by the City through "Stockholm Stadshus AB" SBR is an organization aiming at "developing and promoting Stockholm as a business and tourist location" through the collaboration "with trade and industry, other organisations in the city, other municipalities in the Stockholm Region, academic institutions and government agencies" and with the ultimate goal of creating "growth in the region, with the goal of becoming the leading growth region in Northern Europe by 2010"[1]. In the Stockholm Stadshus AB Group there are several other companies fully owned by the City (see Figure 2) with businesses providing a range of different services to the people and companies in the region such as water and sewage (Stockholm Vatten AB), apartments, buildings and maintenance (several different companies), ICT-infrastructure (AB Stokab) and culture (Stockholms Stadsteater AB). As all other companies in the Stockholm Stadshus AB Group, SBR has a Board of Directors, consisting of politicians representing the democratically elected City Council. Chairing the Board is the Mayor of Stockholm i.e., the highest ranking political official in the city, and as deputy chair, the leader of the second largest political party in local politics is appointed.

Together with its subsidiary Stockholm Visitors' Board (SVB) which is working specifically for and with tourists in cooperation with hotels, travel agencies, etc, SBR had in the fall of 2007 a staff of 79 and a yearly turn over of about 122,4 mSEK.[2]

As head of SBR Jürgen Kleist was appointed, previously chief manager of the Stockholm Economic Development Agency. In his former role Kleist in early 2003 participated in a presentation where the new brand for Arlanda airport was presented. The airport had experienced a big draw back since the terrorist attack on New York on September 11th in 2001 with the pulling out of American Airlines, the falling of passenger numbers and the down-turn in the Swedish economic cycle and had a strong need of doing something about this. The task of developing a new brand strategy was given to Dowell//Stubbs, a brand communications agency situated in Stockholm but with international staff. Julian Stubbs, founder, CEO and Senior Brand Planner at Dowell//Stubbs[3], who led the work for Arlanda airport and who did the presentation which Kleist listened to, thought that the airport had had the wrong focus. Instead of focusing on the airport itself he suggested that the vision should be something bigger: "to make the greater Stockholm as

Figure 2. Organisation chart showing the place of Stockholm Business Region AB in the city organisation. Author's figure based on organization chart of Stockholm Business Region

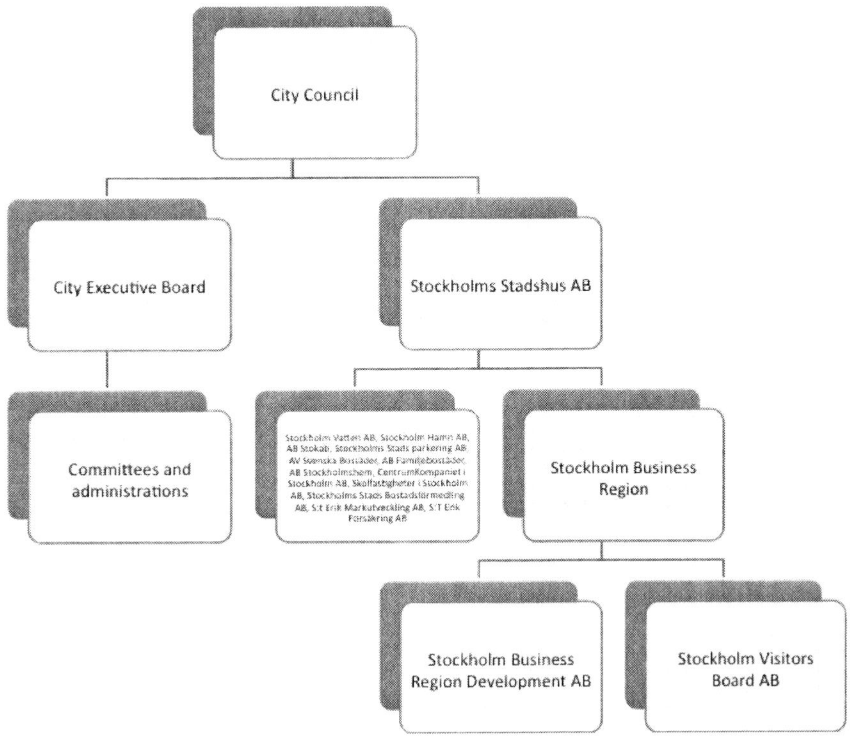

the preferred choice in Scandinavia for business and travellers"[4] With positive connotations like "high tech", "healthy living", "socialist welfare", "non-political", "non-border", to Stubbs, "Scandinavia" was the obvious choice of name to use instead of for example "The Baltic Sea Region" (which Stubbs describes as being "more potato soup than anything else"). Kleist was impressed by the work of Dowell//Stubbs and asked for the company's participation in the formulating of a new brand for Stockholm.

In Stubbs' opinion the city had not been clear in communicating but to get an idea of what the people of the city thought Dowell//Stubbs carried out "vox pops"; mini-interviews with people in the streets and asked "Which is the bigger city: Stockholm or Copenhagen?" The picture that emerged was fuzzy and to Stubbs this was a clear indication of the need for a coherent brand. The principle behind place branding is the same as for product- or company branding according to Julian Stubbs, but to brand a place is more difficult as there are more stakeholders involved. "A place is not there to be sold" Stubbs says, and to make all stakeholders of a city join the same vision is difficult as they all have different and quite disparate goals. Also, a place does not have one owner that can control the brand. Instead, the stakeholders must agree upon which story should be told, which is a challenge when it comes to a city with a multitude of different stakeholders. In fact "the largest challenge has been to show that this project is good for all parties", the SBR communications manager Anne Årneby said in an interview with the media (see Jansson & Power, 2006).

INVENTING "STOCKHOLM: THE CAPITAL OF SCANDINAVIA"

The story of a brand must be told in a pedagogical way and the starting point is always "what's special about us?" Therefore slogans like "Stockholm. Venice of the North", a saying used for Stockholm for at least 80 years (see Anderson, 1953) or "Stockholm. Beauty on Water" are not very good. "Venice of the North" is a weak message saying only that Stockholm is a Nordic copy of another city using a slogan that several other cities also claim such as Amsterdam, Brügge and St Petersburg. A place brand also must be inclusive enough to be able to encompass all the typical qualities of the place which is why "Beauty on water" is problematic. It only refers to one dimension of the city namely its environment. It is also important to be in control of the brand and that is not the case when designing it around the kind of clichés mentioned above. An example of a city that has succeeded in finding a good brand is Glasgow, "the friendly city", where the brand is built around its people with the slogan "Glasgow's miles better". Other cities build their brand on their historical inheritance or on architecture like Paris with the Eiffel Tower, London with the Millenium Dome and the Gurkin and the Swedish city of Malmö with the Turning Torso. It seems as if a city doesn't become great until its architecture is great.

After a lot of consideration, the brand of "Stockholm: The Capital of Scandinavia" emerged. As a word "the capital" signals the centre of power, "the most of everything" and thus a place strong enough to support a multitude of activities. "Scandinavia" also recalls powerful and positive emotions and in contrast to "Sweden" it is less political but still encompasses more territory. Carrying powerful connotations of health, design and architecture, "Scandinavia" was early the preferred choice.

In the spring of 2005 Anne Årneby was employed as *communications manager* for SBR and manager of media and communication at SVB. With a background in brand management at several large Swedish companies, Årneby was considered perfect for the job. Together with Dowell//Stubbs she developed a brandbook to be used by all the stakeholders. Here the arguments for the cheeky proposition "Stockholm: The Capital of Scandinavia" are given:

Central: that Stockholm is the geographical centre

Business: that Stockholm is the business centre, having a stock market bigger than the other Nordic markets put together

Cultural: that Stockholm is the cultural centre, with the Nobel price and festivities, its fashion and design

This three-folded argument was of course given a lot of thought. To Stubbs it was important not to have too many different arguments ("three is a good number") as the aim is that all stakeholders remember them. This ensures effective communication. Therefore arguments concerning the environment such as ecology and "the green city" were discarded, even though this is an important aspect of Stockholm for example recognised by established business sources: Stockholm has been the top city in the European Cities Monitor by Cushman & Wakefiled Healey & Baker for the last years when it comes to freedom from pollution, and not surprisingly, this is also reported on the Stockholm city web.[5] According to Stubbs, however, the green image and the freedom of pollution might fall under one of the other categories, and besides; a common truth needs no changed brand activity. Also it is important that the arguments are not too time-specific when building a brand but that they last over time as it is important to have a longitudinal perspective and to be consistent in the message, Årneby says[6].

FORMING A PARTNERSHIP: STOCKHOLM BUSINESS ALLIANCE

Communications manager Anne Årneby invited all the municipalities in the Stockholm region to join a partnership especially set up for the purpose of branding the region internationally. The rationale behind forming the partnership of the Stockholm Business Alliance (SBA) is that the Stockholm region on the threshold to globalisation increasingly is competing with other city regions in Europe in general, and Northern Europe in particular for inbound investments, visitors and people settling down. The city managers of Stockholm think of the Stockholm region as the natural focal point in Scandinavia, its economical centre and leading cultural city, which is not surprising as about three million of Sweden's nine million live and work in the larger Stockholm region together representing 42 percent of the GNP of Sweden. In 2005, about 25.000 new companies were started in the area, which is almost half of all newly started companies in Sweden. The region thus has an entrepreneurial, strong and diversified business setting with world leading industries such as life sciences, information and communication technology, banking and finance, and environmental technology. Communication to and within the region is well developed with several international airports and an increasing train commuting in the larger Stockholm region. With 28 universities and university colleges education and research is widely spread and deepened often in close cooperation with trade and industry (Stockholm Business Region, 2007b).

In 2007, a year after it was formed, the SBA consisted of 43 municipalities in the Stockholm—Mälardalen region and had the specific aim of deepening and developing the commercial policy and activities in the region. An important element in the partnership is to attract inbound investments to the region and thus create a joint responsibility for branding, processing and receiving business organisations to the region. Members of the partnership are bound for five years with the goals to by 2010 become (Stockholm Business Region, 2007b):

- the leading growth region in Northern Europe
- one of three regions in Europe best in marketing
- the region with the best inbound investment operations

To reach these goals the partnership has the following leading ideas for their organisation: First, municipalities in the larger Stockholm region work together to enhance the local and regional trade and industry. Second, these municipalities develop the local and regional work with commerce in joint responsibility to increase the attractiveness of the region for people, companies and international capital. Third, the partnership appeared on the international arena under the joint brand of "Stockholm: The Capital of Scandinavia".[7]

Each municipality in the partnership contributes with four Swedish Kronor per inhabitant and year to the joint budget for the Stockholm Business Alliance. This adds up to 15,5 million Swedish Kronor for 2007 (Stockholm Business Region, 2007a, 2007b).[8] (see Figure 3)

About half of the 43 partnering municipalities come from the Stockholm county but not all have close proximity in geographical meaning to Stockholm such as Karlskoga in the west (about 240 km from Stockholm centre), Ludvika in the north-west (about 220 km), Gävle in the north (about 170 km), and Oxelösund in the south (about 120 km). Only six municipalities of the 26 in the Stockholm county have not joined the SBA partnership (Norrtälje, Österåker, Danderyd, Vaxholm, Järfälla and Salem) (see Figure 4 and Figure 5).

The municipalities differ widely in structural parameters such as population, commuting patterns and economic structure. The highest populated municipality is Stockholm with more

Figure 3. Budget and activities of the Stockholm Business Alliance (Swedish Kronor) (Source: Stockholm Business Region, 2007a, 2007b)

Income	
Funding by partner municipalities	15,500,000
Costs	
SBA, coordination and development	2,500,000
Marketing activities	1,600,000
Analysis and benchmarking	100,000
International marketing,	
Invest in Sweden Agency	1,830,000
Mess and Stockholm Access	1,025,000
Travel, project and printing	1,335,000
Reception services, evaluation and aftercare	50,000
Staff and administration	7,060,000

Figure 4. List of Municipalities of the Partnership 'Stockholm Business Alliance'

43 municipalities have joined the Stockholm Business Alliance. They come from the following regional counties:

Stockholm county: Stockholm, Lidingö, Sollentuna, Solna, Sundbyberg, Täby, Upplands-Bro, Upplands Väsby, Sigtuna, Vallentuna, Ekerö, Nacka, Värmdö, Haninge, Tyresö, Nynäshamn, Huddinge, Botkyrka, Nykvarn and Södertälje

Uppsala county: Uppsala, Knivsta, Enköping and Heby

Södermanland county: Eskilstuna, Nyköping, Oxelösund, Gnesta, Trosa, Katrineholm, Flen and Strängnäs

Västmanland county: Västerås, Surahammar, Hallstahammar, Arboga, Köping and Kungsör

Örebro county: Örebro and Karlskoga

Gävleborg county: Gävle

Dalarna county: Ludvika and Smedjebacken

than 760.000 inhabitants and the least populated municipality in the partnership is Surahammar municipality with 10.200 inhabitants (year 2004).

In a general classification scheme by the Swedish Association of Local Authorities, all 290 municipalities in Sweden have been divided into nine categories, and those being partners in the SBA are mentioned in Figure 6. Most municipalities in the partnership are considered suburban municipalities i.e. suburbs to the city of Stockholm (20). Seven are considered large cities on their own with a number of inhabitants ranging from 80.000 (Södertälje) to 180.000 (Uppsala), and five are "commuter municipalities" meaning that people might live in these municipalities but that a high number of persons commute to other municipalities (mainly Stockholm) to work.

The suburban municipalities and the commuter municipalities consist of about half of the municipalities in the partnership and it is not strange that these see the need to join such an initiative. What is more interesting are the municipalities in the other half. Here, we find six municipalities defined as "large cities", "manufacturing municipalities" and other municipalities where

Figure 5. A map of member municipalities and their proximity to the larger Stockholm region (© 2007 Stockholm Stad. Used with Permission)

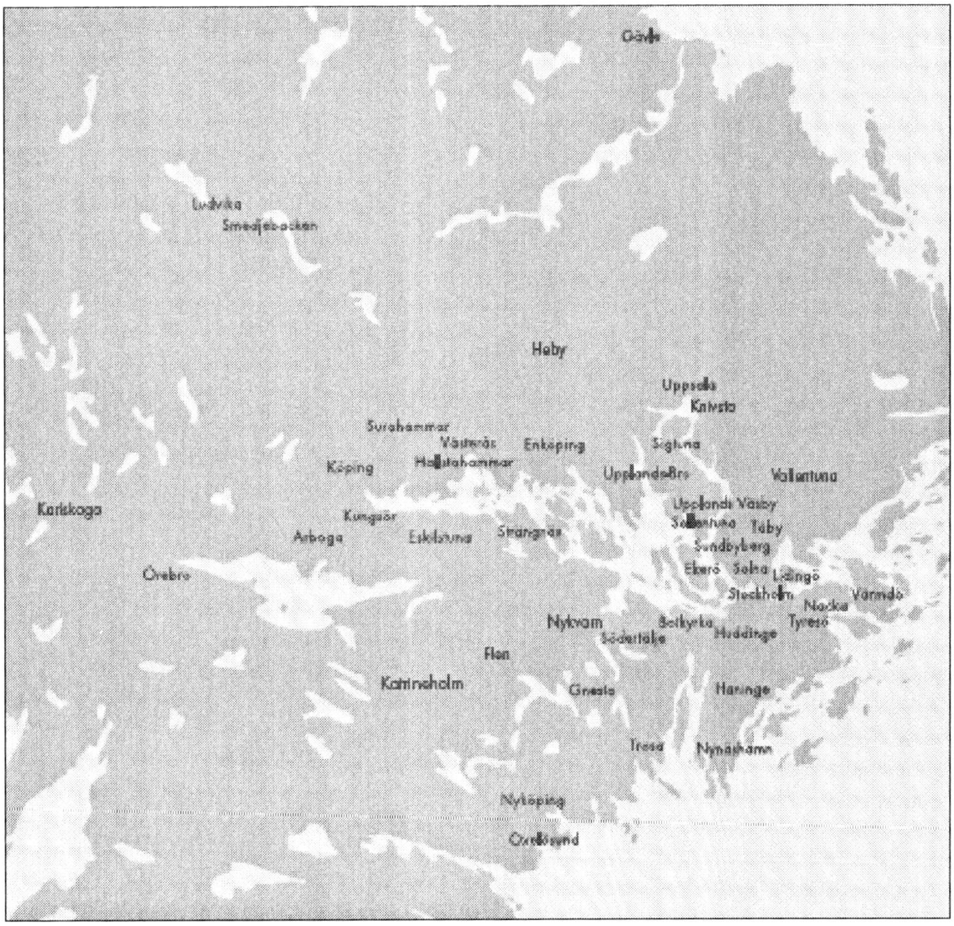

the economic structure evolves around the own municipality in some way or another. Still, they obviously see the need to be part of an initiative such as the SBA.

To start up the partnership Stubbs ran a workshop in October 2006 for all the municipalities involved. The workshop was run in English as English is the language to be used when promoting the area internationally. During the workshop not only was "Stockholm: The Capital of Scandinavia" presented but the representatives for the municipalities were also given exercises to make them think about what is special for their municipality. The SBA has obvious advantages for smaller municipalities: "today every place is looking at how to brand itself. But how shall small municipalities away from the big cities succeed internationally?" Stubbs asks rhetorically and illustrates his answer with a similar case from another country: "people come to the UK because of London and to be close to London is good for a company's business". He adds that Stockholm has the same capacity. "Why not be in Västerås and have access to Stockholm?" Stockholm is seen as a magnet and when international investors knock on the Stockholm door they will have a lot of alternatives when deciding where to set up business, perhaps settling in Västerås if an

Figure 6. Classification of municipalities, January 1st 2005, by the Swedish Association of Local Authorities (Source: the Swedish Association of Local Authorities)

The Swedish Association of Local Authorities made the following classification of municipalities. The municipalities were divided into nine categories on the basis of structural parameters such as population, commuting patterns and economic structure. The categories of municipalities are, listing SBA municipalities:

1. ***Metropolitan municipalities*** (3 municipalities). *Municipalities with a population of over 200,000 inhabitants*: Stockholm
2. ***Suburban municipalities*** (38 municipalities). *Municipalities where more than 50 per cent of the nocturnal population commute to work in another area. The commonest commuting destination is one of the metropolitan municipalities*: Botkyrka, Ekerö, Haninge, Huddinge, Lidingö, Nacka, Sollentuna, Solna, Sundbyberg, Tyresö, Täby, Upplands Väsby, Upplands-Bro, Vallentuna, Värmdö.
3. ***Large cities*** (27 municipalities). *Municipalities with 50,000-200,000 inhabitants and more than 70 per cent of urban area*: Eskilstuna, Gävle, Södertälje, Uppsala, Västerås, Örebro.
4. ***Commuter municipalities*** (41 municipalities). *Municipalities in which more than 40 per cent of the nocturnal population commute to work in another municipality*: Gnesta, Knivsta, Kungsör, Nykvarn, Nynäshamn, Trosa.
5. *Sparsely populated municipalities* (39 municipalities). *Municipalities with less than 7 inhabitants per km² and less than 20,000 inhabitants*: 0.
6. *Manufacturing municipalities* (40 municipalities). *Municipalities where more than 40 per cent of the nocturnal population between 16 and 64 are employed in manufacturing and industry.* (SNI92): Oxelösund, Surahammar.
7. ***Other municipalities, more than 25,000 inhabitants*** (34 municipalities). *Municipalities that do not belong to any of the previous categories and have a population of more than 25,000*: Enköping, Karlskoga, Katrineholm, Ludvika, Nyköping, Strängnäs.
8. ***Other municipalities, 12,500-25,000 inhabitants*** (37 municipalities). *Municipalities that do not belong to any of the previous categories and have a population of 12,500-25,000*: Arboga, Flen, Hallstahammar, Heby, Köping.
9. ***Other municipalities, less than 12,500 inhabitants*** (31 municipalities). *Municipalities that do not belong to any of the previous categories and have a population of less than 12,500*: Smedjebacken.

industrial firm, in Kista if a high tech firm and so on. To be part of the SBA is thus a good insurance for the smaller places in the greater Stockholm area according to this logic.

The next step in the SBA-work, from Stubbs' perspective, is to continue educating the organizations involved and to work on the inhabitants of Stockholm. "Places are built by word of mouth", says Stubbs, but adds that from his perspective as a CEO, place branding is not so lucrative as to work with commercial brands as the former requires convincing politicians to invest money which in some cases can prove difficult.

INTERNET AS A TOOL FOR PROMOTING THE BRAND

All the stakeholders in the SBA-partnership have access to the brandbook through the Internet and can use it within the guidelines. They also have access to an intranet where guidance material of different kinds can be downloaded and added to web pages of the partnering municipalities. There is also detailed information on how to use logo, text segments etc in printed material, ads, and for presence on the Internet. The partnership was formed in 2006 and has had a fixed number of participating municipalities since early 2007, so the municipalities have had the first half of 2007 to

Figure 7. Partner logotype for Stockholm Business Alliance to be used on the Internet by the partnership municipalities (© 2007 Stockholm Stad. Used with Permission)

infuse their web page with SBA information. The brand-book suggested that the logotype showed in Figure 7 is to be used together with some kind of text informing about the partnership.

However, by August 2007 only seven of the 43 municipalities have used this logo on their *Swedish* starting page to communicate the SBA partnership and the brand of "Stockholm: The Capital of Scandinavia" (Stockholm, Upplands-Bro, Upplands Väsby, Värmdö, Södertälje, Trosa and Katrineholm). It is interesting to note that English starting pages overall did not include the partner logotype with one exception only: the Värmdö web page. A few examples of screen dumps of Web sites are given in Figures 8-10. Another 21 municipalities have described the partnership on sub pages of which 16 have HTML-related texts and five refer to downloadable documents related to the partnership like decision protocols, press releases etc. Out of these 21 merely four have used the logo as described. The rest, about half of the municipalities in the SBA partnership, have no information about the partnership on their official web page. Not even after using their own search engines does any information about the SBA appear - one municipality does not even have a search engine.

Figure 8. Screen dump of the Swedish starting page of Södertälje municipality: www.sodertalje.se (Accessed on October 4, 2007; © 2007 Södertälje Kommun. Used with Permission)

Figure 9. Screen dump of the Swedish starting page of Trosa municipality: www.trosa.se (Accessed on October 4, 2007; © 2007 Trosa Kommun. Used with Permission))

Figure 10. Screen dump of the English starting page of Värmdö municipality: www.varmdo.se (Accessed on October 4, 2007; © 2007 Värmdö Kommun. Used with Permission))

A CLOSER LOOK AT THE STOCKHOLM CITY WEB PAGES

As part of the working process with the brand the Stockholm "What's on"-guide, banners and all other kinds of tourist material was redesigned according to the guidelines in the brandbook and Årneby hired a web agency to look through the web pages of the city of Stockholm reducing them to two main portals: a visitor's portal for tourist information and a business portal for business information. Both are integrated into each other by vertical, sideway menus on the right hand side, thus signalling a coherent image in terms of layout and providing good accessibility. Since it can be assumed that either of these portals are the first web pages that a presumptive visitor meet we must take a closer look at these two portals to see how the web site elements as listed by Rowley (see Box 1 above) communicate brand value (Rowley, 2004a, 2004b).

The visitor's portal (www.stockholmtown.com as accessed on October 4, 2007; see Figure 11) is labelled "The Official Visitors Guide" to Stockholm.

Graphics and logotype. The brand logotype is placed in the upper right hand corner with the text slightly tilting to the right in a white circle on top of the green background, placed not in the centre of this green field, but a little bit to the left. To the left of the brand logotype is a colourful photographic image of Stockholm depicting smaller sailboats in full sail on the Riddarfjärden in front of the City Hall. On the right hand side of the screen dump of the portal are banners consisting of photographic images or logotypes for the Wasa Museum (one of the major tourist attractions in the city), the express train between Stockholm Arlanda airport and downtown Stock-

Figure 11. Screen dump of the starting page of Stockholm visitors' portal: www.stockholmtown.com (Accessed on October 4, 2007; © 2007 Stockholm Stad. Used with Permission)

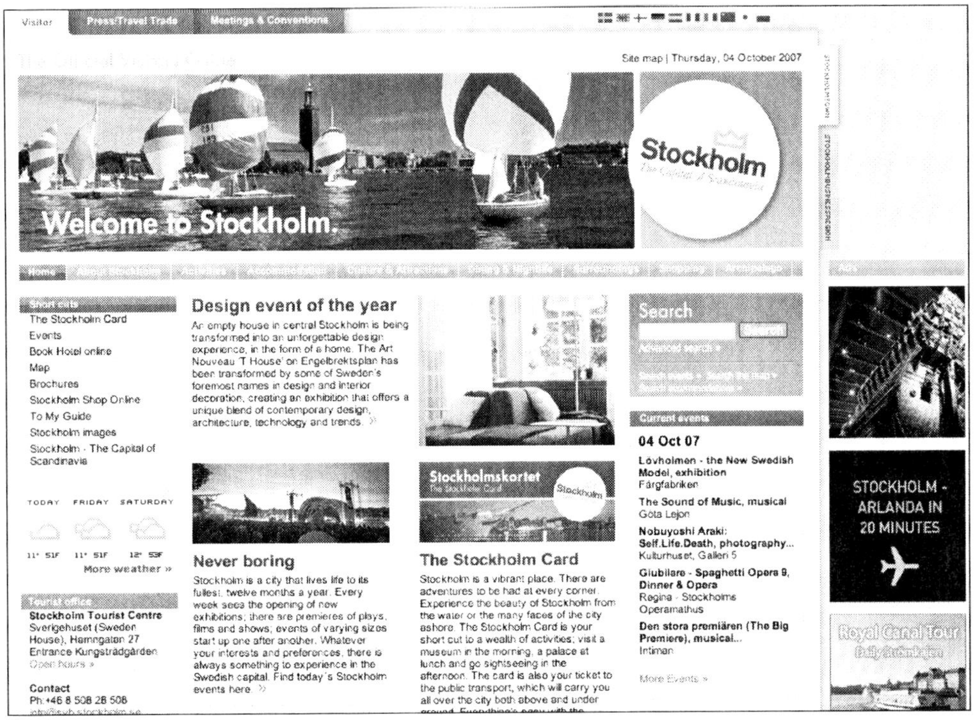

holm and for guided tours around the island of Djurgården. More information is found beneath the bottom part of the screen dump but due to the screen size of our working lap top no more can be viewed for a screen dump. Flags symbolize that the portal is available in ten languages: Swedish, English, Finnish, German, Spanish, French, Italian, Chinese, Japanese and Russian, and weather symbols indicate together with number of degrees the weather conditions in the city, currently and two days ahead.

Text and copy. Under headlines like "Design event of the year" and "Never boring" shorter introductory texts introduce Stockholm and current events in the city using words like "vibrant", "adventures to be had at every corner", "a city that lives life to its fullest" and "unforgettable design experience".

Currency and news. One column of the portal is reserved for news under the heading of "current events". This column presents several events taking place the very day you visit the portal. These events are drawn from a major database in which the organisers themselves have entered events.[9] Apart from this the current date is written in the upper right hand corner above the brand logotype in small, black letters and is of course upgraded automatically on a daily basis. Also, the weather forecast is upgraded daily.

Colour. The colour pallet of this portal is blue, grey, green and white, with occasional details in red. However, the colour pallet changes when you use the menus and enter sub pages of the portal.

Shapes. Shapes are distinct and separated only with fields of different colours. The portal has three flaps on top and two on the right hand side, these flaps have been given shades building the impression of paper folders placed on top of each other.

Layout and combination of images. As described above the page contains a mix of photographic images, logotypes, drawn images, symbols and text, with the brand logotype and the summer picture of Stockholm by the waterside being almost a third of the page. Below this the other elements are arranged in five columns, first a slender column with a menu with short cuts to other kind of information for visitors like "The Stockholm card", "Stockholm Shop Online", the weather forecast and tourist office contact details, then two somewhat wider columns with featured information such as "Design event of the year", "Never boring" and "The Stockholm Card" including text and pictures of the city landscape and modern furniture probably designed in Scandinavia. To the right of these a column with two distinct services: access to a search engine and a board of "Current events". And finally, the ads placed outside of the page on which the other information is placed. For navigating the layout of the portal provides several different ways of entering further web pages on the site. First, on top of the whole web page to the left are three flaps with the texts: "Visitors", "Press/Travel Trade" and "Meetings & Conventions". Second, the menu directly below the photo and the logo has flaps with "Home", "About Stockholm", "Activities", "Accommodation", "Culture & Attractions", "Dining & Nightlife", "Surroundings", "Shopping", and "Archipelago".

Relationship features. All the flaps just mentioned lead to further pages, in the case of the flaps below the photo, six to eleven under pages are found under each category. Further, the many ads on the right hand side, the menu in the first column to the left with options to book hotels and to buy a Stockholm Card take the user beyond the web page. If visitors still have not found what they are looking for there is a search function on the portal.

Figure 12. Brand logotype for "Stockholm: The Capital of Scandinavia" (© 2007 Stockholm Stad. Used with Permission)

Wallpapers and screen savers. On the portal there are no wallpapers or screen savers to download to remind the visitors of the brand. However, on several sub pages there are pictures and images of the brand that can be downloaded particularly for partnering municipalities.

The business portal (www.stockholmbusinessregion.se as accessed on October 4, 2007; see Figure 13 below) is labelled "The Official Business Guide for the Stockholm Region".

Graphics and logotype. Also on the business portal, the brand logotype is placed in the upper right hand corner in the same way as on the visitors' portal, and also here a photographic image of Stockholm can be found to the left, even though here, the image shows the Old town and the hills of Södermalm. No ads are found on this portal and the Swedish and British flags indicate that the portal is available in only two languages.

Text and copy. Three headlines give three perspectives on the City, two of them by referring to rankings of different kinds. "It's official: Stockholm rated World's Best City" summarizes a ranking done my the American magazine *Readers's Digest* and "Stockholm. A green leader

Figure 13. Screen dump of the starting page of Stockholm business portal: www.stockholmbusinessregion.se (Accessed on October 4, 2007; © 2007 Stockholm Stad. Used with Permission)

of the world" tells of how the city is listed as "one of four Green Leaders by the business magazine Fast Company" i.e. that Stockholm is one of four cities in the world which are thought to shape the world's environmental future. Thirdly, "Mobile Future 2007" provides information about an event where "…the future and prospects of the mobile industry" will be discussed.

Currency and news. From the visitors' portal we recognize the heading of "current events" even though the events that are advertised here are business events rather than leisure activities.

Colour. The colour pallet of this portal is blue, grey, and white, with occasional details in mustard yellow, and here the colour pallet does not change depending on the flaps in the menus.

Shapes. Just as for the visitors' portal the shapes are distinct and separated only with fields of different colours, and also here the graphic, colour and layout of these indicate papers on top of each other.

Layout and combination of images. The overall layout of the business portal is similar to the visitors' portal with a mix of photographic images, logotypes, drawn images, symbols and text, with the brand logotype and the summer picture of Stockholm by the waterside. Here, there are only four columns however; a first column to the left with three distinct flaps ("Contact us", "Stockholm region" and "Meet us at") above two logotypes related to Stockholm, the blue logotype for Stockholm Business Region and the logotype with St Erik's yellow head on a blue shield with the text "Stockholm stad" ("The City of Stockholm") written around it. This is the logotype for the Stockholm municipality and is well known to most Stockholmers as the logotype can be found on most municipal information and documents. The second and third column consist of text and photos just as the second and third columns on the visitors' portal. The fourth column consists of three boxes, a search-box, a box providing the possibility of subscribing to three different kinds of newsletters depending on area of interest and a list of current events. For navigating, this layout provides similar ways of entering further web pages as the visitors' portal but here the top menu only corresponds with the two languages that the portal is available in. The flaps in the menu below the photo are "Home", "Why Stockholm", "Key Sectors", "Establishing a Business", "Our Services", "News & Publications", "Press", "Brand Sthlm", and "About Us".

Relationship features. All flaps lead to further pages as does the two Stockholm-related web pages, the news and the list of events. Beyond the flap called "Stockholm region" the portal has extensive information about the Stockholm Business Alliance and the brand of "Stockholm: The Capital of Scandinavia". One page covers the alliance partners in an overview, their geographical proximity, and some basic information like population, distance to Stockholm City, industries and key companies. You can easily access the English web page of the other municipalities guided by an alliance based introduction. This introduction can be skipped but shows the logo in brand-true colours of light grey and blue, and shapes known from the brand. A rough map of Scandinavia is shown and the Stockholm larger region is focused in and expanded. From this map the chosen municipality is enlarged and thereafter first the English web page of the municipality is shown.

Wallpapers and screen savers. No wallpapers or screen savers can be found here either but, just as on the visitors' portal, it is possible to download picture and images on several sub pages.

In line with the conviction to guide web page visitors in the graphical design of the brand, the city has designed a press room on the business portal with ready-to-use materials about Stockholm that are free to use for journalists interested in writing about Stockholm and the Stockholm area including high-resolution images, stock shots, comprehensive facts about Stockholm and interviews with leading representatives within medical research, fashion and industry.

In its branding work the city has cooperated with internationally well-known people and companies interviewing them about why their company has an office in the area. People that have participated are Margareta van den Bosch heading the H&M-design head quarter situated in Stockholm, Jonas Frisén, professor at Karolinska Institute world known for his research on stem cells and Hans Stråhberg, president and CEO of Electrolux which has placed their product development team in the city. Also the Chinese pop star Wei Wei living in Stockholm who sang at the closing ceremony of the Olympic Games in Beijing in 2008 has been involved in this. Årneby and her team regularly work on 500 journalists world wide to inspire them to write about Stockholm and the Stockholm region. This has proven fruitful. In August of 2006, *Business Week* for example had a big article on "Stockholm: Mecca for Affordable Talent" where Sweden and the Stockholm area was depicted in a very favourable way[10].

STOCKHOLM: A MODERN CAPITAL

From this close look at the two Internet portals we can conclude that they are very similar in layout, colours and graphics, which is precisely what was Årneby's intention with the work she initiated when she was appointed Communications Manager. The brand logotype of "Stockholm: The Capital of Scandinavia" is placed centrally, in fact in the place where the Western eye normally first look on the first page of a magazine, newspaper or document. Just like a campaign button it is pinned to the web page bearing similarities to a stamp perhaps a quality stamp. According to Julian Stubbs a lot of time was spent defining the graphic expression of the brand and designing the templates. In the end the colour pallet was decided to be blue ("for water and sky") and grey ("cool colour for Scandinavian design") with a mix of two fonts. "Stockholm" is written in modern serifs and "The Capital of Scandinavia" in an older style together intended to connote Stockholm as an interesting mix of modern and old.[11] In contrast to Sweden which is usually represented by the three crowns "Stockholm: The Capital of Scandinavia" is represented by one crown, representing The Capital and drawn in a modernistic style. This logotype is used on both Stockholm portals but also in printed brochures, and at the Stockholm Arlanda Airport.

Looking at the pedagogical layout of the two portals with their comprehensive navigation, the search functions, the languages in which the portals are available and the vast possibilities of finding more information on transportation, sights, booking hotels etc it is not difficult to draw the conclusion that on the Internet Stockholm stands out as a city interested in branding itself, in wanting to be accessible to its visitors regardless of what kind. It is emphasized that these pages are "the official" web pages of the city and there seem to be an attempt to provide "objective" information as it is clearly indicated what is advertisement and what is not. (Even though the two authors of this text are fully aware of the difficulty with using the word "objective" here...) On the business portal there is no advertisement at all; this is a "neutral" area.

The two closely interconnected portals together indicate that there are many different sides to Stockholm; the more playful, leisurely side and the more sober, serious business side. Regardless which portal the visitor is interested in there are large possibilities of finding customized information through the many relational features, the possibility of searching and subscribing to newsletters. Not surprisingly news of Stockholm's high ranking in different surveys are placed centrally on the business portal probably with the intention of providing weight to the arguments, not only for the City but for the whole region.

The natural element of water returns in several ways on the two portals: through the photos of boats on water, the Wasa ship and is an obvious symbol of Stockholm which is built on many islands situated on both Lake Mälaren and the Baltic Sea. The high quality of the water in Stockholm makes this a good element to enhance working as a symbol for something fresh and fluid and connects to the older brands ""Beauty on Water" and "Venice of the North". But the two portals also give proof of that Stockholm is a city of arts, for example by pushing for design including photos of Scandinavian designed furniture. This is important as the arts generally are seen as projecting "an aura of high quality, civility, creativity [...] and consequently conferring status on its visitors" (Holcomb, 1999, p. 64).

Even though Stockholm is a city with history (for example visible on the visitor's portal through the ad about the Wasa museum) it is first and foremost depicted as a city of modernity; a city on the move which the racing sailboats, the Stockholm-Arlanda express-advertisement, the design-news and the large extent to which the two portals depend on continuous updating indicate. The basic message of these two portals is that, in Stockholm, things are "contemporary" or "future". This way Stockholm on the Internet stands out as a modern capital, an image coherent with how previous research has found that city managers of Stockholm regard the city (Czarniawska, 2002). Even though information about the partnership can be found by going through the portals, the main focus of both portals is the City of Stockholm rather than the region.

ORGANIZING THE CAPITAL

According to the brand specialists one of the biggest challenges with place branding is to have all stakeholders join the common vision (Rowley, 2004a). The Stockholm case seems to have been successful in this respect due to the reorganizing of the city, forming a strong base for the work as described above. Through its design, with all political parties from local government represented on the board, the Stockholm Business Region (SBR) has a good chance of surviving changes in political leadership which was proven in 2006 when the former Social Democratic Mayor changed places with the leader of the Conservative party, the former stepping down from the seat as Chair due to the party's loss in the democratic election but remaining in the Board of SBR as vice chair. This way, SBR can make long term plans and work according to them. Growth politics thus seem to have become non-political in Sweden in the sense that all politicians agree on its importance and to a fairly large extent it seems also to how it shall be done. Another possible interpretation is that this unity regarding the reorganisation of SBR indicates that the issue of goals and methods is a non-question, a question of no or little importance to the politicians, thus accepting the direction proposed by whichever stakeholder.

Still, the forming of the Stockholm Business Alliance (SBA) seems to illustrate the unifying of different stakeholders around a common vision. Despite its name the partnership does not consist of businesses but of municipalities. As described above the partnership's goal is to attract business the greater Stockholm area rather than to have businesses join the partnership even though businesses are welcome to use the SBA logos and the material available on the Internet. To a large extent the SBA works in close cooperation with the Stockholm-based companies in the forming of branding campaigns, for example visible at the web pages of the City of Stockholm. What is amazing with SBA is that 43 out of Sweden's 290 municipalities have joined the SBA and agreed to market "Stockholm: The Capital of Scandinavia" and the numbers given above indicate the area's economic importance for the country as a whole. The three specific goals for the partnership should make it possible to measure the outcomes in 2010 and to see whether the partnership has been successful in their endeavour or not.

By organizing the work as the responsibility of the municipality of Stockholm (through SBR) and a partnership of municipalities (SBA) makes it easier to develop a brand from the public sector perspective which might not be the case had the businesses been involved in the work as equal partners. To involve the business community in the development of the brand makes it more difficult to find a brand which is of direct benefit to them which is a lesson to be learned for example from Bradford where the city suggested that the brand of Bradford should be "a city of culture"; a suggestion that was not met with approval by the Bradford business community which questioned how they would profit from this (Trueman et al., 2004). In the case of Stockholm the brand developed is not developed only for the benefit of the businesses but for the region as a whole, making it an inclusive but still distinct statement with the power to attract all possible stakeholders in the greater Stockholm area.

Compared to other countries the Stockholm-model where so many public sector actors have joined forces, also financially, to promote the area in order for the area to grow seems fairly unique. In Amsterdam, for example, a public-private-partnership has been formed with the main task of coordinating the branding efforts of the city. The public-private-partnership called "Amsterdam Partners" consists of actors from the public sector as well as from the private sector and also for this organization the Internet was considered a major tool of developing the city's image. Here, though, the Internet was seen as one branding activity next to five others which is a difference from the Stockholm case where the Internet is seen as a tool not limited to an activity (Kavaratzis & Ashworth, 2006). This means that in Stockholm the Internet is regarded a natural communication and distribution channel through which information about the city and the brand itself is disseminated and the work is set up to support this channel continuously. To form public-private-partnerships to plan and implement promotional campaigns seem to be quite common in an international perspective but the difficulty of knowing who is responsible for what often leads to the local government taking at least a coordinating role in the work (Holcomb, 1999). As is obvious from our case study Stockholm has taken a much more proactive role than that.

Through SBA the municipalities has established a forum for regional collaboration and coordination (even though it should be added that the SBA is not the only initiative in this area) to maintain the strong economic position the region has today in the future. At the same time it is somewhat surprising that the common brand is not visible to a larger extent on the different web pages of the partners since one of the goals set up for the partnership is to become "one of three regions in Europe best in marketing" by 2010. A simple explanation could be the division of resources; the City of Stockholm has by far the largest budget for these kinds of activities, regardless the partnership[12], as the City of Stockholm is the largest municipality in Sweden.[13] The smaller municipalities might not have the resources to work with their web page to the extent that the City of Stockholm does.

But this also reveals another aspect of the relationship. Even though the SBA is a partnership the power balance is obviously not equal. Stockholm being the only metropolitan municipality in the partnership has the most people, the most money and the most of all other kinds of resources compared to all the other municipalities. It could be argued that the Stockholm Business Alliance is an imperialistic attempt of taking over the neighbouring area which is also expressed in the brand "*Stockholm*: The Capital of Scandinavia" (italics is our emphasis) clearly telling us what is the centre (and thus what is the periphery). From this perspective the SBA can be seen as a smart way of extending Stockholm's own budget for branding. At the same time, the lack of the brand's presence on most of the partnership municipalities' Internet pages indicate that they rest in the belief of the power of Stockholm to reach the goals set for 2010. By not using the designated logo on their web page they seem to trust in Stockholm to do the job.

REDRAWING THE MAP

Historically, cities have been seen as organisms; spatially bounded entities with strict spatial and social divisions (Amin & Thrift, 2002) but as this case study illustrates this metaphor is no longer

Figure 14. Geographical distribution of the 43 regional partners in the Stockholm Business Alliance, August 2007[14] (© 2007 Stockholm Stad. Used with Permission)

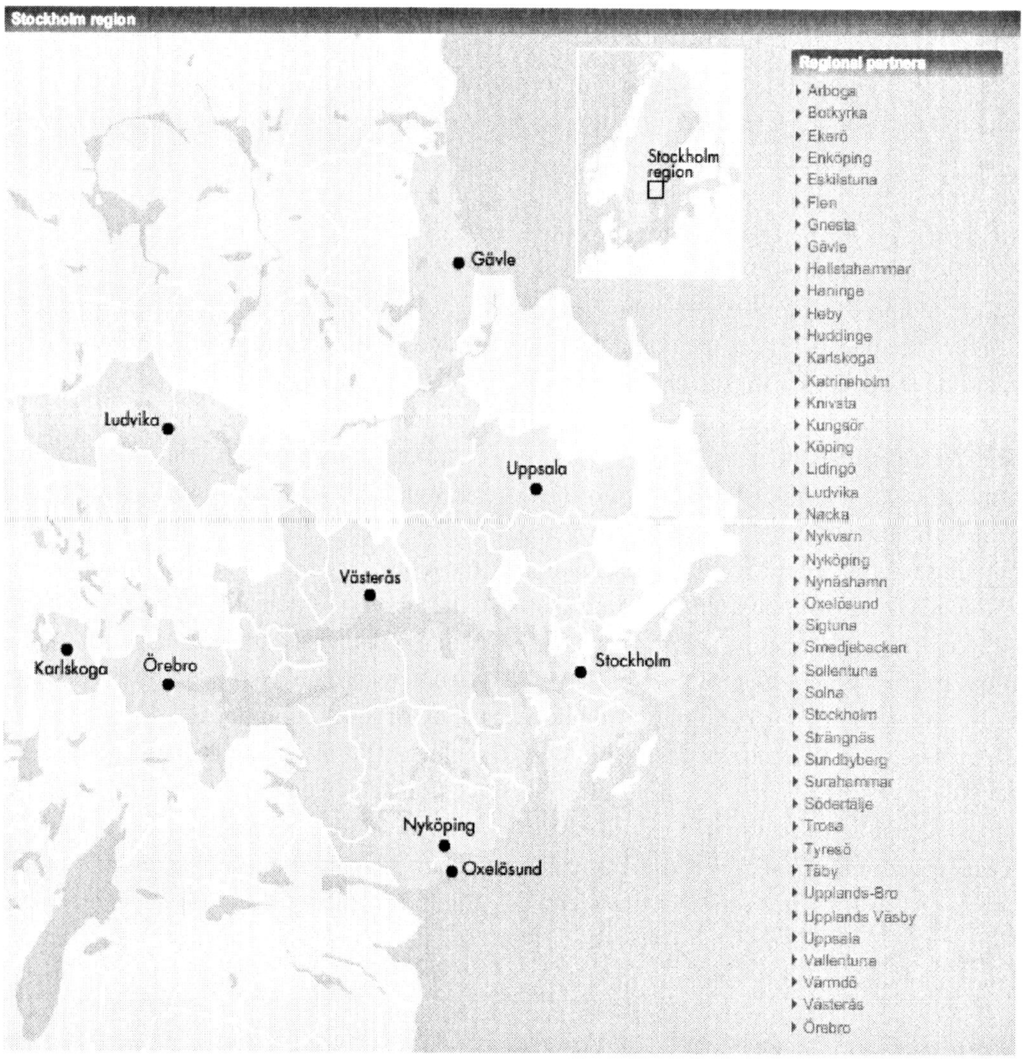

Figure 15. Virtual distribution of the seven regional partners in the Stockholm Business Alliance, who by August 2007 had published the partner-logotype on the first page of their web pages (this figure is edited by us to highlight these seven partners) [15] *(© 2007 Stockholm Stad. Used with Permission)*

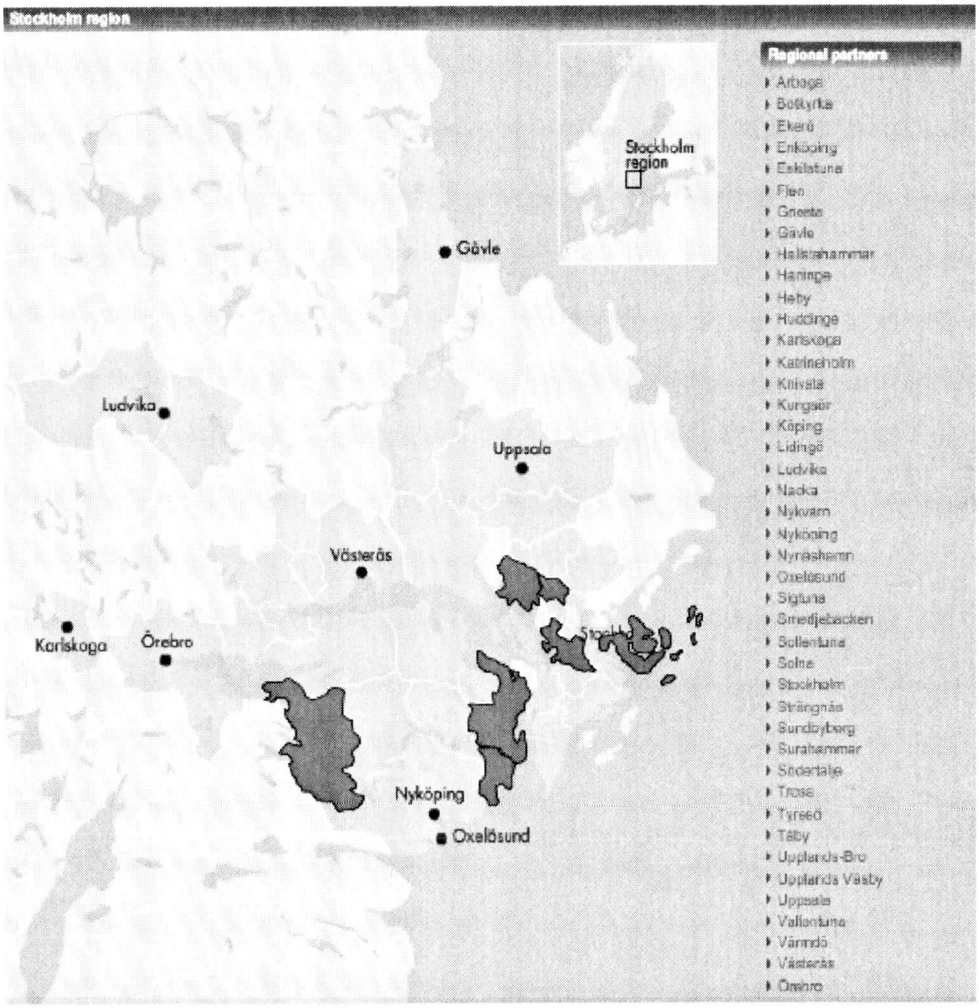

enough. "The Capital of Scandinavia" is not only a place but a network of places, a *space of places* to use Castells' (1997) term which illustrates another trend in today's urban reality: the blurring of boundaries between what counts as a city and what is its suburbs and countryside (Amin & Thrift, 2002). Where are the boarders of "The Capital of Scandinavia"? Are they geographical regarding *all 43* regional partners (see Figure 14) or are they virtual regarding those *few seven* partners who by August 2007 had published brand-related logotypes on their web pages: Stockholm, Upplands-Bro, Upplands Väsby, Värmdö, Södertälje, Trosa and Katrineholm (see Figure 15). Does this depend on who is drawing the borders or who is looking? And will Stockholm and the whole region become "The Capital of Scandinavia" in the eyes of its inhabitants and those working there?

Figures 14 and 15 illustrate the question of how visible and present "Stockholm: The Capital of Scandinavia" is on the Internet and how the group of contractually connected partners is different from the group of logotype connected partners. This is an important observation since the logotype connected partners and their logotype presence on the Internet might be the first or perhaps only impression an international visitor to the municipality web pages ever see of the Stockholm Business Alliance and the brand of "Stockholm: The Capital of Scandinavia".

It could be argued that if becoming a natural part of everyday discourse "The Capital of Scandinavia" could actually come into existence. This is backed up by Stuart Hall who has conducted studies on how British cities in the 1980's and 1990's were socially reconstructed through discourse in media and space. Hall argues that "representation is a process whereby cultural systems of space are envisioned into being" (S. Hall, 1997, p. 203). According to Hall, by stating that Stockholm *is* the capital of Scandinavia and also giving evidence of which places are closest to this centre (the municipalities which are members of the SBA), the map could be redrawn as a certain pattern of centre and periphery will be established, at least on a symbolic level. Whether it will have an effect on the material/economic level is yet to be seen. What can be seen is the impact of the press room. Here, the Internet obviously plays an important role and is an example of an important source of communicating the brand to people all over the world who are in the position of formulating the discourse.

FROM CLEAR BRAND TO FUZZY IMAGE?

It is true that the naming of a place contributes to its identification (Amin & Thrift, 2002) and that we through names and slogans have images of places even though we haven't been there. But in the Stockholm case, despite the launching of a brand, people obviously have not embraced the bold statement it builds on to encompass it as an internalized identity yet. For instance when *Svenska Dagbladet*, a large Swedish Daily Newspaper, asked its online readers in June 2007 which city they preferred as the capital of Scandinavia Copenhagen came out first with 53,6 percent of the votes, Stockholm second with 43,3 percent and Oslo third with 3,2 percent of the votes.[16] Political scientist Jacob Christensen currently at the Department of Political Science, Umeå University, Sweden, comments upon another interesting survey. In his blog about the Stockholm initiative to brand itself as The Capital of Scandinavia he refers to a Mastercard ranking of cities as commerce centres.[17] Here, Copenhagen ranked 14 and Stockholm ranked 16. Copenhagen is ahead of Stockholm on most sub issues such as "legal and political framework", "economic stability", "ease of doing business", "financial flow" and "business centre". Thus, to many it seems that Stockholm is *not* the natural capital of Scandinavia. From the perspective of Kotler, Haider and Rein the element of "developing a strong and attractive positioning and image for the community" has not been fulfilled (Kotler et al., 1999, p. 25) which indicates that the brand albeit its conceptual clarity might end up in fuzzy images once we pay attention to the receiver side.

Perhaps a double approach to avoid this is necessary. On the one hand the city must continue to work with its physical appearance since branding is often most powerful when linked to and coordinated with physical transformation of the city. In fact, to build or refurbish signature buildings is a quite common strategy for branding cities. Branding through advertising as in the case of Stockholm is another and branding through planning strategies and urban redevelopment is a third (Jansson & Power, 2006). Cities that lack powerful symbols or historical or architectural signifiers even devise them as several examples in the USA shows (Judd & Fainstein, 1999) and the aim is of course to construct a landmark which puts the city on the mental map of people (remember how Paris is connected with the Eiffel Tower or Sydney with the Opera House) (Holcomb, 1999).

This may lie outside the scope of SBR or SBA though and illustrates the need for an overall view of the city's activities.

On the other hand a continuous work with the nontangible images is also necessary through the work with traditional and Internet media. That SBA views the Internet as an important channel of affecting and changing people's image of Stockholm is obvious through the pedagogical web page with functions like the web based press room. Another example is the official brand-related movie of four minutes named "Stockholm: The Capital of Scandinavia" which was produced in January 2007 and soon found its way to YouTube™. This official YouTube movie is introduced with a helicopter film guided by the brand logo. Happy upbeat music accompanies MTV-style movie clips of well-known buildings and places in the larger Stockholm region, people in everyday situations, stores and many facets of how it feels to be in Stockholm; in summer as in winter times. [18] The official movie has not been viewed as many times as another movie called "Stockholm—The Capital of Sweden" posted in December 2006 though,[19] which is a movie produced by a private person in Norway who calls herself "jentemor".[20] This low-budget slide show of still pictures taken downtown Stockholm in the late fall of 2006 mostly accompanied by music by Vollenweider[21] supports the image of Stockholm as a nice place but there are also examples of forums on or available through the Internet where the claim that Stockholm is the Capital of Scandinavia is questioned.

According to the Norwegian daily *Dagsavisen* which had an article about the Stockholm brand in March 2006 the Copenhagen tourist office manager Lars Bernhard Jørgensen commented that the Stockholm brand prevents cooperation like that between Copenhagen and Stockholm, two cities that once used the slogan of "Capitals of Scandinavia".[22] Another critical voice is the organisation promoting the Øresund region.[23] Representatives for the organization have published a statement in which they express their surprise that Stockholm has taken this step clearly alienating Copenhagen and Oslo. The Øresund Region uses the slogan of "Øresund. The *Human* Capital of Stockholm" and feel trampled upon.[24] These examples illustrate the difficulty of controlling the image of a city regardless the branding budget and the efforts put into work. On a contemporary online marketing channel like YouTube™ visitors can find several different videos presenting Stockholm in different ways which might contribute to an unwanted variety of how Stockholm is imagined and by browsing the Internet a range of images of Stockholm are encountered.

Also, one could not help but wondering whether the political climate in Sweden is ready for the kind of work that would aim at developing the conceived identity of the people in the area covered by the Stockholm Business Alliance towards encompassing "Stockholm: The Capital of Scandinavia". Having a strong history of fighting for decentralization and emphasizing the importance of "a living country side" local politicians might find it difficult having to explain the fee paid to the SBA (out of tax payers' money) even though potential positive effects in the future. The work of "inward branding" (Jansson & Power, 2006) i.e. the branding which aims at building the identity of the residents of the Stockholm Business Arena-region might clash with the "inward branding"-activities of the smaller towns or municipalities.

CONCLUSION

The case study of "Stockholm: The Capital of Scandinavia" as we have presented it here illustrates the challenge that many city managers face today: to work beyond the mere construction of new productive infrasystems and to attract the unconditional love, attraction and commitment of many disparate actors (Dobers & Strannegård, 2001). Even though Stockholm has been successful in organizing the SBA it could be argued that this "top-down"-approach will not be as far-reaching as the "bottom-up"-approach visible for example in the case of the branding of the "SoFo"-area in Stockholm, a gentrified area in the Southern parts of the City where the branding process has been part of building a neighbourhood identity (Jansson & Power, 2006). "Stockholm - The Capital of Scandinavia" does not seem to be part of the identity of neither local nor regional residents and for reasons that we only have speculated on here the Internet presence of "Stockholm: The Capital of Scandinavia" seems to deal more with the City of Stockholm rather than the region of SBA. To develop a deeper understanding of the rationales behind the commitment of the SBA partners and to follow the case until 2010 when the SBA will be evaluated could be the obvious next step and should provide new insights into the relationship between image, substance and brand identity. On a general level we believe that the role of Internet in raising awareness of brands and in promoting identification with the brand (Rowley, 2004b) is an interesting area of future research.

Another question is what happens to the city when it is reduced to a commodity. City branding is per definition an approach that centres around the conceptualization of the city as a brand (Kavaratzis & Ashworth, 2006) and the aim of the branding process is to establish a close link between the character of the object and its branded image of form (Jansson & Power, 2006). How then does this affect the city? On the one hand the commodification process involves the transformation of the city into a product that can be exchanged somehow (Fainstein & Gladstone, 1999) emphasizing the importance of souvenirs, photos, brochures and other artefacts that can symbolize the commodified city. To some this is to destroy "the soul" of the city (Holcomb, 1999) as the city is made attractive and accessible "through processes of reduction and simplification" (Judd & Fainstein, 1999: 37). On the other hand the commodification of the city is also interesting to study in relation to channels like the Internet. Through the Internet, the investor, the knowledge worker and the tourist (all those whose interest the SBA hope to catch) have the possibility of looking not only at Stockholm but at all the cities and regions that SBA compete with. This way the place as a product is always sold to the consumer before she sees it making branding more significant than for conventional products where the consumer can see, test, and compare the product (Holcomb, 1999). In this the Internet today plays an important role. And the possibility of Internet portals leading to an endless number of sub pages and hypertext providing a limitless stream of continuously updated information, photos, moving images etc makes Internet a different kind of forum for representing the place compared to other sources. Could it be argued that the Internet is a way to represent the multifacetted complexity of the city compared to traditional forums and that this is a way to balance the reduction of the city which is inevitable in branding endeavours? And which role does Internet play in the commodification process? Obviously, several questions seek answers.

ACKNOWLEDGMENT

We are thankful to the participants of a text seminar at the Royal Institute of Technology in October 2007 for discussing and commenting on a draft to this chapter. Special thanks goes to Assistant Professor Hans Rämö, Stockholm University.

REFERENCES

Adolfsson, P. (1999). *Environment as a part of the city.* Paper presented at the Managing Big Cities Conference (August 26-28). Gothenburg, Göteborg.

Adolfsson, P. (2003). *Miljö och dess många ansikten i staden. Om kvalitetsmätningar och organisering i Stockholm*. Göteborg: BAS.

Amin, A., & Thrift, N. (2002). *Cities. Reimagining the urban*. Cambridge: Polity Press.

Anderson, B. (1953). *Stockholm. Capital and crossroads*. Stockholm: The Swedish Institute.

Ashworth, G. J., & Voogd, H. (1990). *Selling the city. Marketing approaches in public sector urban planning*. London: Belhaven.

Beckerson, J. (2001, March/April). *Marketing British tourism. Early state stimulation of a service sector, 1890-1950*. Paper presented at The Economic History Society, 75th Anniversary Conference, Kelvin Conference Centre, University of Glasgow.

Bennett, F. C., & Nathanson, J. (1997). Metropolitan alliances. *Economic Development Commentary, 21*(1), 34-38.

Bergström, O., & Dobers, P. (2000). Organizing sustainable development. From diffusion to translation. *Sustainable Development, 8*(4), 167-179.

Boyle, M. (1997). Civic boosertism in the politics of local economic development. 'Insitutional positions' and 'strategic orientations' in the consumption of hallmark events. *Environment and Planning A, 29*(11), 1975-1997.

Brotchie, J., Batty, M., Hall, P., & Newton, P. (Eds.). (1991). *Cities of the 21st century. New technologies and spatial systems*. New York, NY: Longman Cheshire.

Castells, M. (1989). *The informational city. Information technology, economic restructuring, and the urban-regional process*. Oxford: Blackwell Publishers.

Castells, M., & Hall, P. (1994). *Technopoles of the world. The making of twenty-first-century industrial complexes*. London: Routledge.

Chevrant-Breton, M. (1997). Selling the world city. A comparison of promotional strategies in Paris and London. *European Planning Studies, 5*(2), 137-161.

Cushman & Wakefield, H. B. (2006). *European Cities Monitor 2006*. London: Cushman & Wakefield, Healey & Bakero.

Czarniawska, B. (2002). *A tale of three cities. Or the glocalization of city management*. Oxford: Oxford University Press.

Dobers, P. (2001). Netting the information infrasystem of Stockholm. An idea spreads throughout the world. In H. Glimell & O. Juhlin (Eds.), *The social production of technology. On the everyday life with things* (pp. 189-206). Göteborg: BAS Publishers.

Dobers, P. (2002). Broadband—boom and bust in the new economy. In I. Holmberg, M. Salzer-Mörling & L. Strannegård (Eds.), *Stuck in the future? Tracing 'the new economy'* (pp. 79-104). Stockholm: Bookhouse.

Dobers, P. (2006). Empty spaces or illusionary images? "Stockholm as a Mobile Valley". In S. R. Clegg & M. Kronberger (Eds.), *Space, organization and management* (pp. 287-304). Malmö / Copenhagen: Liber / Copenhagen Business School Press.

Dobers, P., & Strannegård, L. (2001). Loveable networks. A story of affection, attraction and treachery. *Journal of Organizational Change Management, 14*(1), 28-49.

Erickson, B., & Roberts, M. (1997). Marketing local identity. *Journal of Urban Design, 2*(1), 35-59.

Fainstein, S., & Gladstone, D. (1999). Evaluating urban tourism. In D. R. Judd & S. Fainstein (Eds.), *The touristCity* (pp. 21-34). New Haven and London: Yale University Press.

Fishman, R. (1987). *Bourgeois utopias. The rise and fall of suburbia*. New York: Basic Books.

Graham, S., & Marvin, S. (1998). *The richness of cities. Urban policy in a new landscape* (Net effects: Urban planning and the technological future of cities No. Working Paper 3). Newcastle: Centre for Urban Technology, Department of Town and Country Planning, University of Newcastle.

Graham, S., & Marvin, S. (2001). *Splintering urbanism. Networked infrastructures, technological mobilities and the urban condition*. London: Routledge.

Hall, P. (1977). *The world cities* (2nd ed.). London: Weidenfeld and Nicolson.

Hall, S. (Ed.). (1997). *Representation: Cultural representations and signifying practices*. London: Sage / Open University Press.

Hall, T. (1998). Introduction to selling the entrepreneurial city. In T. Hall & P. Hubbard (Eds.), *The entrepreneurial city. Geographies of politics, regime and representation* (pp. 27-30). Chichester: John Wiley.

Hannigan, J. (1998). *Fantasy city. Pleasure and profit in the postmodern metropolis*. London: Routledge.

Hepworth, M. E. (1989). *Geography of the information economy*. London: Belhaven Press.

Hepworth, M. E. (1990). Planning for the information city. The challenge and response. *Urban Studies, 27*(4), 537-558.

Holcomb, B. (1999). Marketing cities for tourism. In D. R. Judd & S. Fainstein (Eds.), *The tourist-City* (pp. 54-70). New Haven and London: Yale University Press.

Jansson, J., & Power, D. (Eds.). (2006). *Image of the city. Urban branding as constructed capabilities in Nordic city regions*. Oslo: Nordic Innovation Centre.

Judd, D. R., & Fainstein, S. S. (Eds.). (1999). *The tourist city*. New Haven, CT.

Kapferer, J.-N. (1997). *Strategic brand management. Creating and sustainaing brand equity long term* (2nd ed.). London: Kogan Page.

Kavaratzis, M., & Ashworth, G. J. (2006). Partners in coffeeshops, canals and commerce. Marketing the city of Amsterdam. *Cities, 24*(1), 16-25.

Kearns, G., & Philo, C. (1993a). Culture, history, capital. A critical introduction to the selling of places. In G. Kearns & C. Philo (Eds.), *Selling places. The city as cultural capital. Past and present* (pp. 1-32). Oxford: Pergamon.

Kearns, G., & Philo, C. (Eds.). (1993b). *Selling places. The city as cultural capital. Past and present*. Oxford: Pergamon.

Kiang, M. Y., Raghu, T. S., & Shang, K. H.-M. (2000). Marketing on the internet. Who can benefit from an online marketing approach? *Decision Support Systems, 27*, 383-393.

Kotler, P., Asplund, C., Rein, I., & Haider, D. H. (1999). *Marketing places Europe. How to attract investments, industries, residents and visitors to cities, communities, regions and nations in Europe*. London: Pearson Education.

Kotler, P., Haider, D. H., & Rein, I. (1993). *Marketing places. Attracting investment, industry, and tourism to cities, states, and nations*. New York, NY: Free Press.

OECD. (1992). *Cities and new technologies*. Paris: OECD.

Porsander, L. (2000). Translating a dream of immortality in a (con)temporary order. *Journal of Organizational Change Management, 13*(1), 14-29.

Rainisto, S. (2003). *Success factors of place marketing. A study of place marketing practices in northern Europe and the United States*. Helsinki: Helsinki University of Technology.

Rowley, J. (2004a). Online branding. *Online Information Review, 28*(2), 131-138.

Rowley, J. (2004b). Online branding. The case of McDonald's. *British Food Journal, 106*(3), 228-237.

Sadler, D. (1993). Place marketing, competitive places and the construction of hegemony in Britain in the 1980s. In G. Kearns & C. Philo (Eds.), *Selling places. The city as cultural capital. Past and present* (pp. 175-192). Oxford: Pergamon Press.

Shostak, A. (1982). Seven scenarios of urban change. In G. Gappert & R. V. Knight (Eds.), *Cities in the 21st century* (pp. 69-93). Beverly Hills: Sage.

Smyth, H. (1994). *Marketing the city. The role of flagship developments in urban regeneration.* London: E and FN Spon.

Stewart, M. (1996). Competition and competitiveness in urban policy. *Public Money and Management, 16*(3), 21-26.

Stockholm Business Region. (2007a). *Marknadsplan. Investment Promotion.* Stockholm: City of Stockholm.

Stockholm Business Region. (2007b). *Verksamhetsprogram och budget Stockholm Business Alliance 2007.* Stockholm: City of Stockholm.

Trueman, M., Klemm, M., & Giroud, A. (2004). Can a city communicate? Bradford as a corporate brand. *Corporate Communications: An International Journal, 9*(4), 317-330w.

Urban, F. (2002). Small town, big website? Cities and their representation on the internet. *Cities, 19*(1), 49-59.

Ward, S. V. (1998). *Selling places. The marketing and promotion of towns and cities, 1850-2000.* London: E & FN Spon.

Ward, S. V., & Gold, J. R. (Eds.). (1994). *Place promotion. The use of publicity and marketing to sell towns and regions.* Chichester: John Wiley.

Wells Jr, L. T., & Wint, A. G. (2000). *Marketing a country. Promotion as a tool for attracting foreign investment* (revised edition ed.). Washington: Foreign Investment Advisory Service.

Wolf, M. (1999). *The entertainment economy. The mega-media forces that are re-shaping our lives.* New York, NY: Penguin Books.

ENDNOTES

1. http://www.stockholmbusinessregion.se/templates/page____19284.aspx?epslanguage=EN (Accessed on October 10, 2007)
2. http://www.stockholmbusinessregion.se/templates/page____19570.aspx?epslanguage=EN (Accessed on October 5, 2007)
3. www.dowellstubbs.com
4. All quotes from Julian Stubbs are from an interview made with him November 23, 2006
5. http://www.stockholmtown.com/templates/page____14737.aspx. (Accessed on August 30, 2007)
6. Our interview with Anne Årneby was carried out October 31, 2006
7. The partnership is further described in Swedish and in English at www.thecapitalofscandinavia.com, where you for instance can find a 16 page business plan and budget for the Stockholm Business Alliance for 2007 (in Swedish), and a marketing and investment promotion plan for 2007 (in Swedish).
8. A detailed description of planned activities etc can be found in the brochure "Market plan. Investment promotion 2007 (In Swedish) (Stockholm Business Region, 2007a).
9. Interview with Anne Årneby, October 31, 2006

[10] Business Week, August 21, 2006

[11] These are quotes from the interview with Stubbs, November 23, 2006.

[12] Some figures gives an idea of the size of the City of Stockholm-organization: In 2006, the City of Stockholm had 50 195 people on the monthly pay-rolls (including the city-owned companies) and the budget for the City's all activities 2007 was 34,7 billion SEK.

[13] In Sweden the inhabitants pay income taxes to the municipality. This means that a municipality with many high-income citizens is more wealthy than a municipality with low-income citizens. In addition to this, though, there is a "Robin Hood"-like system by which the Swedish state redistributes some of the financial resources to municipalities in parts of the country which have difficulties surviving on the income taxes of their own citizens.

[14] http://www.stockholmbusinessregion.se/templates/page____18554.aspx?epslanguage=EN (Accessed on September 20, 2007)

[15] http://www.stockholmbusinessregion.se/templates/page____18554.aspx?epslanguage=EN (Accessed on September 20, 2007)

[16] http://www.svd.se/dynamiskt/stockholm24/did_15853848.asp (Accessed on August 31, 2007)

[17] http://blog.jacobchristensen.name/2007/06/20/the-capital-of-scandinavia/ (Accessed on September 20, 2007)

[18] http://www.youtube.com/watch?v=TbNnw5Z0V6k (Accessed on August 22, 2007)

[19] On August 22, 2007, the YouTube movie had been seen about 950 times, and the other movie about 4.400 times.

[20] "mother of girls" or "young, girlish mother"

[21] http://www.youtube.com/watch?v=sqjPn0fRda4&mode=related&search= (Accessed on August 22, 2007)

[22] http://www.dagsavisen.no/innenriks/article266610.ece (Accessed on September 20, 2007)

[23] http://www.oresundsregionen.org/2fa92ef4 (Accessed on September 20, 2007)

[24] http://www.oresundsregionen.org/42c90029 (Accessed on September 20, 2007)

Chapter XIV
Marketing the mCity: How a City Based ICT-Project Can Make Sense

Anette Hallin
Royal Institute of Technology, Sweden

ABSTRACT

Information and communication technologies (ICTs) cannot only be used practically in marketing efforts, but also as symbols, due to the images and associations they provoke of for example modernity and speed. The marketing of a city through the use of ICT-images however, also involve risks, as ICTs among certain people also bring about negative associations. Therefore, marketers need to be aware of what happens with the marketing material after it has been developed and sent out. The main argument of this chapter is that sense making emerges through a dialogic process. By analyzing semiotically a marketing leaflet for the Stockholm-based ICT-project mCity, and two ads for Nokia phones that appeared in Europe at about the same time as mCity, this chapter challenges the traditional cybernetic sender-receiver model of communication, and proposes that when the sender has sent the message, the message becomes a speaker on its own, interacting with the listener through a dialogic process set in the mind of the listener. When understanding this, marketers should benchmark the use of ICTs in other contemporary media in order to ensure the success aimed for with the city marketing material using ICT-imager.

INTRODUCTION

Throughout history, images of places have been of interest to different actors. As early as in the Middle Ages, places like Jerusalem, Santiago de Compostela and Nidaros attracted pilgrims due to their images of holiness, and in the early 20th century, cities like Paris and Trieste attracted poets and artists due to their images of modernity and cultural creativity (T. Hall, 1997). Since the mid-1970's, though, it can be argued that the development of strategic "place-selling" has been affected by the process of globalization, and in order to compete on a global market, there is a clear attempt to build an imagery suited for that competitive purpose (Harvey, 1989).

A more positive attitude among official policy-makers towards the market-forces can also be traced (Ward, 1998), perhaps because the upgrading of the image of a city can cast a "seemingly beneficial shadow" over the whole metropolitan region (Harvey, 1989, p. 8). At the same time, the responsibility of place marketing has moved from the national to the local or regional levels (Ward, 1998), and therefore, the branding, marketing and selling of the city have become important issues to city managers (Jessop, 1998). This is not the place of discussing the differences between these three concepts, but to only briefly say that they all deal with connecting positive images with the city. As "marketing" most often is used as the term denoting the practical work, this is the term that will be used in this chapter.

Different cities use different marketing-strategies depending on institutional, political and cultural factors (Chevrant-Breton, 1997), and so far, more research needs to be done concerning how cities work practically with these issues; which tools are used, how they work and their result. Due to the growing awareness of cities'—and the regions'—roles in innovation and regional growth, a renewed interest in how cities create attractive urban spaces can be traced, though, and this interest is directed, not only towards urban redevelopment through the building up of hard infrastructure, or the supporting of new firms entering the area, but also towards how cities work with tangible measurements and focus on constructing advantages. (Jansson & Power, 2007)

Traditionally, the cities of the world evolved as places of commerce and trade, and as centers of religious, military, cultural and political power. But today, cities—especially if they do not possess this historical heritage—need to construct their own images. In this process, one would expect cities to distinguish themselves from their competitors, but so far, many cities seem to have chosen similar strategies: they have built their story around promotional events such as the Olympics, the World Cup or other sports events; or exhibits and fairs, like the Cultural Capital, thus relating themselves to other cities which have adopted the same strategy. (Fainstein & Judd, 1999; Porsander, 2000)

To yet other modern cities, information and communication technologies (ICTs) have played a central role, making ICTs not only important in a material way, building the infrastructure of eGovernment, providing citizens and city-based companies with the means to communicate and carry out their business, but through their image qualities, lending to the cities all the notions associated with them. This is nothing new though, technology has always been connected to the idea of modernity, through its associations with energy and speed (Löfgren, 2001). Place marketing is full of metaphors of movement, speed and tempo, because to have a high level of energy has become quite important in a globalized world, where cities compete for investors, entrepreneurs and visitors. A place of "energy" is also a "modern" place i.e. a place of the future, and of future promises—associations which have been used in the marketing not only of cities, but of regions, where ICTs sometimes has been used since the area has a lot of ICT-related activities, such as Silicon Valley, and sometimes as the result of a strategic governmental intervention, as in the case of the Multimedia Super Corridor in Malaysia (Bunnell, 2003), and there are obviously many more examples.

Even though ICTs play a vital role, carrying associations of modernity and success, they must be used in a different way than earlier, as today, the interest of technologies seem to have moved from the technologies themselves to their content and effect. First of all, ICTs not only carry positive connotations. During the last few years, research has been carried out on the non-technical aspects of mobile technology, based on the idea that mobile technologies are not only technological but social objects (Brown, 2002), raising issues of privacy and integrity, the blurring of borders between work-life and private-life, the speed of modern society, the redefinition of communication and other not so positive effects of ICTs on modern society. (Myerson, 2001; Cooper, 2002; Gant & Kiesler, 2002; Sherry & Salvador, 2002). And even though the digital revolution has only begun, somehow, ICTs seem to have slipped into darkness, becoming part of the every day life rather than interesting in themselves (Dobers & Hallin, 2006), so cities interested in marketing themselves as technologically advanced face a challenge in how to do this in a good way in order to avoid being connected to the negative associations of ICTs.

Also, since people make sense of something new in relation to previous experiences, the marketing of cities cannot be seen in isolation, but must be related to the society at large. Meaning is produced in a "circuit of culture", where the understanding of something new emerges through individual interpretation (S. Hall, 1997), as the individual relates the new to all that which seems relevant in his or her particular case. This can be described as a dialogical process between the listener and the speaker (Bakhtin, 1934-35/1981), when, in the case of for example marketing material, the speaker is no longer the sender of the message, but the marketing material in itself, and the listener is the audience which takes in the marketing material.

Based on Stuart Hall's and Michail Bahktin's thinking, the main argument of this chapter is that the marketing material of a city, for example for an ICT-related project, enters into a dialogical process in what Hall called "the circuit of culture" (Hall, 1997), which means that the interpretation of the marketing material cannot be seen in isolation. Instead, the sense making process of this involves references (consciously as well as unconsciously) to a variety of other images and verbal representations of ICTs. It is argued that this process takes place in the mind of the listener, through a dialogical process between the speaker's verbal and visual expressions and the listener's associations to these.

The aim of this chapter is to show how this dialogic process can take place, and to thereby contribute to the understanding of how ICTs can be used in city marketing. For this purpose, a semiotic analysis of a marketing leaflet for a Stockholm-based project called "mCity" will be carried out, and the focus will be on how the project has entered two types of dialogue, verbally, through its name, and visually, through the marketing material. By connecting the semiotic analysis of the marketing leaflet with a similar analysis of two Nokia-ads which appeared in Europe at about the same time, the purpose is to develop an understanding of how the representation of ICTs in relation to cities can make sense when engaging in dialogue through the "circuit of culture". Even though the case is a city-related case, it could be argued that the same process takes place in the marketing of regions or countries, as the dialogic process is the same. It should be pointed out though, that the practicalities of the marketing of regions or countries might differ from the marketing of cities—more actors might be involved, and the marketing process itself might thus be different compared to the marketing of a city.

The outline of the chapter is as follows: after a few words on semiotics, a theoretical section follows which aims at explaining how sense making happens through dialogue. Then, after an introduction to the mCity-project, an analysis of the verbal representation of the project (its name) is carried out, followed by the semiotic analysis of a marketing leaflet for mCity and two Nokia-ads, chosen as they were issued at about the same time as the mCity leaflet, and spread in Europe at the same time. Through the semiotic analysis, the intention is to show that the marketing leaflet and the Nokia-ads belong to the same circuit of culture, and the theoretical and practical implications of this are finally discussed.

A FEW WORDS ON SEMIOTICS

In order to understand how the marketing leaflet of mCity creates meaning I will do a semiotic analysis. Dealing with the study of "signs", the basic idea of semiotics is that language is made up of signs which communicate meaning, and that our understanding of reality is constructed by the words and other kinds of signs which we use—not only by that which the words or the signs denote.

Different schools of semiotics have explained the relation between the sign and that which it denotes in different ways. In the early 20th century, the French linguist Ferdinand de Saussure introduced the idea that the sign consists of two sides; the "signifier" and the "signified", where the "signifier" is the perceptual side and the "signified" is the conceptual side of the sign (Selden & Widdowson, 1993). To Saussure, meaning thus resides in the image/sign/symbol itself, just as the value of money resides in the money itself, even though this value is depending on a specific context with a specific time and a specific place. The literary critic Roland Barthes explained the difference through the terms "denotation" and "connotation", denotation being the analogy between the image and its object, and connotation being the associations the image provokes (Barthes, 1977). In the United States, Charles Pierce, about the same time as Saussure, introduced a third dimension by talking about "the sign", "the referent"/"object" and "the interpretant", arguing that the semiotic process is a dynamic relation between the three, resulting in the creation of meaning (Christensen & Askegaard, 1999).

Regardless the terminology, by recognizing the ideas of semiotics, one see that words and signs also may carry associations beyond that specific object to which the word refers, connotations, which are constructed in a specific cultural context. The linguistic sign "Rolls-Royce" for example, not only denotes a car, but in our times carries connotations of luxury and wealth (Bignell, 2002). This way, language is central in our process of making sense of the world, and meaning is social and relational, rather than idiosyncratic (Tietze, Cohen & Musson, 2003). By analyzing the bits and pieces of a visual representation in order to find a hidden meaning, one tend to leave out the context in which the image was produced and read, which not only would add valuable perspective to the analysis (Leeuwen, 2001), but is necessary if the ideas on dialogicity as described briefly above, and in more detail below, are recognized. This is why I will not only study an mCity-leaflet, but also two Nokia-ads which appeared at about the same time as the mCity-leaflet (spring of 2003), and that can be said to be examples of two ICT-ads with which mCity enters into dialogue in my interpretation. The main reason for this is that they are similar to me in a very apparent way; both represent photos of mobile phones, which is why it could be argued that they belong to the same circuit of culture.

In order to understand how the leaflet and the ads create meaning, I will apply semiotic terminology and use the methodology presented by Jonathan Schroeder which include three steps: description, interpretation and analysis (Schroeder, 2003; Schroeder, 2004). Through these steps, my intention is to point to the codes that connect meanings to specific graphic patterns in the ads. These have to do with techniques of "photogenia", i.e. techniques of lighting, exposure and printing (Barthes, 1977), as well as disposition and organization of the images, for example vectors (lines in the pictures) expressing point of view, placement of objects/subjects, expressing position, and salience, i.e., the use of eye-catching details in order to accentuate something special (Jewitt & Oyama, 2001).

Even though the chapter uses semiotics as a tool to analyze how images create meaning, this does not mean that this meaning, the one expressed here, is the *only* meaning, the *"true"* meaning which everyone engaging in dialogue with the images experience the same way. Rather, the meaning is the result of a sense making process which is not only dependent on the cultural and historical context, but on individual experiences. Therefore, my claim is neither that these interpretations are general, or objectively true, nor that they reveal the intention of those who developed the material, but rather that these interpretations are only a few, possible interpretations. The point here is not to reveal the true meaning of the images, but to show that by representing the same object (in this case a mobile phone), they place themselves in the same circuit of culture, and when recognizing this it then becomes interesting to see how they engage in dialogue with each other in one listener's mind (mine), in order to see how the interpretation of the one is related to the interpretation of the other. The fact that the reader of this chapter might—and probably will—interpret the visual and verbal representations of mCity and the ads differently is the basic evidence that the dialogic process actually takes place, which is the very point of the chapter. What is interesting is not if the imagery of mCity or the two Nokia-ads studied here provoke positive or negative associations in my mind, but that they in fact can be related to each other through the dialogic process in the way described in this chapter—something which exemplfies how different but related impressions can be (and often is) related to each other; they interact in the sense making process. The semiotic exercise, then, is only necessary in order to show that the dialogic process actually takes place. But what, then, is the "dialogic process"?

THE DIALOGIC PROCESS AND THE CIRCUIT OF CULTURE

The argument in this chapter is that the images of Stockholm as mCity has emerged through a dialogic process in the minds of people conciously or unconciously, as they have engaged in dialogue by linking to the associations in this case provoked by two Nokia-ads which appeared as posters in the city, as ads in magazines and new papers at about the same time as the marketing leaflet for mCity was issued—just as everything new is made sense of in relation to whatever seems (consciously and unconsciously) relevant. In this process of meaning production, language plays a crucial role through its ability to transmit meaning, operating as a representational system, being a medium through which we 'make sense' of things (S. Hall, 1997).

Often, the cybernetic communication-model is used to describe the communication-process. But this simplifies the process by reducing the aspects of the communication-situation to the sender, the receiver, the message and the medium (a channel), possibly making it a good model for the transmission of information in mechanical systems, but not for human communication, as the complexity of the process is not taken into account (Czarniawska, 2000). To 'make sense' of things is not only a passive incorporation of the outside, but an active re-organizing, re-writing and re-formulating of it (Weick, 1995), and in this process, language—in all its forms—plays a crucial role. Language is not only a system of signs, or abstract grammatical categories, but ideologically saturated, intertwined with worldviews, opinions and ideas (Bakhtin, 1934-35/1981), and "meaning" is the effect of an interaction between speaker and listener. This idea was introduced by the Russian Literary scholar Mikhail Bakhtin in the 1930s (Bakhtin & Volosinov, 1929/1994), and is today encompassed by several scholars of different disciplines, for example, the British cultural studies researcher Stuart Hall.

According to Hall, "meaning" is produced in a "cultural circuit", where language plays a crucial role. Hall argues that it is because of its representational function that language creates meaning when it is used:

It is by our use of things, and what we say, think and feel about them—how we represent them—that we give them a meaning [...] we give things meaning by how we use them, or integrate them into our everyday practices [....] we give things meaning by how we represent them—the words we use about them, the stories we tell about them, the images of them that we produce, the emotions we associate with them, the ways we classify and conceptualize them, the values we place on them. (S. Hall, 1997, p. 3)

Bakhtin goes even further by stating that interaction between people—dialogue—is a basic element of human existence. To live is to engage in dialogue and to find oneself imbursed in never-ending dialogues according to Bakhtin (in Dysthe, 1996), who also points out that dialogue not only takes place face-to-face between people as a vocalized verbal action but also between people through printed material of various kinds (Bakhtin & Volosinov, 1929/1994). Thus, humans are constantly engaged in dialogue.

There are of course a vast number of theoreticians which have discussed the complexity of dialogue, but Michail Bakhtin is of special interest here because of his ideas that the object produced by—to use cybernetic terminology—the 'sender', is not only inscribed with the 'message', but that the object actually becomes a 'sender' of its own, when the 'sender' has 'sent' it to the 'receiver'. Bakhtin points to several features of dialogue of which I have chosen four which I will discuss further below. I believe this can spread light on the sense making process by functioning as theoretical tools for interpretation: the unimportance of time, the distinction between internal and subjective dialogism, authoritative and internally persuasive discourse and the importance of the speech genre.

Dialogue, argues Bakhtin, is always involved in a dialogic relationship to utterances in history and in the future, making time irrelevant (Bakhtin, 1934-35/1981). Making a distinction between "internal dialogism", where the speaker breaks through the conceptual horizon of the listener, constructing the utterance against her perception of the speaking partner, making dialogue itself more important than the object of discussion, and "subjective dialogism", where the utterance is created with respect to what is talked about, leading to a larger risk of conflict of understanding between the speaking partners, Bakhtin points out that dialogue can have different orientation, depending on the speaker, the object of dialogue and the situation itself. (Bakhtin, 1934-35/1981)

Every utterance is impregnated with centrifugal and centripetal forces. Centrifugal forces are decentralizing and dis-unifying, whereas centripetal forces are centralizing and unifying. The result of these forces is the "dialogic orientation" of language, creating new and significant potential in discourse—both linguistically and in terms of meaning (Bakhtin, 1934-35/1981). Language always exists between the creative border-zones of human consciousnesses, and it is this very responsive interaction between speakers that makes it possible for language to produce new meaning (Bakhtin, 1952-54/1994). In fact, responsive understanding is … a fundamental force, one that participates in the formulation of discourse, and it is moreover an active understanding, one that discourse senses as resistance or support enriching the discourse" (Bakhtin, 1934-35/1981, pp. 280-281). The participants of dialogue take in what is said and relate it to their "apperceptive background" which is not a linguistic background but a background composed of specific objects and emotional expressions (Bakhtin, 1934-35/1981, p. 281) In dialogue, people always place themselves in relation to what is said, filling that which is said with various content from their own associations and beliefs, making language everything but neutral. This is the result of social life creating a multitude of concrete worlds and verbal-ideological social belief systems, within which language is filled with various content. Thus, language is never unitary, but heteroglot, representing the co-existence of contradictions between the present and the past, between different socio-ideological groups, different ideas etc (Bakhtin, 1934-35/1981).

Bakhtin distinguishes between two types of discourse: authoritative and internally persuasive. The authoritative discourse (for example religious dogma, acknowledged scientific truths, current fashionable ideas) enters our consciousnesses as a compact mass, which one must either totally encompass, or totally reject. The internally persuasive discourse, on the other hand, is the slow process of assimilating other people's worlds/ideas into one's own discourse. The sense making process, according to Bakhtin, then, is a continuous struggle among various available verbal and ideological points of view, approaches, directions and values and this process can take whatever route possible through the dialogic interaction with either of these options. (Bakhtin, 1934-35/1981, p. 346)

Of course, discourses can be fused with the image of the speaking person, affecting our idea of what is said—in fact, when attempting to understand and create meaning in everyday life, we do not separate discourse from the personality speaking it, because the personality is so materially present—but eventually, one's own discourse will begin to liberate itself from this and from the authority of other people's discourse (Bakhtin, 1934-35/1981).

For dialogue to be successful, i.e. for the members of the group to understand each other, the speaker and the listener must belong to the same language community (Bakhtin & Volosinov, 1929/1994). People in different language communities use different speech genres, depending on the speaker's speech will or speech plan. The speech will or speech plan is, according to Bakhtin, crucial to the choice of subject of speech and also provides the boundaries for that which is said. In all kinds of communication, the participants will try to orient themselves by trying to make sense of the speaker's speech plan. (Bakhtin, 1963/1994, p. 83).

As each utterance not only expresses the speaker's attitude towards the object of speech, but also towards other people's utterance, the utterance can be said to be filled with dialogic overtones, which must be taken into account in order to create meaning (Bakhtin, 1963/1994). Thus, if the listener does not have command of the generic forms of the speech genres used by the speaker, she will fell helpless and left out, disregarding the extent of her general knowledge of the language used, because she will not be able to hear the dialogic overtones.

Within literary theory, several scholars have presented similar ideas as Bahktin; "tout texte est absorption et transformation d'une multiplicité d'autres textes", wrote for example Julia Kristeva (Hallberg, 1992, p. 186) and even though both she and Bahktin referred to literature and words, today, the expanded definition of a "text" also includes every form of human expression; photos, drawings, pictures, etc. In order to understand how images—created linguistically or visually—create meaning, the focus of study should be dialogue. But what dialogue an image engages in depends on the associations and references of the listener and the speaker. Of course, these associations can never be completely documented by an outsider, but the intertextual frames to which the listener/speaker relates the image she encounters somehow contain similarities, despite difference in details (Rimmon-Kenan, 1983).

In order to understand the creation of meaning of "mCity", I will explore two dialogues in which the mCity-project can be said to engage. First, the verbal dialogue in which "mCity" can be said to engage through its name, and secondly the visual dialogue of the project through which the project engages through its promotion leaflets with images similar to those in the leaflets. But first, a background to the mCity-project will be provided.

ICTS IN THE CITY OF STOCKHOLM AND THE MCITY-PROJECT

Having a history of dominance in the field of telecommunications, it is not surprising that the Swedish capital of Stockholm has had a specific interest in mobile technologies. In fact, ICTs in the form of broadband and mobile technology have been particularly important in building up the image of the city of Stockholm as a high-tech city (Dobers, 2003). This has been seen as part of a global trend of cities competing with each other, raising the importance of sustaining, effecting and developing the city images (Czarniawska, 2000).

Stockholm has put much effort into joining the group of "IT-cities", through initiatives like the founding of the broadband communication company STOKAB in the mid 1990's and the starting of the IT-competition Stockholm Challenge where IT-projects from all over the world compete and are to work as sources of inspiration to each other (Dobers, 2003). At the turn of the millennium, Stockholm was also perceived an IT-capital in international media, for example through a special coverage article in Newsweek on "Hot IPOs and Cool Clubs in Europe's Internet Capital" (McGuire, 2000), and scored high in rankings on cities (Wirtschaftswoche, 2002), to the great pleasure of the Stockholm management, as they found support for their strive to promote Stockholm as the IT-capital of Europe.

During 2002, the image of Stockholm as an IT-city was questioned though, at least in domestic media, as the ICT-industry experienced a set back with people being laid off, companies going bankrupt etc. Headlines like "The IT-crisis—worst in Stockholm" were found in the newspapers, and numbers were presented to support the alarming news: the unemployment-rate in Stockholm rose with 58% during 2001 and 2002, compared to 7% in Gothenburg (the second largest city in Sweden) and 2% in Skåne (the most southern part of the country). (Computer Sweden 2002-10-21, quoting county governor Mats Hellström) Voices arguing for governmental intervention to help IT-companies were raised (Göran Johnsson, chairman of the union "Metall" in an article in the Swedish daily Dagens Nyheter 2002-10-18) and Stockholm was described as an important economic driver for the whole country (Mats Hellström, county governor of Stockholm region in an article in the Swedish daily Svenska Dagbladet, 2002-11-03).

At the same time, city managers in Stockholm were very aware of the rapid development of 3G and mobile services in the Asian countries, especially Japan, and the news about I-Mode and other innovative applications were used as an example of the advances against which Sweden might be regarded as lagging behind. As an answer, the head of the Stockholm Economic Development Agency, Christer Asplund, in January 2002 launched the idea of the "mCity-project", a project aiming at organizing "the mobile city" through the implementation of relevant ICTs, the project consisting of several small pilot-projects, focusing on identifying needs in the community and creating solutions to these. The pilot projects evolved around five different segments: tourists, students, SMEs, commuters and city employees and a few examples of pilot projects included a website for tourists with information on statues, art objects and buildings of interest available to tourists via mobile or fixed Internet; up-to-date traffic information available on an internet site through for example WAP and internet; absence management systems in Stockholm-schools, and scheduling services within the care sector.(Hallin & Lundevall, 2007)

The Stockholm Economic Development Agency, an organization within the City of Stockholm, initiated the mCity project together with the management of one of the city's local City Districts, and a number of Swedish ICT-companies. Even though it was originally planned as a two-year project, starting in January 2002, ending in January 2004, the project continued after that, becoming a part of the city's strategic work with

ICTs. When the final report was written in 2006, marking the closing of the project, it had engaged a large number of actors, locally and nationally, and the word about mCity had spread all over the world, making it interesting as a case for other city managers interested in "m Government", i.e. the work with providing citizens the possibility of accessing the public services regardless their location (Hallin & Lundevall, 2007). The project had been presented at a number of different conferences, and had attracted the attention of media and ICT-people all over the world.

Even though mCity was perceived as big, in numbers it was small (Hallin, forthcoming). It had a very limited budget of about 2-4 mSEK per year (about 320.000-640.000 USD/year), but was situated within the city organization and benefited form the help of overhead functions such as marketing staff as well as the involvement of the different companies. Between 2002 and 2006, three marketing leaflets about mCity were produced and distributed at different events where mCity was presented. Still, it could be argued that the fame of mCity was mainly due to word-of-mouth, rather than a result of a conscious marketing strategy. With a staff of 1-3 people, mainly involved in the actual running of the pilot projects, it is surprising that mCity became so broadly known among city people interested in ICTs. One reason for this was probably the good timing—there were not so many other ICT-projects of this kind around at the time, and the Swedish ICT-industry were having problems after the burst of the IT-bubble a year or so after the turn of the millennium.

The "m"-City

The name "mCity" was born in January of 2001, at a workshop in Stockholm to which Christer Asplund had invited different actors of the European telecom-industry to discuss the future possibilities of 3G. "mCity" is the primary verbal representation of the mCity-project, often even used alone, without "-project", opening up for several interpretations: the mobile city, the city where people are mobile, the city where mobile technology is used, etc. To abbreviate "mobile" or "mobility" into "m" and put it in front of a noun builds on a tradition started by the abbreviation of "atomic bomb" into "a-bomb" and continued by for example "e-mail" (electronic mail). Not only does the acronym stand as an adjective explaining the following noun, but there is also a trend of using it for many things related to the Internet, and today, several words are modeled in the same way. A quick search at the Internet provides a few examples: eBay, eTrade, eMuseum, eMedicine, eInclusion, eCards, eLoan, ePals, eLearning, eMusic, eNature, eGold, eStrategy/mStrategy, eCommerce/ mCommerce, eGovernment/mGovernment, etc. These acronyms seem to flourish, especially in marketing, probably because marketers believe in the images they create; images of technology and modernity, images of products and services at the edge of the latest ideas and discoveries. The assumption is that the latest and the most modern is what is best, according to the Western belief in linear evolution. The "mCity" can thus be interpreted as the modern city. Today, when the difference between traditional information technology and communication technologies is diminishing, there are indications that "e" is being replaced by "m", perhaps as an attempt to move away from the more technically denoting "e"—electronic"—to "m", meaning "mobile", but not necessarily referring to technology. The use of prefixes, "prefix-management", can thus be said to create legitimacy to a normal concept—in this case "city"—which is through the prefix made different enough to be interesting (Catasús & Lundgren, 1999).

"mCity" as "the mobile city" or "the city of mobilities" could also lead to associations on "mobility". Before the age of mobile technology, to be "mobile" denoted someone or something that could easily be moved physically. The perpetuum mobile (the machine that keeps on running without the addition of energy), the mob (the unpredictably moving crowd), the automobile (the car), the mobile home (caravans for permanent living) and to mobilize (to summon) are examples of how the Latin stem (mobilis) has been used to denote mobility in different ways. With the development of mobile radio communication and the introduction of the cell-phone, or the "mobile phone", the word "mobile" has received new meanings. "Mobility" today, thus implies both physical movement of people and things, and the interaction of people *without* physical movement, through ICT-devices. Through the usage of "mobile technologies" people become mobile in that they can go wherever they want and still have access to any information they want and the possibilities of communicating with whom they want at any time. Not only is the gap between time and place comprised or reduced (Urry, 2000), but "mCity" as the "city of mobile technology" is also a city of high-tech.

To be "mobile" is often regarded as something positive. The free movement of people, goods and labor, regardless of boundaries and borders, is an important idea within the EU, realized through the Schengen-agreement, and it is thought to be of vital importance of the development for the European economy. On the level of the individual, the idea of travel, the meeting of new cultures, ideas and values, is always thought to broaden individual's minds and understanding of others. "Mobility" is also regarded as a human right; even though you are physically disabled, you should have equal opportunities of moving around and of going where you want. The creation of "mCity" as "the city of mobilities" can thus be interpreted as the creation of a city accessible to everyone, an interpretation made by the project managers in the early stage of the project when a pre-study was carried out regarding how to improve the life for the handicapped through mobile technology, or in the later stage of the project in the traffic-surveillance pilot project.

As the discussion above indicates, the "mCity" can be interpreted as an example of an internally persuasive discourse, where the indefiniteness of what mCity is leads to the possibility of assimilating it into a number of different discourses, depending on who is entering the dialogue and thus making sense of the verbal representation of the project.

THE MOBILE PHONE CITY

Images of the "mCity" have also been created by visual representations of the project. Throughout its history, the project has been presented visually mainly in two ways: power point presentations and leaflets, the power point presentation having been developed by the project managers themselves, changing them depending on the audience and the occasion, and the Marketing leaflets having been produced by a professional

Image 1. Front page of the mCity-leaflet developed during the fall of 2002

Swedish PR-agency, Blomquists. In total, there were three Marketing leaflets about mCity produced and the first, developed during the fall of 2002, was written in Swedish and English, while the two later ones—developed during the end of May 2003 and in the beginning of 2004—were produced in two versions, one in Swedish and one in English. The two first leaflets had photographic images of mobile phones on the first page (see Images 1-3).

The front page of the first leaflet also had eight smaller photographic images of happy people in different ages speaking or using mobile phones (see Image 1). But while the first page of the first leaflet was dominated by a photo of a mobile phone—which phone was not clear—the first page of the second leaflet had a large image of a Sony Ericsson P800, with a heart with the inscription "Stockholm loves mobile services" (see Image 2).

The third leaflet, produced after a reorganization resulting in the moving of mCity from the Stockholm Economic Development Agency to the Stockholm Competence Development Fund, did not have any images on its first page, but followed the graphic concept developed for the organization to which it had moved. The otherwise white first page had the text "mCity Improving mobile solutions" written in large, green letters (see Image 3).

Nowhere in the third leaflet are there any pictures on phones, except for a photo of an anonymous woman in sunglasses, who under a statue talks in a mobile phone (see Image 6). On the previous pages, there are two photos of people in the city being involved in the project: Ednah Lord, supervisor for the care of the disabled and participant in the care-sector pilot project (Image 4), Josef Danell and Kristina Petterson, mStudent project managers (Image 5). Next to each photo, the pilot-project is presented in text and a stylized graphic illustration, illustrating the technology implemented—an image of a computer connected to a mobile phone with a

Image 2. Front page of the mCity-leaflet developed in May 2003

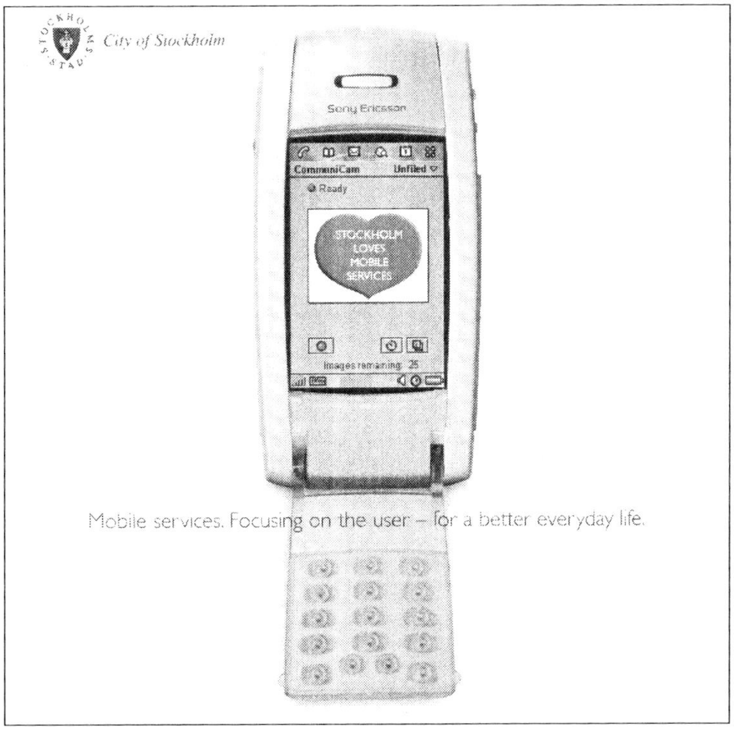

Image 3. Front page (page 1) of the mCity-leaflet developed in the beginning of 2004

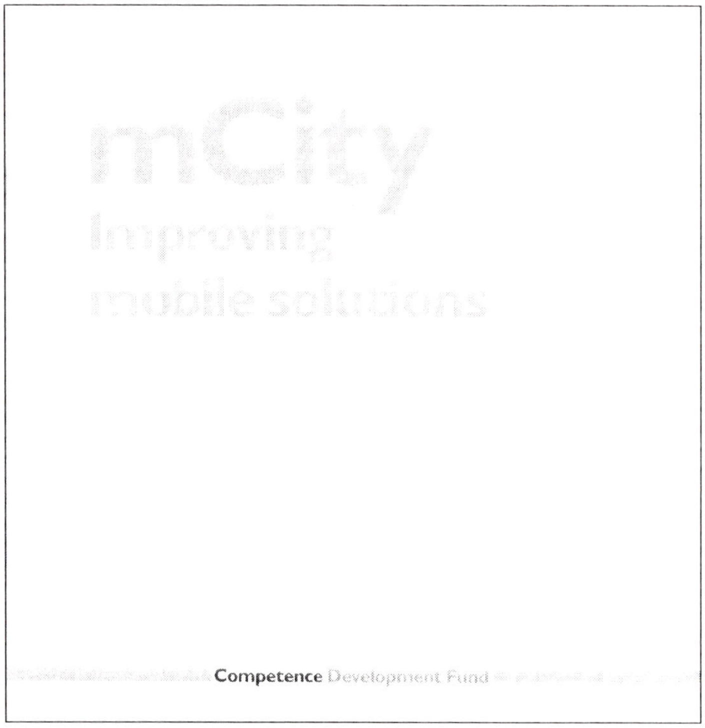

Image 4. Pages 2 and 3 of the mCity-leaflet developed in the beginning of 2004

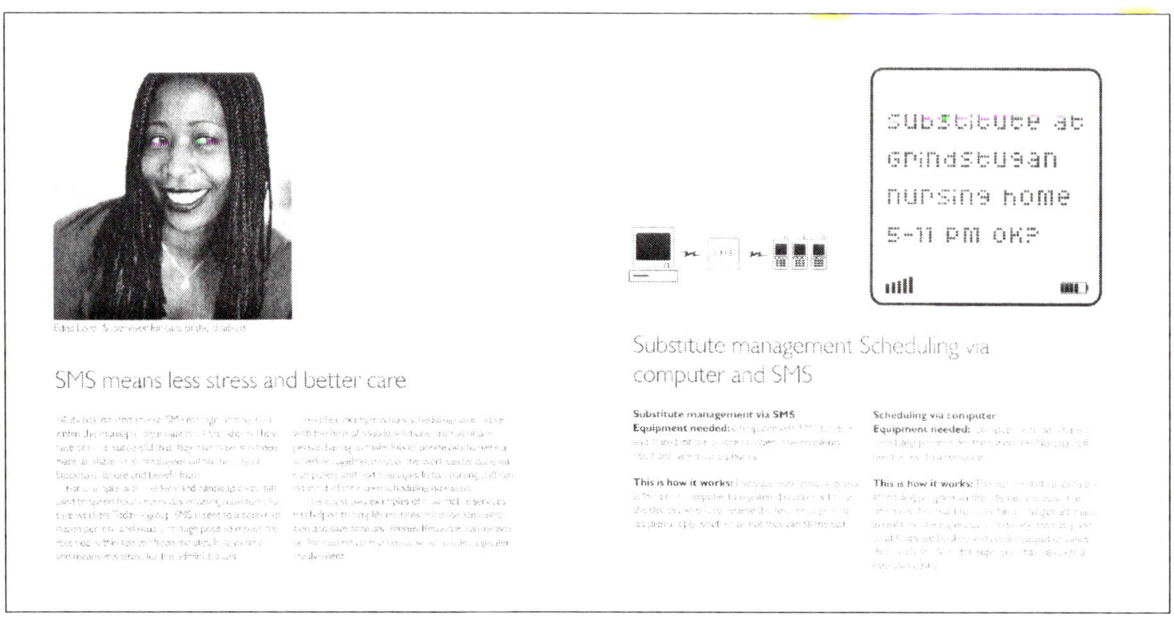

Image 5. Pages 4 and 5 of the mCity-leaflet developed in the beginning of 2004

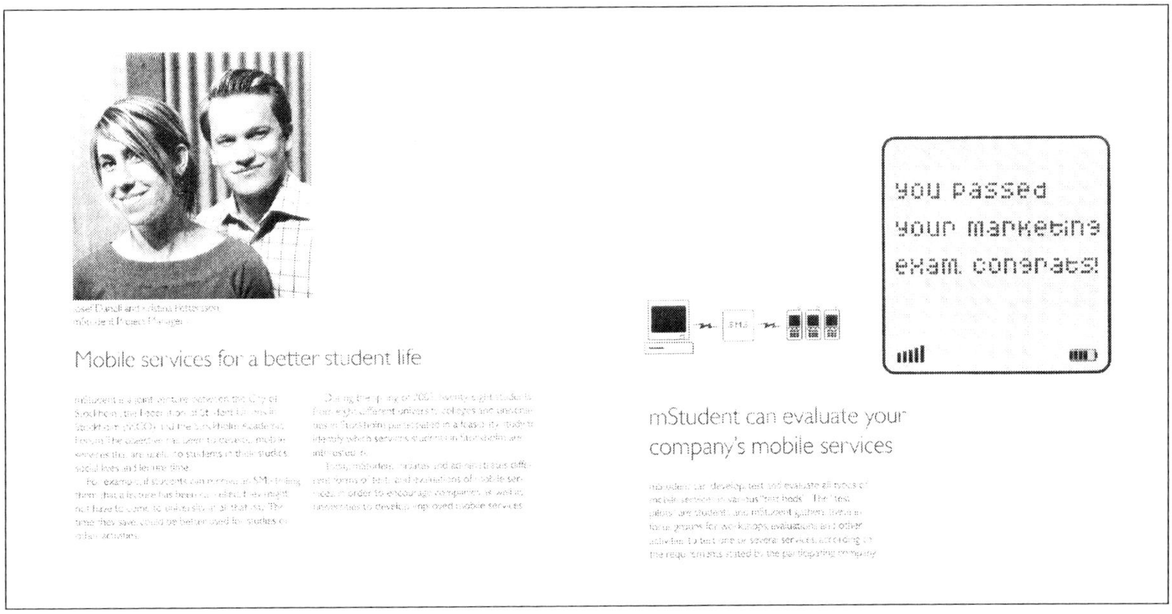

Image 6. Pages 6 and 7 of the mCity-leaflet developed in the beginning of 2004

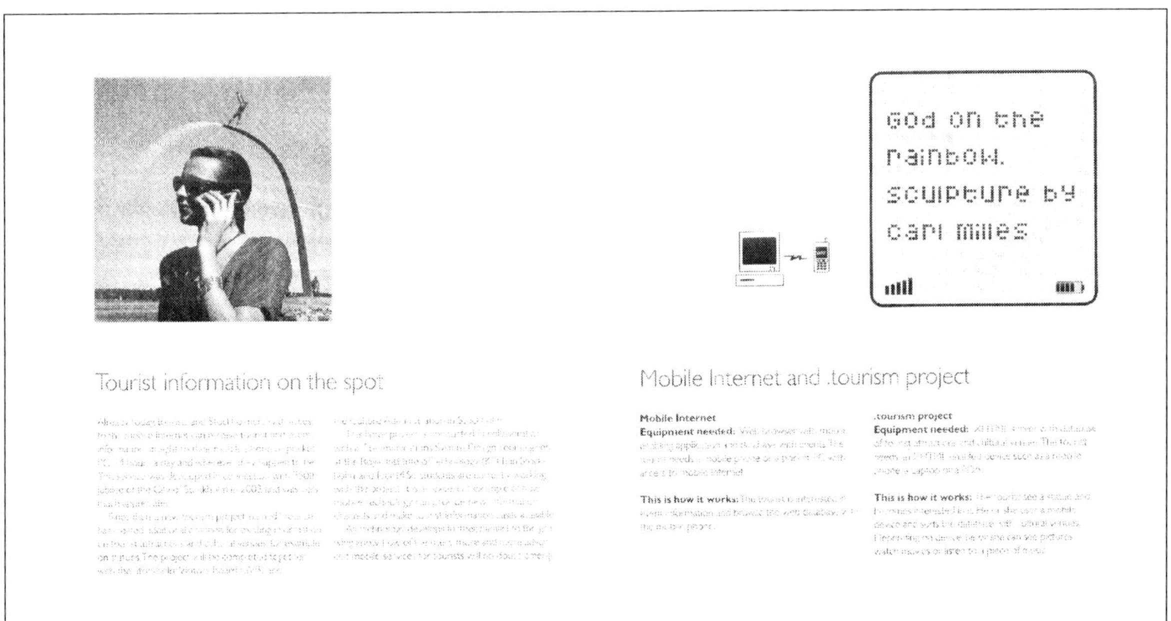

symbol looking like lightning. To the right of this there is a square with text as it could be seen in a mobile phone, exemplifying the kind of information possible through the mobile application that has been implemented. The lack of images of mobile phones might support the observation that ICTs seem to have "slipped into darkness", becoming natural parts of everyday life, rather than objects of interest in themselves (Dobers & Hallin, 2006).

Bringing People Together

Now, a semiotic analysis of one of the marketing leaflets for mCity (the one from May 2003) will be carried out, according to the methodology of description, interpretation and analysis (Schroeder, 2003; Schroeder, 2004). The aim is to understand the sense making process, or—put in other words—to understand how the interpretation happens through the dialogic process. The particular marketing leaflet was chosen since it was more widely distributed than the first one and since it was developed without having to adhere to any graphic concept, in contrast to the third leaflet.

The leaflet can be described as a four page-brochure, square, with text and photos. On the front page (see Image 7) is a Sony Ericsson P800-phone, centrally placed and standing up in front of a white background. In the top, left corner is the logotype of the City of Stockholm, next to the text "City of Stockholm" in italics. The mobile phone is open; i.e. the hatch is folded down, and across it, there is a text saying: "Mobile services. Focusing on the user –for a better everyday life." The menu of the phone shows different symbols of the mode "CommuniCam"—the camera-function—which displays an image in the window of the phone, there is a red heart, with the white text "Stockholm loves mobile services".

Turning the page (see Image 8), there is a photo to the left, showing a middle-aged woman in pony-tales, smiling and looking at an old woman, who seems to laugh with the younger woman, while holding hands. Across the picture text reads: "There are several trial projects in Stockholm to help facilitate everyday life for citizens, visitors, companies and municipal operations." Above the photo is some more information about the mCity-project; giving examples of some of the pilot-projects that have been carried through.

The page to the right is built the same way; a black-and-white photo, depicting four people in a crowd, barely visible, facing each other, their

Image 7. Front page (page 1) of the mCity-leaflet developed in May 2003

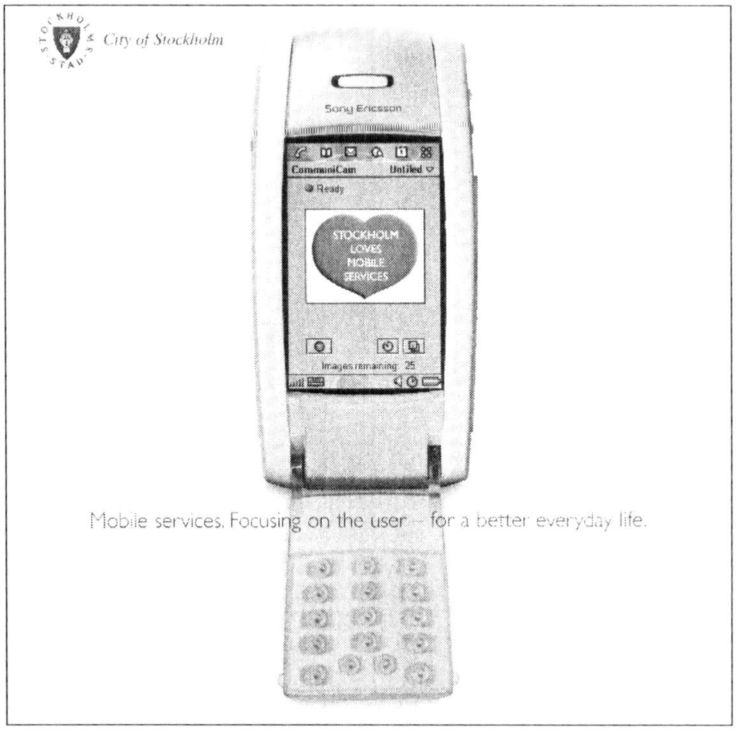

Image 8. Pages 2 and 3 of the mCity-leaflet developed during the end of May 2003

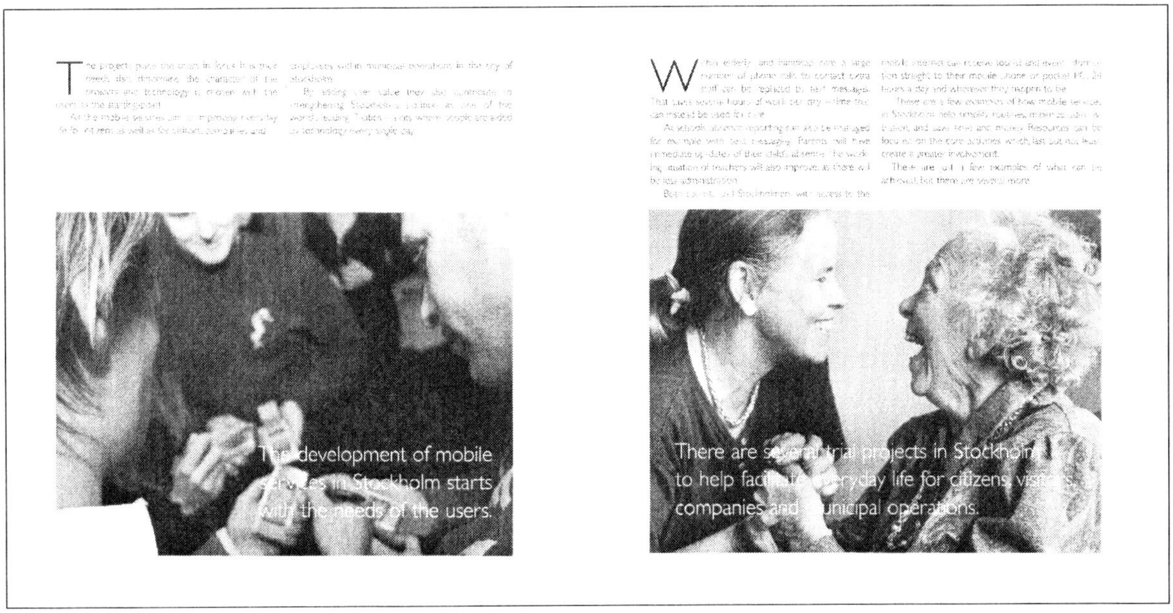

heads bent, using what appears to be devices very similar to the mobile phone depicted on the front page of the leaflet, with smiles on their faces. They are grouped together in two's, holding their devices close, looking into one. The picture is taken from over the shoulders and in between one pair. The text across this photo says: "The development of mobile services in Stockholm starts with the needs of the users."—a statement which the text above the photo elaborates on.

On the last page of the leaflet, the call "Find out more!" in large letters is printed centrally; above a text, providing details about "the mCity concept" and inviting anyone interested to contact the project-manager. Under it, the Stockholm city-logo appears again, and below, the name "City of Stockholm". Beneath it, "MCITY" is printed in block letters, the same size as the following text, stating an address, telephone- and fax number, and a web page address. The most eye-catching thing on the front of the leaflet is the red heart. The heart is placed centrally on the page, in the center of the mobile phone. The heart has a salient position through its placement and its color red, always a salient color (Jewitt & Oyama, 2001), emphasizing its traditional symbolic value of human love. In my interpretation, the red heart stands out as a contrast to the technical device—the mobile phone—a contrast which is further added to by the photos on pages 2 and 3; especially on page 2, showing two people enjoying each others company, laughing or smiling at each other. In the photo of page 3, the human and technological sides of the leaflet are brought together through the photo depicting people interacting through their mobile phones.

The mobile phone on the front page is shot from a frontal angel, showing an open hatch, which invites the viewer to interact and increases viewer's identification and involvement; whereas the point of view created in the photo on page 2 detaches the viewer as the picture is taken from the side; depicting the people at eye-level, which indicates a relation of symbolic equality (Jewitt & Oyama, 2001). The photo on page 3 is also shot from eye-level, but here the angle is from in between two people standing next to each other but opening up to the viewer; as if the viewer is standing close to them, looking over their shoulders; thus not being directly involved, but close enough to see what they are doing, being part of the group. Interestingly enough, the most dominant vectors in the two photos on pages 2 and 3, created by the actors, i.e. the people in the pictures and their eye lines (Jewitt & Oyama, 2001), contrast each other, as the dominant vector in the photo on page 2 is between two people, looking intently at each other, whereas the dominant vectors in the photo on page 3 are between the people, looking at their mobile phones. But at the same time there are other vectors in the photo of page 3 between the two people with their backs towards the camera and the person facing the camera, indicating inter-human contact.

Placing things in the center is a typical way of indicating centrality; holding marginal elements together (Jewitt & Oyama, 2001) and a comparison of what is at the center of the first three pages show that on page 1 it is the heart, on page 2, the two smiling mouths—almost the laughter we cannot hear, and on page 3, as the center is black, the eye is drawn to the mobile devices in the peoples' hands. Also on page 4, the call urging the reader to "Find out more!", is centrally placed at the top, indicating its importance, and at the bottom, the essential details; the down to earth information, are given for the reader who decides to follow up on this invitation.

So what is the analysis of this semiotic analysis of the leaflet? As I see it, there is a contrast in the leaflet between a "soft", human side, and a "hard", technological side, but that the leaflet brings the two sides together; on page 1 by the salience of the red heart in the mobile phone and by the dominant vector on page 3. Through the open hatch of the mobile phone on page 1, the point of view created through the angels from witch the photos on pages 2 and 3 are shot, and finally the urge on page 4, the viewer is invited, but not forced, to interact. Interestingly enough, the name "mCity" only appears in the text of page 4, and at the bottom of page 4, where the size of "mCity", is smaller than "City of Stockholm", indicating a lower degree of importance.

CREATING BALANCE

Even though the leaflet in different ways tries to bring humans into the setting of mCity, the mobile phone—a Sony Ericsson—on the front page is dominant, due to its size and central position. In this way, the leaflet in a very obvious way relates to all ads for mobile phones—ads that can be seen everywhere today. During the last few years, the mobile phone industry has grown steadily; for 2004, the year after the production of this mCity-leaflet, the Business Intelligence provider Gartner predicted that over half a billion mobile terminals would be sold (http://www4.gartner.com/research/spotlight/asset_55593_895.jsp, 2004-08-09). A the time, the largest single manufacturer was Nokia with some 35% of the market-shares world wide, but the competition was getting harder as several Chinese companies tried to enter the European and American markets, making it even more important for other manufacturers to keep and attract new customers. As today, advertisements and marketing-campaigns were important tools in this work.

I will scrutinize two ads which appeared on posters in the streets and airports of several large European cities in April and May of 2003, and on full pages of Swedish business-magazines like "Veckans Affärer" and "Dagens Industri" at about the same time (see images 9 and 10). They are for Nokia 6610-mobile phones and were originally produced in Finland at the Nokia head quarter. One of the ads was obtained from the PR-company used by the Swedish branch of Nokia (Grey Momentum, see http://www.grey.se) (image 9) and the other one is taken from the supplement to "Dagens Industri" in mid-May, 2003 (image 10). Also here I will use the methodology introduced by Schröder as previously described.

My description starts with the observation that both ads use a standing-up format and have a similar overall-composition: the left part consisting of a photo and the right part being mostly white. The photos represent everyday scenes of

Image 9. First Nokia 6610-ad, published in European business media, on airports etc during the spring of 2003

human life; one showing what can be part of the inside of someone's clothes-closet: a rack of neatly hung men's neck-ties and an equally neatly hung red scarf and one showing what appears to be a living room furnished with modern pieces of furniture, but with black footprints of an animal's feet leading across the floor and the rug. Across the photos a white or black text says "Balans är en

Image 10. Second Nokia 6610-ad, published in European business media, on airports etc during the spring of 2003

befriande känsla", which translates into something like: "Balance is a liberating feeling".

More specifically, the photo on the first ad (image 9) shows a tie-rack mounted on a wall or a door and with ties of different patterns; checked, lined and single-colored, but all within the same color-scale: light blue to light gray. The scarf, which is knitted in red yarn with white letters on it, occupies the last third of the tie-rack: the letters 'F' and 'C' most visible. The photo corresponds with the photo on the screen of the phone to the right, showing a photographic image of a football lying on the grass of a football-field below a blue sky with light, white clouds. Behind the phone, a white shirt on a hanger hanging in some kind of closet is barely discernible. The pictures of the tie-rack and the telephone are in the foreground, whereas the faint images of the shirt and the closet are in the back. The background picture is represented with shadows as if the light is coming in from the right, placing the tie-rack, and especially the red scarf, in focus.

The wooden, polished floor and the white, square rug, neatly placed in front of a green sofa with two orange-checked cushions and a white coffee table of the kind available in more expensive stores for furniture, add up to the impression of this being a photo of a living room in the other ad. The black, perhaps muddy, footprints have the shape and size of those of a larger dog—an impression supported by the photographic image of the face of a Dalmatian-dog on the screen of the mobile phone to the right. The square coffee table, which actually seem to come apart into a larger and a smaller one, and the sofa, modern through its square shape and steel frame, are placed in front of a white wall, with no pictures or paintings on it. On the coffee table there are two books entitled "Paris Interiors" and Portraits", and on top of them another book, or perhaps a magazine, is placed a bit carelessly, not in line with the books.

Whereas the tie-rack photo contains two main signifiers: the ties and the scarf, the living-room attributes of the other photos together make up the signifier of the minimalist living-room. These signifiers, though different, work in a similar way, leading to interesting interpretations.

Ever since it was introduced in the 1890's, the tie was worn together with jacket-suits. Soon, it became part of the dressing-code for "white-collar-labor"; i.e. people working with tasks sitting down at desks (book-keeping, banking, insurance and other office-jobs) as opposed to "blue-collar-workers" whose work included physical activity. Today, the tie is part of a common iconography signifying a formal occasion of some kind. It is still often worn on formal occasions with a white shirt of the kind represented in the ad. Here, the ties on the tie-rack are neatly hung, signaling an orderly owner leading a structured life. But not only do the ties work as signifiers for middle-class/upper-class work and a higher standard of life. They also represent something that literally ties you; that restrains you in certain ways. When wearing a tie there are a lot of things you don't do; you don't run, you don't play, but you sit still and behave in a controlled manner, according to social rules and expectations. The notion of restriction and restraint is even present in the English word "tie". The tie, when tied around the neck, alludes to the noose of the gallows, and is actually also used sometimes by people committing suicide by hanging. The ghostly image of the shirt in the background enforces the associations of life and death—are these the shirts of a living or a dead person?

The scarf is a signifier that works in stark contrast to the ties. The central position of the scarf in the ad as a whole and its red color draws attention to the ad (Jewitt & Oyama, 2001). The material of the scarf also differs from the fabric and the fabrication of the ties. Whereas the ties are made of materials like silk, wool or perhaps polyester, the scarf is knitted of wool or cotton yarn. This is a scarf made to be worn in rough settings, when moving about, and to manage frequent washing. The appearance of the scarf and the letters 'F' and 'C'—probably standing for "Football Club"- confirms the idea that this is the kind of scarf commonly worn by football-fans. The picture on the screen of the phone, showing a football lying in mid-court of a football-field as if ready to be kicked off, further enhances the football-imagery.

The furniture in the photo in the second ad (Image 10) forms the signifier of the living room, a room to relax and socialize in. The objects in this particular living room are carefully chosen to complement each other in a minimalist style, both in shape and color. The square with slightly rounded corners ads to the impression of stylistic unity and is a shape that reoccurs in several pieces of furniture, the sofa, the two fabrics of the cushions on the sofa, and the coffee table. That the inhabitant of this room is interested in interior decorating and art is confirmed by the books lying on the coffee table, entitled "Paris interiors" and "Portraits", and the magazine on top of the books has probably been nonchalantly placed deliberately—as everything else seems selected and put together on purpose in this living room. The sofa is not the kind you huddle up in with a nice book—it has more probably been chosen because of its design appeal, just as the coffee table has been chosen to match the sofa in shape and style. This living room, thus, is not primarily decorated for the purpose of relaxation, but rather for socialization an identity-creation. Many young Europeans—both men and women—might associate with the inhabitant of this living room, as indulging in "the pornography" of interior design magazines has become an important leisure-time activity to many of us during the last few years. This photo speaks to the ones who dream of composing the ultra-designed home which will convey the images of a selective, aware and modern consumer.

Even though this living room might feel impersonal to some people (depending on whether you like the minimalist style or not), the living room as a signifier works in the same way as the clothes closet with its tie-rack and shirts; it signifies a private, personal side of a human individual's life (Image 9). Not only are these people orderly, they are also aware of what is modern and have the skill and financial ability to choose and combine attributes in a sophisticated way. This impression is further enhanced by the color composition of the two photos. The discrete colors of the ties—bluish and grayish—go well with dark suits of gray, blue and perhaps black, build the image of a user who is not interested in showing a particular personality, or perhaps is restricted in expressing personality in the setting where the ties are mostly worn. The user has or likes to wear clothes that are representative and not intrusive. The color of the ties are the colors often used for uniforms of different kinds; for policemen, air-hostesses and others, as these colors commonly are thought to express calm, diplomacy and stability—in the Nordic countries even conservatism and aristocracy ("blue blood" being the sign of nobility). The ties represented through the photo thus do not convey anything about the user, as ties otherwise can do through logotypes or specific patterns of companies, schools or clubs, these can be the ties of anybody.

The red color of the scarf heavily ads to making it a contrast to the ties, being the color of contrast commonly used on signs of warning of different kinds and here also catching the spectator's eye straight away. The denotation of the red color of "VIP"-persons (cf. "the red mat") or politically radical ideas (cf. the red color used on socialistic emblems and flags) further ads to the contrast of the gray/blue ties. The red color enforces the differences of how the scarf and the ties respectively are worn in totally different settings and at totally different times, being attributes of two different identities. The red color and its association to blood perhaps also ads to the life-and-death-imagery touched upon above.

The other ad (Image 10) is mainly dominated by the white and the green colors. White is traditionally in Western symbolism seen to symbolize light, purity and virginity and is often the color for wedding-dresses, confirmation-dresses/suits etc. It is worldwide seen as the color of peace (cf the white flag) and is a popular basic color in Nordic interior decorating and in the minimalist style. In this picture, green is used as a contrast color. Green, etymologically connected to the verb 'grow', symbolizes the growing—both in nature and in human life. The green color is also used to indicate, "go", for example on green traffic lights, in customs, etc. Both ads, thus, create associations through the choice of contrasting colors, even though the color-combinations of the two photos differ.

Another similarity is the play with commodity codes: football and the owning of a dog. In the first ad (Image 9), football says something about the lifestyle of the tie and scarf-owner. Originally, football was an upper-class sport, primarily played in private schools in GBR. But in the late 19th century, football gradually became a working-class sport, since it was one of the first sports with professional players, i.e. players making money on sports. To make money on playing sports was not thought to be acceptable in upper classes. Today, football is one of the world's largest sports, and in Britain and in many other countries it is extremely important to many people, not only because they are playing, but also because they are spending a lot of money and time watching games. Football thus not only denotes activity, but social life with the drinking of beer, being part of a large group cheering for the team, etc.

Compared to other sports, like tennis or golf, football is a fairly playful sport. "To kick ball" is something both young and old can do, regardless of skill and knowledge of rules. This playfulness is also conveyed through the dog-ad, where the muddy footprints of a semi-large dog on the white carpet bear evidence of a playful and fun time out in the fields. The owning of a dog is just like football a commodity code; it tells us something about the inhabitant of the living room (presuming it is the same person who owns the dog) (see Image 10). This is a person who has another side in his or her life, which contrasts to the minimalist, to some people cold and impersonal, style of his or her home. The fact that the dog has been allowed to—or at least not successfully prevented from—entering the living room with its dirty paws further gives the impression of a living room, and dog-owner with a love for living creatures. It is also important to notice what kind of dog it is. The photo on the screen of the mobile phone gives us the answer: a Dalmatian. This race, commonly known from the Disney-movie "The 101 Dalmatians", is a popular race for guarding and hunting, but also for company. It is not the kind of small, harmless dog one might expect in a living room like the one of this photo, but rather a fairly large and clumsy dog, needing a lot of outdoor exercise and play.

So far, I have pointed to the contrast of the two photos used in the ads making up the associations of human individuals with "real lives": work and leisure. The individuals owning the ties/the scarf, living in the living room together with the dog are pretty orderly, but not over-orderly. Are they male or female? Neither of the two photos conveys gender through the use of commodity codes. Both men and women can play and/or be interested in football, and both men and women can own and enjoy the company of a dog. But neckties and shirts are mostly worn by men, which is why it is reasonable to argue that the tie-rack-ad aims at men. Does this mean that the target group for the other ad is women? This is difficult to tell, as interior decorating today is popular among both women and men, and the look of this living room neither gives feminine nor masculine associations in my opinion. But perhaps this works differently in other parts of Europe?

Either way, my final analysis is that the two ads both consist of elements of different parts of /wo/men's lives—work and leisure or order and playfulness. This duality is also indirectly referred to in the text written across the photos, speaking of balance. In our time of stress, burn-outs and over-worked people, most of us would agree upon the idea that balance between different parts of life; work-leisure, sitting still-moving around, intellectual activity-physical activity, is important for mental and physical health. The idea of balance is also reflected in the photos—the tie-rack photo showing ties and a scarf neatly hung/balanced in a proportionate way, and the living room-photo representing neatly arranged, very clean furniture balanced by the muddy footprints of the Dalmatian. The text saying that balance is/gives you a feeling of relief; delivering you from outer demands/constraints/worries is metaphorical. Originally the Swedish word "befriande" was used to denote the physical setting free of people from prison or captivity in a literal sense. The ghost image of the shirt perhaps makes the idea of balance even stronger through a hidden threat: if you don't achieve balance in life, your shirts will become the property of a dead man...

THE MOBILE SOLUTION AS THE COMMON SOLUTION

Implicitly, the Nokia ads propose the Answer to the question haunting many people of today's Western world; how can I create balance in life, the Answer of course being "through the Nokia 6610 mobile phones". This is just as important, if not more important than the technical features of the phones of which none is presented in the ads, confirming the conclusion drawn by several researchers of advertisements who have shown that consumers not only buy products because of their utilitarian aspects, but because of their symbolic qualities. In this way, the consumer produces identity through the consuming process (Lury, 1998; Miller, 1998; Churchill & Wakeford, 2002; Townsend, 2002; Schroeder & Borgerson, 2003). Advertisements—being part of "the circuit of culture"—help us construct and reconstruct our identities, and thus, the consuming subject and the consumed object are tied together in a dialectic relationship, where both are "I" in order to constitute the other (Miller, 1998). In this process, the object and brand itself is unimportant; what is important is what the object and the brand expresses to other people. The item a person goes shopping for is not a commodity that intervenes in the relationship between a person and discourse, but also reveals itself as discourse, as the semiotics it carries feeds into the imagination of different possibilities for the self (Lury, 1998; Miller, 2001). The meaning created in the circuit of culture is what gives us a sense of identity: "... of who we are and with whom we 'belong'..." (S. Hall, 1997:3).

Over the last decade, anthropologists have realized that the more people see themselves as global the more they understand themselves as local. It also has been argued that shopping is a way of tackling the gap between the ideal space and the physical place (Miller, 2001); and in this process, markers of identity, attributes like clothes, watches, mobile phones etc, also carry the function of confirming the subject in the physical realm, signaling to other people who I am/want to be, what I stand for and to what group I /want to/ belong. It can even be argued that the attributes when presented in advertisements like the ones analyzed here challenge the identity of the individual, which no longer can be seen as natural connected to the individual, but the result of the "prosthetic culture" in which we live, where things like mobile phones become extensions of ourselves (Lury, 1998)—we are what we have, one could argue (Belk, 1989).

The message of the two ads studies here is that the Nokia 6610 mobile phone can help create balance between different parts of an individual's life. Even though there are no people in the ads, the personal attributes (the ties, the scarf, the shirt, the living room with its furniture—and the dog!) help personalize them, closing the gap between the potential customer and the technological device, which in fact also becomes personal. The mobile phone is presented to be as basic in a person's life as his ties or his football scarf, or as personal as his dogs. It becomes the missing piece of life's jigsaw puzzle, completing the picture, functioning as a junction between all parts of life. Almost like a prothesis, it helps people do things they couldn't do otherwise; keep track of football-results while traveling, being available for work-calls on the football stadium, keep in touch with the loved ones (even the dog!) while at work. Thus, the mobile phone is presented as a necessary link connecting and completing different parts of human life; helping to create the balance needed for a sound, healthy and human life, which connects well to the observation that today, play and having fun seem to be important in the companies of "the new economy" (Strannegård & Friberg, 2001).

In my interpretation, the mCity-leaflet communicates the same message as the Nokia ads, namely that of mobile technology being the tool which brings the 'soft' and the 'hard' sides of human life together, thus 'humanifying' mobile technology. The slogan "Focusing on the user—for a better everyday life" on page 1 raises a number of questions, to which the answers can be found explicitly or implicitly on the different pages of the leaflet:

- What does the user use?—The mobile phone (Image 7, page 1)
- Who is the user?—Any one; different ages, different settings (Image 8, pages 2 and 3)
- Focusing on the user, how?—Observing, participating (Image 8, pages 2 and 3)
- Better everyday life, how?—More time to laugh, to interact (Image 8, page 2 and 3)
- Who does this?—The City of Stockholm (Image 7, page 1, and 4)
- Where can I find more information?—Address, telephone number, contact person, web page (page 4)

The mCity leaflet can be said to join the authoritative discourse created by advertisements for mobile phones, here exemplified by the two Nokia-ads, which, at least in Sweden at the turn of the millennium, were seen as fashionable tools, necessary for the modern person. But even though the leaflet and the ads through their way of communicating belong to the same language community, having the same speech plan (to make more people interested in mobile solutions, the mCity and the Nokia phones), they engage in an internal dialogism with the speaking partner, rather than a subjective dialogism, departing from a perceived image of the speaking partner as a human being with needs of balance in life between for example work and leisure. The common solution, given both by the mCity-leaflet and the Nokia-ads, is the mobile phone. Thus, the dialogue is oriented towards the situation of the speaking partner, rather than the object of discussion or the speaker, having a centripetal force, aiming at unifying the world of the perceived speaking partner and the speaker.

By relating to the speech genre of the ICT-industry and to people familiar with the tradition of abbreviations like "mCity", mCity creates meaning both through its verbal and visual representations. The leaflet studied here proposes a closing of the gap between humans and technology, relating to a dialogue which takes place on a broader arena through ads in public places addressing the subject of how to create balance in times of stress and burn-outs. By absorbing and relating to different discourses, together, the verbal and visual representations form images of "mCity" within the speaking partner. One could for example envisage "Stockholm as an mCity" as a modern, fast-moving and quickly adapting, trend-sensitive city, as it is a city of high-tech, but also a city of free(-ly moving) people where technology is in the service of humans.

CONCLUSION

It has been argued that in today's society, the visual impressions we get create memories which in turn create new identities (Lury, 1998). This way, everything is connected in a huge and complex circuit of culture, which needs to be taken into account also when designing the strategy for city marketing. This chapter has been an attempt to show how, in the circuit of culture, linguistic expressions (be they verbal or images) engage in dialogue, by interpreting the verbal and visual expressions of the Stockholm-based mCity-project and relating this to the interpretation of two ads for mobile phones, which appeared in Europe at about the same time.

Initially, mCity was created through a Swedish translation of I-Mode, and against the background of the Swedish telecom-sector crisis. mCity then entered into dialogue both through its verbal and visual representations, and as showed in this chapter, it is possible to interpret these representations as relating to a more general discourse, illustrated by two of the contemporary ads for mobile phones. Through the use of the image of the mobile phone, mCity and the ads belong to the same circuit of culture, and the purpose with this chapter has been to show how the sense making of the one can be related to the other. Again, it must be pointed out that the interpretations here are dependent on the interpreter's view (mine), but the purpose has not been to establish the "true meaning" of the marketing leaflet or the ads, but rather to show how sense making can take place through the dialogic process in the circuit of culture.

Now, then, how does this study contribute to the work of city marketers or to the theories of city marketing? As has been made clear, the purpose here has not been to take the receivers' side into account in the traditional way, by for example carrying out a study of how people have perceived the marketing material of mCity. This would of course be an interesting approach, but different in its basic assumptions, departing from the idea that such a study would, in fact, reveal the "truth" about how people perceive this material. Such a study would certainly also lead to interesting implications.

Instead, this chapter points to the powerful symbolic qualities of ICTs, and that the interpretations of these change all the time, which is why it is a challenge for city marketers interested in using representations of ICTs in their marketing efforts to design the marketing material in such a way so that it on the one hand "fits" with what is perceived as positive with ICTs, but on the other hand so that it stands out as special. A benchmark of how ICTs are used in marketing at the particular time and in the context of the cultural setting where the marketing effort is planned should thus be carried out by those interested in using ICT images in city marketing.

One implication of the present chapter is that communication is much more complex than what the classic cybernetic model reveals. A message, also in the form of a marketing campaign, a marketing leaflet as studied here, engages its receiver in dialogue in an intricate way, which involves all possible associations. This way, the marketing material becomes involved in the circuit of culture, and in order for the marketing effort to be "successful" in the sense that the interpretations made are not too different from those that were intended, I would argue that an awareness of how this process takes place on a theoretical level is necessary among marketing people. This chapter has been an attempt of shedding light upon this process.

ACKNOWLEDGMENT

Thanks to Prof. Jonathan Schröder at the University of Exeter for helpful comments on an early version of this text.

REFERENCES

Bakhtin, M. (1934-35/1981). Discourse in the novel (C. Emerson & M. Holquist, Trans.). In M.

Holquist (Ed.), *The dialogic imagination: Four essays* (p. 259-422). Austin: Univ. of Texas P.

Bakhtin, M. (1952-54/1994). Speech genres and other late essays (V. W. McGee, C. Emerson & M. Holquist, Trans.). In P. Morris (Ed.), *The Bakhtin reader. Selected writings of Bakhtin, Medvedev, Voloshinov* (pp. 81-87). London: Edward Arnold.

Bakhtin, M. (1963/1994). Problems of Dostoevsky's poetics (C. Emerson, Trans.). In P. Morris (Ed.), *The Bakhtin reader. Selected writings of Bakhtin, Medvedev, Voloshinov* (pp. 89-96). London: Edward Arnold.

Bakhtin, M., & Volosinov, V. (1929/1994). Marxism and the philosophy of language (L. Jatejka & I. R. Titunik, Trans.). In P. Morris (Ed.), *The Bakhtin reader. Selected writings of Bakhtin, Medvedev, Voloshinov* (pp. 26-37). London: Edward Arnold.

Barthes, R. (1977). The Photographic Message (S. Heath, Trans.). In *Image, music, text*. New York, NY: Hill and Wang.

Belk, R. W. (1989). Extended self and extending paradigmatic perspective. *Journal of Consumer Research, 16,* 129-132.

Bignell, J. (2002). *Media semiotics: An introduction* (2nd ed.). Manchester: Manchester University Press.

Brown, B. (2002). Studying the use of mobile technology. In B. Brown, N. Green & R. Harper (Eds.), *Wireless world. Social and interactional aspects of the mobile age* (pp. 3-15). London: Springer.

Bunnell, T. (2003). *Malaysia, modernity and the multimedia super corridor: A critical geography.* Singapore: Routledge Pacific Rim Geographies.

Catasús, B., & Lundgren, M. (1999). Coupling the environmental issue: The environmental managers and their allies. *Global Focus, 11*(2), 21-36.

Chevrant-Breton, M. (1997). Selling the world city: A comparison of promotional strategies in Paris and London. *European Planning Studies, 5*(2), 137-161.

Christensen, L., & Askegaard, S. (1999). Corporate identity and corporate image revisited. A semiotic perspective. *European Journal of Marketing, 35*(3/4), 292-315.

Churchill, E. F., & Wakeford, N. (2002). Framing mobile collaborations and mobile technologies. In B. Brown, N. Green & R. Harper (Eds.), *Wireless world. Social and interactional aspects of the mobile age* (pp. 154-179). London: Springer.

Cooper, G. (2002). The mutable mobile: Social theory in the wireless world. In B. Brown, N. Green & R. Harper (Eds.), *Wireless world. Social and interactional aspects of the mobile age* (pp. 19-31). London: Springer.

Czarniawska, B. (2000). The European capital of the 2000s: On image construction and modeling. *Corporate Reputation Review, 3,* 202-217.

Dobers, P. (2003). Image of Stockholm as an IT city: Emerging urban entrepreneurship. In C. Steyaert & D. Hjort (Eds.), *New movements in entrepreneurship* (pp. 200-217). Aldershot: Edward Eldgar Publishing.

Dobers, P., & Hallin, A. (2006). Slipping into darkness: A study of the role of ICTs in the making of Stockholm's image. *Journal of Urban Technology, 13*(3), 119-122.

Dysthe, O. (1996). *Det flerstämmiga klassrummet. Att skriva och samtala för att lära.* (B. Nilsson, Trans.). Lund: Studentlitteratur.

Fainstein, S. S., & Judd, D. R. (1999). Global forces, local strategies and urban tourism. In D. R. Judd & S. S. Fainstein (Eds.), *The tourist city* (pp. 1-17). New Haven and London: Yale University Press.

Gant, D., & Kiesler, S. (2002). Blurring the boundaries: Cell phones, mobility, and the line between work and personal life. In B. Brown, N. Green & R. Harper (Eds.), *Wireless world. Social and interactional aspects of the mobile age* (pp. 121-131). London: Springer.

Hall, S. (1997). The work of representation. In S. Hall (Ed.), *Representation. cultural representations and signifying practices* (pp. 1-11). London, Thousand Oaks, New Delhi: Sage.

Hall, T. (1997). (Re)placing the city. Cultural relocation and the city as centre. In S. Westwood & J. Williams (Eds.), *Imagining cities. Scripts, signs, memory* (pp. 202-218). London and New York: Routledge.

Hallberg, P. (1992). *Litterär teori och stilistik* (4. uppl. ed.). Göteborg: Akademiförl.

Hallin, A. (forthcoming). *Size matters: The problem of organizational size and the case of mCity.* PhD thesis, The Royal Institute of Technology Stockholm.

Hallin, A., & Lundevall, K. (2007). mCity—User focused development of mobile services within the city of Stockholm. In I. Kushchu (Ed.), *Mobile government: Emerging directions in E-Government*: Hershey, PA: Idea Group publishers.

Harvey, D. (1989). From managerialism to entrepreneurialism: The transformation in urban governance in late capitalism. *Geografiska Annaler, 71 B*(1), 3-17.

Jansson, J., & Power, D. (Eds.) (2007). *The image of the city—Urban branding as constructed capabilities in Nordic city regions.* Uppsala: Nordic Innovation Centre/Dept of Social and Economic Geography.

Jessop, B. (1998). The narrative of enterprise and the enterprise of narrative: Place marketing and the entrepreneurial city. In T. Hall & P. Hubbart (Eds.), *The Entrepreneurial city. Geographies of politics, regime and representation.* Chichester & New York: Wiley.

Jewitt, C., & Oyama, R. (2001). Visual meaning: A social semiotic approach. In T. V. Leeuwen & C. Jewitt (Eds.), *Handbook of visual analysis* (pp. 134-156). London, Thousand Oaks, New Delhi: Sage Publication.

Leeuwen, T. v. (2001). Semiotics and iconography. In T. V. Leeuwen & C. Jewitt (Eds.), *Handbook of visual analysis* (pp. 92-118). London, Thousand Oaks, New Delhi: Sage Publication.

Lury, C. (1998). *Prosthetic culture: Photography, memory, and identity.* London: Routledge.

Löfgren, O. (2001). Urbana koreografier—rum och rörelse i 1900-talets städer. In R. Solli & B. Czarniawska (Eds.), *Modernisering av storstaden: Marknad och management i stora städer vid sekelskiftet* (1. ed., pp. 15-34). Malmö: Liber.

McGuire, S. (2000, February 7th). Shining Stockholm. *Newsweek,* 52-59.

Miller, D. (1998). *A theory of shopping.* Oxford: Polity Press.

Miller, D. (2001). *The dialectics of shopping.* Chicago; London: University of Chicago Press.

Myerson, G. (2001). *Heidegger, Habermas and the mobile phone.* Cambridge: Icon books & Totem books.

Porsander, L. (2000). *Titt-skåp för alla, en berättelse om hur Stockholm blev kulturhuvudstad.* Göteborg: BAS.

Rimmon-Kenan, S. (1983). *Narrative Ficion: Contemporary poetics.* London and New York: Routledge.

Schroeder, J. (2003). Att marknadsföra identiteter och konsumera skillnader. In Á. Sverrisson, P. Aspers & P. Fuehrer (Eds.), *Bilderna i samhällsanalysen.* Lund: Studentlitteratur.

Schroeder, J. E. (2004). Produktion och konsumtion av reklambilder. In P. Aspers, P. Fuehrer & Á. Sverrisson (Eds.), *Bild och samhälle. Visuell analys som vetenskaplig metod.* (pp. 75-96). Lund: Studentlitteratur.

Schroeder, J., & Borgerson, J. L. (2003). Dark desires: Fetishism, ontology, and representation in contemporary advertising. In T. Reichert & J. Lambiase (Eds.), *Sex in advertising. perspectives on the erotic appeal* (pp. 65-87). Mahwah, N.J & London: Lawrence Erlbaum Associates.

Selden, R., & Widdowson, P. (1993). *A reader's guide to contemporary literary theory* (3rd ed.). New York & London: Harvester Wheatsheaf.

Sherry, J., & Salvador, T. (2002). Running and grimacing: The struggle for balance in mobile work. In B. Brown, N. Green & R. Harper (Eds.), *Wireless world. Social and interactional aspects of the mobile age* (pp. 108-120). London: Springer.

Strannegård, L., & Friberg, M. (2001). *Already elsewhere—Play, identity and speed in the business world.* Stockholm: Raster förlag.

Tietze, S., Cohen, L., & Musson, G. (2003). *Understanding organizations through language.* London, Thousand Oaks, New Delhi: Sage.

Townsend, A. M. (2002). Mobile communications in the Twenty-first Century City. In B. Brown, N. Green & R. Harper (Eds.), *Wireless world. Social and interactional aspects of the mobile age* (pp. 62-77). London: Springer.

Urry, J. (2000). *Sociology beyond societies, mobilities for the Twenty-first Century.* London & New York, NY: Routledge.

Ward, S., V. (1998). *Selling places. The marketing and promotion of towns and cities 1850-2000.* New York, NY: E & Fn Spon.

Weick, K. E. (1995). *Sensemaking in organizations.* Thousand Oaks, CA: Sage.

Wirtschaftswoche. (2002). Europas Städte im Vergleich. *Wirtschaftswoche, 33*, 18-24.

Chapter XV
Walled City to Wireless City

Sandra Moffett
University of Ulster, Magee Campus, Northern Ireland

T. M. McGinnity
University of Ulster, Magee Campus, Northern Ireland

M. Callaghan
University of Ulster, Magee Campus, Northern Ireland

J. Harkin
University of Ulster, Magee Campus, Northern Ireland

D. N. Woods
University of Ulster, Magee Campus, Northern Ireland

M. Paris
University of Ulster, Magee Campus, Northern Ireland

ABSTRACT

This chapter outlines the journey that the city of Londonderry (Derry), Northern Ireland, undertook when converting a traditional walled city to a technology-enhanced wireless city. The chapter presents an overview of the three project strands, namely wireless city (civic aspect), wireless walls (tourism aspect), and wireless campus (educational aspect), along with the contribution made by each partner organization. A detailed case study of the educational element is presented, employing a dual qualitative/quantitative research approach. The case study focuses on the experience of one academic member of staff in using the wireless initiative via Tablet PC and SMART classroom. Quantitative analysis to gain insight into usability of wireless networking services and wireless technologies is presented from both a staff and student viewpoint and finally conclusions are drawn on the overall project experience. The project successfully, completed in December 2006, has received a number of awards for its innovative approach.

INTRODUCTION

In August 2004 an exciting and innovative project, the Flagship Broadband Project was launched in Northern Ireland. The aim of this project, which attracted funding to the value of £1.38 million sponsored by the Department of Enterprise, Trade and Investment (DETI) Northern Ireland, under the EU Building Sustainable Prosperity Programme, was to create a wireless city in Londonderry (Derry), a city situated in the North West of Northern Ireland http://encarta.msn.com/map_701514266/londonderry.html. The project entitled "Walled City to Wireless City" was an innovative partnership between three partners, namely Derry City Council, the Intelligent Systems Research Centre (ISRC), University of Ulster and the North West Institute of Further and Higher Education (NWIFHE). The aim of the Flagship project was to create an innovative 'wireless city' incorporating a range of civic, tourism and educational aspects. There were three aspects to the project, namely, wireless city (civic aspect), wireless walls (tourism aspect) and wireless campus (educational aspect). The project completed in December 2006. As the authors of this chapter are researchers within ISRC, the focus presented is from an educational perspective based on a case study of wireless campus implementation.

This chapter will commence by presenting an overview of the partners involved in the project (Section 2) and briefly detailing the contribution of each to the project (Section 3). The chapter continues by focusing on the educational aspect of the project by presenting a case study of wireless campus implementation. This case includes three distinct but complementary components, namely wireless networking services on and off campus, the piloting of a Tablet PC mandate initiative and the development of SMART classrooms. Practical application of educational tools to support the implementation is presented (Section 4). Research undertaken to assess the impact and success of this initiative is presented (Section 5) and conclusions are drawn on the overall experience (Section 6).

This project has been a very successful achievement for each of the partners involved, and has won a number of awards including the Best Wireless Project in the Communication in Business Awards 2005 and two categories (namely Education, sponsored by Hewlett Packard and Mobile Technology (WiFi), sponsored by BT) in the 2006 BT GoldeneyeT Awards.

PARTNERS

The Walled City to Wireless City Project works with its stakeholders to realise the full potential of Derry as a Regional City, driving forward economic change in the North West. To incorporate the various strands of the project (wireless city, walls and campus) three project partners were involved, these are outlined below:

Derry City Council

Realising Derry's potential is the central aim of the Economic Development Strategy of Derry City Council http://www.derrycity.gov.uk/economicdevelopment/strategy.htm. The Strategy identified that the City and its region require significant local and external intervention to build its physical, human and capital resources. The Council's Economic Development Strategy identified twelve strategic priorities for intervention. These priorities underpin the development of all the sectors of the local economy. Addressing these priorities is essential to the development of Derry as the key strategic employment location of the North West and as the central focus of the North-West Development Corridor between Letterkenny and Coleraine. The Strategy also identified six "strategic development zones", the functions of which underpin the City's capability to drive forward economic change in the North West. This demonstrates that the Strategy is made up of actions which are not only integrated thematically but also spatially. Advancement of the 'Digital Economy' is one of the priority actions, further information is available from http://www.derrycity.gov.uk/economicdevelopment/digital.htm.

The objective of the digital economy theme is to create new business opportunities in providing integrated services for e-business, e-government, e-tourism and e-learning. As part of the 'Walled City to Wireless City' project Derry City Council was involved in two elements, firstly the creation of an e-government spectrum which would take advantage of the broadband and wireless technology access and secondly the promotion of tourism within the City of Derry through the wireless walls concept.

The Intelligent Systems Research Centre, University of Ulster

The Intelligent Systems Research Centre (ISRC) (http://isrc.infm.ulst.ac.uk) is located within the School of Computing and Intelligent Systems (SCIS) (http://www.infm.ulst.ac.uk/scis/) on the Magee Campus of the University of Ulster in Londonderry http://www.multimap.com/motoring/?&hloc=GB|BT48%207JL#t=l&map=55.00624,-7.32403|16|4&loc=GB:55.00624:-7.32403:16|BT48%207JL|BT48%207JL. Research within ISRC is focused on computational intelligence, including research on wireless sensor networks and ambient intelligence. The University of Ulster is concerned with the educational aspect of the project including the creation of a wireless campus, as detailed later in the chapter.

The North West Institute of Further and Higher Education

The North West Institute of Further and Higher Education (NWIFHE), now the North West Regional College (NWRC) (http://www.nwrc.ac.uk/) is one of the fastest growing colleges in Northern Ireland with approximately 22,000 students undertaking study in a range of full and part-time programmes. The College aims to encourage and facilitate participation in education, provide a wide choice of high-quality learning opportunities for people aged over sixteen, of varying abilities and backgrounds, meet efficiently the needs of corporate clients and public organisations for training services and consultancy and enable those who study and work with the Institute to achieve success, fulfilment and progress. NWIFHE (NWRC) is concerned with the educational aspect of the project linking into the wireless technologies available in the Magee Campus, University of Ulster, which is located adjacent to their premises.

PROJECT STRANDS

This section provides a brief overview of the three project strands, the civic aspect, the tourism aspect and the educational aspect.

The Wireless City

The Wireless City element of the project built on and developed existing e-government services from both the Council Offices on the Strand Road of Londonderry and the Council Chamber in Guildhall Street, Londonderry. In practical terms this involved the provision of wireless infrastructure on both sites. As Derry City Council offices are located adjacent to both the University and NWRC Campuses this provides a seamless wireless network between the civic quarter on the riverside and the city's 'knowledge corridor' extending from the riverside towards the border.

One initiative undertaken as part of the project to develop the digital presence of the Council (outlined as key in their Economic Development Strategy) was to create an e-government system, known as the Records and Document Management Systems (ERDMS). This technological system provided the blueprint for the other 25 local authorities in Northern Ireland as an exemplar system for information sharing and storage. To promote the use of this system Derry City Council provided all senior management and elected members with Laptop/Tablet PCs enabling council business to be carried out electronically, permitting more flexible access to computing resources and the Internet in designated broadband and wireless areas. This e-government initiative resulted in cost savings in terms of staff time, travel allowance and reduction in paper wastage. In addition, the Council utilised technological equipment including Blackberry-type devices, Pocket PCs and wireless enabled video cameras for Council demonstration, training and testing purposes. Throughout the duration of the project a number of Council events were televised as part of the project promotion. Being part of a Wireless City also encouraged local businesses to take advantage of broadband/wireless capabilities. Examples of application can be referred to in the following documentation.

- Empowering the Mobile Society": A Digital Economy Strategy for the North West Crossborder Region
- Mason Telecommunications Report (Executive Summary): A Review of broadband infrastructure and Services within the Derry City Council Area
- Case studies of best practice in the use of broadband technologies by local SMEs

The Wireless Walls

The Wireless Walls strand focused on the implementation of a wireless/mobile network based upon a Wi-Fi networking standard in the historic walled area of Derry. The Walled City is approximately one mile in circumference. Within this area Wi-Fi services were established with almost total coverage. The Wireless City service provides free Internet access to anyone with a wireless enabled device such as a laptop or personal digital assistant (PDA). This is the first service of its kind in Ireland and one of the first in the UK. Network access is provided free for one hour time slots, after which the user needs to logout and login again. Further information is available from http://www.wirelessderry.org/

Visitors have the facility to access the network through the rental of handheld electronic tourist guides. The Visitor Location Based Services use new technologies in the provision of enhanced location dependant information services to visitors. The delivery of these services are facilitated through wireless hotspots based around the boundaries of the historic walled city and manifested by the use of electronic guides in the form of GPS/WLAN enabled PDAs. These handheld systems offer content generally available at tourist offices but augmented by comprehensive interactive personalised location based dynamic information and services. This approach improves on current travel guides by using GPS/WLAN technology to allow visitors to access geographical location specific information and services, similar to the techniques adopted by De Donatis (2006). In addition previous issues related to the limited data storage capacity of current hand held devices have been addressed by allowing information to be downloaded from a central server when needed. This ensures that information is current, up to date and accurate.

Planned additional location based services include:

- Information on nearby hotels, restaurants, amenities, shopping and entertainment centres including event listings, current/special offers and daily menus.
- Guides to locating nearby attractions, landmarks or sights with downloadable maps identifying routes to desired locations.

- Updated listing of local events, museum information, historical information and sporting events.
- Medical assistance guide as well as the location of medical services in the area.
- Information on safety and security and up to date parking tips.
- Information on basic travel needs including public transport schedules, airport, train and bus services and car rental locations.

The Wireless Campus

The wireless campus focuses on the creation of an educational precinct in the Strand Road area of the City, which in practical terms will involve the provision of a wireless network covering large areas of the Magee Campus, University of Ulster and the adjacent NWRC Campus. The education strands also sees the creation of enhanced teaching and learning environments in the University and NWRC which will facilitate greater, more flexible access to the institutions computing and educational resources.

Another aspect of the University's element of this project saw the creation of a Wireless Technology Demonstration Centre. The Wireless Technology Demonstration Centre provides independent information, advice and consultancy to businesses on wireless sensor systems, broadband applications and content. The Centre has a number of objectives, which include facilitating the research, development, prototyping and manufacture of cutting edge innovative wireless hardware/software solutions for integration into existing or new products for businesses and the development of novel, innovative and localised applications/content for delivery over the extensive wireless infrastructure provided by the project. This will include innovative e-learning and e-University applications to accelerate the utilisation of the Tablet PCs provided by the project. Further information is available at http://wtrc.infm.ulst.ac.uk/

The creation of the wireless campus had three distinct components, namely:

- Wireless Networking Services On and Off Campus
- Piloting a Tablet PC Mandate Initiative
- Development of SMART Classrooms

Wireless Networking On and Off Campus

The first strand of the initiative involved the provision of a wireless network for wireless connectivity covering large sections of the Magee Campus, University of Ulster and the NWRC Campuses. The objective here was to allow enhanced, more flexible access to computing resources and the Internet in the designated areas. In January 2005 an interim wireless network was deployed on the Magee Campus, this was deployed and tested over the next few months with complete functionality achieved by October 2005. This network now covers the majority of the Magee Campus and in particular the larger lecture theatres, laboratories, external and internal public areas including canteens and the Learning Resource Centre (LRC). Access to the network is freely available to all staff and students on the campus upon the completion, submission and validation of a brief application form available in the LRC.

Tablet PC Mandate Initiative

To take full advantage of these resources and the accompanying infrastructure a pilot project was undertaken to identify issues related to the development of a 'Tablet PC mandate' approach in both Institutions. The Tablet PC is a lightweight, portable machine similar to a fully-functional laptop PC which offers wireless connectivity and advanced features such as pen based computing and text to digital manipulation. The Tablet PC selected for the project is the Toshiba Portege M200 (see Waldock, 2006 for recent Tablet PC review). The Tablet PC is powered by the Windows XP Tablet PC Edition operating system, an augmented and functionality enhanced version of Windows

XP and comes equipped with a sensitive screen designed to interact with a complementary pen. The pen can be used directly on the screen (Figure 1) in a similar fashion to a mouse to select, drag, and open files or in place of a keyboard to create handwritten notes and annotations (Figure 2).

In February 2005 all teaching related staff (n=30) in SCIS received a Tablet PC for personal and teaching/work related use. The staff also attended a number of short induction and training sessions related to the practical use of the Tablet PC and accessing the wireless network. In October 2005, 120 first year undergraduate students in SCIS received a Tablet PC for the duration of their academic career. The only pre-requisite for receiving the Tablet PC was attendance on a

Figure 1. On screen pen function

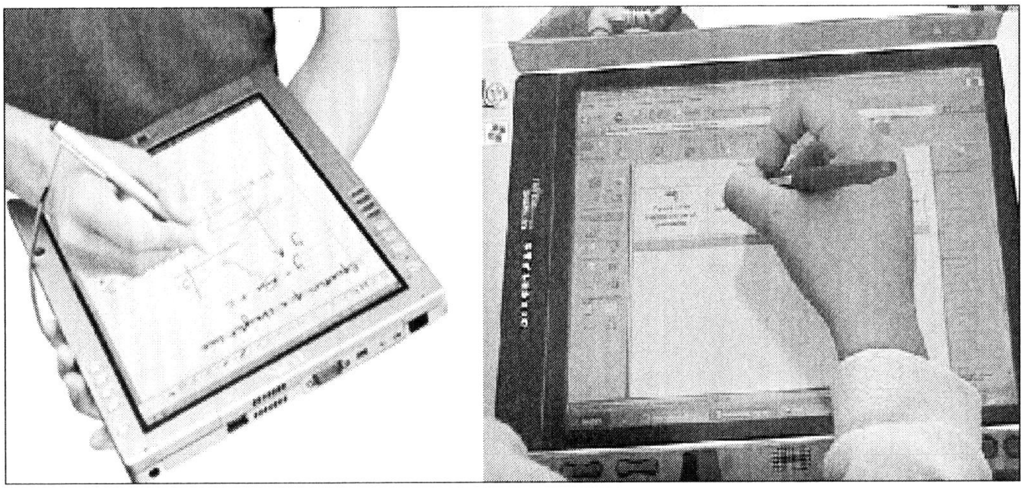

Figure 2. Annotation and creation of documents using digital ink (Microsoft product screen shot(s) reprinted with permission from Microsoft Corporation)

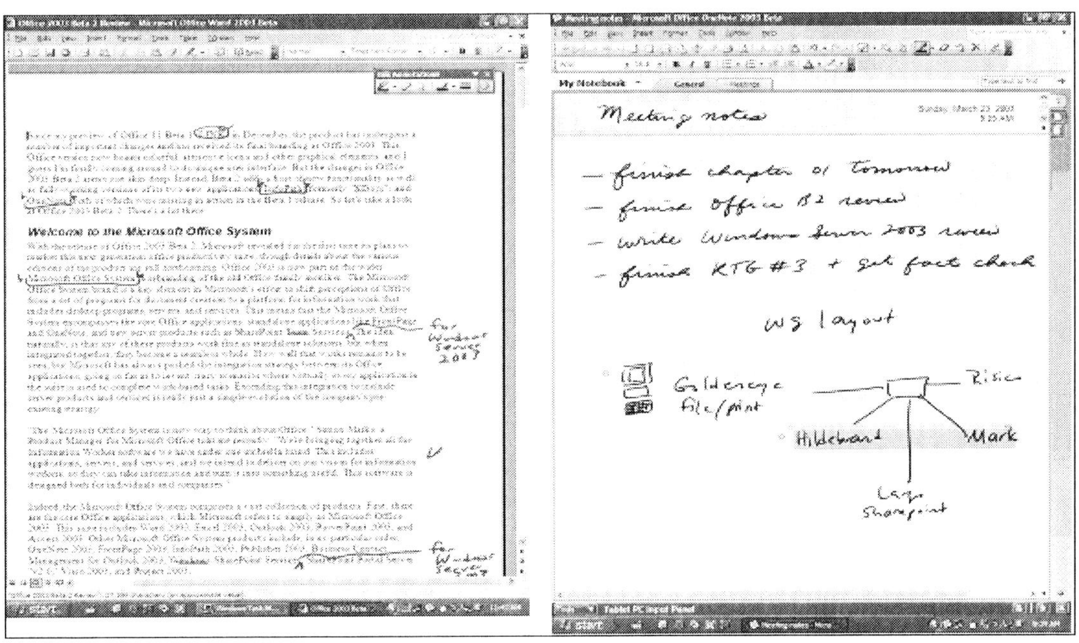

designated course approved by the project and the one-off compulsory payment of £100 to provide extra insurance cover for 2 years for the student in the event of damage or theft to the computer. On the day of distribution the students attended an induction session on the practical use of the Tablet PC and accessing the wireless network. This distribution process was repeated in October 2006 when a further 100+ Tablet PCs were given to students. At the time of writing the Tablet PC user base is approximately 300.

SMART Classrooms

The last strand of the initiative involves enhancing the capabilities of a number of existing lecture halls on the Magee and NWRC Campuses by equipping them according to the SMART classroom model. SMART classrooms are defined as technology-enhanced multimedia learning environments designed for instructional purposes to allow the presentation of complex information to students and encourage advanced levels of interaction in the lecture hall. During the summer of 2005 the computer laboratory MF124/5 was redeveloped using a SMART classroom model and equipped with Plasma TVs, rear projection SMART board and data projectors (Figure 3). The project team chooses a unique cluster design for the laboratory to facilitate collaborative working among students and to allow the Lecturer easy access to the students. The focus point of the room is a 70 inch interactive rear projection SMARTBoard which allows the Lecturer to use a diverse range of teaching material including the Internet, DVD, video, audio and facilitates the demonstration of software packages. A number of other classrooms and laboratories have also had wireless data projectors installed. This allows the Lecturer to connect to the projector using the Tablet PC from anywhere in the room giving added mobility and flexibility.

Figure 3. SMART Classroom in use

PROJECT IMPLEMENTATION AND PRACTICAL APPLICATION

Upon receiving the Tablet PCs staff within SCIS were encouraged to use them in all teaching/related work duties. Initially, most users experimented by utilisation in tasks with low public exposure (more administrative than teaching) until confidence and competence in the use of the device was achieved. The main application areas were at meetings, during school visits, to correct and revise final year projects, to note comments during supervision meetings, to facilitate learning in small group tutorials, and during lectures and laboratory sessions. The latter is presented in more detail as a case study example of application.

Lectures and Laboratory Sessions

Each academic member of staff has been encouraged to use the Tablet PC during lectures and practical computing laboratory sessions. This section outlines the experience of one member of staff who was responsible for teaching the module COM314M1 Multimedia Authoring, a first semester module which consisted of 70+ second year students in the academic year 2005/6. While this example relates to the teaching of multimedia content other cases of application within the Computing and Engineering fields are available (see Enriquez et al., 2006; Chang Liu & Young, 2006; Charlevoix et al., 2006; Ambikairajah, 2007). In this example module delivery was broken down into two sessions, lecture (2 hours) and hands-on laboratory time (3 hours). The lecture time focused on the theoretical material underpinning the practical aspects of multimedia design while the laboratory session facilitated student familiarity in practical and applied use of multimedia packages. The laboratory is equipped with a high specification desktop machine and fixed data projector so there was no real advantage in using the Tablet PC here but there was ample opportunity to do so in the lecture theatre.

Multimedia is a visually rich medium and requires the use of a diverse range of examples in a variety of media (graphical, audio, video and animation) to reinforce the theoretical aspects of the discipline. Previously the Lecturer had felt restricted during module delivery in that he did not have access to hardware/software to highlight/demonstrate this aspect of the discipline in the lecture theatre where the typical setup was a very low specification, fixed desktop PC and data projector capable of delivering basic Microsoft PowerPoint slides. The Lecturer felt he could rectify this shortcoming using the Tablet PC platform with the necessary software installed and used with the complementary wireless data projector and wireless access.

Module notes for each lecture would normally be available for students to digitally download from the Lecturers' module website in advance of the lecture. Before each lecture the Lecturer reviewed the content available online to see where additional multimedia material could be added and/or new functionality could be incorporated. Where appropriate a range of video, animation, sound files and web links (both inside and separate to) the Microsoft® PowerPoint slides were added, these were then accessible to students during the lecture. The Lecturer tried out this new approach in a live lecture environment.

During the lectures there was some teething problems such as when the data projector stalled or Microsoft® PowerPoint failed to launch. The Lecturer felt that this was probably more related to his inexperience and lack of awareness of the limitations of the technology than the devices themselves. On a positive note the Lecturer was quickly able to rectify the situation and felt that the novelty of the Tablet PC and the accompanying curiosity in this previously unknown platform helped by making the students more tolerant and forgiving of technological mistakes.

The Lecturer expressed that the flexibility offered by the technology was impressive. One instance success was the annotation of slides in response to student queries, for example when setting the first assignment relating to the creation of a promotional CDROM for the traditional music band, Dervish. In response to student queries on the assignment the Lecturer annotated the slides with additional information as requested by the students, redesigned the slides to incorporate video and web links (Figure 4) and then uploaded the augmented slides straight to the module website for the students to subsequently access. This facilitated the capturing, recording and storing of feedback from students which was subsequently available immediately online.

Figure 4. Microsoft PowerPoint slide annotated with student feedback from lecture (Microsoft product screen shot(s) reprinted with permission from Microsoft Corporation)

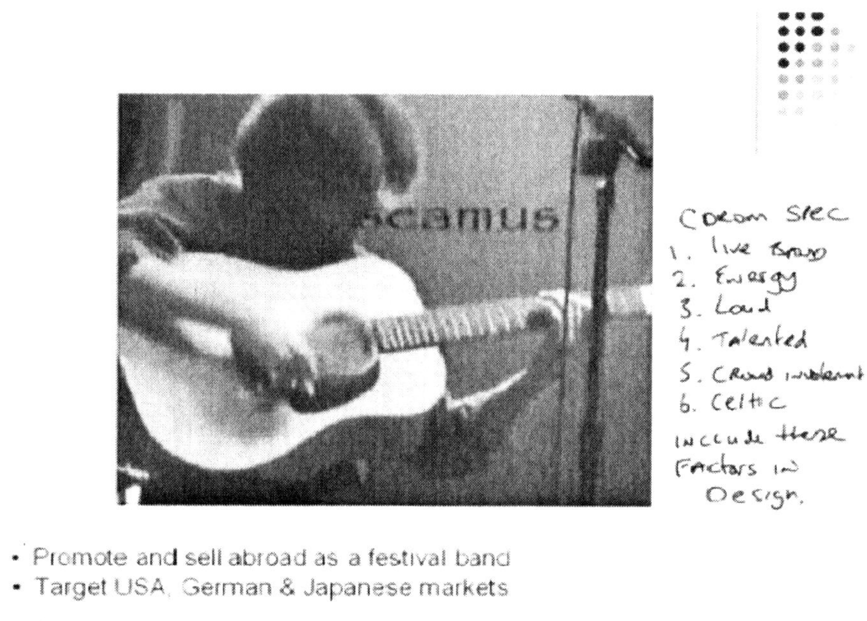

Figure 5. Dynamic slide to clarify issue related to CDROM structure (Microsoft product screen shot(s) reprinted with permission from Microsoft Corporation)

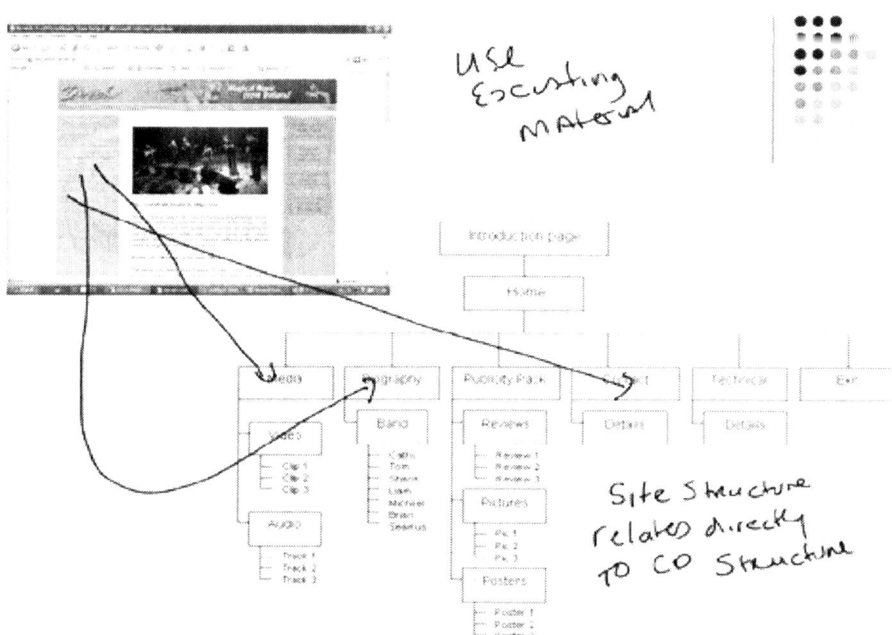

Figure 5 shows a further example of this process where the Lecturer had opened the Dervish website in a web browser, captured a screenshot, pasted this into an existing Microsoft® PowerPoint slide and combined both with a series of annotations.

In addition the Lecturer found that when demonstrating software packages the ability to capture screenshots of the package in use (again directly related to student queries), add the screenshot into Microsoft® PowerPoint, annotate it where necessary and finally save to the module website (Figure 6) was invaluable. This functionality alone justified the investment in terms of Lecturer time and resource in the adoption of the technology. The additional benefits of full web access where appropriate and a more intuitive and natural lecturing style facilitated by the form factor and design of the Tablet PC made the experience a worthwhile and rewarding undertaking.

The examples provided in this section outline the potential for the Tablet PC, wireless access and digital inking for teaching and related duties in an educational environment. The flexibility of the platform is obvious and the number of possible areas of use extensive. However, there does seem to be a steep learning curve associated with becoming competent using the Tablet PC and its' associated applications. The time needed to experiment with the technology may be an inhibitor to user acceptance as outlined in the Technology Acceptance Model (TAM) (Davis, 1986; 1989) where factors such as limited ability, time, environmental or organisational limits, and unconscious habits limit the freedom to act. While students may have full intention to use the system it may be perceived as difficult to use and could contribute to user rejection. In light of this deficiency the next section of this chapter seeks to quantify the experiences of staff and students who participated in the first year of the DETI funded wireless campus. This study is similar to that conducted by El-Gayar and Moran (2007) where student acceptance of the Tablet PC was investigating using TAM. In our case research employing a quantitative approach was under-

Figure 6 Annotated slide of software demonstration in response to student queries (Adobe product screen shot(s) reprinted with permission from Adobe Systems Incorporated)

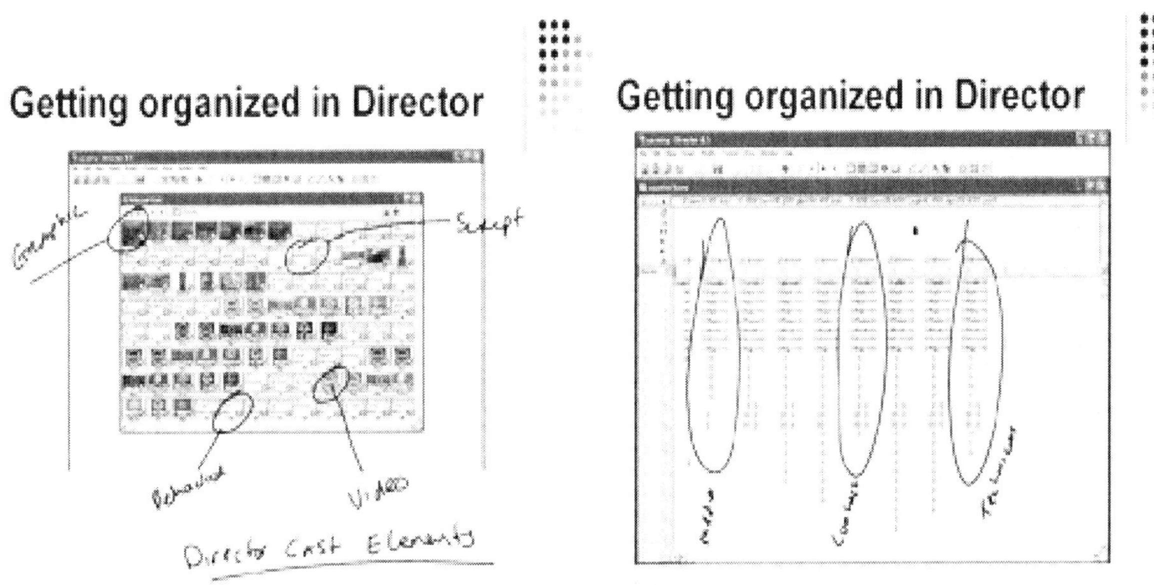

taken by conducting two surveys, one with staff and one with students as user groups.

QUANTITATIVE RESEARCH METHODOLOGY

This section summarizes the results of two surveys carried out with the co-operation of teaching staff and students in an attempt to catalogue and quantify their experiences in the use of the Tablet PC and wireless infrastructure. The survey looks at the level and type of Tablet PC use among the users groups, the impact the DETI project has had on the teaching and learning environment in general, any perceived user identified deficiencies in the implementation approach taken and suggestions for the future directions in the project. An example of the survey questionnaires are available in appendix 1 (staff survey) and appendix 2 (student survey).

Context for Staff Survey

Thirty Tablet PCs were distributed to Lecturers, Teaching Fellows, Computing Training Officers and Graduate Demonstrators in SCIS in February 2005. Within this group there are 21 male and 9 female members of staff with ages ranging from 20+ to over 50 years of age (Table 1).

Given the School profile, research and teaching expertise it would be reasonable to assume that the users are technically orientated and familiar with computing related hardware and software. The survey was carried out in April 2006 (approximately 15 months after the Tablets were distributed) ensuring ample time for familiarity with the platform and for use patterns to emerge. Respondents were asked to comment on their use of the Tablet, the context in which they used it, perceived student use as well as the overall effect on the Lecturer/student teaching experience. 24 responses from a possible 30 respondents were received giving a high representative sample (80%).

Table 1. Age profile of school April 2006

Age Group (Years)	Number Of Staff
20-30	6
30-40	14
40-50	6
50 +	4

Staff Use

In this series of questions respondents were asked to comment on the general level of Tablet PC use. 45% of respondents used the Tablet PC as their primary computer replacing their desktop PC. 75% of respondents indicated that they used their Tablet PC on a daily basis while the remainder indicated frequent use (2-3 times a week). 87% of respondents used the pen driven functionality of the Tablet PC. 80% of users utilized the hand writing recognition functionality of the Tablet PC in a range of applications (Figure 7). 16% hand wrote emails, 30% regularly used Microsoft OneNote and 55% used Sticky notes. 80% used the pen driven mode of the Tablet PC in other Microsoft® Office applications including Microsoft® PowerPoint, Excel and Access. A small percentage used the Tablet PC to sign documents.

Staff Activities

In this series of questions respondents were asked to comment on where they used the Tablet PC. The range of activities where the respondents used the functionality of the Tablet PC was widespread and diverse covering all aspects of teaching and administrative duties, such as running presentations, annotating notes, demonstrating projects, and downloading notes. Unsurprisingly general use for access to the wireless network was high (over 80%). However it is interesting to note that the percentage use for administrative related

Figure 7. Staff use (Applications)

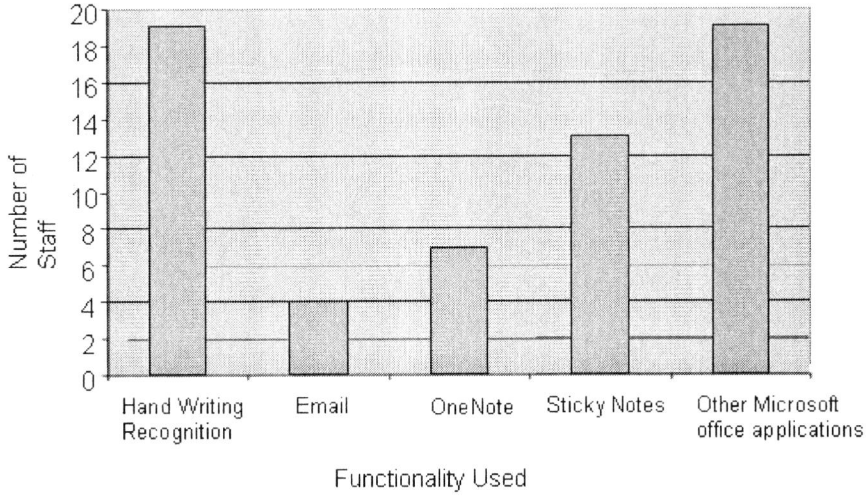

tasks is slightly higher than the percentage use for teaching related activities. Further informal discussions with respondents indicated that this was due to a number of factors including:

- Exposure levels related to level of competence. Respondents had some reservations about using new technology which they had not fully mastered in front of large groups of students.
- The majority of lecture rooms or laboratories have permanent audio visual equipment installed.
- The weight and size of the Tablet PC and accompanying carry case along with lecture related materials are awkward to carry around.
- Lack of wireless access in some classrooms discouraged some users (20%).
- 16% of respondents indicated that the Tablet PC interfered with their teaching style while 40% felt that it complimented it.
- 37% indicated that they did not use the Tablet PC because the students in their classes did not have one, for example if a student was registered on a split degree programme they did not participate in the scheme (refer to list of eligible courses in Table 2).

The majority of respondents were happy with the level of training received to use the Tablet PC (87%). 40% of respondents used the Tablet PC at home and have wireless broadband connectivity at home. A number of staff had general difficulties configuring the Tablet PC for dial up connections (23%).

Student Use (Lecturer Perspective)

In this section respondents were asked to comment on the level of student use in a range of teaching related activities including classrooms, practicals, tutorials and studies advice sessions. This percentage of student use needs to be set in context, as only 33% of the possible respondents would have Lecture/Laboratory contact with the Foundation and First year students eligible to receive the Tablet PC under this years' scheme and also that only approximately 75% of first year students received Tablet PCs. 37% of respondents indicated that students used Tablet PCs in their lectures and practicals. 41% of respondents indicated that students used Tablet PCs in their module tutorial sessions and over 58% indicated that students used Tablet PCs in the first year studies advice/tutorial sessions.

Main Student Activities

The main student Tablet PC use observed by the respondents in teaching related activities include downloading, annotating and inking notes and completing tutorials (Figure 8). In addition a number of students were accessing websites related to the content under discussion.

Wireless Network Access

All respondents were aware of the wireless network and how to access it. 40% of respondents used the wireless network on a daily basis and a further 20% use it more than 3 times a week. 54% of respondents used the wireless network for email, internet and network access. 25% used the Tablet PC for Internet access, 16% for email and the remainder for local network access (Figure 9).

Wireless Network: Coverage and Reliability

75% of respondents rated the level of satisfaction with wireless campus coverage at 5 or greater (Figure 10) while 63% of respondents rated the reliability of the wireless network connection at 5 or greater (Figure 11).

Wireless Network: Ease of Use

Over 50% of respondents rated ease of use of the wireless network as 8 or greater (Figure 12).

Wireless network: Technical support

80% rated technical support for the wireless network as 6 or greater. 35% of users had com-

Figure 8. Students Tablet PC use during teaching related activities

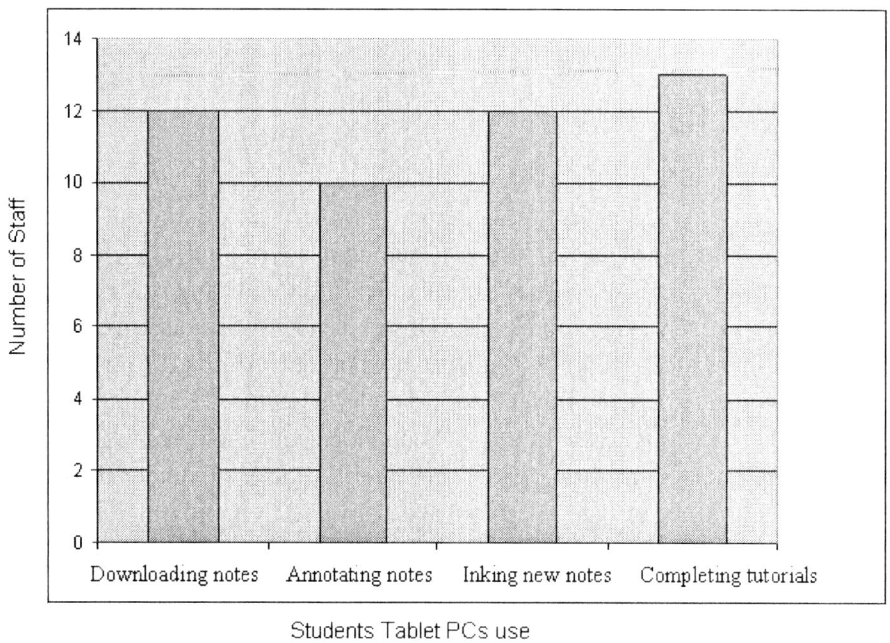

Figure 9. Staff wireless network related activities

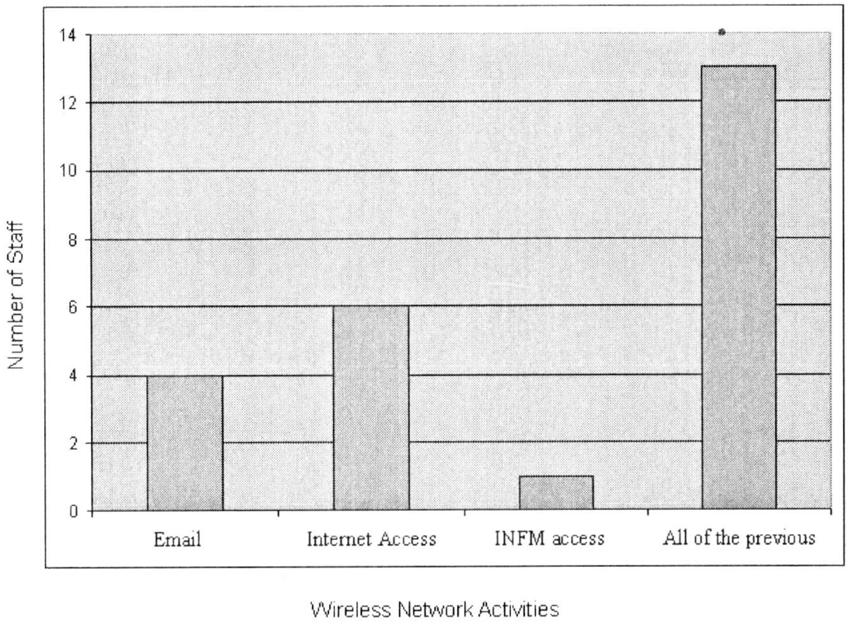

Figure 10. Wireless coverage (Scale 1-10: 1-Poor, 10-Excellent)

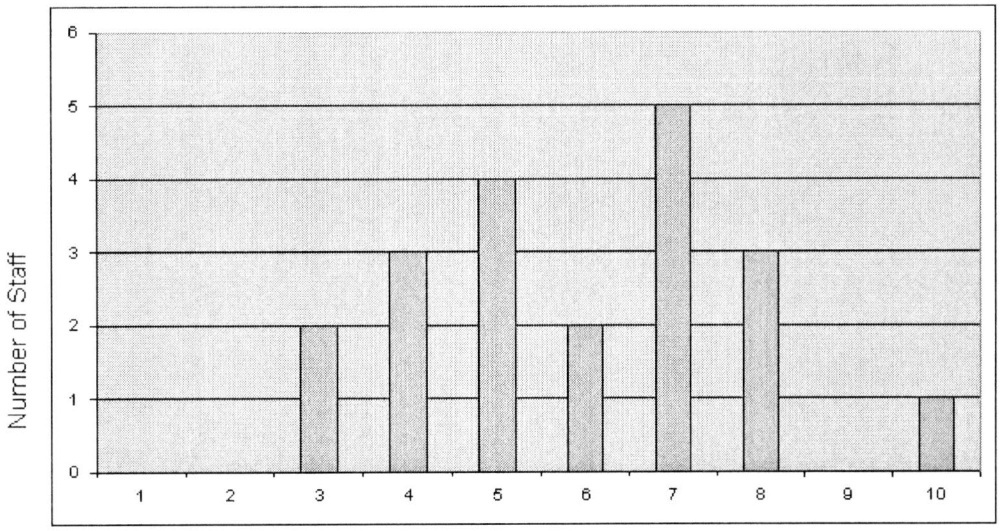

Figure 11. Wireless reliability (Scale 1-10: 1-Poor, 10-Excellent)

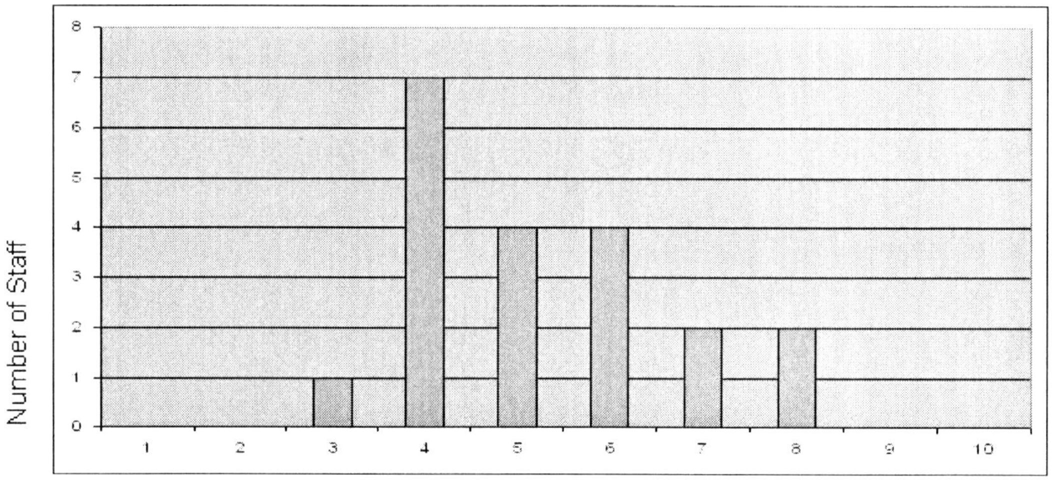

Figure 12. Ease of use (Scale 1-10: 1-Poor, 10-Excellent)

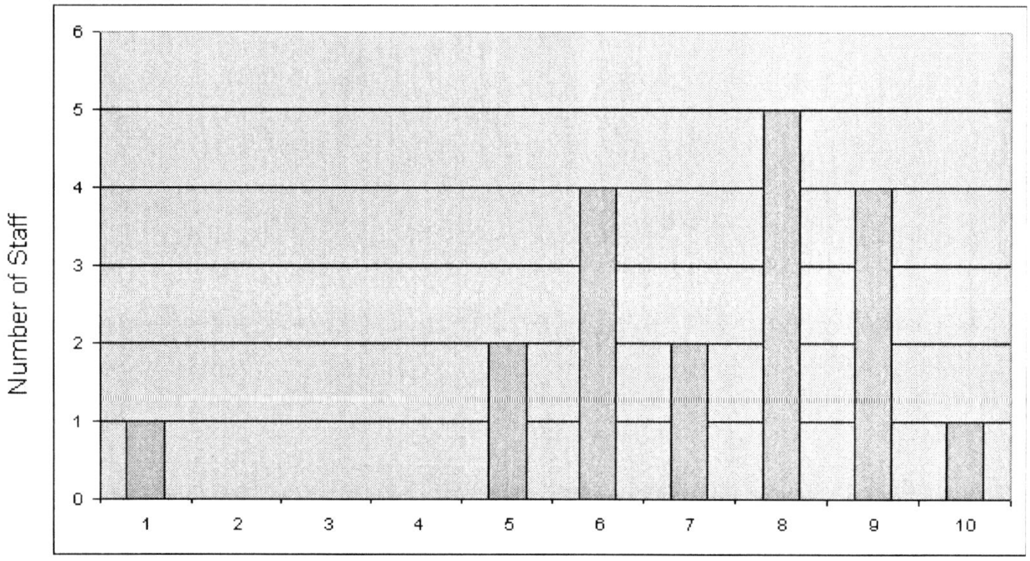

plained about an aspect of the wireless network but stated that in most cases it was quickly resolved (Figure 13).

Overall Impact on Overall Lecturer/ Student Teaching Experience

When asked to comment on the overall usefulness of the Tablet PC for teaching and impact on the Lecturer/student experience, 80% of respondents indicated that the use of this technology had a positive impact on the Lecturer/student experience (Figure 14). The vast majority of respondents indicated that they had adapted very quickly to this technology and used it in a range of teaching and administrative related tasks. The general feeling was that the pilot scheme carried out among students should be expanded. Many respondents

Figure 13. Technical support (Scale 1-10: 1-Poor, 10-Excellent)

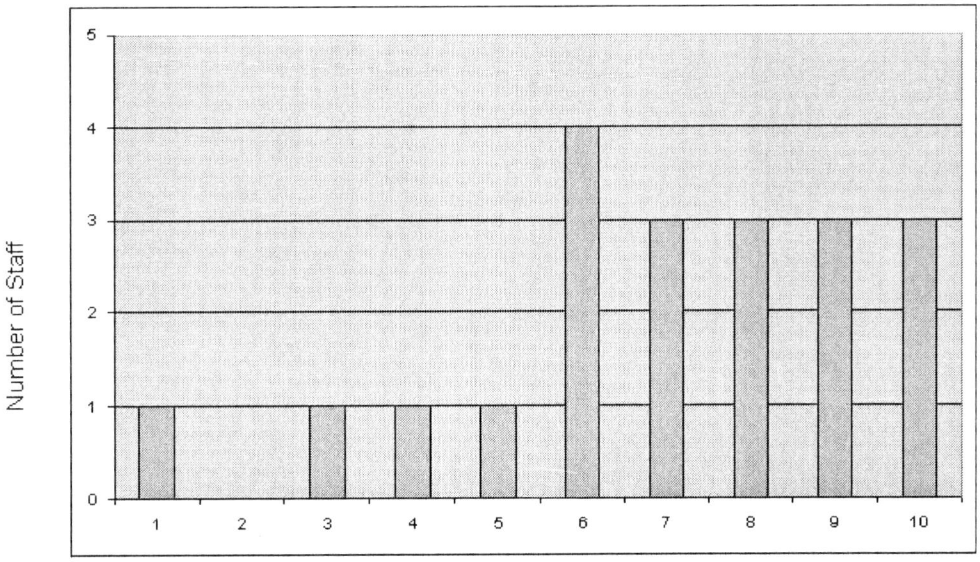

observed large numbers of students using their Tablets PCs as their main PC in laboratories and a high level of student use was also noted outside of formal scheduled teaching activities e.g. group work sessions in the LRC and in on-campus catering provision areas.

Context for Student Survey

Over 100 Tablet PCs were distributed to first and foundation year students attending designated courses in SCIS during semester one (October) of the 2005/6 academic year (Table 2). This group contained approximately 69% male and 31% female students with typical age profile ranging from 18 to 20 years of age.

The students are attending computing related courses so in this context it would be reasonable to assume that the recipients are technically minded and technology aware (moderate/expert users). The survey was carried out in May 2006 (approximately 8 months after the Tablet PCs were distributed) ensuring ample time for familiarity with the platform and for use patterns to emerge.

Respondents were asked to comment on their use of the Tablet PC, the context in which they used it, perceived teaching staff use and the effect they felt it had on their overall University experience. 53 responses (over 50%) were received giving a good representative sample of opinion.

Student Use

In this series of questions respondents were asked to comment on their general level of Tablet PC use. Tablet PC use was extensive with over 82% of respondents reporting over 3 hours of use daily and 54% over 5 hours (Figure 15). 63% of respondents had wireless Internet access at their term time address.

When asked how they used the Tablet PC and for what applications 90% of respondents used the Tablet PC in pen driven mode, 87% regularly used the handwriting recognition functionality of the Tablet PC, 83% have sent handwritten emails and

Figure 14. Assessment of Impact of Tablet PC on Lecturer/Student experience

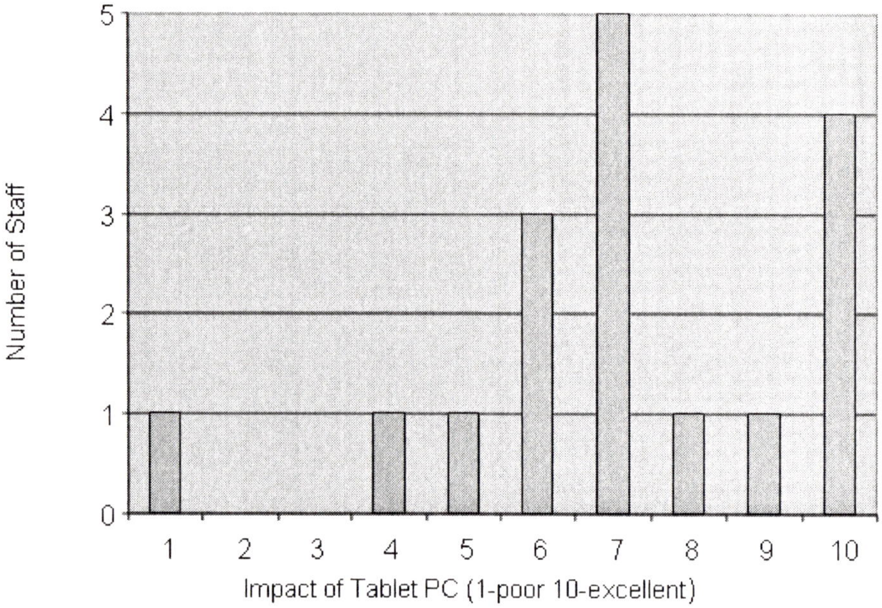

Table 2. Courses eligible to receive Tablet PC in 2005/06

Programme	Course code
Computer Science	E411UM
Computer Games Development	E138UM
Multimedia Computer Games	E139UM
BSc Hons. Computing with Accounting BSc Hons. Computing with Entrepreneurship BSc Hons. Human Resource Management BSc Hons. Computing with Marketing BSc Hons. Computing with Psychology BSc Hons. Computing with French BSc Hons. Computing with German BSc Hons. Computing with Irish BSc Hons. Computing with Spanish BSc Hons. Computing with E-Business	E140UM E132UM E143UM E135UM E137UM E141UM E142UM E144UM E145UM
Integrated Foundation Year BSc (Hons.)/BEng (Hons.) BEng Hons. BSc Hons.	E134UM E409UM
Electronics and Computer Systems	E133UM

made handwritten Microsoft Outlook tasks. 47% use Sticky notes regularly to take quick notes and record memos. 75% use the pen and ink functionality in a range of Microsoft applications including Word and Microsoft PowerPoint. These results are similar to those found by Leeds, (2005).

In relation to teaching and learning activities 100% of respondents indicated they had regularly used the Tablet PC in a lecture environment, 90% in tutorials, 75% in laboratories, 73% in first year tutorials/studies advice and 94% indicated use in completing assignments (Figure 16).

When asked how they used the Table PC in these environments 94% had annotated the Lecturers' notes in a lecture context, 78% had inked new notes during lectures and tutorials, 63% had used the Tablet PC to access module related resources during class and 36% had used it for messaging. In addition, 98% of respondents had used the Tablet PC to work collaboratively with other students and activities included completing coursework and assignments (92%), sharing resources found online (62%) and emailing fellow students or members of staff (82%).

Wireless Campus

84% of respondents have used the campus wireless network. 91% rated the campus wireless network coverage as good or better. 81% of respondents rated the reliability of connection as good or better. 88% of respondents rated ease of use of the campus wireless network as good or better while 73% rated the technical support services for wireless network as good or better. 92% of respondents felt that wireless access had improved and enriched their learning experience at the University allowing mobile flexible access to learning resources, reduced printing costs and facilitating collaborative working with other students.

Figure 15. Typical daily student use

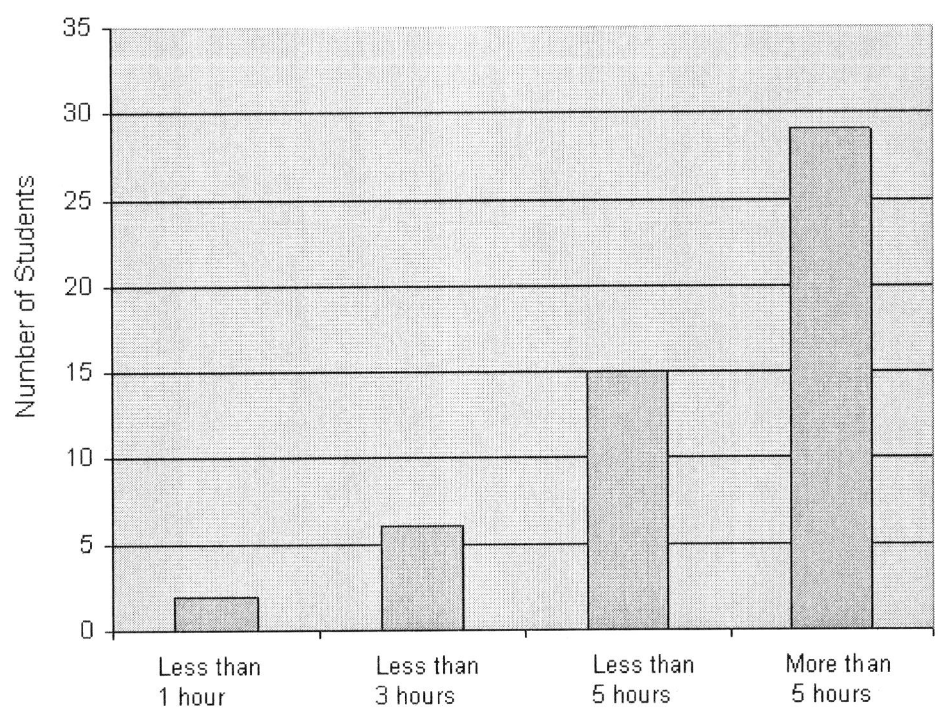

Figure 16. Student Tablet PC use: activities

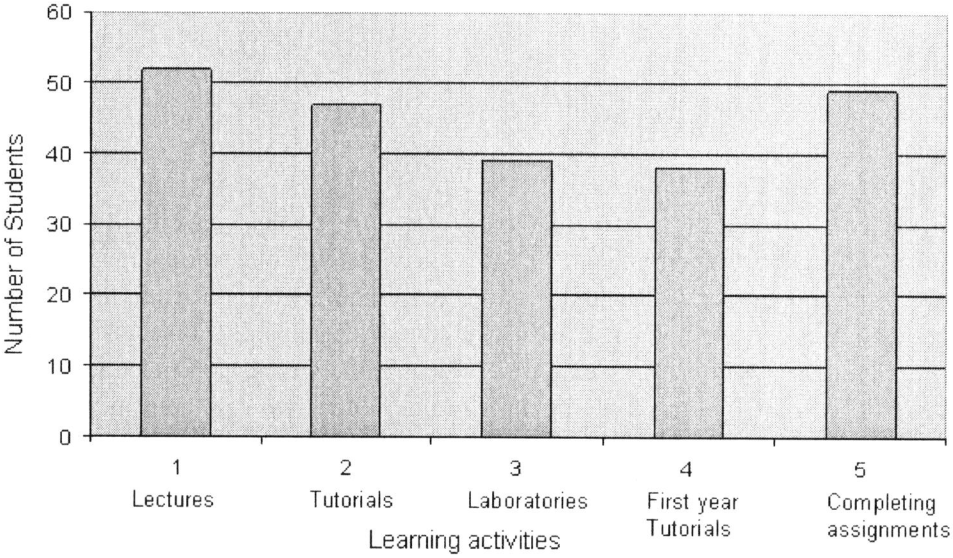

General Issues Related to Tablet PC Use

- **Battery life and charging issues:** Only 47% of students felt that the battery life of the Tablet PC was satisfactory and 80% of respondents charged the Tablet PC on the University premises. A variety of locations were used to charge the Tablet PC including lecture theatres (47%), laboratories (60%), student residences (30%) and ad hoc outlets (28%) in the University (e.g. corridors, canteen). 67% of respondents are in favour of secure on- campus lockers for storage and as a place to charge the Tablet PCs but were overwhelming against paying a charge for this service. There was some support for a battery swap facility and casual drop in help area.
- **Compulsory student insurance scheme:** 80% of respondent felt that the Tablet PC insurance scheme was a good idea and that they would purchase insurance themselves if it was not compulsory. 92% felt that the £100 cost levied was reasonable and acceptable.
- **Induction session:** 81% of respondents had attended the Tablet PC induction session and 93% felt the induction session was useful. When asked on ways to improve the induction session 65% would prefer smaller induction groups and 30% requested more sessions with increased technical content.
- **Technical support:** 70% of respondents had found the Tablet PC website helpful but requested more technical material on the site and felt that a user forum would be beneficial. 70% of respondents were aware of on campus technical support offered by Computer Officers and Technicians. 38% of respondents had used the support services with the average use of less than 5 visits. 80% of the problems were software based problems with the majority related to wireless access. 90% of support services users rated the support received as good or better.

- **Staff use:** 86% of respondents encountered staff that used the Tablet PC for teaching related activities in lectures, 55% in tutorials and 25% in first year tutorial sessions. 73% of respondents indicated that staff Tablet PC utilization in these environments was competent and appropriate.

Impact on Overall Student Experience

94% of respondents felt that the use of the Tablet PC and wireless access improved the quality of their overall University learning experience. When asked how it improved their University experience respondents replied: it encourages attendance (25%), it encourages study (60%), it facilitates access to online resources for assignments (97%), it encourages and facilitates collaboration with fellow students (46%), it allows flexibility by not having to queue for fixed laboratory resources (90%), it provides a convenient storage and management facility for student work (93%), it allows flexible management and annotation of lecture material. 90% of respondents felt that the removal and absence of the Tablet PC would have a detrimental effect on their studies. Reasons cited included reduced flexibility by restricting the ability to work at home or on the move, loss of benefits of note annotation stored on the Tablet PC and reduced access to campus based and complementary online resources. When asked for general feedback on the project and opportunities for improvement respondents suggested; onsite facility to carry out repairs on damaged machines, the inclusion of a CDROM drive on the Tablet PC, improved battery life for the Tablet PC or better general access to power sockets, the need for a secure storage and charging facilities, a battery swap facility, smaller carrying cases and better wireless access in some black spots.

Survey Conclusions

The results of the two surveys clearly indicate that a high percentage of teaching related staff and students have embraced the use of the Tablet PC and wireless access in the context of their teaching, work and study. The general consensus is that the use of this technology has had a positive impact on teaching and on overall University experience. From a teaching staff perspective the Tablet PC initiative offers flexibility, mobility and opportunities for new modes of teaching and work related activities. The student perspective is even more positive where the technology was very quickly integrated into the overall student experience becoming indispensable in a very short time period. Even though the students were considered to be moderate/expert users they still found the Tablet PC experience rewarding and felt using the tool enhanced participation and learning.

CONCLUSION

This chapter presents an overview of the Flagship Broadband Project which enabled the city of Londonderry (Derry), Northern Ireland to take advantage of wireless capabilities. Three partners, namely Derry City Council, the ISRC at SCIC, University of Ulster and NWRC, were successful in attracting funding to the value of £1.38 million from the Department of Enterprise, Trade and Investment (DETI) Northern Ireland, under the EU Building Sustainable Prosperity Programme. The project had three strands focusing on civic, tourism and educational avenues.

The main focus of this chapter has been on the third strand. Research has shown that information and communications technologies (ICT) can have a powerful impact on teaching and learning (Sandholz et al., 1991, Knight & Knight, 1995). Opportunities for technological driven transformation in approaches to teaching and learning exist in a range of subject areas and have the potential to contribute to the development of intellectual and related skills in students by increasing interest, participation and concentration levels (Selinger, 2002). The importance of this process is recognized in a recent UK government initiative which seeks to embed ICT in everyday teaching practice regardless of course or subject matter (DfES report, 2003). The critical question however is the appropriate use of technology in the teaching and learning experience. The authors believe that the infrastructure described in this chapter, and the review of its implementation, have contributed to that debate.

This chapter has investigated the usefulness of advanced technologies as a tool for teaching and learning from both a personal perspective and from the perspective of teaching staff and students involved in the Flagship project. The two complementary technologies focused on are the Tablet PC and wireless network access, quantitative research proves the success of the undertaking. Both staff and students alike expressed benefit in having these new technologies incorporated into the teaching/learning experience to the extent that withdrawal of the resources and facilities would have a detrimental effect on the entire Lecturer/student experience.

While the overall project has been successful, the research team has noted some issues for consideration. Firstly, the provision of wireless facilities on the Magee and NWRC Campuses are temperamental and costly to maintain in terms of technical support and firefighting mechanisms. As the ongoing success of the project is dependant on full provision of the wireless architecture it is important to maintain the network so it is available 24 hours a day, 7 days a week, 365 days a year. Also as students have become dependant on the wireless capabilities provided on the two Campuses' there is expectation that these will be available in other University/College locations, such as halls of residence. Work is currently ongoing to meet this demand. Secondly, the success of the project has also been its downfall to some extent. During the pilot and implementation phases of the project, the Tablet PC and SMART classroom provision were used as a strong marketing and recruitment tool for our courses. Student numbers on the eligible courses increased to the cost of some other courses not participating in the scheme. Increased numbers on the eligible courses led to a demand for the Tablet PC provision and wireless facilities which in turn led to costs and technical provision as previously outlined. Thirdly, as the scheme is now in its third year the Tablet PCs first provided in 2005/06 are outdated and need to be replaced and upgraded. Costs of maintaining the initial standards set by the project now have to be met by SCIS as initial funding has expired. Finally, we are catalysts to our own success and are always striving for the next new and innovative teaching/research ideas and applications.

Overall the Flagship Broadband project has been a very successful achievement and has won a number of awards to date including the Best Wireless Project in the Communication in Business Awards and two categories (namely Education, sponsored by Hewlett Packard and Mobile Technology (WiFi), sponsored by BT) in the 2006 BT GoldeneyeT Awards. It has helped to establish the City as an area of high technological achievement and created strong links between the project partners. Future collaborative technological projects are currently being considered by the project team and the challenge to develop the Wireless City of Derry is met with much enthusiasm. This is particularly so within the University environment as student expectation is becoming more technically orientated. Internet-based learning, increased use of digital content and library resources, increased application of e-texts, real-time discussions and online collaboration tools are becoming essential for the educational experience (Toshiba, 2007). With innovative features such as the transflective LED backlit monitors and visible, illuminated touch pads (Toshiba, 2007b) being injected into the Tablet PC and new technologies such as the mobile web, ultra mobile PCs, IPTV and mobile VoIP (BBC News, 2008) the future looks bright for the newly digitised Walled City of Derry.

ACKNOWLEDGEMENT

The authors acknowledge the financial support of the Department for Enterprise, Trade and Investment, Northern Ireland under the European funded Building Sustainable Prosperity program.

REFERENCES

Ambikairajah, E. (2007). Tablet PC and electronic whiteboard use in signal processing education. *IEEE Signal Processing Magazine,* 24(1), 130-133.

BBC News, (2008). *Technologies on the rise in 2008.* Retrieved February 14, 2008, from http://news.bbc.co.uk/2/hi/technology/7147804.stm.

Chang L., & Young, V. (2006). Student self-reported use of wireless Tablet PCs in classrooms. *Frontiers in Education 36th Annual Conference*, San Diego, USA.

Charlevoix, D. J., Jackman, S. K., & Twine, T. E. (2006), Conference notebook: Tablet PCs: A welcome aid to lecture-based meteorology courses. *Bulletin of the American Meteorological Society, 87*, 737-738.

Davis, F. D. (1986). *A technology acceptance model for empirically testing new end-user information systems: Theory and results.* Doctoral dissertation, Sloan School of Management, Massachusetts Institute of Technology.

Davis, F. D. (1989). Perceived usefulness, perceived ease of use, and user acceptance of information technology. *MIS Quarterly, 13*(3), 319-339.

De Donatis, M. (2006). Map IT: The GIS software for field mapping with Tablet PC. *Computers & Geosciences, 32*(5), 673-680.

Department for children, schools and families (DfES) (2003). *Towards a unified E-Learning strategy*, http://www.dfes.gov.uk/, Retrieved November 1, 2003, from http://publications.teachernet.gov.uk/default.aspx?PageFunction=productdetails&PageMode=publications&ProductId=DfES+0455+2003&

El-Gayar, O., & Moran, M. (2007). Examining student acceptance of Tablet PC using TAM. *Issues in Information Systems, VIII*(1), 167-172.

Enriquez, A., Gunawardena, A., Kowalski, F., Kowalski, S., Millard, D., & Vanides, J. (2006). Innovations in engineering education using Tablet PCs – panel discussion with four institutions, *Frontiers in Education 36th Annual Conference*, San Diego, USA.

Knight, C., & Knight, B. A. (1995). Cognitive theory and the use of computers in the primary classroom. *British Journal of Educational Technology, 26*, 141-148.

Leeds, K. (2005). *Technology: Fad or fixture? A study on students' perceptions of using Tablet PCs during class.* Retrieved June 6, 2006, from http://www.cccone.org/scholars/05-06/Kelvin-LeedsExecutiveSummary.pdf

Sandholz, J. H., Ringstaff, C., & Dwyer, D. C. (1991). *Teaching with technology: Creating pupil centred classrooms.* New York, NY: Teachers College Press.

Selinger, M. (2002). Learning computer technology skills in the subject context of learning. *Journal of Information Technology for Teacher Education, 10*(1-2), 143-157.

Toshiba (2007). *The Classroom and beyond,* Retrieved September 28, 2007, from http://eu.computers.toshiba-europe.com/cgi-bin/ToshibaCSG/download_whitepaper.jsp?service=EU&

WHITEPAPER_ID=2007-09-Classroom-and-Beyond-EN

Toshiba (2007b). *Trends and technology 2007.* Retrieved March 15, 2007, from

http://eu.computers.toshiba-europe.com/cgi-bin/ToshibaCSG/download_whitepaper.jsp?service=EU&WHITEPAPER_ID=2007-01-Trends-to-watch-out-for-EN

Waldock, L. (2006). Travel light with a tablet [Tablet PC review]. *Personal Computer World, 29*(4), 94-96.

APPENDIX 1: ORIGINAL STAFF SURVEY

Questionnaire Examining Use of the Tablet PC by Staff Date:__/__/__

General Operation

1. Is your Tablet PC used as your primary computer? YES / NO

2. On average how many times do you use your Tablet PC per week ________

3. Do you utilise the pen driven mode of the Tablet PC? YES / NO

4. Did you know that you can work mouse-free thanks to the Tablet PC pen? YES / NO

5. Have you used handwriting recognition on your Tablet PC? YES / NO

   If you answered YES have you:

   a. Created hand written emails or just added a handwritten word or drawing to personalize your typed text e-mails? YES / NO
   b. Used OneNote 2003 to create Microsoft Outlook tasks from items in your notes, and send as email? YES / NO

6. Have you used Sticky Notes YES / NO

   If you answered YES have you used Sticky Notes for:

   a. Writing notes down? YES / NO
   b. Creating quick gestures? YES / NO
   c. Recording quick memos? YES / NO

7. Have you used right click on your pen? YES / NO
   If you answered YES did you find it easy to use? YES / NO

8. Did you use pen and digital ink in any of the Office applications? YES / NO

WORK RELATED USE

9. Do you use your Tablet PC for:
   a. Meetings YES/ NO
   b. School visits YES / NO
   c. External UU business YES / NO

Questionnaire continued on following page

10. Do you use your Tablet PC for any of the following teaching activities

a.	Lectures	YES / NO
b.	Tutorials	YES / NO
c.	Practical Classes	YES / NO
d.	Project Supervision	YES / NO
e.	First Year Tutorials	YES / NO

If you answered YES to any of the above, do you use it for

a.	Running a presentation?	YES / NO
b.	Annotating slides during a presentation?	YES / NO
c.	Inking new notes?	YES / NO
d.	Demonstrating particular software?	YES /NO
e.	Other purposes (outline briefly below)	YES / NO

____

____

____

If you answered NO to any of the above, do you NOT use it because:

a.	None of the students you teach use the Tablet PC	YES /NO
b.	You **do not** find the Tablet PC useful	YES / NO
c.	You find the Tablet PC interferes with your teaching style	YES / NO
d.	The teaching rooms do not support Tablet PC use	YES /NO
e.	You find it awkward to carry around	YES /NO
f.	You were never shown how to use it properly	YES /NO
g.	There was no wireless connectivity	YES /NO
h.	Any other reasons (outline briefly below)	YES / NO

____

____

____

11. Do your students use their Tablet PCs during:

a.	Lectures	YES / NO
b.	Tutorials	YES / NO
c.	Practical Classes	YES / NO
d.	First Year Tutorials	YES / NO

Questionnaire continued on following page

If you answered YES to any of the above do they use it for

a.	Downloading your notes	YES/ / NO
b.	Annotating your notes	YES / NO
c.	Inking new notes	YES / NO
d.	Working through tutorial questions	YES / NO
e.	Any other purpose (outline briefly below)	YES / NO

12. If you answered YES to any part of Q11 then approximately what percentage of students used the Tablet PC during:

a.	Lectures	_______
b.	Tutorials	_______
c.	Practical Classes	_______
d.	First Year Tutorials	_______

13. On a scale of 1-10, (1-poor 10-excellent) how useful do you rate the Tablet PC in terms of helping improve the student/Lecturer experience _______

NETWORK USE

14. Do you know how to logon to the INFM network from your Tablet PC? YES/NO

15. Do you know that you can log onto the INFM network wirelessly? YES/NO

16. Have you used your Tablet PCs wireless capability at home YES/NO

17. Do you utilise the wireless campus network YES/ NO

    If YES then on a scale of 1-10 (1-poor, 10-excellent) how do you rate the wireless network in terms of:

a.	Coverage	_______
b.	Reliability of connection	_______
c.	Ease of Use (configuration/login etc)	_______
d.	Technical support for wireless service	_______

Questionnaire continued on following page

18. If you answered YES to Q17 then:

    a. On average how often do you use the wireless network (times per week) _______
    b. Is your main purpose of use (1) Email (2) Internet access (3) INFM network access (4) All _______
    c. Do you know where your closest wireless hotspot is? YES/NO
    d. If your coverage is poor have you ever made a complaint? YES/NO
    If the answer to (d) was YES who did you raise the issue with and what was the response?

Miscellaneous

19. If you have any other comments on the Tablet PC and its use, please specify below:

Thank you for your participation

APPENDIX 2: ORIGINAL STUDENT SURVEY

Tablet PC / Wireless Campus Student Questionnaire   Date:__/__/__

General Operation

1. Do you utilise the pen driven mode of the Tablet PC? YES / NO

2. Did you know that you can work mouse-free thanks to the Tablet PC pen? YES / NO

3. Have you used handwriting recognition on your Tablet PC? YES / NO

   If you answered YES have you:

   a. Created hand written emails or just added a handwritten word or drawing to personalize your typed text e-mails? YES / NO
   b. Used OneNote 2003 to create Microsoft Outlook tasks from items in your notes, and send as email? YES / NO

4. Have you used Sticky Notes? YES / NO

   If you answered YES have you used Sticky Notes for:

   a. Writing notes down? YES / NO
   b. Creating quick gestures? YES / NO
   c. Recording quick memos? YES / NO

5. Have you used right click on your pen? YES / NO

   a. If you answered YES did you find it easy to use? YES / NO

6. Did you use pen and digital ink in any of the Office applications? YES / NO

7. Do you find the battery life of the Tablet PC satisfactory? YES / NO

8. Do you charge your Tablet PC in the University? YES / NO

Questionnaire continued on following page

If your answer was YES, do you charge your Tablet PC in any of the following locations?

a.	Lecture Theatre	YES / NO
b.	Laboratory	YES / NO
c.	Student Residences	YES / NO
d.	Ad-hoc power outlets in the University (e.g. corridors, canteen)	YES / NO

9. Would you like the use of secure lockers on campus where you could store and re-charge your Tablet PC during the day? YES / NO

   If you answered YES, would you be prepared to pay an hourly locker use fee? YES / NO

10. What other provisions or facilities would you like to see made available in relation to on-campus use of the Tablet PC? Please tick all that you feel are desirable.

a.	Recharging stations	_______
b.	Secure storage	_______
c.	Battery swap facility	_______
d.	Drop in casual help area	_______

Student Experience

11. Typically how many hours per day do you use your Tablet PC?

a.	Less than 1 hour	_______
b.	Less than 3 hours	_______
c.	Less than 5 hours	_______
d.	More than 5 hours	_______

12. Do you use your Tablet PC for any of the following **learning** activities:

a.	Lectures?	YES / NO
b.	Tutorials?	YES / NO
c.	Practical Classes?	YES / NO
d.	First Year Tutorials?	YES / NO
f.	Assignments (individual or group based)?	YES / NO

Questionnaire continued on following page

If you answered YES to any of the above, do you use it for:

a.	Annotating lecture notes during a lecture?	YES / NO
b.	Inking new notes during lectures/tutorials?	YES / NO
c.	Browsing the web during lectures/tutorials?	YES / NO
d.	Messaging?	YES / NO

If you answered NO to any of the above, do you NOT use it because:

a.	You **do not** find the Tablet PC useful?	YES / NO
b.	The teaching rooms do not support Tablet PC use?	YES /NO
c.	You find it awkward to carry around or use?	YES /NO
d.	You were never shown how to use it properly?	YES /NO
e.	There was no wireless connectivity?	YES /NO
f.	Your Tablet PC is broken or unusable?	YES / NO
g	Staff **do not** encourage it?	YES / NO
h.	Staff **do not** seem familiar in the use of the various features?	YES / NO
i.	The battery life is not long enough?	YES / NO
j.	There is no safe secure storage area provided?	YES / NO

13. Do any of your Lecturers use the Tablet PC during any of the following teaching activities:

a.	Module Lectures?	YES / NO
b.	Module Tutorials?	YES / NO
c.	First year tutorial classes?	YES / NO

14. If you answered YES to any part of question 13 do you feel that:

a.	The Lecturer was familiar with the use of the Tablet PC?	YES / NO
b.	The Lecturer made good use of the Tablet PC?	YES / NO

15. Do you think having the Tablet PC has improved your learning experience at the University? YES / NO

Questionnaire continued on following page

If you answered YES, can you list some examples where you feel it has helped you as a student and in particular in relation to your studies? Please tick all that apply:

a. It encourages me to attend ________
b. It encourages me to study ________
c. It helps me access online resources for my assignments ________
d. It allows me to collaborate with my fellow students ________
e. I find it easier to understand computing related concepts ________
f. I do not have to wait for a desktop computer to become available in the lab ________
g. I find it helps me manage the storage of all my work ________
h. I can easily write notes during lectures and change later ________

NETWORK USE

16. Do you use the Tablet PC wireless connection to logon to the INFM network? YES/NO

17. Have you used your Tablet PCs wireless capability at home? YES/NO

18. How do you rate the campus wireless network in terms of:

a. Coverage

Poor	Good	Very good	Excellent

b. Reliability of connection

Poor	Good	Very good	Excellent

c. Ease of Use (configuration/login etc)

Poor	Good	Very good	Excellent

d. Technical support for the wireless network

Poor	Good	Very good	Excellent

Questionnaire continued on following page

19. Do you think having wireless access to the Internet has improved your learning experience at the University? YES / NO

    If your answer is YES, can you list some examples where you feel it has helped you as a student. Please tick all that apply:

    a. Access to online resources from different locations _______
    b. Email fellow students in relation to coursework and group activities _______
    c. You feel that you are always contactable wherever you are on campus via email _______
    d. You can access resources outside of the traditional lab environment _______
    e. You don't have to access lecture notes prior to attending the lecture hall – they can be downloaded when you arrive at the lecture _______
    f. The Tablet PC saves you from printing lecture notes as you can read them from the Tablet PC in the lecture hall _______

20. If the Tablet PC was taken from you tomorrow would it have a negative effect on your studies? YES / NO

    If you answered YES, outline how it would affect your studies

    a. No access to PC resources _______
    b. No online access _______
    c. Others (comment) ____

21. Have you used the Tablet PC for collaborative working with other students? YES / NO

    If your answer was YES, how did you use the Tablet PC and was it useful?

    a. Completing coursework and assignments _______
    b. Sharing resources found online _______
    c. Emailing fellow students or members of staff _______

Support Services

22. Are you aware of the Tablet PCs technical support website? YES / NO

    If your answer was YES and you used the site, did you find the Tablet PC support website useful in relation to the Tablet PC? YES / NO

Questionnaire continued on following page

23. If you have any comments on the site or on how it could be improved please make them here. Please tick all that apply.

    a. More technical information is needed ________
    b. Less technical information is needed ________
    c. More basic help is needed ________
    d. Less basic help is needed ________
    e. A user forum where students can post comments and upload new links ref software downloads ________

24. Are you aware of the Tablet PCs technical support services provided by computer officers and technicians? YES / NO

25. Approximately how many times have you used the University technical support services in relation to issues you have had with your Tablet PC? Please tick whichever one applies:

    a. Never ________
    b. Less than 5 times ________
    c. Less than 10 times ________
    d. More than 10 times ________

26. If you have used the technical support services, have your Tablet PC problems been hardware or software related or both? Please tick all applicable

    a. Hardware ________
    b. Software ________
    c. Both hardware and software ________

27. Were the problem(s) resolved? YES / NO

28. How would you rate the Tablet PC support service?

Poor	Good	Very good	Excellent

Tablet PC Insurance

29. Do you feel the Tablet PC insurance is necessary and would you purchase it if it was not compulsory? YES / NO

30. Do you think the £100 cost is reasonable? YES / NO

Questionnaire continued on following page

Induction Session

31. Did you attend the induction session for the Tablet PC? YES / NO

If your answer was YES, did you find it useful? YES / NO

32. How do you think it could be improved? Please tick all that apply:

a. Make the duration of the session longer _______
b. Make the duration of the session shorter _______
c. Smaller induction groups allowing for more question to be answered _______
d. Make the induction more technical _______
e. Make the induction less technical _______
f. Have more help sessions _______
g. No induction required—easy to use _______

MF124/5 Classroom

33. Have you had any classes in the MF124/5 laboratory? YES / NO

34. If your answer to question 28 was YES, do you think the facilities in MF124/5 have improved your learning experience at the University? YES / NO

If your answer was YES, can you list some examples where you feel it has helped you as a student and in particular in relation to your studies? Please tick all that apply.

a. The clustered layout of the room encourages small group working _______
b. The visual facilities in the room helps me follow the Lecturer better _______
c. I can hear the Lecturer more clearly _______
d. The software available on the desktops allow me to communicate with the lecture more easily _______
e. I feel more comfortable in the lab environment _______
f. The temperature in the room is adequate during lecture/laboratory sessions _______

Questionnaire continued on following page

35. Do you have any general comments or feedback on the Tablet PC, wireless campus or MF124/5.Can you think how the facilities have helped you as a student or things you would like to see improved?

Thank you for taking the time to complete the questionnaire

Chapter XVI
Using Information Communication Technology to Decentralize City Marketing: Challenges and Opportunities

Bantu L. Morolong
University of Botswana, South Africa

ABSTRACT

This chapter introduces the reader to the idea of city marketing. This idea has developed over time, globally, as cities continue to grow rapidly. The chapter assesses how city marketing as a concept and a practice has generally developed with particular reference to Botswana, Southern Africa. Using evidence from documented material this chapter critically reviews city marketing as perceived and applied by planners and policy makers. The chapter looks at how Information Communication Technologies (ICTs) can be used to market cities such as Gaborone, the capital city of Botswana. Literature and experiences from other parts of the world are used to underscore city growth as a serious development issue. City dwellers are identified as key stakeholders in city marketing because they have a more permanent association with the city in contrast to those who come into the city to visit or for business, tourism, and other purposes.

INTRODUCTION

By focusing on the locals therefore, the chapter upholds the need to decentralize city planning and management arguing that participation by locals is central to city marketing processes and that effective ICT use could facilitate this. Views around what could contribute to development of sustainable cities in the context of the developing world using Botswana as a point of reference are advanced. Ideas that this chapter generates are expected to serve as useful springboards for research and analyses of the practice of city marketing in Africa and other parts of the developing world. The chapter serves as a contribution to the promotion of the concept of city marketing as it continues to gain currency with special interest in the role of ICT in urban development.

The chapter observes that the idea and practice of city marketing does not seem to be fully developed in the context of the developing world. Yet as a practice it seems to hold great potential for positive influences on the development of cities and their sustainability, that is, if it is applied early enough in the lives of cities. Exploitation of this potential seems to be particularly possible in these regions of the world where most of the current cities seem to be relatively new cities. In this context cities which are relatively young such as Gaborone might not need to reinvent the wheel.

As cities grow at an alarmingly high rate they make city growth a daunting development phenomenon. In the information era however, growing cities have a strong fall back in the form of ICTs to face the challenges of city growth which involves a need for communication about the processes and their impacts on people. The thesis of this chapter is that there is limited awareness about the importance of city marketing for systematic city growth in this context. It is argued that this lack of awareness has significantly influenced the ways in which cities are viewed, planned and utilized. This also seems to have impacted the extent to which participation in city development and the way the benefits of city development processes are distributed among the members of society with a special concern about the locals. This chapter therefore highlights the dynamics of city growth in the context of national development and the connections between that and the position and role of cities in the global system. It serves to fill the gaps in information about the magnitude of urban growth as both a desirable development and at the same time a frightening prospect if not well integrated into the broader development landscape. The chapter points to the existing paucity of data about trends of city growth due to the erratic attention given to this and challenges planners and researchers to intensify efforts in this area. Without such data which the chapter suggests could be effectively generated through effective application of ICTs in city planning and city marketing, the rapidly growing cities will not achieve any significant levels of sustainability. Posing a set of questions about sustainability of the new cities such as Gaborone, the chapter characterizes a sustainable city as one that is inclusive in its planning, marketing and management processes. Such a city builds its growth on the sound principles of decentralized decision making and democratic governance. It uses Information Communication Technologies to be proactive and also responsive to some of the biggest challenges of this era which are poverty, diseases, unbridled depletion of the environment, inequity in the sharing of the benefits of development, globalization and of course the biggest ever challenge to face humanity which is climate change (C 40 Climate Summit, New York, 2007)

This chapter briefly introduces the idea of city marketing against the background of the unprecedented growth of cities. The chapter outlines the challenges that come with this growth particularly in the context of the developing world. Statistical evidence which is described as frightening is used to demonstrate the magnitude of this problem and a pertinent question is asked about awareness, preparedness and concern by most countries in this context about this phenomenon. Focusing on city development in Africa, a brief contrast is made between old and new cities pointing to the fact that very much against expectations, growth in the new cities is not accompanied by any significant rates of economic growth and increases in the quality of life. Evidence for this is the failure by these cities to meet the four principles of livability, good governance, bankability, and competitiveness in the global systems. Instead, new cities are showing signs of lack in sustainability. Using Gaborone City as a unit of analysis the chapter points to implications of uncontrolled city growth, and inefficiently planned and managed cities. In this context, the chapter appraises some of the attempts of democratic city governance which holds potential for sustainable cities. Following a brief critical analysis of these attempts, the chapter is concluded by pointing to the potential role of Information Communication technologies in the marketing of Gaborone city. ICTs are presented as a tool for inclusive city planning and management in Botswana where it is observed that the ICT development landscape is significantly being developed. With the use of ICTs, enough and current data on the needs, aspirations and dynamics of the city's growth could be generated. And this could serve as a useful basis for effective planning and management of a sustainable city.

CITY GROWTH AS AN ISSUE IN CURRENT DEVELOPMENT DISCOURSE

Literature surveyed for this chapter shows that cities have moved to the forefront of global socio-economic change, mainly because half of the world's population is now reported to be living in urban areas. According to UN projections, cities and urban areas are gaining an estimated 60 million people per year - over 1 million every week. In many developing countries cities are growing two or three times faster than the overall population. As urban areas - particularly smaller towns and cities - continue to grow in size, about 5 billion people are expected to live in cities by 2030. This is about 61 per cent of the global population of 8.1 billion. The other half of the world's population is said to be increasingly dependent upon cities for economic, social and political progress. Factors such as globalization and democratization of governance processes have also increased the importance of cities which serve as hubs of most national development activities. Evidently, the phenomenon of city growth is also increasingly becoming a complex challenge to societies across the globe

The World Bank is one of the key players in the quest for solutions to some of the challenges that the rapid growth of cities brings in its train and the broad issues of social development, of which cities and their growth is one. Other authors point to the fact that city growth itself can be a daunting challenge that needs to be effectively managed because it has implications for political and social structural processes, resources , environmental factors, cultural dynamics, production arrangements, policies and for the lives of those who inhabit them (Rakodi, 1997; Hicks, 1998; Helten & Fischer, 200; Maundeni, 2004). A strongly held view by the bank is that as the world becomes more urban and governments are decentralizing responsibility and authority for governance there is also an increasing need to closely work with cities and municipalities to address urban development. The bank has as a result declared this as its urgent priority through its corporate strategy for urban development (World Bank, 2006).

Anchoring its discussion on how cities and towns could respond to the needs, concerns and welfare of their residents and businesses, the World Bank through its urban development strategy suggests that; livability, competitiveness, good governance and management, and bankability are the key pillars of city sustainability. In looking at city growth in the context of the developing world, therefore, there are fundamental questions with regard to the extent to which these pillars are recognized in those contexts .This chapter determines the relevance of ICT in achieving these pillars.

The Magnitude of Global Urban Expansion

Globally, the annual urban growth rate is projected at 1.8% in contrast to the rural growth rate of 0.1%. Further it is predicted that growth will be even more rapid in the urban areas of less developed regions, averaging 2.3% per year, with a doubling time of 30. Simply put, cities and urban areas are said to be gaining an estimated 60million people per year or over one million every week. .This makes urban growth in the form of cities one of the most important development issues in this era, with far reaching implications. This also seems to be much more of an important issue for the developing world than it is for the developed world where the rates of urban growth seem to be declining and so are the challenges that urbanization brings with it (UNEP and UNESCO Facts and Figures, 2006)

The key dimensions of the policy debate on the expansion of cities are certainly not new. The age-old question underlying this debate is still whether expansion should be resisted, accepted, or welcomed. At one extreme, there are those who have fought to limit the growth of cities by any and all means. At the other, there were those who welcomed it and actively prepare cities for absorbing the oncoming waves of new migrants (Shlomo et al., 2005). Evidently there are different views held by development analysts and practitioners of different persuasions about urbanization and city growth. UNESCO further notes that, it is generally accepted that cities not only pose potential threats to sustainable development but also hold promising opportunities for social and economic advancement and for environmental improvements at local, national, and global levels. (UNEP/UNESCO, 2006)

URBANIZATION AND CITY GROWTH IN THE AFRICAN CONTEXT

In the last twenty years urbanization has become one of the most important trends in human settlement development. The prospects of equitable development in all developing countries continue to be deeply affected by rapid urban population growth (Mosha, 2004). Africa's urban development is closely connected with this continent's political history of colonization. However, some African historians report on cities as old as between 700 AD and the 1600th century pointing to some cities which came much earlier even though then they were only known as kingdoms, chiefdoms and empires. Rakodi (1997) however, notes that some cities and kingdoms were ephemeral, linked to the reign of a particular ruler; while others, such as the Yoruba towns of south-west Nigeria, prospered through many regimes and continue to be important today (Mabogunje, 1968 cited by Rakodi, 1997). It is further noted that during the colonial period however, cities developed not as industrial centers, but to facilitate the extraction of commodities and the politico-administrative system on which this depended (p. 21)

New Cities

Many new cities were established in the post-colonial period, but not for the same reasons as in the colonial period. The new cities were meant to give the nations a 'fresh start and to be the beginning of a new golden future (Stock, 1995 as cited by Rakodi, 1998). For those with a history of liberation movements they seem to have been part of what was promised by the liberation politicians. And therefore cities served to induce nationhood and as centers from which countries could source their national identity as they are seats of national governments' power and authority (p. 5). This is the context within which the growth of Gaborone as discussed in this chapter should be understood. As a result of the somewhat divergent views about the history of city development in the African context it seems most appropriate in this chapter to refer to new and old cities. The former being the post colonial cities and the latter being for many African countries the post independence towns most of which have grown to become cities.

It is on record that since gaining independence, as new African countries actively planned their economies, governments also took full responsibility to develop the countries by building infrastructure and developing social services; developing industry and creating employment. Most of these investments were taking place in the cities through centralized planning approaches. And for many, centralization seemed to symbolize a strong and, powerful state. Centralization of the nature described here also meant that cities became hubs of business and international interaction as investors found it most practical and convenient to establish themselves in the growing capitals. These are closest to power, better infrastructural services, the legal system and other amenities which are needed for effective engagement in business (Rakodi, 1997; Mosha, 2004; Maundeni, 2004; World Bank, 2006; UNEP/ UNESCO, 2006; Wong, Millar & Choi, 2006).

Rates and Implications of African City Growth

The Africa region is experiencing the fastest rate of urbanization in the world in spite of the fact that it is still a predominantly rural continent with only 37.3% of its population living in urban areas in 1999 (Hicks, 1998; UNEP/ UNESCO, 2006). However, it is further noted that urbanization is occurring in this context without generating the resources (public or private) and employment opportunities to accommodate the surge in urban population. Today, up to two thirds of African urban dwellers live in informal settlements with inadequate transport, water, sanitation facilities, electricity, and health services. Housing finance systems are said to be non-existent or limited to upper income sectors. Crime is a major problem in many African cities, and worsening with increased poverty and deteriorating living conditions. Hicks suggests that as more and more African countries achieve some success in macro fundamentals, it is urgent to identify concomitant micro policies and measures that promote growth and sustainable poverty reduction. Thus, as heavy investments are made on city growth, the legitimate concern should be about balancing the benefits between the urban and rural dwellers and even across different sectors of the population in the urban centers (Mosha, 2004; Maundeni, 2004; World Bank, 2006; UNEP/ UNESCO, 2006)

In an analysis of the fiscal situation of Gaborone city, Mosha (2004) concludes that contrary to the common trend in the African context the financial strength of Gaborone City Council (GCC) is fairly sound, in that it nearly meets most of its needs. This is achieved either through its own resources or from the central government which provides most of the city's recurrent and capital funds. This is in contrast to a number of municipalities in various parts of Africa, which are reeling under heavy debt burdens.

Urban Poverty

Urban poverty is one of the major consequences of unplanned and uncontrolled city growth. According to the World Bank (2006), due to the dynamics of rapid urban growth, the nature of urban poverty is more than an income or employment issue. It is also characterized by squalid living conditions; risks to life and health from poor sanitation, air pollution, crime and violence, traffic accidents, natural disasters; and the breakdown of traditional familial and communal safety nets. Urban populations are also particularly hard hit by macro-financial shocks, such as the recent crises in East Asia and Russia. Urban environmental degradation has the most immediate effects on poor urban residents but also has serious national and global impacts (p. 36) In the case of Gaborone and other fast growing cities, the concept of urban dwellers needs to be unpacked to answer the question of whose benefit is urban growth and city marketing.

Furthermore, it is increasingly becoming evident that, unless economic growth accelerates substantially, there will be insufficient resources to fund the backlog of investments in cities, let alone to meet future requirements. In the financially constrained economies it is also evident that the already overstretched central governments budgets are unlikely to carry the added burden of supporting the increased demands for services and business investments in the cities. This may render cites which are faced by these challenges virtually dysfunctional. While different authors emphasize different root causes of this urbanization without growth, many agree that the causes are also unclear. They also note that there is no clear explanation of why cities have much less economic dynamism in Africa than elsewhere. This lack of economic dynamism in the urban centers makes relevant in this chapter, the concept of sustainable cities which is briefly discussed in the next section. From this presentation, it is clear that if issues of urban growth are not given enough attention in the overall development discourse of the developing world, this could have significant, economic, social structural and environmental impacts. The most adversely affected by these are the poor and already vulnerable groups who come to cities with hopes to gain in some what cities offer.

The Rapid Growth of Cities as a Development Concern

Cities are essentially hubs of economic activity but they are at the same time in this era of globalization, national and regional engines of socio economic growth. It is an adequately documented fact that this growth has far reaching implications such as urban poverty, crime, overcrowding, unemployment, poor social services, landlessness, ineffective city management and challenges of good governance. Cities as entities also need a positive image through capacity to respond to the demands of their inhabitants and those of the business and industrial sector that they serve. While this chapter does not attempt to fully answer these questions, it teases out areas of concern around city growth with particular reference to the relevance of the potential use of ICT to address those concerns (Brockerhoff, 2000)

Cities are also centers of technological and cultural creativity, homes of the poor and deprived, and sites and sources of environmental pollution". And yet there is a generally held belief by those who live in the remote areas that life in the cities is comfortable and the most desirable (Fuchs, 1994, cited by Rakodi, 1997). Hence the high rates of migration into the cities. And this increases the need for cities to be effectively planned and managed to accommodate these expectations. Further Rakodi (1997) raises a concern about the fact that in the context of the developing world, understanding of the socioeconomic and political dynamics of cities and the urban systems of which they are part is limited. And so is capacity to manage cities effectively. Unprecedented, often unplanned and seemingly unanticipated city growth leaves the stakeholders with many burdens and many unanswered questions.

Beyond concerns about the extent to which cities as they grow are able to cope or fail to cope with the demands of their inhabitants, for many authors the biggest challenge for cities and city planners lies with being able to integrate the fast growing cities into the global economy. One of the most pertinent questions which are asked by this chapter is how much of the city growth processes in the context of the developing world are taking this issue into account. This question makes relevant for this chapter the issue of sustainable cities.

SUSTAINABILITY OF CITIES: THE CASE OF AFRICA

Concern about sustainability of cities centre around ensuring that the current changes in the ways cities are managed, built, and lived in tips the balance towards a healthier and more peaceful urban future (State of the World 2007: Our Urban Future). In the foreword to this report *Anna Tibaijuka*, Executive Director, UN-HABITAT observed that;

Though cities are important engines of growth and provide economies of scale in the provision of services, most of them are environmentally unsustainable. In addition, in this age of increasing insecurity, with more than 50 percent of their residents living in slums without adequate shelter or basic services, many cities are rapidly becoming socially unsustainable

Largely, issues of great concern about urbanization include its impact of human settlements, chaotic cities surrounded by slums and squatter settlements, marginalization of the bulk of the world population. However, on a positive note the director said that many of the targets of the Millennium Development Goals address the living conditions of the urban poor, in particular. Targets 9 and 11 within Goal 7 on environmental sustainability are particularly relevant as indicators that urbanization is being taken increasingly seriously. For Africa she said even though the promise of independence has given way to the harsh realities of urban living mainly because these countries were ill prepared for their urban future, the leaders are now taking note.

Recent analyses of the economy of African urban areas have shown that they are characterized by fragmentation of their private sector: a very "extroverted" modern sector with limited linkages with the local economy. The fact that the urban informal sector plays a key role in the local economy is acknowledged as well as the fact that it primarily caters to its own needs and demands in the areas of food, housing, transport to mention but a few, while the small and medium enterprise sector is underdeveloped (World Bank, 2006). The World Bank's Urban and Local Government Strategy proposes an agenda for working with both national and local governments to develop sustainable cities, which are first and foremost livable, competitive, well governed and managed and are bankable. Of the four principles of sustainable cities, this chapter chooses to emphasize the issue of a well managed and well governed city saying that the role of ICT in this regard would be most strategic. The chapter also focuses on city governance viewing it as central to effective city management and city sustainability. This focus on the human face of city growth is a deliberate departure from the World Bank's and other authors concerns as mentioned elsewhere in the chapter who give particular attention to economic aspects of cities. (Retrieved June 20, from 2008 http://www.worldwatch.org/node/4853)

City Governance as a Factor in the Sustainability of Cities

In its broad sense the concept of governance is simple and complex at the same time. It is used in the context of organizations to mean the processes through which these systems facilitate the meeting of their goals in a patterned fashion. As a process governance may operate in an organization of any size from a single human being to all of humanity, and to nation states, and corporations engaged in purposeful activities. The World Bank defines Governance as the exercise of political authority and the use of institutional resources to manage society's problems or affairs and observes that it is often used interchangeably with government. From a mainly political perspective, governance has been defined as the rule of the political to solve conflicts between actors and adopt a decision. For others it is used to describe the proper functioning of institutions and their acceptance by the public. This concept becomes even more elusive when qualified as good governance. Good performance of government institutions is pivotal to good governance. It is usually reflective of the workable relationships between government institutions business industry which are comprised of city dwellers, their representatives, community organizations and local and non-governmental organizations (Discover Botswana, 2000).

It is apparent that discussions of 'good governance' in general, have gained currency among researchers, political scientists and in the agenda of international development agencies such as the World Bank. This seems a rather belated recognition that the quality of service offered by organizations depends heavily on the way they are governed. Very unfortunately discussions of good governance all too often remain at the national level, perhaps because international agencies tend to deal directly with national governments. The importance of good local governance is usually forgotten. And yet the success of local institutions such as cities also depends on good local governance. With this note a focus is now made on Gaborone the capital city of Botswana with a purpose to analyze how the ideas as presented thus far about city growth and sustainability could be applied in this context

Botswana

Botswana, a country in Southern Africa adopted its new name upon independence in 1966 to become a parliamentary republic. Four decades of uninterrupted civilian leadership, progressive social policies, and significant capital investment have created in Botswana, one of the most dynamic economies in Africa. Mineral extraction, principally diamond mining, dominates economic activity. However, tourism is a growing sector due to the country's conservation practices and extensive nature reserves. Botswana has one of the world's highest known rates of HIV/AIDS infection. Even though this threatens its very small population of a little over one million and which occupies a very vast land area with rich cultural diversity, Botswana also has one of Africa's most progressive and comprehensive programs for dealing with this disease (Investors Guide to Botswana, 2007)

Botswana is reported to have maintained one of the world's highest economic growth rates since independence in 1966. Through fiscal discipline and sound management, Botswana has transformed itself from one of the poorest countries in the world to a middle-income country with a per capita GDP of more than $11,000 in 2006. Its economy is described as stable and so is its political climate making it one of the most desirable foreign investment climates in the continent (Investors Guide to Botswana, 2007, Jenman, Africa Safaris, 2007). While the government of Botswana adopts a policy of zero tolerance to corruption, it has flexible and generous investment incentives for business industry manufacturing, tourism, the knowledge economy, financial services and trading for domestic and foreign investors. On the downside, the government of Botswana must deal with high rates of unemployment and poverty. Unemployment officially was 23.8% in 2004, but unofficial estimates place it closer to 40%. HIV/AIDS infection rates are the second highest in the world and threaten Botswana's impressive economic gains. An expected leveling off in diamond mining production overshadows long-term prospects. (Investors, Guide, 2007). This is the country whose capital city Gaborone is a unit of analysis for this chapter.

Gaborone

Gaborone is the capital city of Botswana. In 2005, the estimated population of Gaborone was 208,411. Gaborone is a post independence city, a new city, with the same age as the country which is 43 years old this year. Therefore this city is part of the achievements of independence. It was a result of the relocation of the capital town from Mafikeng which is more than hundred kilometers away from Gaborone and is now part of the Republic of South Africa. This had been an odd administrative arrangement that dated back to the early colonial period. When Botswana which was a British protectorate became independent in 1966, it needed a capital city within its territory. The centre of the city was constructed in three years, and the basic infrastructure was in place for Independence Day on 30 September 1966, when Bechuanaland was the eleventh British colony in Africa to become independent. Gabs" is still a common abbreviation for Gaborone, sometimes used in casual speech while a newer nickname for Gaborone is 'GC' or 'Gabz (Wikipedia, the free encyclopedia, 2008)

Gaborone is located in the far south of the country and it has a fascinating history having its roots in a village setting in the South East of the country. By far, this has been the most important of the five towns in Botswana because it is the capital administrative centre which has also only recently (1986) been elevated to the status of city, with the other smaller towns having mainly specialized functions Gaborone is reported to have been the least populous urban settlement in the country at the time it was designated the capital of Botswana. Even though Gaborone is said to have grown in leaps and bounds over the last forty years, by international standards it is still a very small city with a population of close to 300, 000 inhabitants but which has grown rapidly from a population of 3855 in 1963, to 18 799 in 1971, 59 657 in 1981, 133 468 in 1991 and186 007 in 2001 (Discover–Botswana, 2006).

The enormous population increase in Gaborone has resulted in the physical expansion of the city to 169 square kilometers, eating up most of the developed land. Furthermore, the growth of Gaborone has always been characterized by population growth beyond projections. This has led to the city swallowing a significant portion of the land and the cultural and economic make-up of the outlying villages and incorporating some of these within its borders (Maundeni, 2004; Mosha, 2004). Therefore by description, Gaborone is a village city. This means that this city is in touch with its local rural roots. This is accentuated by the fact that many of Gaborone's dwellers also have a close connection with their rural origins and a good number of them only come to work in Gaborone and commute between the city and their home villages where they permanently reside.

The broader national context of Gaborone is a predominantly rural environment with the population sparsely distributed over a large land area that includes the Kalahari Desert. This is a big veritable wilderness in the heart of the African Continent and whose wide horizons reveal Africa in all its Majesty. The Botswana wild life is one of the biggest and the richest in Africa. Botswana does everything to conserve its diverse ecological system which includes the sun baked Makgadikgadi saltpan, waterways, swamps, lagoons reed plains and the largest inland delta in the world, the Okavango which is said to feature some of the best wild life viewing anywhere in Africa (Discover Botswana, 2006),

GABORONE CITY PLANNING, MANAGEMENT: THE ISSUE OF LOCAL PARTICIPATION

Fundamentally, cities are products of plans which also serve as frameworks for city development which in turn is based on policies set up by established structures of national and local governance such as city Councils and municipalities. City planning is one of the services that are offered by the local authority. Cities as part of the local government system are increasingly becoming important. This fact effectively justifies the general call to which this chapter subscribes, for renewed approaches and a different set of objectives to city planning. Some of the renewed approaches are covered by for example, the World Bank's new assistance strategy which is geared to serving national, as well as corporate objectives of cities. These objectives according to the bank include reducing poverty, promoting broad-based growth, protecting the environment, and supporting effective institutions. In this strategy the bank underlines the fact that in order to achieve these objectives there is need for a closer working relationship with the local government structures. These structures assume greater responsibility for delivering services that have direct impacts on people's daily lives. They do this in partnership with national governments and the communities themselves as demonstrated below for Gaborone city.

According to literature on participatory development planning, public participation at grassroots level is at the core of sustainable development in general. In particular Keiner (2003) posits that it is an indispensable part of urban development. He further notes that any measures aimed at achieving sustainable urban development in Africa and Latin America imply prerequisites on the national level, such as decentralization, democratization, and "good governance", encouragement of community participation and integration of women. For Botswana it is a generally shared view that the government recognizes the importance of involving its people in the planning process by using participatory planning and consultation structures. As a result of this, there are officially defined, structures at all levels of society.

Participatory Planning Structures in Gaborone

Some of the structures are more permanent and are made up of elected representatives while others are formed on an ad hoc or voluntary basis. Formal structures which include Village Development Committees (VDCs), Council Committees and the City Council itself, Land Boards, Tribal Administration, District Administration and Parliament are already taking part in the planning process. Perhaps most important amongst these is the kgotla which is a traditional instrument for articulating public and individual interests in Botswana. Its functions are similar to those of a local parliament. The Kgotla simply defined is a meeting place in a ward. Wards are subsets of Councils political areas, and there are 26 wards in Gaborone. The Kgotla is a legitimate institution for public decision-making. Once a decision has been made at the Kgotla, it is considered binding for the entire community. The Kgotla holds public meetings for development activities concerning the city, in addition to its customary court functions. It is also an important forum that brings urban development issues to the attention of the government. It is the starting point for bottom up development and forms a basis for involving the people in decision-making processes (Keiner, 2003, p. 27).

An Assessment of Gaborone's Planning and Management Processes

In July, 2007 a workshop was held in Gaborone under the theme: "Making Gaborone a livable, safe, functional and economically sustainable city." This workshop was part of an ongoing consultation exercise on the review of the city plan for the period 1997-2012. The workshop was jointly hosted by the Gaborone City Council (GCC) and consultants from Planted Africa and Mosienyane & Partners International. Officiating at this workshop the Mayor of Gaborone said that the Gaborone Central Business District (CBD) is not attracting the right type of development initiatives. The mayor said, consequently, there is need to assess reasons for the slow development of the area and to come up with ways of attracting the right developers. The mayor also emphasized the need to revitalize the Gaborone Main Mall and to make it look attractive even after normal business hours (Botswana Daily News, July 24th, 2007)

At the same workshop Mr. Ike Iloanya an official of Planted Africa underlined the general perception that Gaborone lacks a local character and a defining image of a city that is welcoming. This seems to corroborate the view which had been expressed by the mayor on the previous day who said that Gaborone currently acts only as a stopover for tourists flying to the Okavango Delta which is the centre of tourism in the northern part of the country. Mr. Iloanya further said Gaborone did not satisfy the livability, safety, and functionality criteria from an urban planning and design standpoint. He said these shortfalls needed to be addressed by the revised plan. Thus, he said, one of the tasks of the revised plan is to enhance

Gaborone's comparative advantage and competitive edge over other African cities. It could be observed that even internally the city might need to increase this edge over other parts of the country which feature more than it does in the media and tend to overshadow its image in the eyes of both the locals and visitors to Botswana.

Towns and cities are part of a constantly evolving relationship between people, land, culture and the wider environment because as it is observed by most authors, good connections in cities enhance choice, support social cohesion, make places lively and safe and facilitate contact among people. The current context of Gaborone according to Mosienyane (2007) does not consider the impact on the health of the population which lives and works in the city. It also does not observe the social, cultural and economic context as well as physical elements and relationships (p. 4).

These views evidently suggest that the current city plan should use and include collaborative approaches to the design of Gaborone. Through inclusive planning multiple perspectives that those involved can bring could enrich the processes and make them more holistic. This would also facilitate the making of meaningful contributions by the various city communities in decision making through good communication and coordinated actions. These role players could include decision-makers from central and local government, professionals, transport operators, developers and users. This would be democratic and participatory approaches to the city in its planning and these approaches could easily spill over to other aspects of the city's life. The sentiment of participatory city planning is shared by authors who subscribe to the idea of sustainable cities chief among whom is the World Bank through its Urban Assistance Programme which is guided by four main principles of livability, competitiveness, well governed, and managed cities and bankability. While this chapter upholds all of these principles in the context of Gaborone city, it also aims to give special attention to issues of democracy and participation of locals in the life of the city through the use of ICTs.

Contrary to the view by Keiner (2003) about the process of city planning in Gaborone being democratic, in his mapping exercise of democracy in Gaborone, Maundeni (2004) reports that, Gaborone right from the outset was primarily the seat of government that subsequently attracted business activities. It was unlike some cities in Southern Africa neither a mining, nor, an agricultural place. In his discussion of institutional arrangements in Gaborone and their implications for local democracy, Maundeni (2004) says that official records therefore, show extensive central government involvement in the running of Gaborone City. As a result, the city council is not responsive to the needs of the city dwellers because there is very little interaction between them and the councilors. Furthermore, this author observes that there are very few mechanisms through which city dwellers can present their complains, except to attend meetings addressed by the Minister of local Government where they complain about council's delayed or failed delivery of services (p. 37). The institutions that the city would use to push the agenda of participatory planning are varied and very crucial. However, in the case of the city of Gaborone, Maundeni (2004) observes that, most of the structures that are employed for this purpose are closely associated with party politics and most specifically with the ruling party and therefore with the government.

Maundeni concludes by saying that the scenario that he describes has potential to be divisive. And this seems to be the point where ICTs could come in to be to serve a unifying role. There is evidence that the media has served in other parts of the world to sell the brand of cities and in the promotion of their image to the internal and external users of the city (Kavaratzis, 2007).These experiences show that where there is awareness of the potential to use ICTs and opportunities were opened up to do so, some of the concerns city dwellers have about the city and its management were effectively communicated through those tools. Observations are that with many types of media which are now technology driven, people tend to be upfront with their views thereby effectively contributing to the life of the city whether in its planning or management.

The Benefits of Decentralized Democratic City Planning and Management

City governments that have involved their inhabitants are reported as having found themselves benefiting immensely from the experiences that those most associated with the city can bring to the table for discussion and implementation (Ashworth & Voogd, 1990). Those who live in the city are expected to be more able to be resourceful in terms of the type of information that they may have about the city, its physical and social history, problems and challenges and trends of thought about its future. It seems in this case that the basic principles of participatory development such as ownership of development processes demand driven needs based programming and decentralized trickle up of ideas can be applied to city development. The institutions that the city would use to push this agenda are varied and very crucial.

However, in the case of the city of Gaborone, Maundeni (2004) observes that most of the structures that are employed in the planning and management of this city are closely associated with party politics and most specifically with the ruling party and therefore with the government. This has a potential to be divisive and the ICTs could be used to serve a unifying role. For example the media has served in other parts of the world to sell the brand of cities and in the promotion of their image to the internal and external users of the city and what it has to offer. City populations (and businesses) depend on local institutions for most forms of infrastructure and services. For city dwellers the basic principles of their involvement in city processes is not different from local and decentralized government as has been promoted as one of the fundamental features of democratic, efficient and sustainable development. Therefore, in any city, good governance should be about inclusive decision making processes on, service provision , policy changes , physical developments, so that city dwellers can assume responsibility and their proper stakeholder role in the life of their city.

One of the most important aspects of the city in which its dwellers could be effectively involved is its marketing. There is evidence which suggests that the marketing processes of every city are central to the city's life right from its inception throughout all the stages of its growth. It is argued in this chapter, that this is most likely going to contribute to the city being better understood and appreciated and in turn to its offering a balanced set of benefits to all of its dwellers. Before a look is made at how city marketing is relevant for Gaborone, a brief conceptual framework of city marketing is given for the benefit of readers who might not have thought about it in this context (Ashworth & Voogd, 1990).

CITY MARKETING: A TOOL FOR INCLUSIVE CITY GROWTH

City marketing as a phrase generated about 850,000 Google hits in August 2006 (Kavaratzis, 2007) and lacks a clear and unanimous set of definitions of its components. As an idea it conjures up images of earnest applications of marketing strategies and principles to cities around the world. As a field of study Kavaratzis (2007) observes that city marketing has grown over the years into an established sub discipline and an area of research, in spite of its underdeveloped state in terms of theory and application (p. 695). It is also said to have attracted interest of many academics and commentators from various disciplines resulting in a substantial body of publications on it as a process (Ashworth & Voogd, 1990).

According to Kavaratzis and Ashworth, (2006), city marketing is a complex process that requires planning with communication as a central feature if this is to succeed. In this information era therefore, ICT use seems the most relevant pivot around which city marketing should revolve. City marketing has also been employed as a response to economic, political and social changes in cities resulting in its accelerated use as cities position themselves strongly against fierce competition.

In many contexts, city marketing seems to be a tool and a means to other ends. One of the most important of these ends is promotion of the key philosophy of a city and its image. For all of the functions that it serves, City marketing requires careful and systematic planning for positive results (Ashworth & Voogd, 1990, Kotler, 1999 cited in Kavaratzis & Ashworth, 2006). Kavaratzis (2004) and Short and Kim (1999, cited in Kavaratzis, 2007) note that city marketing has developed through distinct and discreet phases, overtime.

Cities are an integral part of the urban local government system. According to Kellogg and Lillquist (1999), historically, local government and other providers of professional services did not think it necessary to get close to the customer through the use of sophisticated communication and marketing skills and techniques. However, these authors observe that the environment has changed and it is much more competitive. There is also increasing evidence that city marketing and strong communication links between cities and their internal and external stakeholders contribute to an improved city climate in terms of business and the social aspects of a city such as its image and its civic pride, hence the need for city marketing (p. 17). This in the view of Kellogg and Lillquist (1999) is not about selling the city but it is a process that uses communication to discover citizens' needs and build continuing relationships for the benefit of the city dwellers (Preface).

Writing on Globalization and Africa and the challenge of urban growth, (Rakodi, 1997) said that it is increasingly becoming true that the planet's future is an urban one and that the largest and fastest-growing cities are primarily in developing countries. She goes on to observe that even in Africa, which for a long time was regarded as one of the least urbanized continents, it is projected by the UN that over half of Africa's population will be urban by 2020 (UN, 1993).The largest cities "serve simultaneously as national and regional engines of economic growth, centers of technological and cultural creativity, homes of the poor and deprived, and the sites and sources of environmental pollution" (Fuchs, 1994, cited by Rakodi, 1997). Therefore, it should be unsettling that there is still limited understanding of the dynamics of these cities and the urban systems of which they form part, and our capacities to manage them effectively, also so limited (p. 2).

The central thesis of this chapter is that cities in most parts of the developing world are marketed in a centralized fashion, to an extent that the local people do not know their cities and what they can offer and vice versa. It is posited in this chapter that ICTs can serve to move the focus of city marketing from emphasis on the economic and tourist aspects of the city to a local focus by decentralizing the processes. Such decentralization would add to the benefits that locals would receive from their cities as the cities in turn benefit from being the business services and tourist transit points that most are at the present. Gaborone city is the case in point on this. As the city is the centre for public service activity, government and business, it is also a stop over for example, for tourists en-route to the North of the country where most of the tourism activities are concentrated.

Application of the concept of marketing to cities is made a special challenge by the fact that marketing is often considered to be the domain of private business whose purpose was the successful sale of commercial products (p. 2). According to Kellogg and Lillquist (1999) marketing is increasingly being recognized not only as a set of methods for selling goods but as a body of professional skills and techniques for bringing the producers of goods and services together. This is happening in a context where consumers of goods and services are also increasingly developing high expectations for quality. Therefore, in city marketing, progressive governments are reaching beyond the traditional mandate of service provision to provide customers with what they need or want (p. 2). This chapter strongly advances the view that ICTs have a central role to play in city marketing and are strategic in this information age.

INFORMATION, COMMUNICATION TECHNOLOGIES AND CITY MARKETING

Information and Communication Technologies (ICTs) is used here to refer to all the hardware, software, networks and media for the collection, storage, processing, transmission and presentation of information as well as related services. Communication technologies consist of a range of communication media and devices, including print, telephone, fax, radio, television, video, audio, computer and the Internet. ICT can be split into ICI and IT (World Bank, 2006).

It is an adequately documented fact that marketing requires the use of appropriate techniques to enhance effectiveness in service delivery. Professional communicators are central to this process which is where ICTs become strategic. Citing the views of Benest the city manager for Brea in California, Kellogg and Lillquist (1997) further note that most cities are public entities and that given the consumers' experiences within the private sector they demand timelines and high quality service, and low cost but responsive service.

Like in any enterprise, the marketing of cities is an image builder. In view of the suggested steps for city marketing which are needs analysis, marketing plan, plan implementation, getting feedback, and evaluation of results. ICT use in City marketing seems very crucial for all of these processes. It is an established fact that for any marketing activity to be successful there has to be a marketing plan which in this information era is usually technology mediated. Like all plans a marketing plan, be it for a city, a business or a particular product, is based on facts and information. It is a research based product whose aim is to help the city, to seek answers to questions around for example, the city's achievements, business climate particularly its investment possibilities; its state of tourism, cultural factors and cultural heritage; infrastructural services, Safety and security; everyday life practices and events and Labor market issues including rates of unemployment

Research on a city generates data bases that in this era are mostly electronic, and technology mediated. Reports are prepared, presented and could be stored electronically in the internet and websites of different city stakeholders. For those coming into the city particularly from far a-field, answers to some of their questions about the city, its people, business and social landscape can be searched electronically from different sources. In a city context like in modern day public and private organizations, teleconferences have also served to elicit data and publicize information about cities as entities. Recording Instruments are used extensively by researchers and businesses for various purposes including for marketing the cities.

The pertinent question that this chapter is raising in the context of the developing world is how prepared are cities in this context to apply the concept of city marketing in the promotion of awareness about them. This question is posed in view of the fact that evidently in the long term cities such as Gaborone might find themselves having to compete for resident business people, investors, tourists, customers and service users and providers. This might even be more so in a context of Botswana where there are several mining towns that offer some of what people come to cities for such as jobs, business and other services The history of cities in some parts of the world shows that sometimes this competition comes from internally and might become so fierce that the core city gets such as Gaborone may get out competed by cities that were otherwise not the main cities causing relocation of the capital city to new sites.

A big part of the city marketing process is advertising which in this era takes place through in addition to the already mentioned technologies, radio, TV and other tools such as CD's and cassettes. This is media coverage of the city. The research results that are used to develop city marketing plans are usually intensively communicated through electronic mediums such as press releases, use of billboards, video presentations, speakers corners using PR systems, seminars, meetings and conferences at which information is technology mediated, slide shows, power point and again through on-line communication (Kellogg & Lillquist, 1999).

The Potential and Challenges of Gaborone City Marketing Through Information, Communication Technologies

The Botswana Telecommunications Act of 1996 established the Botswana Telecommunications Authority (BTA) and embarked on a liberalization process, leading to a mobile phone duopoly and open competition in the ISP arena in 1998 (James & Wild, 2005). According to a report by this team of authors, Botswana is a prominent implementer of ICT for economic development and in the carrying out of several significant ICT processes and projects. But despite this, it has only been in 2004 that this country commenced a formal ICT Policy process. Significantly, rather than relying on external support, the country is funding the process out of its own funds (p. 67).

This report further notes that with regard to infrastructure, Botswana stands out in the region by enjoying a modern infrastructure with a high penetration of fibre in urban areas. The rural telephone and power programs are reported to be continuing their expansion to remote villages. And the high growth rate of mobile users suggests a willingness to use technology to communicate. There are multiple radio and TV stations available and a growing number of private networks are being deployed. The Government Data Network (GDN) and police private networks serve the bulk of government requirements (p. 60).

Public internet access is available through residential dial up modems, and Internet cafés in urban areas, with fifteen ISPs and a variety of plans to choose from. Business access can be achieved through packet switched managed services, ISDN and private line services. Improvement steps include increasing the bandwidth for Internet connectivity to 45Mbps. With the Botswana Telecommunications Company (BTC)'s plan to use the submarine cable connection through South Africa (SAT3/WASC, Southern Africa/Western Africa Submarine Cable) Botswana's connection to Europe via South Africa is envisaged with improved internet capacity. This is a good picture on a macro scale. In the context of a city such as Gaborone, however, the issue of access and unequal access to some if not all of the technologies poses as a big challenge in the use of ICT for city marketing (p. 64).

Implementation of a city marketing plan has budgetary implications that need careful planning particularly so if such implementation is going to be technology mediated. Budget preparation is demanding and requires skills that might not always be available. This is particularly so in the context of the developing world given that usually the steering officials of city plans such as mayors are not always from the technologically inclined fields. But instead have a political orientation Maundeni (2004). This means that staff capacities are crucial for any city marketing exercise. Beyond capacities, Kellogg and Lillquist (1999) say that the success of city marketing plans depends upon the support of a well informed and enthusiastic staff (p. 22). They further emphasize the need to evaluate implementation of those plans so that strategic decisions can be made that could impact the next plan.

There is strong evidence that for ICTs to deliver on their promise of economic and social development, it is critical that countries adopt enabling legal and regulatory environments that support e-development. "An enabling environment is that which has effective policy frameworks, legal, market, and social considerations that interact both at domestic and global levels to create fertile conditions for ICT-led growth. The importance of this enabling environment was recognized in the Declaration and Action Plan of the first phase of the World Summit on the Information Society (WSIS), which emphasized that a trustworthy, transparent, and non-discriminatory environment was essential for the use and growth of ICTs in the developing world (Fust, 2005). A major obstruction to effective urban policymaking in sub-Saharan Africa is identified as the lack of recent, reliable data on the region's urban populations, poverty levels, and growth rates. Evidently, most data is collected at the national level not the city level mainly because most of the planning processes are centralized.

In the context of Southern Africa, there are many identified bottlenecks to effective city planning and integration of processes that could enhance this, such as use of ICTs for city marketing. According to Hanson (2007) one of those bottlenecks is that much of the attention on poverty alleviation in Sub Saharan Africa was on rural populations and yet over one-third of sub-Saharan Africans currently live in urban areas, and in the next thirty years that figure may swell to over half the continent's population. Further, noted is the fact that some 72 percent of sub-Saharan Africa's urban population currently lives under slum conditions (UN Population Fund's *State of World Population*, 2007). These indicators can be used to support the strong arguments that without improvements in local governance and infrastructure, as well as coordination between local, national, and international groups, the urban areas of Africa in general, and those of sub Saharan Africa in particular will remain impoverished (Retrieved June 20, 2008, from (http://www.cfr.org/publication/14327/urbanization_in_subsaharan_africa.html#).

CONCLUSION

Information and communication technologies (ICTs) are increasingly being recognized as essential tools of development, that can also empower poor people, enhance skills, increase productivity, and improve governance at all levels, However, the success of ICT-enabled development (or e-development) will thus not be measured by the diffusion of technology, but by advances in development itself and, ultimately, in the achievement of the Millennium Development Goals (World Bank, 2006; E-development is defined as development that makes use of ICTs or ICT applications to provide information and knowledge services to enhance productivity, efficiency and quality of life (Wikepedia, 2008). Within the broader context of e-Government which refers to the use of internet technology as a platform for exchanging information, providing services and transacting with citizens and industry or business. This chapter introduces the concept of e- city to denote the use of ICTs to facilitate efficiency in the marketing and delivery of public services, or processes in the city.

Successful application of Information Communication Technologies seems to require systematic reforms that take into account the fundamental principles of change management bearing in mind the nature of the socio economic context. From the foregoing account, the city of Gaborone does not seem to have quite gotten there in its response to these challenges. There is evidence to suggest that Gaborone city has not like other cities in the world developed into an information city by taking advantage of the potential offered by ICTs. In view of the fact that the winds of change that sweep across urban areas and local governments underscore the importance of integrating urban development to national goals, the city of Gaborone is challenged by this chapter to liaise with technology providers to create synergy for its image building. There is great potential for Gaborone to capitalize on the strength of its rather promising ICT adoption environment and that of its advanced neighbor, South Africa, in its efforts to join the information highway.

Mosha (2004) observes that in the developing world cities are seen as engines of national development, offering job opportunities and a higher quality of life to new urban dwellers as a way of inducing faster national development. Therefore, they have to be managed properly and funds provided to run them effectively. The consequences of shortfalls in this area have been highlighted in this chapter. While cities such as Gaborone are relatively new and their circumstances unique to their contexts, it seems timely to suggest that they align their growth processes with current global concerns about city growth. It is suggested that such aligning of priorities would require a better understanding of the key dimensions of this expansion, the forces that are driving it globally, regionally and locally, so as to put in place appropriate policies that are likely to be effective, efficient, equitable and sustainable (p. 12).

Based on documented evidence this chapter has sought to illuminate issues that have been identified by many authors as areas for possible future research on cities in the developing world with a special focus on Africa. This has been done bearing in mind the fact that, for a continent as large as Africa, it is usually very difficult to develop a common frame of reference on issues of development. The same applies for the issues of city growth and city marketing. However, it has been possible to identify common threads that link experiences from the different parts of the continent. Thus, pointing to the fact that on this subject Africa like other parts of the developing world could be studied and understood holistically.

According to Maundeni (2004), a large part of dwellers of many cities are migrant populations. This has negative consequences for local democracy because normally, in-migrants see themselves as visitors and therefore, are less likely to be committed to local democracy. Instead they are more likely to be committed to the local democracy in their original home places. Gaborone is no exception in this regard, meaning that an extra mile is needed for all of the city's inhabitants to take part in city processes of governance. Literature surveyed for this chapter also strongly confirms the fact that in most cases, creating the "right" environment for ICT use is a daunting task for policymakers. While best practices are emerging from countries that have successfully crafted policies to facilitate digital opportunities, it is a generally held view that, there is no single blueprint that has been developed and could be followed in every case. One of the biggest challenges that countries of the developing world face in this information age is lack of policies on ICT

In the case of Gaborone Andrew Sesinyi, deputy permanent secretary in the ministry of Communications, Science and Technology giving a keynote address at the Botswana Council of Non Government Organizations (BOCONGO) ICT forum in Gaborone said technology is important to attain the pillar of An Educated and Informed Nation as espoused in the country's vision 2016. Therefore, its importance in our daily lives can never be over emphasized. Sesinyi assured the participants of his ministry's recognition that Botswana like other countries is in a knowledge based era. And therefore most of that information and knowledge should be stored, retrieved, manipulated, analyzed, etc., in one electronic form or another. It is against this background that Botswana has developed a draft ICT policy, called ***Maitlamo*** through which, Botswana aims to be a globally competitive, knowledge and information society where lasting improvements in social, economic and cultural development is achieved through effective use of ICT. This is a bold commitment to take this country beyond Vision 2016. In his view, the wealth of any country, as much as it may be dependent on its minerals or manufacturing capability, comprises the ideas and knowledge that are generated. Sesinyi further noted that evidently, knowledge management has become central to governance, and as such it is the ministry's intention to make ICT an important engine of economic growth by liberalizing the telecommunications industry.

This speech laid emphasis on the role of Civil Society Organizations in driving the ICT use agenda and strongly underlined the fact that although some Botswana CSOs are already active users of ICTs, much more needs to be done to build the capacity of the sector to capitalize on the potential of ICTs. Calling on society activists to be always at the cutting edge of the latest technologies as they are advocating for social economic change, Sesinyi noted that nowadays, there is even online activism whereby petitions can be signed online. He, however, went on to say that more should be done to build the capacity of Civil Society Organizations (CSOs) to capitalize on the potential of technologies. Sesinyi added that the use of ICTs in support of social change and development has already seen an impressive evolution, and that the Internet has the potential to revolutionize the way CSOs conduct their operations.

An expression of this degree of commitment and by-in by government into the idea of ICT as a tool for development is a very positive indicator of a conducive environment for experimenting with technology mediated city marketing for Gaborone. The Government of Botswana is currently in the early stages of a move towards e-Government and has a central government website. While Botswana acknowledges the fact that ICT use is not a panacea for all of its development challenges, it has taken seriously the potential that this tool holds for driving socio economic change. It also duly recognizes some of the challenges brought by the sector which can even be harmful to society.

It is evident from the foregoing presentation that a better understanding of Africa's urban economies and urbanization process is urgently needed, in order to identify their role and position them well in national development. Within this framework and given its economic and geopolitical characteristics, Africa is still bargaining from a position of disadvantage in its efforts to enter the information age within a discourse that identifies application of ICTs almost as a panacea for sustainable development. This is an era in which every country and every region of the world is under pressure to integrate ICTs into their development processes. This has certain preconditions one of the most important of which is human capacity to drive the ICT in the development agenda (Nijkamp & Jonkhoff, 2001).

While the proposals made for sustainable cities as espoused by the World Bank strategy are appreciated, this chapter argues that part of the responsibility to sustain cities should be shared with the city government structures through the locals. Maundeni (2004) writing on democracy in Gaborone observed that attempts to achieve good governance in this context have been made up of processes of consultation. However, as some authors observe, even if regular public consultations are held (for instance, city fora), it is easier for city governments to listen than to act on what they hear. In the view of Maundeni this difficulty can get compounded if the city structures are not enjoying some degree of autonomy from the central government structures which are part of the political landscape and therefore not non partisan. In addition, many of the suggestions and requests voiced by citizens through open channels of communication as would be provided by ICTS may be the responsibilities of national or state/provincial agencies, not local governments. But city governments that provide room for involving their citizens and civil society organizations in the 'governance' of the city are said to get much in return. Suddenly, solutions appear to what previously seemed impossible or too expensive.

The World Bank on the other hand is of the strong view that urban transition offers significant opportunities to improve the quality of life for all individuals. However, whether this potential is realized depends critically on how cities are managed and also on the national and local policies affecting their development. Across all countries over time, urbanization should accompany sustained economic improvement and, when well-managed, can be an important contributor to broad-based social welfare gains. The development of urban areas is also closely linked to the rural economy through the exchange of labor, goods, services, information and technology, capital, and social transactions that benefit residents in both locations (p. 42). It is further noted that Africa is the worlds most rapidly growing market for mobile telephony and also home to the fastest growing fixed telephony markets in the world even though. This is the case despite the fact that Africa still has the world's lowest penetration rates and basic telephony provision remains a major need in many parts of its parts, particularly in rural areas. There is a strong belief by the bank that the surge in demand for Internet access and broadband capabilities will drive these developments further in the coming years concluding that overall, Africa's telecommunications s future looks very promising and offers great opportunities to service providers, equipment vendors and investors. This optimism is to be an important part of the city marketing agendas in this context and the planners and managers of the city of Gaborone are challenged to take advantage of these opportunities early enough in the life of this city.

In the view of (Keiner & Salmerón 2003) although for Gaborone some "good governance" policies that create opportunities for sustainable development are already in use, they are not sufficient as a basis for sustainable urban development. And therefore, no proper vision for sustainable urban development exists yet. Currently, there is still a quest to define Gaborone's sustainable identity as a whole and for the city's constitutive parts. For ICT use these authors argue and I concur, nowadays, the opportunities that the Internet offers should not be disregarded. And for Botswana it is further observed that the potential for effective ICT use for development is high because Botswana's youth are very familiar with the Internet and other communication technologies (Sairosse, 2003). While evidently, there is still a lot of work to be done in this area, in this context, there seems to be a strong basis to experiment with ICT use for which a useful framework is provided by the: *Vision 2016.* As McGranahan and Satterthwaite (cited in Keiner & Salmeron, 2003) point out:

The goal is not sustainable cities but cities that contribute to sustainable development within their boundaries, in the region around them and globally (p. 10). And as this aptly puts it "A small city poorly managed is bad, a large city poorly managed may be as good (or as bad) as a small city well managed, but a large city well managed is definitely best." (Prud'homme 1996, p. 3)

REFERENCES

Ashworth, G. J., & Voogd, H. (1990). *Selling the city.* London: Belhaven Press.

Botswana Daily News (July 24th, 2007). *Botswana government information services.*

Brockerhoff, M. P. (2000). An urbanizing world. *Population Bulletin, 55*(3), 1.

Cavric, B. I., & Mosha, A. C. (2003). *Towards better urban development and management in Botswana*. Gaborone. University of Botswana.

Discover–Botswana, travel and tourism magazine (2006). Retrieved June 27, 2007, from http://www.discover-botswana.com/investors-guide.php

Guermazi, B., & Satola, D. (2005, November). Creating the "right" enabling environment for ICT. In Schware, R. (Ed.), *E-Development: From excitement to effectiveness*. A report prepared for the World Summit on the Information Society held in Tunis. Global, Information and Communication Technologies Department. World Bank Group, pp. 23-47.

Fust, W. (2005). *Knowledge society, knowledge management and ICT:* Editorial contribution to a publication ed. by the Community Development Library Dhaka/Bangladesh.

Hanson, S. (2007). *Urbanization in Sub-Saharan Africa*. Retrieved June 20, 2008, from http://www.cfr.org/publication/14327/urbanization_in_subsaharan_africa.html

Hicks, J. F. (1998, July). *Sub-Saharan Africa: Enhancing the productivity of urban Africa.* Paper presented at the International Conference on Research Community for the Habitat Agenda Forum of Researchers on Human Settlements. Geneva, Italy.

Investors Guide to Botswana Magazine (2007). Retrieved August 10, 2007, from http://www.discover-botswana.com/investors-guide.php

James, T., & Wild, K. (2005). *Report of a feasibility study for an information society program for the African, Caribbean and Pacific (ACP) countries; ANNEX IV: Regional Report – Africa*.

Jenman Safaris Africa (2007). Retrieved July 15, 2007, from http://www.jenmansafaris.com/

Kavaratzis, M., & Ashworth, G. J. (2007). *Partners in coffee shops, canals and Commerce: Marketing the city of Amsterdam. Netherlands*. University of Groningen, Urban and Regional Studies Institute.

Keiner, M. (2003). *Sustainability oriented urban development: A general introduction with case studies from Gaborone, Johannesburg and Santiago de Chile*. Zurich. Institute for Spatial and Landscape Planning.

Kellogg, C., & Lillquist, R. (1999). *How to market your city*. City, County Communications Association. Washington. Municipal Research & Services Center.

Mayor Bloomberg, Mayor of New York City in a Keynote Address at the C40 Large Cities Climate Summit (2007, May). New York. Retrieved June 10, 2008, from http://www.nycclimatesummit.com

Maitlamo (2005). *National policy for ICT development*. Ministry of Communication, Science and Technology. Government of Botswana. Retrieved June 10, 2008, from http://www.maitlamo.gov.bw/ict-vision.asp

Maundeni, Z. (2004). *Mapping local democracy in Gaborone city.* Gaborone, Botswana, Botswana Association of Local Authorities – BALA.

Molebatsi, C. O. (1996). Towards a sustainable city. Gaborone, Botswana. *Ambio, 25*(2), 126-133.

Mosha, A.C. (1996). The city of Gaborone, Botswana – Planning and management. *Ambio, 25*(2), 118-125.

Mosha, A. C., & Cavric, B. (2001). Incorporating urban agriculture in Gaborone city planning. *Urban Agriculture, 4*, 25–27.

Mosha, A. C. (2004, September). *Challenges of municipal finance in Africa. Gaborone City, Botswana.* Paper presented at Habitat's 2nd World Urban Forum for Professionals on Land and Urban Poverty: The Challenges of City Financing: Organized by the Habitat Professionals Forum and the UN-HABITAT. Barcelona, Spain.

Nijkamp, P., & Jonkhoff, W. (2001). The city in the information and communication technology age: A comparative study on path dependency. *International Journal of Technology Policy and Management, 1*(1), 78-99.

Obaid, T. A. (2007, December). *Population and human development: The key connections*. The Millennium Development Goals Report 2007, UN.

Press Conference (2007, July 11), ITU Secretary-General Dr. Hamadou Touré for the *Connect Africa* Summit held in Kigali, Rwanda, 29-30 October 2007 held jointly with the UN Global Alliance for ICT and Development (GAID).

Prud'homme, R. (1996). Size, sprawl, speed and the efficiency of cities. *Urban Studies Journal, 36*(11), 1849-1858.

Rakodi, C. (1997) (Ed.). *The urban challenge in Africa: Growth and management of its large cities*. New York, NY: United Nations University Press.

Rakodi, C. (1998). *Globalization: Trends and Sub-Saharan African cities*. In F.C. Lo and

Y. Yeung (Eds.), *Globalization and the world of large cities (pp. 314-351)*. Tokyo, New York; Paris: UNU Press.

Sairosse, M. T. (2003). *Economic and social impact of the Internet: A study of cybercafés in Gaborone, Botswana*. University of Botswana, Gaborone. MLIS dissertation.

Satterthwaite, D. (1996). Toward healthy cities. *People and the Planet, 5*(2), 12-15.

Satterthwaite, D. (2002, December) *City Governance for and with Children*. Paper presented at the Conference on Children and the City; Learning from international experiences, creating local solutions. Patronage of Her Majesty Queen Rania Al-Abdullah. Amman, Jordan.

Satterthwaite, D. (2002*). Why the cities have grown*: Commissioned Paper by the International Institute for Environment and Development (IIED). London

Schware, R. (Ed.) (2005, November). *E-Development: From excitement to effectiveness.* Paper presented at the World Summit on the Information Society. Tunis, Tunisian Republic.

Schware, R. (Ed.) (2005, November). *Global forces, urban change and urban management in Africa:* The urban Challenge. In Africa: Growth and Management of its Large Cities. Paper presented at the World Summit on the Information Society. Tunis, Tunisian Republic.

Shlomo, A., Stephen, C., Sheppard, S. C., Civco, D. L., Buckley, R., Chabaeva, A., Gitlin L., Kraley, A., Parent, J., & Perlin, M. (2005). *Transport and urban development department.* Washington, D.C. International Bank for Reconstruction and Development. World Bank Group.

Stavrou, A. (2005, October). *Productive cities: Information needs of the urban poor.* Paper presented at the 4th International Forum on Urban Poverty Conference, Marrakech, Morocco.

UNEP/UNESCO (2006). *Facts andfigures.* New York, NY: United Nations.

United Nations Development Programme's Regional Project on Local Governance for Latin America (2005). New York, NY: United Nations.

UNFPA, (2007). World population Report: Unleashing the potential of Urban Growth. New York, NY: United Nations. http://www.unfpa.org/swp/2007/english/chapter_1/smaller_cities.htm. Retrieved February 16, 2008.

Wong, C. Y. L., Millar, C. J. M., & Choi, C. J. (2006). Singapore in transition: From technology to culture hub. *Journal of Knowledge Management, 10*(5), 79-91.

World Bank (2006). *Information and communications for development (IC4D): Global trends and policies.* Washington D.C. Global Information and Communication Technologies Department. World Bank Group.

World Bank (2006). *Cities in transition: A strategic view of urban and local government issues.* International Bank for Reconstruction and Development, Washington, D.C.: World Bank Group.

World Bank (1998). *Reports of findings on ongoing operational, economic and sector work carried out by the World Bank and its member governments in the Africa Region* Published periodically by the Africa Technical Department on behalf of the Region. A report of findings on the Africa Region. Number 7.

Compilation of References

20 World's Twenty Best (2004). Television program that selects the top 20 destinations in the World in several sections, including the top 20 cities. RDF Media production for channel broadcasting Ltd.

Aaker, D. (1991). *Managing brand equity*. New York, NY: Free Press.

Aaker, D. (2001). *Strategic market management*. (6th ed.). New York, NY: John Wiley and Sons.

Achrol, R. S., & Kotler, P. (1999). Marketing in the network economy. *Journal of Marketing, Special Issue*, *63*, 146-163.

Adam, J., Cobos, X., & Liu, S. (2007). *Travel 2.0: Trends in Industry Awareness and Adoption*. New York University and PhoCusWright Inc.

Adolfsson, P. (1999). *Environment as a part of the city*. Paper presented at the Managing Big Cities Conference (August 26-28). Gothenburg, Göteborg.

Adolfsson, P. (2003). *Miljö och dess många ansikten i staden. Om kvalitetsmätningar och organisering i Stockholm*. Göteborg: BAS.

Aguilera, M., & Perales, A. (1994, March). La Imagen de las ciudades en el marketing urbano. *Marketing y Ventas*, *79*, 10-14.

Ahmad, A. (2005, March 28). Intel chief puts spotlight on digital era: Barrett wants GCC states to invest more in information communication technology. *Gulf News*, A7.

Amarsaikhan, D., Lkhagvasuren, T., Oyun, S., & Batchuluun, B. (2007). Online medical diagnosis and training in rural Mongolia. *Distance Education*, *28*(2), 195.

Amin, A., & Thrift, N. (2002). *Cities. Reimagining the urban*. Cambridge: Polity Press.

Amis, R. (2007, May). You can't ignore social media: How to measure Internet efforts to your organization's best advantage. *Tactics*, 10.

Andersen, D. F., Belardo, S., & Dawes, S. S. (1994). Strategic information management: Conceptual frameworks for the public sector. *Public Productivity and Management Review*, *17*(4), 335-353.

Andersen, P. H. (2005). RM and brand involvement of professionals through Web-enhanced brand communities: Coloplast case. *Industrial Marketing Management*, *34*(3), 285-297.

Anderson, B. (1953). *Stockholm. Capital and crossroads*. Stockholm: The Swedish Institute.

Anholt, S. (2000, November/December). The nation as brand. *Across the Board*, 22-27.

Anholt, S. (2004). Nation-brands and the value of provenance. In N. Morgan, A. Pritchard, & R. Pride (Eds.), *Destination branding: Creating the unique destination proposition*. Oxford: Elsevier Butterworth-Heinemann.

Anton Claver, S. (2004). La presencia en Internet de los principales destinos turísticos del litoral mediterráneo español. *Actas del Congreso TURITEC 2004*. Universidad de Málaga, 2004. Retrieved June 1, 2008, from http://www.turismo.uma.es/turitec/turitec2004/index.htm

Archdale, G. (1993). Computer reservation system and public tourist offices. *Tourism Management*, *14*, 3-14.

Armstrong, A., & Hagel, J. (1997). *Net gain: Expanding markets through virtual communities*. Boston, MA: Harvard Business School Press.

Arnold, M. (2007, Spring). *Die elektronische Behörde*. Pictures of the Future—The Magazine for Research and Innovation, Siemens, (pp. 41-42).

AS EMT (n. d.). Retrieved August 13, 2007, from http://www.emt.ee/wwwmain?screenId=mainpage.private&language=ENG

Ashling, J. (2008). Market growth in Asia, Africa, and Middle East set to tip the balance. *Information Today, 25*(2), 24-25.

Ashworth, G. J., & Voogd, H. (1990). *Selling the city. Marketing approaches in public sector urban planning*. London: Belhaven.

Awad, N. F., & Zhang, S. (2007, January). *Stay out of my forum! Evaluating firm involvement in online ratings communities*. Paper presented at the 40th HICSS, Waikoloa, Big Island, Hawaii.

Baeumerth, (2002). *Handbuch mobile commerce: technische Grundlagen, Marktchancen und Einsatzmoeglichkeiten* (pp. 225-230). Berlin/Heidelberg, Germany: Springer.

Bailor, C. (2007, January). When disaster does not strike. *Customer Relationship Management*, 42-43.

Baker, B. (2007). Places: The new brand frontier. *Total destination management*. Retrieved June 1, 2008, from www.DestinationBranding.com

Baker, S., & Green, H. (2005). Blogs will change your business. *Business Week, 3931*, 56-67.

Bakhtin, M. (1994). Problems of Dostoevsky's poetics (C. Emerson, Trans.). In P. Morris (Ed.), *The Bakhtin reader. Selected writings of Bakhtin, Medvedev, Voloshinov* (pp. 89-96). London: Edward Arnold.

Bakhtin, M. (1994). Speech genres and other late essays (V. W. McGee, C. Emerson & M. Holquist, Trans.). In P. Morris (Ed.), *The Bakhtin reader. Selected writings of Bakhtin, Medvedev, Voloshinov* (pp. 81-87). London: Edward Arnold.

Bakhtin, M., & Volosinov, V. (1994). Marxism and the philosophy of language (L. Jatejka & I. R. Titunik, Trans.). In P. Morris (Ed.), *The Bakhtin reader. Selected writings of Bakhtin, Medvedev, Voloshinov* (pp. 26-37). London: Edward Arnold.

Baldrige National Quality Program. (2008). Criteria for performance excellence. *National Institute of Standards and Technology, Department of Commerce*. Retrieved May 31, 2008, from http://baldrige.nist.gov/PDF_files/2008_Business_Nonprofit_Criteria.pdf.

Baloglu, S., & McCleary, K. W. (1999). A model of destination image formation. *Annals of Tourism Research, 26*(4), 808-889.

Baloglu, S., & Pekan, Y. A. (2006). The website design and Internet site marketing practices of upscale and luxury hotels in Turkey. *Tourism Management, 27*, 171-176.

Barney, J., & Hesterly, W.S. (2006). *Strategic management and competitive advantage*. Upper Saddle River, NJ: Pearson Prentice Hall.

Barthes, R. (1977). The Photographic Message (S. Heath, Trans.). In *Image, music, text*. New York, NY: Hill and Wang.

Barwise, P., & Farley, J. U. (2005). The state of interactive marketing in seven countries: Interactive marketing comes of age. *Journal of Interactive Marketing, 19*(3), 67-80.

Bastain, L. (2007). Web surfing for great locations. *Area Development Magazine*. Retrieved September 5, 2006, from http://www.angeloueconomics.com/areadevelopment_websearching.html

Bateman, I., Lovett, A., & Brainard, J. (2005). *Applied environmental economics: A GIS approach to cost-benefit analysis*. New York, NY: Cambridge University Press.

BBC News, (2008). *Technologies on the rise in 2008*. Retrieved February 14, 2008, from http://news.bbc.co.uk/2/hi/technology/7147804.stm.

Becattini, G. (1979). Dal sttore industriale al distretto industrialle. Alcune considerazione sull'unità dell'economia industriale. *Revista di Economia e Politica Industriale, 1*.

Beck, E., Madon, S., & Sahay, S. (2004). On the margins of the 'Information Society': A comparative study of mediation. *The Information Society, 20*, 279-290.

Beckerson, J. (2001, March/April). *Marketing British tourism. Early state stimulation of a service sector, 1890-1950*. Paper presented at The Economic History Society, 75th Anniversary Conference, Kelvin Conference Centre, University of Glasgow.

Beerli, A., & Martín, J. D. (2004). Factors influencing destination image. *Annals of Tourism Research, 31*(3), 657-681.

Belk, R. W. (1989). Extended self and extending paradigmatic perspective. *Journal of Consumer Research, 16*, 129-132.

Bellman, S., Johnson, E., Lohse, G., & Mandel, N. (2006). Designing marketplaces of the artificial with consumers in mind. *Journal of Interactive Marketing, 20*(1), 21-33.

Bellmann, B. (2007). *Schicken Sie ihre Gaeste auf Schatzsuche*. Retrieved August 17, 2007, from http://www.bjoernbellmann.de/pdf/pdf_infotrail-geocaching.pdf

Bennett, F. C., & Nathanson, J. (1997). Metropolitan alliances. *Economic Development Commentary, 21*(1), 34-38.

Benyoucef, M., & Verrons, M. H. (2008). Configurable e-negotiation systems for large scale and transparent decision making. *Group Decis Negot, 17*, 211-224.

Berger, P. L., & Luckmann, T. (1968). *Social construction of reality*. Amorrortu: Buenos Aires.

Bergström, O., & Dobers, P. (2000). Organizing sustainable development. From diffusion to translation. *Sustainable Development, 8*(4), 167-179.

Berkowitz, E., Crane, F., Kerin, R., Hartley, S., & Rudelius, W. (2003). *Marketing* (5th Canadian ed.). Toronto: McGraw-Hill Ryerson.

Bessa, M., Coelho, A., Cruz, J. B., & Chalmers, A. (2006). Selective presentation of perceptually important information to aid orientation and navigation in an urban environment. *International Journal of Pattern Recognition, 20*(4), 467-482.

Bickart, B., & Schindler, R. M. (2001). Internet forums as influential sources of consumer information. *Journal of Interactive Marketing, 15*(3), 31-40.

Biel, A. (1997). Discovering brand magic: The hardness of the softer side of branding. *International Journal of Advertising, 16*, 199-210.

Bignell, J. (2002). *Media semiotics: An introduction* (2nd ed.). Manchester: Manchester University Press.

Black, J. (2002). *Oxford Dictionary of Economics* (2nd ed.). New York, NY: Oxford University Press.

Blain, C., Levy, S. E., & Brent Ritchie, J. R. (2005). Destination branding: Insights and practices from destination management organizations. *Journal of Travel Research, 43*, 328-338.

Blitzer, W. (2006, March 14). The situation room transcripts. *CNN.com*. Retrieved August 4, 2007, from http://transcripts.cnn.com/TRANSCRIPTS/0603/14/sitroom.02.html

Blood, R. (2000). Weblogs: A history and perspective. Retrieved October 15, 2007, from http://www.rebeccablood.net/essays/weblog_history.html

Blood, R. (2004). How blogging software reshapes the online community. *Communications of the ACM, 47(12)*, 53-55.

Bloomfield, B., & Coombs, R. (1992). Information technology, control and power: The centralization and decentralization debate revisited. *Journal of Management Studies, 29*(4), 459-484.

Blume, L. (2006). Economic policies as determinants of the local business climate: Empirical results from a cross-section analysis among East German municipalities. *Regional Studies, 40*(4), 321–333.

Book, B. (2003, July). *Traveling through cyberspace: Tourism and photography in virtual worlds*. Paper presented at the conference Tourism & Photography: Still Visions - Changing Lives in Sheffield, UK.

Borgatti, S. P., Everett, M. G., & Freeman, L. C. (2002). *Ucinet 6 for Windows: Software for social network analysis*. Harvard: Analytic Technologies.

Botswana Daily News (July 24th, 2007). *Botswana government information services*.

Boyle, M. (1997). Civic boosertism in the politics of local economic development. 'Insitutional positions' and 'strategic orientations' in the consumption of hallmark events. *Environment and Planning A, 29*(11), 1975-1997.

Bozinis, A. I. (2007). Internet politics and digital divide issues: The rising of a new electronic aristocrats and electronic meticians. *Journal of Social Sciences, 3*(1), 24-26.

Brady, M. (2003). Managing information technology assimilation: A marketing perspective. *Irish Journal of Management, 24*(1), 125-138.

Brady, M., Fellenz, M. R., & Brookes, R. (2008). Researching the role of information and communications technology (ICT) in contemporary marketing practices. *Journal of Business & Industrial Marketing, 23*(2), 108-114.

Brady, M., Saren, M., & Tzokas, N. (2002). Integrating information technology into marketing practice—The IT reality of contemporary marketing practice. *Journal of Marketing Management, 18*, 555-577.

Brandenburger, A. M., & Nalebuff, B. J. (1996). *Coopetition*. New York, NY: Currency Doubleday.

Bregenz (n/d.). *Herzlich Willkommen in der Mobile City Bregenz!* Retrieved February 14, 2008, from http://www.bregenz.ws/mcity/

Brockerhoff, M. P. (2000). An urbanizing world. *Population Bulletin, 55*(3), 1.

Brodie, R.J., Coviello, N.E., & Winklhofer, H. (2008). Contemporary marketing practices research program: A review of the first decade. *Journal of Business & Industrial Marketing, 23*(2), 84-94.

Brookes, R.W., Brodie, R.J., Coviello, N.E., & Palmer, R.A. (2004). How managers perceive the impacts of information technologies on contemporary marketing practices: Reinforcing, enhancing or transforming? *Journal of Relationship Marketing, 3*(4), 7-26.

Brotchie, J., Batty, M., Hall, P., & Newton, P. (Eds.). (1991). *Cities of the 21st century. New technologies and spatial systems*. New York, NY: Longman Cheshire.

Brown, B. (2002). Studying the use of mobile technology. In B. Brown, N. Green & R. Harper (Eds.), *Wireless world. Social and interactional aspects of the mobile age* (pp. 3-15). London: Springer.

Brown, M. & Brudney, J. L. (1998). A "smarter, better, faster, & cheaper" government: Contracting and geographic information systems. *Public Administration Review, 58*(4), 335-345.

Bryson, J. M. (2004). *Strategic planning for public and nonprofit organizations: A guide to strengthening and sustaining organizational achievement* (3rd ed.). San Francisco, CA: Jossey-Bass.

Bryson, J. M. (2004). What to do when stakeholders matter: Stakeholder identification and analysis techniques. *Public Management Review, 6*(1), 21-53.

Bryson, J.M., Ackermann, F., & Eden, C. (2007). Putting the resource-based view of strategy and distinctive competencies to work in public organizations. *Public Administration Review, 67*(4), 702-717.

Bughin, J. R. (2007, August). How companies can make the most of user-generated content. *The McKinsey Quarterly*, Research in Brief, Web Exclusive.

Buhalis, D. (1998). Strategic use of information technologies in the tourism industry. *Tourism Management, 19,* 409-421.

Buhalis, D. (2000). Marketing the competitive destination of the future. *Tourism Management, 21*, 97-116.

Buhalis, D. (2002). *eTourism. Information technology for strategic tourism management*. UK: Prentice Hall.

Buhalis, D., & Cooper, C. (1998). Competition or cooperation? Small and medium sized tourism enterprise at the destination. In E. Laws, B. Faulkner & G. Moscardo (Eds.), *Embracing and managing change in tourism* (pp. 307-323). London: Routledge.

Buhalis, D., & Costa, C. (2006). *Tourism, management dynamics. Trends, management and tools.* Oxford: Elsevier Butterworth Heinemann.

Buhalis, D., & Licata, M. C. (2002). The future E-Tourism intermediaries. *Tourism Management, 23*, 207-220.

Bull, A. (1991). *The Economic of travel and tourism.* Melbourne: Pitman.

Bunnell, T. (2003). *Malaysia, modernity and the multimedia super corridor: A critical geography*. Singapore: Routledge Pacific Rim Geographies.

Burr, C., Patterson, R., Rolland, E., & Ward, K. (2007). Integration of E-CRM in healthcare services: A framework for analysis. *International Journal of E-Business Research, 3*(2), 1-12.

Business World (2004, January 20). *ICTs can help fight poverty*. Manila: Author.

Butler, R. W. (1980). The concept of a tourist area cycle of evolution. *Canadian Geographer, 24*, 5-12.

Byrne, B. M. (2001). Structural equation modeling with AMOS, EQS, and LISREL: Comparative approaches to testing for factorial validity of a measuring instrument. *International Journal of Testing, 1*(1), 55-86.

Cai, L. (2002). Cooperative branding for rural destination. *Annals of Tourism Research, 29*, 720-742.

Caldwell, N., & Freire, J. (2004). The differences between branding a country, a region, and a city: Applying the brand box model. *Brand Management, 12*(1), 50-61.

Carrier, C., Raymond, L., & Eltaief, A. (2004). Cyberentrepreneurship: A multiple case study. *International Journal of Entrepreneurial Behaviour and Research, 10*(5), 349-363.

Castells, M. (1989). *The informational city. Information technology, economic restructuring, and the urban-regional process*. Oxford: Blackwell Publishers.

Castells, M. (1996). *The rise of the network society.* Cambridge, MA: Blackwell Publishers.

Castells, M. (1998). *The information age: Economy, society and culture, I*, & *The rise of the network society, II*. Cambridge, MA: Blackwll Publishers Inc.

Castells, M. (2001). *The internet galaxy: Reflections on internet, business and society*. Oxford: Oxford University Press.

Castells, M., & Hall, P. (1994). *Technopoles of the world. The making of twenty-first-century industrial complexes*. London: Routledge.

Catasús, B., & Lundgren, M. (1999). Coupling the environmental issue: The environmental managers and their allies. *Global Focus, 11*(2), 21-36.

Cavric, B. I., & Mosha, A. C. (2003). *Towards better urban development and management in Botswana*. Gaborone. University of Botswana.

Cegarra, J. G., & Córdoba, J. R. (2006). Assessing and developing E-Government use by SMEs. *Business School Research Memoranda, 56*, 1-21.

Cell Broadcast ist bald Geschichte - GSM-"Textrundfunk" beim UMTS-Mobilfunk nicht vorgesehen (n/d.). Retrieved August 25, 2007, from http://www.teltarif.de/i/cellbroadcast.html

Chan, Y. E., Sabherwal, R., & Thatcher, J. B. (2006). Antecedents and outcomes of strategic IS alignment: An empirical investigation. *IEEE Transactions on Engineering Management, 53*(1), 27-47.

Chang L., & Young, V. (2006). Student self-reported use of wireless Tablet PCs in classrooms. *Frontiers in Education 36th Annual Conference*, San Diego, USA.

Charlevoix, D. J., Jackman, S. K., & Twine, T. E. (2006), Conference notebook: Tablet PCs: A welcome aid to lecture-based meteorology courses. *Bulletin of the American Meteorological Society, 87*, 737-738.

Chaves, N. (2004, April). *La marca destino turístico, cinco estrategias gráficas.* Paper presented at the XIII Simposi Internacional de Turisme i Lleure ESADE-Fira de Barcelona. Retrieved April 30, 2007, from www.esade.es/cedit2004/cat/est_estudios.php

Checkland, P., & Holwell, S. (1998). *Information, systems and information systems: Making sense of the field.* Chichester, UK: John Wiley and Sons.

Chen, J., & Ching, R. K. H. (2004). An empirical study of the relationship of IT intensity and organizational absorptive capacity on CRM performance. *Journal of Global Information Management, 12*(1), 1-17.

Chen, L. D., & Wells, W. D. (1999). Attitude toward the site. *Journal of Advertising Research, 39*(5), 27-38.

Chen, L. D., Gillenson, M. L., & Sherrell, D. L. (2002). Exciting online consumers: an extended technology acceptance perspective. *Information and Management, 39*, 705-719.

Cheverst, K., Davies, N., Mitchell, K., Friday, A., & Efstratiou, C. (2000). *Developing a context-aware electronic tourist guide: Some issues and experiences.* Paper presented at the SIGCHI Conference on Human Factors in Computing Systems. Retrieved June 20, 2008, from http://www.guide.lancs.ac.uk/CHIpaper.pdf

Chevrant-Breton, M. (1997). Selling the world city. A comparison of promotional strategies in Paris and London. *European Planning Studies, 5*(2), 137-161.

Chias, J. (2005). El negocio de la felicidad. *Desarrollo y marketing turístico de países, regiones, ciudades y lugares.* Madrid: Prentice-Hall.

Cho, C-H. & Cheon, H. J. (2005, Summer). Cross-cultural comparisons of interactivity on corporate Web sites. *Journal of Advertising, 43*(2), 99-115.

Choi, S., Lehto, X. Y., & Morrison, A. M. (2007). Destination image representation on the web: Content analysis of Macau travel related Web sites. *Tourism Management, 28*,118-129.

Chon, K. S. (1991). Tourism destination image modification process: Marketing implications. *Tourism Management, 12*(1), 68-72.

Chow, J. (2005). The new frontier. *National Post Business Magazine*, 40.

Christensen, L., & Askegaard, S. (1999). Corporate identity and corporate image revisited. A semiotic perspective. *European Journal of Marketing, 35*(3/4), 292-315.

Churchill, E. F., & Wakeford, N. (2002). Framing mobile collaborations and mobile technologies. In B. Brown, N. Green & R. Harper (Eds.), *Wireless world. Social and interactional aspects of the mobile age* (pp. 154-179). London: Springer.

Ciborra, C. (1994). The grassroots of IT strategy. In C. Ciborra & T. Jelassi (Eds.), *Strategic information systems: A European perspective* (pp. 3-24). Chichester, UK: John Wiley.

Clout, H. (2006). Vive la Bretagne! *Modern and Contemporary France, 14*(1), 79-83.

CNNMoney (2007). Money best places to live 2007. Retrieved August 1, 2007 from http://money.cnn.com/magazines/moneymag/bplive/2007

Collins, C., & Buhalis, D. (2003). Implementation of a new strategic framework for survival of destination management systems In A. Frew, P. O'Connor & M. Hitz (Eds.), *Information and communications technologies in tourism* (pp. 202-211). Wien: Springer-Verlag.

Cooper, C., Fletcher, J., Gilbert, D., Shepherd, R., & Wanhill, S. (1998). *Tourism: Principles and practices.* England: Addison-Wesley, Longman.

Cooper, G. (2002). The mutable mobile: Social theory in the wireless world. In B. Brown, N. Green & R. Harper (Eds.), *Wireless world. Social and interactional aspects of the mobile age* (pp. 19-31). London: Springer.

Córdoba, J. R. (2005). Communities and evaluation of E-Government services. In S. Clarke & E. Coakes (Eds.), *Encyclopaedia of communities of practice in information and knowledge management* (pp. 32-34). Hershey, PA: Idea Group Publishing.

Córdoba, J. R. (2006). Using Foucault to analyze ethics in the practice of problem structuring methods. *Journal of the Operational Research Society, 57*(9), 1027-1034.

Couper, M. P., Baker, R. P., Clark, C. Z. F., Martin, J., Nicholls, W. L., & O'Reilly, J. M. (1998). *Computer assisted questionnaire information collection*: John Wiley & Sons Inc.

Coviello, N., Miller, R., & Marcolin, B. (2001). Understanding IT-enabled interactivity in contemporary marketing. *Journal of Interactive Marketing, 15*(4), 18-33.

Criado, J., & Ramilo, M. C. (2003). E-government in practice: An analysis of web site orientation to the citizens in Spanish municipalities. *International Journal of Public Sector Management, 16*(3), 191-218.

Crompton, J. L. (1979). An assessment of the image of Mexico as a vacation destination and the influence of geographical location upon the image. *Journal of Travel Research, 18*(4), 18-23.

Crotts, J. (1999). Consumer decision making and prepurchase information search. In Y. Mansfield & A. Pizam (Eds.), *Consumer behavior in travel and tourism* (pp. 149-168). Binghamton, NY: Haworth Press.

Cueto, J. (2005, January 23). Ciudades lentas. *El País Semanal, 1478*, 8.

Cushman & Wakefield, H. B. (2006). *European Cities Monitor 2006*. London: Cushman & Wakefield, Healey & Bakero.

Czarniawska, B. (2000). The European capital of the 2000s: On image construction and modeling. *Corporate Reputation Review, 3,* 202-217.

Czarniawska, B. (2002). *A tale of three cities. Or the glocalization of city management.* Oxford: Oxford University Press.

Damianos, L., Cuomo, D., Griffith, J. Hirst, D., & Smallwood, J. (2007, January). *Adoption, utility, and social influences of social bookmarking.* Paper presented at the 40th HICSS, Waikoloa, Big Island, Hawaii.

Davies, J. S. (2005). Local governance and the dialectics of hierarchy, market and network. *Policy Studies, 26*(3/4), 311-335.

Davis, F. D. (1986). *A technology acceptance model for empirically testing new end-user information systems: Theory and results.* Doctoral dissertation, Sloan School of Management, Massachusetts Institute of Technology.

Davis, F. D. (1989). Perceived usefulness, perceived ease of use, and user acceptance of information technology. *MIS Quarterly, 13*(3), 319-339.

Davison, R. M., Wagner, C., & Ma, L. C. K. (2005). From government to e-government: A transition model. *Information Technology & People, 18*(3), 280-299.

Dawes, S.S., Pardo, T.A., Simon, S., Cresswell, A.M., LaVigne, M.F., Andersen, D.F., & Bloniarz, P. A. (2004). *Making smart IT choices—Understanding value and risk in government IT investments*, 2nd edition. Center for Technology in Government, University at Albany, SUNY.

Day, G. S. (1998). Organizing for interactivity. *Journal of Interactive Marketing, 12(1)*, 47-53.

Day, G. S., & Montgomery, D. B. (1999). Charting new directions for marketing. *Journal of Marketing, 63*, 3-13.

DCCED. (2007). Alaska visitor statistics program (Fall/Winter 2006-07): Department of Commerce Community and Economic Development.

De Certeau, M. (1988). *The practice of everyday life.* Berkeley, CA: University of California Press.

De Chernatony, L., & Dall´Olmo, R. F. (1998). Definig a brand: Beyond the literature with experts´ interpretations. *Journal of Marketing Management, 1*(1), 94-105.

De Chernatony, L., & Dall'Olmo Riley, F. (1997). Modelling the components of the brand. *European Journal of Marketing, 32*(11/12), 1074-1090.

De Chernatony, L., & McWilliam, G. (1990). Appreciating brands as assets through using a two-dimensional model. *International Journal of Advertising, 9*(2), 111-119.

De Donatis, M. (2006). Map IT: The GIS software for field mapping with Tablet PC. *Computers & Geosciences, 32*(5), 673-680.

De Elizagárate, V. (2003). *Marketing de ciudades*. Madrid: ESIC Pirámide.

Deighton, J. (1996). The future of interactive marketing, *Harvard Business Review*, *74*(6), 151-152.

Dellarocas, C. (2003). The digitization of word-of-mouth: Promise and challenges of online feedback mechanisms. *Management Science*, *49*(10), 1407-1424.

Department for children, schools and families (DfES) (2003). *Towards a unified E-Learning strategy*, http://www.dfes.gov.uk/, Retrieved November 1, 2003, from http://publications.teachernet.gov.uk/default.aspx?PageFunction=productdetails&PageMode=publications&ProductId=DfES+0455+2003&

Der Spiegel (2007). Bad Air in Beijing? Pollution dangers cast shadow over 2008 Olympics. Retrieved June 28, 2007, from Der Spiegel Online at http://www.spiegel.de/international/world/0,1518,491184,00.html

DeSanctis, G., & Poole, M. S. (1994). Capturing the complexity in advanced technology use: Adaptive structuration theory. *Organization Science*, *5*(2), 121-147.

DiMaggio, P. J., & Powell, W. W. (1983). The Iron Cage Revisited: Institutional isomorphism and collective rationality in organizational fields. *American Sociological Review*, *48*, 147-160.

Discover–Botswana, travel and tourism magazine (2006). Retrieved June 27, 2007, from http://www.discover-botswana.com/investors-guide.php

Dobers, P. (2001). Netting the information infrasystem of Stockholm. An idea spreads throughout the world. In H. Glimell & O. Juhlin (Eds.), *The social production of technology. On the everyday life with things* (pp. 189-206). Göteborg: BAS Publishers.

Dobers, P. (2002). Broadband—boom and bust in the new economy. In I. Holmberg, M. Salzer-Mörling & L. Strannegård (Eds.), *Stuck in the future? Tracing 'the new economy'* (pp. 79-104). Stockholm: Bookhouse.

Dobers, P. (2003). Image of Stockholm as an IT city: Emerging urban entrepreneurship. In C. Steyaert & D. Hjort (Eds.), *New movements in entrepreneurship* (pp. 200-217). Aldershot: Edward Eldgar Publishing.

Dobers, P. (2006). Empty spaces or illusionary images? "Stockholm as a Mobile Valley". In S. R. Clegg & M. Kronberger (Eds.), *Space, organization and management* (pp. 287-304). Malmö / Copenhagen: Liber / Copenhagen Business School Press.

Dobers, P., & Hallin, A. (2006). Slipping into darkness: A Study of the role of ICTs in the making of Stockholm's image. *Journal of Urban Technology, 13*(3), 119-127.

Dobers, P., & Strannegård, L. (2001). Loveable networks. A story of affection, attraction and treachery. *Journal of Organizational Change Management, 14*(1), 28-49.

Doel, M., & Hubbard, P. (2002). Taking world cities literally: Marketing the city in a global space of flows. *City, 6*(3), 351-368.

Doolin, B. (2004). Power and resistance in the implementation of a medical management information system. *Information Systems Journal, 14*(4), 343-362.

Doyle, S. (2001). Software review: How is geography supporting marketing in today's commercial organizations? *Journal of Database Marketing, 9*(1), 85-89.

Du, H. S., & Wagner, C. (2006). Weblog success: Exploring the role of technology. *International Journal of Human-Computer Studies*, *64*, 789-798.

Duncan, T., & Moriarty, S. E. (1998). A communication-based marketing model for managing relationships. *Journal of Marketing, 62*, 1-13.

Dunleavy, P., Margetts, H., Bastow, S., & Tinkler, J. (2006). *Digital era governance: IT corporations, the state, and e-government*. Oxford: Oxford University Press.

Dwyer, C. (2007, January). *Digital relationships in the 'MySpace' generation: Results from a qualitative study*. Paper presented at the 40th HICSS, Waikoloa, Big Island, Hawaii.

Dysthe, O. (1996). *Det flerstämmiga klassrummet. Att skriva och samtala för att lära.* (B. Nilsson, Trans.). Lund: Studentlitteratur.

Eagle, K. (2004). Translating strategy: Public sector applications of the balanced scorecard. *Government Finance Review*, 16-27.

Eden, C., & Ackermann, F. (2000). *Making strategy: The journey of strategic management*. London: Sage Publications.

Eden, C., & Ackermann, F. (2001). A mapping framework for strategy making. In A. S. Huff and M. Jenkins (Eds.), *Mapping strategic knowledge* (pp. 173-195), London: Wiley.

Editorial. (2005, January 6). Fibre optic licenses awarded. *The Jamaica Observer*. Retrieved August 5, 2007, from http://www.jamaicaobserver.com/news/html/20050105T220000- 0500_72709_OBS_FIBRE_OPTIC_LICENCES_AWARDED.asp

Eggleston, K., Jensen, R., & Zeckhauser, R. (2002). Information and communication technologies, markets, & economic development. *Discussion Papers Series, Department of Economics, Tufts University 0203, Department of Economics, Tufts University*, 62-74.

Ekinci, Y., & Hosany, S. (2006). Destination personality: An application of brand personality to tourism destinations. *Journal of Travel Research, 45*, 127-139.

El-Gayar, O., & Moran, M. (2007). Examining student acceptance of Tablet PC using TAM. *Issues in Information Systems, VIII*(1), 167-172.

eMarketer (2007). *Reviews boost e-commerce conversions*. Retrieved May 25, 2007, from http://www.eMarketer.com.

eMarketer (2007). *The rising roar of word-of-mouth*. Retrieved June 29, 2007, from http://www.eMarketer.com.

eMarketer (2007). *UGC users outnumber creators*. Retrieved July 2, 2007, from http://www.eMarketer.com.

eMarketer (2007). *Web 2.0 sites draw more visitors*. Retrieved May 2, 2007, from http://www.eMarketer.com.

Enriquez, A., Gunawardena, A., Kowalski, F., Kowalski, S., Millard, D., & Vanides, J. (2006). Innovations in engineering education using Tablet PCs – panel discussion with four institutions, *Frontiers in Education 36th Annual Conference*, San Diego, USA.

Epps, S. R. (2007). *Demystifying tagging for travel sellers*. Forrester Research Report.

Epps, S. R., Harteveldt, H. H., & McGowan, B. (2007). *Executive Q&A: Social tagging for ebusiness. Answers to E-Business professionals' common questions about social tagging*. Forrester Research.

Erat, P., Desouza, K., Schafer-Jugel, A. & Kurzawa, M. (2006). Business customer communities and knowledge sharing: studying the critical issues. *European Journal of IS, 15*, 511-524.

Erbek, F. S., Ulubay, A., Maktav, D., & Yagiz, E. (2005). The use of satellite image maps for urban planning in Turkey. *International Journal of Remote Sensing, 26*(4), 775-784.

Erickson, B., & Roberts, M. (1997). Marketing local identity. *Journal of Urban Design, 2*(1), 35-59.

ESADE, & BDDO Consulting (2004, April). *Evolución del posicionamiento de los destinos turísticos españoles: De lugares de vacación a marcas de turismo.* Paper presented at the XIII Simposi Internacional de Turisme i Lleure ESADE-Fira de Barcelona. Retrieved April 30, 2007, from www.esade.es/cedit2004/cat/est_estudios.php

Esbjornsson, M., & Vesterlind, D. (2003). Mobility and social spatiality. In M. Hard, A. Losch, & D. Verdicchio (Eds.), *Proceedings from transforming spaces: The Topological Turn in Technology Studies Conference*. Darmstadt, Germany. Unpublished.

Etchner, C. M., & Brent Ritchie, F. (1991). The meaning and measurement of destination image. *Journal of Tourism Studies, 2*(2), 2-12.

Europaeische Kommission (2002). *GALILEO—Das europaeische Satellitennavigationsprojekt. Information zu dem Projekt*. Retrieved June 20, 2008, from http://ec.europa.eu/dgs/energy_transport/galileo/doc/galileo_info_note_2002_03_26_de.pdf

European Commission (EC) (2006). *ICT and E-Business in the tourism industry. ICT adoption and e-business activity in 2006. European Commission*. Retrieved February 20, 2008, from http://www.ebusiness-watch.org/studies/sectors/tourism/documents/Tourism_2006.pdf

European Travel Commission (ETC) (2004). *European tourism insight 2004. European Travel Commission.* Retrieved February 20, 2008, from http://www.etc-corporate.org/resources/uploads/ETC_EuropeanTourismInsights_2004.pdf

Fainstein, S. S., & Judd, D. R. (1999). Global forces, local strategies and urban tourism. In D. R. Judd & S. S. Fainstein (Eds.), *The tourist city* (pp. 1-17). New Haven and London: Yale University Press.

Fainstein, S., & Gladstone, D. (1999). Evaluating urban tourism. In D. R. Judd & S. Fainstein (Eds.), *The touristCity* (pp. 21-34). New Haven and London: Yale University Press.

Farmer, J. (2004). Communication dynamics: Discussion boards, Weblogs and the development of communities of inquiry in online learning environments. In R. Atkinson, C. McBeath, D. Jonas-Dwyer, & R. Phillips (Eds.), *Beyond the comfort zone* (pp. 274-283). *Proceedings of the 21st ASCILITE Conference.*

Fernández Güell, J. M. (1997). *Planificación estratégica de ciudades*. Barcelona: Gustavo Gili.

Fesenmaier, D. R. (2007). Introduction: Challenging destination promotion. *Journal of Travel Research, 46*, 3-4.

Fesenmaier, D. R., Werthner, H., & Woeber, K. (2006). *Destination recommendation system: Behavioral foundation and applications*. Oxford: CABI Publishing.

Fisher, M., Raman, A., & McClelland, A. (2000). Rocket science retailing is almost here—are you ready? *Harvard Business Review, 78*(4), 115-24.

Fishman, R. (1987). *Bourgeois utopias. The rise and fall of suburbia*. New York: Basic Books.

Florida, R. (2005). The flight of the creative class. *The new global competition for talent*. New York, NY: Harper Collins Business.

Font, J. (2003). *Competencia y colaboración de ciudades: La aportación del benchmarking*. Retrieved July 10, 2008, from http://www.rosario.gov.ar/per/Acti/Ponen2.html

Forbes (2007). *Best places for business and careers*. Retrieved August 1, 2007, from http://www.forbes.com/lists/2007/1/07bestplaces_Best-Places-For-Business-And-Careers_land.html

Forrester Research (2006). *Social computing*. Retrieved September 14, 2007, from http://www.forrester.com.

Foucault, M. (1977). *The history of sexuality volume one: The will to knowledge* (Vol. 1). London: Penguin.

Foucault, M. (1980). Truth and power. In P. Rabinow (Ed.), *The Foucault reader: An introduction to foucault's thought* (pp. 51-75). London: Penguin.

Foucault, M. (1984). The ethics of the concern of the self as a practice of freedom (R. e. a. Hurley, Trans.). In P. Rabinow (Ed.), *Michel Foucault, ethics subjectivity and truth: Essential works of Foucault 1954-1984* (pp. 281-301). London: Penguin.

Fountain, J. E. (1995). *Enacting technology: An institutional perspective*. Cambridge, MA: John F. Kennedy School of Government, Harvard University.

Fountain, J. E. (2001). *Building the virtual state. Information technology and institutional change*. Washington, D.C.: Brookings Institution Press.

Fraunhofer IIS (2008). *Testumgebung WLAN-Lokalisierung Nürnberg: Vor-Ort-Information und ortsabhängige Dienste mit starken Partnern*. Retrieved June 20, 2008, from http://www.iis.fraunhofer.de/pr/Presse/Presseinformationen_2008/WLAN_Testumgebung.jsp

Frederickson, H. G. (2004). All public administration is local. *PA Times, 27*(11), 11-12.

Fretter, A. (1993). Place marketing: A local authority perspective. In G. Kearns & C. Philo (Eds.), *Selling places: The city as cultural capital, past and present* (pp. 163-174). New York, NY: Pergamon Press.

Fust, W. (2005). *Knowledge society, knowledge management and ICT*: Editorial contribution to a publication ed. by the Community Development Library Dhaka/Bangladesh.

Galí, N., & Donaire, J. A. (2005). The social construction of the image of Girona: A methodological approach. *Tourism Management, 26*, 777-785.

Gallarza, M. G., Gil, I., & Calderón, H. (2002). Destination Image – Towards a conceptual framework. *Annals of Tourism Research, 29*(1), 56-78.

Galliers, R. (2004). Reflections on information systems strategizing. In C. Avgerou, C. Ciborra & F. Land (Eds.), *The social study of information and communication technology: Innovation, actors and contexts* (pp. 231-262). Oxford: Oxford University Press.

Gant, D., & Kiesler, S. (2002). Blurring the boundaries: Cell phones, mobility, and the line between work and personal life. In B. Brown, N. Green & R. Harper (Eds.), *Wireless world. Social and interactional aspects of the mobile age* (pp. 121-131). London: Springer.

Gardberg, N. A., & Fombrun, C.J. (2002). The global reputation quotient project: First steps towards a cross-nationally valid measure of corporate reputation. *Corporate Reputation Review, 4*(4), 303-307.

Gartner, W. C. (1986). Temporal influence on image change. *Annals of Tourism Research, 13*, 635-644.

Gartner, W.C. (1993). Image formation process. *Journal of Travel and Tourism Marketing, 2*(2), 191-215.

Gartner,W., & J. Hunt (1987). An analysis of state image change over a twelve year period (1971-1983). *Journal of Travel Research, 16*(2), 15-19.

Gartrell, R. B. (Ed.). (1988). *Destination marketing for convention and visitor bureaus*. Dubuque: Kendall/Hunt.

Gascó Hernández, M. (2003). New technologies and institutional change in public administration. *Social Science Computer Review, 21*(1), 6-14.

Gascó Hernández, M. (2004). E-gobierno en Bolivia y Paraguay. In R. Araya Dujisin & M. A. Porrúa Vigon (Eds.), *América latina puntogob: Casos y tendencias en gobierno electrónico* (pp. 125-150). Santiago: FLACSO-Chile/ AICD-OEA.

Gentile, A. (2007, September). Help! 311 and CRM systems generate quick responses to residents' ordinary problems. *American City & County*, 32-36.

George, M. L. (2003). *Lean six sigma for service: How to use lean speed and six sigma quality to improve services and transactions*. New York, NY: McGraw-Hill.

Gibbert, M., Leibold, M., & Probst, G. (2002). Five styles of CKM, and how smart companies use them to create value. *European Management Journal, 20*(5), 459-469.

Giddens, A. (1984). *The constitution of society*. Berkeley and Los Angeles, CA: University of California Press.

Gilbert, D. (1990). Strategic marketing planning for national tourism. *The Tourist Review, 1*, 18-27.

Gil-Garcia, J. R. (2005). *Enacting state websites: A mixed method study exploring e-government success in multi-organizational settings*. Unpublished Doctoral Dissertation, University at Albany, State University of New York, Albany, NY.

Gil-Garcia, J. R., & Luna-Reyes, L. F. (2006). Integrating conceptual approaches to e-government. In M. Khosrow-Pour (Ed.), *Encyclopedia of e-commerce, e-government and mobile commerce* (pp. 636-643). Hershey, PA: Idea Group Inc.

Gillenson, M. L., Sherrell, D. L., & Chen, L. (1999). Information technology as the enabler of one-to-one marketing. *Communications of the AIS, 2*(18), 43.

Glasgow (2007). *Scotland with style*. Retrieved August 16, 2007, from http://www.seeglasgow.com

Gnoth, J. (1998). Conference reports: Branding tourism destinations. *Annals of Tourism Research, 25*, 750-760.

Goldman, C. A., Gates, S. M., & Brewer, D. J. (2001). Prestige or reputation: Which is a sound investment? *Chronicle of Higher Education, 48*, 13-15.

Gommans, M., Krishnan, K.S., & Scheffold, K. B. (2001). From brand loyalty to e-loyalty: A conceptual framework. *Journal of Economic and Social Research, 3*(1), 43-58.

Gordon, S. (2006, April 12). Flow is neither new nor different say competitors. *The Jamaica Gleaner*. Retrieved August 5, 2007, from the http://www.jamaica-gleaner.com/gleaner/20060412/business/business2.html

Govers, R., & Go, F. M. (2004). Cultural identities constructed, imagined and experienced: A 3-gap tourism destination image model. *Tourism, 52*(2), 165-182.

Grabow, B., & Hollbach-Groemig, B. (1998). *Stadtmarketing—eine kritische Zwischenbilanz*. Berlin, Germany: Deutsches Institut fuer Urbanistik.

Graham, S. (2002). Bridging urban digital divides? Urban polarization and information and communication technologies. *Urban Studies, 39*(1), 33-56.

Graham, S., & Marvin, S. (1998). *The richness of cities. Urban policy in a new landscape* (Net effects: Urban planning and the technological future of cities No. Working Paper 3). Newcastle: Centre for Urban Technology, Department of Town and Country Planning, University of Newcastle.

Graham, S., & Marvin, S. (2001). *Splintering urbanism. Networked infrastructures, technological mobilities and the urban condition*. London: Routledge.

Granovetter, M. S. (1985). Economic action and social structure: The problem of embeddedness. *American Journal of Sociology, 91*, 481-510

Gregorio, D. D., Kassicieh, S. K., & De Gouvea Neto, R. (2005). Drivers of E-business activity in developed and emerging markets. *IEEE Transactions on Engineering Management, 52*(2), 155-166.

Grenier, R., & Metes, G. (1995). *Going virtual*. Upper Saddle River, NJ: Prentice Hall.

Grenny, J., Maxfield, D., & Shimberg, A. (2007). How project leaders can overcome the crisis of silence. *MIT Sloan Management Review, 48*(4), 46-52.

Gretzel, U., Fesenmaier, D. R., Formica, S., & O'Leary, J. T. (2006). Searching for the future: Challenges faced by destination marketing organizations. *Journal of Travel Research, 45(2)*, 116-126.

Gretzel, U., Yuan, Y., & Fesenmaier, D. R. (2000). Preparing for the new economy: Advertising strategies and change in destination marketing organizations. *Journal of Travel Research, 39*, 146-156.

Gruen, T. W., Osmonbekov, T., & Czaplewski, A. J. (2006). eWOM: The impact of customer-to-customer online know-how exchange on customer value and loyalty. *Journal of Business Research, 59*, 449-456.

Guenther, K. (2002). *Stadtmarketing—Grundlagen und konzeptionelle Gestaltungsansaetze fuer eine raeumlich ganzheitliche Versorgungspolitik unter besonderer Beruecksichtigung des stationaeren Einzelhandels*. Goettingen, Germany: GHS.

Guermazi, B., & Satola, D. (2005, November). Creating the "right" enabling environment for ICT. In Schware,

R. (Ed.), *E-Development: From excitement to effectiveness.* A report prepared for the World Summit on the Information Society held in Tunis. Global, Information and Communication Technologies Department. World Bank Group, pp. 23-47.

Guia, J., Prats, L., & Comas, J. (2006). The destination as a local system of innovation: The role of relational networks. In L. Lazzeretti, & C. Petrillo (Eds.), *Tourism local system and networking.* Elseivier: Amsterdam.

Gummesson, E. (2002). Relationship Marketing in the new economy. *Journal of Relationship Marketing,* 1(1), 37-57.

Gunn, C. A. (1972). *Vacationscape: Designing tourist regions.* Washington DC: Taylor and Francis/University of Texas.

Gwinner, K. P., Gremmler, D. D., & Bitner, M. J. (1998). Relational benefits in services: The customer's perspective. *Journal of the Academy of Marketing Science, 26*(2), 101-114.

Hagel, J. (1999). Net gain: Expanding markets through virtual communities. *Journal of Interactive Marketing, 13(1),* 55-65.

Håkansson, H., & Snehota, I. J. (Eds.) (1995). *Developing relationships in business networks.* London: Routledge.

Halaris, C., Magoutas, B., Papadomichelaki, X., & Mentzas, G. (2007). Classification and synthesis of quality approaches in E-Government services. *Internet Research, 17*(4), 378-401.

Hall, C. M. (2000). *Tourism planning: Policies, processes and relationship.* Harlow: Prentice Hall.

Hall, P. (1977). *The world cities* (2nd ed.). London: Weidenfeld and Nicolson.

Hall, S. (Ed.). (1997). *Representation: Cultural representations and signifying practices.* London: Sage / Open University Press.

Hall, T. (1997). (Re)placing the city. Cultural relocation and the city as centre. In S. Westwood & J. Williams (Eds.), *Imagining cities. Scripts, signs, memory* (pp. 202-218). London and New York: Routledge.

Hall, T. (1998). Introduction to selling the entrepreneurial city. In T. Hall & P. Hubbard (Eds.), *The entrepreneurial city. Geographies of politics, regime and representation* (pp. 27-30). Chichester: John Wiley.

Hall, T., & Hubbard, P. (Eds.). (1998). *The entrepreneurial city: Geographies of politics, regime and representation.* Chichester, UK: John Wiley.

Hallberg, P. (1992). *Litterär teori och stilistik* (4. uppl. ed.). Göteborg: Akademiförl.

Hallin, A. (forthcoming). *Size matters: The problem of organizational size and the case of mCity.* PhD thesis, The Royal Institute of Technology Stockholm.

Hallin, A., & Lundevall, K. (2007). mCity—User focused development of mobile services within the city of Stockholm. In I. Kushchu (Ed.), *Mobile government: Emerging directions in E-Government*: Hershey, PA: Idea Group publishers.

Han, J. H., & Mills, J. E. (2006). Zero acquaintance benchmarking at travel destination Web sites: What is the first impression that national tourism organizations try to make? *International journal of Tourism Research, 8,* 405-430.

Hankinson, G. (2004). The brand images of tourism destinations: A study of the saliency of organic images. *Journal of Product & Brand Management, 13*(1), 6-14.

Hankinson, G. (2005). Destination brand images: A business tourism perspective. *Journal of Services Marketing, 19*(1), 24-32.

Hannigan, J. (1998). *Fantasy city. Pleasure and profit in the postmodern metropolis.* London: Routledge.

Hanson, S. (2007). *Urbanization in Sub-Saharan Africa.* Retrieved June 20, 2008, from http://www.cfr.org/publication/14327/urbanization_in_subsaharan_africa.html

Harloe, M. (2001). Social justice and the city: The new 'liberal' formulation. *International Journal of Urban and Regional Research, 25*(4), 889-897.

Harteveldt, H., Johnson, C. A., Epps, S. R., & Tesch, B. (2006).*Travelers embrace social computing technologies. Guidelines for travel e-commerce and marketing executives and managers.* Cambridge, MA, USA: Forrester Research.

Harvey, D. (1989). From managerialism to entrepreneurialism: The transformation in urban governance in late capitalism. *Geografiska Annaler, 71 B*(1), 3-17.

Hashim, H. H., Murphy, J., & Muhamad Hashim, N. (2007). Islam and online imagery on Malaysian tourist destination Web sites. *Journal of Computer-Mediated Communication, 12*(3), article 16.

Hassan, S., & Gil-Garcia, J. R. (2007). Institutional theory and e-government research. In G. D. Garson & M. Khosrow-Pour (Eds.), *Handbook of research on public information technology*. Hershey, PA: IGI Global.

Haven, B. (2007). *Making podcasts work for your brand*. Cambridge, MA, USA: Forrester Research.

Heath, E., & Wall, G. (1992). *Marketing tourism destinations, a strategic planning approach*. New York, NY: John Wiley and Sons.

Heeks, R. (2005). E-government as a carrier of context. *Journal of Public Policy, 25*(1), 51-74.

Heidelberg-Mobile (n/d). Retrieved May 23, 2007, from http://www.heidelberg-mobil.de

Heijden, H. (2003). Factors influencing the usage of web sites: The case of a generic portal in The Netherlands. *Information and Management, 40*, 541-549.

Heise Mobil (2008). WLAN-Navigationssystem leitet Nürnberger Fußgänger. Retrieved February 20, 2008, from http://www.heise.de/mobil/newsticker/meldung/print/101765

Henderson, J. C., & Venkatraman, H. (1999). Strategic alignment: Leveraging information technology for transforming organizations. *IBM Systems Journal, 38*(2/3), 472-484.

Henderson, J., & Venkatraman, N. (1999). Strategic alignment: Leveraging information technology for transforming organizations. *IBM Systems Journal, 38*(2&3), 472-484.

Hepworth, M. E. (1989). *Geography of the information economy*. London: Belhaven Press.

Hepworth, M. E. (1990). Planning for the information city. The challenge and response. *Urban Studies, 27*(4), 537-558.

Hicks, J. F. (1998, July). *Sub-Saharan Africa: Enhancing the productivity of urban Africa*. Paper presented at the International Conference on Research Community for the Habitat Agenda Forum of Researchers on Human Settlements. Geneva, Italy.

Hirschheim, R., & Klein, H. (1989). Four paradigms of information systems development. *Communications of the ACM, 32*(10), 1199-1216.

Hirschheim, R., & Klein, H. (1994). Realizing emancipatory principles in information systems development: The case for ETHICS. *MIS Quarterly, 18*(1), 83-105.

Hirschheim, R., Klein, H. K., & Newman, M. (1991). Information systems development as social action: Theoretical perspective and practice. *Omega, 19*(6), 587-608.

Hite, J. M. (2003). Patterns of multidimensionality among embedded network ties: A typology of relational embeddedness in emerging entrepreneurial firms. *Strategic Organization, 1*(1), 9-49.

Hjalager, A. M. (2000). Tourism destinations and the concept of industrial districts. *Tourism and Hospitality Research, 2*(3), 199-213.

Hoffman, D. L., & Novak, T. (1996). Marketing in hypermedia computer-mediated environments: Conceptual foundations. *Journal of Marketing, 60*, 50-68.

Hoffmann-Martinot, V. (1998). Urban Political Parties: Role and Transformation. In V. Hoffmann-Martinot, T. N. Clark (Eds) & M. Gromala, *The new political culture* (pp. 195-218). Oxford: Westview Press.

Holcomb, B. (1994). City make-overs: Marketing the post-industrial city. In J. Gold & S. Ward (Eds.), *Place promotion: The use of publicity and marketing to sell towns and regions* (pp. 115-131). Chichester, UK: John Wiley.

Holcomb, B. (1999). Marketing cities for tourism. In D. R. Judd & S. Fainstein (Eds.), *The touristCity* (pp. 54-70). New Haven and London: Yale University Press.

Holland, C., & Naude, P. (2004). The metamorphosis of marketing into an information-handling problem. *Journal of Business and Industrial Marketing, 19*(3), 167-177.

Holley, L. M., Dufner, D., & Reed, B. J. (2002). Got SISP? Strategic information systems planning in U.S. state governments. *Public Performance & Management Review, 25*(4), 398-412.

Holley, L. M., Dufner, D., & Reed, B. J. (2004). Strategic information systems planning in U.S. county governments: Will the real SISP model please stand up? *Public Performance & Management Review, 27*(3), 102-126.

Hollis, N. (2005, June). Ten years of learning on how online advertising builds brands. *Journal of Advertising Research*, 255-268.

Holmen, E., Pedersen, A. C., & Torvatn, T. (2005). Building relationship for technological innovation. *Journal of Business Research, 58*(9), 1240-1250.

Hong Kong Trade and Investment Council. Retrieved August 4, 2007, from http://www.tdctrade.com/.

Hosany, S., Ekinci, Y., & Uysal, M. (2006). Destination image and destination personality: An application of branding theories to tourism places. *Journal of Business Research, 59*, 638-642.

Huang, C. D., & Hu, Q. (2007). Achieving IT-business strategic alignment via enterprise-wide implementation of balanced scorecards. *Information Systems Management, 24*(2), 173-184.

Hubbard, P. (1996). Urban design and city regeneration: Social representations of entrepreneurial landscapes. *Urban Studies, 33*(8), 1441-1461.

Hubbard, P., & Hall, T. (1998). The entrepreneurial city and the 'new' urban politics. In T. Hall & P. Hubbard (Eds.), *The entrepreneurial city: Geographies of politics, regime and representation* (pp. 1-23). Chichester, UK: John Wiley and Sons.

Hudson, S., & Ritchie, J. R. B. (2006). Promoting destinations via film tourism: An empirical identification of supporting marketing initiatives, *Journal of travel research, 44*, 387-396.

Hull City Council. (2007). *Hull City Council website.*

Hull. (2007). Flood-hit Hull a 'forgotten city'. *BBC news, available at http://news.bbc.co.uk/1/hi/england/humber/6270236.stm.*

Hulm, P. (2007). Tunisia's boom in ICT. *International Trade Forum, 1*, 24.

Hwang, Y., Gretzel, U., Xiang, Z., & Fesenmaier, D. (2006). Information search for travel decisions. In D. Fesenmaier, H. Werthner & K. Wöber (Eds.), *Destination recommendation systems: Behavioral foundations and applications* (pp. 3-16). Cambridge, MA: CAB International.

Investors Guide to Botswana Magazine (2007). Retrieved August 10, 2007, from http://www.discover-botswana.com/investors-guide.php

Ishida, T. (2002). Digital city Kyoto. *Communications of the ACM, 45*(7), 76-81.

J.D. Power (2008). *Business ratings.* Retrieved February 23, 2008, from http://www.jdpower.com/business

Jackson, J., & Murphy, P. (2002). Tourism destinations as clusters: Analytical experiences from the World. *Tourism and Hospitality, 4*(1), 36-52.

Jamaica Digiport. Retrieved August 4, 2007, from http://www.jadigiport.com

Jamaica Trade and Invest. Retrieved August 4, 2007, from http://www.jamaicatradeandinvest.org/.

James, T., & Wild, K. (2005). *Report of a feasibility study for an information society program for the African, Caribbean and Pacific (ACP) countries; ANNEX IV: Regional Report – Africa.*

Jang, H. Y, Ko, I. S., & Koh, J. (2007, January). *The influence of online brand community characteristics on community commitment and brand loyalty.* Paper presented at the 40th HICSS, Waikoloa, Big Island, Hawaii.

Janssen, M., & Cresswell, A. M. (2005, January). *The development of a reference architecture for local government.* Paper presented at the 38th Hawaii International Conference on System Sciences, Hawaii.

Jansson, J., & Power, D. (Eds.) (2007). *The image of the city—Urban branding as constructed capabilities in Nordic city regions.* Uppsala: Nordic Innovation Centre/Dept of Social and Economic Geography.

Jenkins, O. H. (1999). Understanding and measuring tourist destination images. *International Journal of Tourism Research, 1*, 1-15.

Jenman Safaris Africa (2007). Retrieved July 15, 2007, from http://www.jenmansafaris.com/

Jessop, B. (1998). The narrative of enterprise and the enterprise of narrative: Place marketing and the entrepreneurial city. In T. Hall & P. Hubbart (Eds.), *The Entrepreneurial city. Geographies of politics, regime and representation.* Chichester & New York: Wiley.

Jessop, B., & Sum, N.-L. (2000). An entrepreneurial city in action: Hong Kong's emerging strategies in and for (inter)urban competition. *Urban Studies, 37*(12), 2287-2313.

Jewitt, C., & Oyama, R. (2001). Visual meaning: A social semiotic approach. In T. V. Leeuwen & C. Jewitt (Eds.), *Handbook of visual analysis* (pp. 134-156). London, Thousand Oaks, New Delhi: Sage Publication.

Johnson, G., & Scholes, K. (1999). *Exploring corporate strategy.* Harlow, UK: Prentice Hall.

Judd, D. R., & Fainstein, S. S. (Eds.). (1999). *The tourist city.* New Haven, CT.

Kabbaj, M. Y. (2003). *Strategic and policy prospects for semantic web services adoption in US online travel industry.* Doctoral dissertation. Retrieved December 20, 2008, from http://ebusiness.mit.edu/bgrosof/paps/kabbajmasters-thesis-travel+sws.pdf

Kapferer, J.-N. (1997). *Strategic brand management. Creating and sustainaing brand equity long term* (2nd ed.). London: Kogan Page.

Kapferer, J-N. (1991). *La marca, capital de la empresa.* Bilbao: Deusto.

Kaplan, R. S. (1999). *City of Charlotte (B).* Boston, MA: Harvard Business School Press.

Kaplan, R. S., & Norton, D. P. (1992). The balanced scorecard--Measures that drive performance. *Harvard Business Review, 70*, 71-79.

Kaplan, R. S., & Norton, D. P. (1996). Using the balanced scorecard as a strategic management system. *Harvard Business Review, 74*, 75-76.

Kaplan, R. S., & Norton, D. P. (2004). *Strategy maps: Converting intangible assets into tangible outcomes.* Boston, MA: Harvard Business School Press.

Kaplanidou, K., & Vogt, C. (2006). A structural analysis of destination travel intentions as a function of web site features, *Journal of Travel Research, 45(2)*, 204-216.

Karavatzis, M. (2004). From city marketing to city branding: Towards a theorical framework for developing city brands. *Journal of Place Branding, 1*(1), 58-73.

Kavaratzis, M. (2005). Place branding: A review of trends and conceptual models. *The Marketing Review, 5*(4), 329-342.

Kavaratzis, M. (2007). City marketing: The past, the present and some unresolved issues. *Geography Compass, 1*(3), 695-712.

Kavaratzis, M., & Ashworth, G. J. (2005). City branding: An effective assertion of identity or a transitory marketing trick? *Tijdschrift voor Economische en Sociale Geographie, 96*(5), 506-514.

Kavaratzis, M., & Ashworth, G. J. (2006). City branding: An effective assertion of identity or a transitory marketing trick? *Place Branding, 2*(3), 183-194.

Kavaratzis, M., & Ashworth, G. J. (2006). Partners in coffeeshops, canals and commerce. Marketing the city of Amsterdam. *Cities, 24*(1), 16-25.

Kavaratzis, M., & Ashworth, G. J. (2007). *Partners in coffee shops, canals and Commerce: Marketing the city of Amsterdam. Netherlands.* University of Groningen, Urban and Regional Studies Institute.

Kearns, G. S., & Sabherwal, R. (2007). Strategic Alignment between business and information technology: A knowledge-based view of behaviors, outcome, and consequences. *Journal of Management Information Systems, 23*(3), 129-162.

Kearns, G., & Philo, C. (1993). Culture, history, capital. A critical introduction to the selling of places. In G. Kearns & C. Philo (Eds.), *Selling places. The city as cultural capital. Past and present* (pp. 1-32). Oxford: Pergamon.

Kearns, G., & Philo, C. (Eds.). (1993). *Selling places. The city as cultural capital. Past and present.* Oxford: Pergamon.

Keen, M., & Mahanty, S. (2006). Sustainability assessment and local government: Achieving innovation through practitioner networks. *Local Environment, 11*(2), 201-216.

Keiner, M. (2003). *Sustainability oriented urban development: A general introduction with case studies from Gaborone, Johannesburg and Santiago de Chile.* Zurich. Institute for Spatial and Landscape Planning.

Keller, K. (1993). Conceptualizing, measuring, and managing customer-based brand equity. *Journal of Marketing, 57*, 1-22.

Kellogg, C., & Lillquist, R. (1999). *How to market your city.* City, County Communications Association. Washington. Municipal Research & Services Center.

Kellogg, C., & Lillquist, R. (1999). *How to market your city.* Retrieved July 16, 2007, from http://www.3cma.org/members/resourcecenter/How%20To%20Market%20Your%20City.pdf.

Kelly, P. (ed.) (2005). *Economies of the world.* London: Routledge.

Kenny, C. (2006). The Internet and economic growth in LDCs: A case of managing expectations. In A. D'Costa (Ed.), *The new economy in development: ICT challenges and opportunities* (pp. 67-88). New York, NY: Palgrave Macmillan.

Kiang, M. Y., Raghu, T. S., & Shang, K. H.-M. (2000). Marketing on the internet. Who can benefit from an online marketing approach? *Decision Support Systems, 27*, 383-393.

Kim, C. K. (2007). A cross-national analysis of global e-government. *Public Organization Review, 7*(4), 317-329.

Kim, J. (2008). A model and case for supporting participatory public decision making in E-Democracy. *Group Decis Negot, 17*, 179-193.

Kim, W. G., Lee, C., & Hiemstra, S. J. (2004). Effects of an online virtual community on customer loyalty and travel product purchases. *Tourism Management, 25*(3), 343-355.

Klein, L. R. (2003). Creating virtual product experiences: The role of telepresence. *Journal of Interactive Marketing, 17*(1), 41-55.

Knauth, B. (2006). *Tourism and the Internet in the European Union*. Luxembourg: European Communities.

Knight, C., & Knight, B. A. (1995). Cognitive theory and the use of computers in the primary classroom. *British Journal of Educational Technology, 26*, 141-148.

Ko, H., Cho, C-H., & Roberts, M. S. (2005). Internet uses and gratifications. A structural equation model of interactive advertising. *Journal of Advertising, 34*(2), 57-70.

Koelmel, B., & Porak, A. (2003). *Real Life Scenarios of Location Based Advertising*. Eurescom Summit 2003.

Kolb, B. M. (2006). *Tourism marketing for cities and towns: Using branding and events to attract tourists*. Burlington, MA: Elsevier Butterworth-Heinemann.

Konecnik, M., & Gartner, W.C. (2006). Customer-based brand equity for a destination. *Annals of Tourism Research, 34*(2), 400-421.

Kotler, P. (2000). *Marketing Management*. Upper Saddle River, NJ: Prentice-Hall.

Kotler, P., & Gertner, D. (2004). Country as brand, product and beyond: A place marketing and brand management perspective. In N. Morgan, A. Pritchard & R. Pride (Eds.), *Destination branding: Creating the unique destination proposition* (pp. 40-56). Oxford: Elsevier Butterworth-Heinemann.

Kotler, P., & Levy, S. (1969). Broadening the concept of marketing. *Journal of Marketing, 33*(1), 10-15.

Kotler, P., & McDougall, G. (1983). Principles of marketing (Canadian ed.) Scarborough: Prentice-Hall Canada.

Kotler, P., Asplund, C., Rein, I., & Haider, D. H. (1999). *Marketing places Europe. How to attract investments, industries, residents and visitors to cities, communities, regions and nations in Europe*. London: Pearson Education.

Kotler, P., Bowen, J., & Makens, J. (1996). *Marketing for hospitality and tourism*. New Jersey: Prentice Hall

Kotler, P., Gertner, D., Rein, I., & Haider, D. (2007). *Marketing internacional de lugares y destinos: Estrategias para la atracción de clientes y negocios en Latinoamérica*. Mexico City: Pearson Prentice Hall.

Kotler, P., Haider, D. H., & Rein, I. (1993). *Marketing places: Attracting investment, industry and tourism to cities, states and nations*. New York, NY: Free Press.

Kotler, P., Haider, D. H., & Rein, I. (1994). *Marketing places*. New York, NY: Free Press.

Kotler. (1997). *Marketing management. Analysis, planning, implementation and control*. New Jersey: Prentice Hall.

Krugman, P. (1996). Making sense of the competitiveness debate. *Oxford Review of Economic Policy, 12*, 17–25.

Krugman, P., & Venables, A. J. (1993). *Integration, specialization and adjustment*. Discussion Paper No. 886. Centre for Economic Policy Research (CEPR), London.

Kuehnberger, P. (2004). *Mobile City Bregenz*. Retrieved June 20, 2008, from http://wko.at/bshandel/Zahlungsverkehr/Veranstaltung-Bezahlen-im-Internet/Praes-Filme-Unterl/mobilkom/BREGENZ.ppt#336,6,Umsetzung

Kumar, R., Novak, J., Raghavan, P., & Tomkins, A. (2004). Structure and evolution of blogspace. *Communications of the ACM, 47(12)*, 35-39.

LA Times (2007). *Clearing the air for Beijing Olympics—China is moving factories, reducing traffic and even changing the weather to limit pollution at the '08 Games*. Retrieved August 11, 2007, from http://www.latimes.com/news/printedition/asection/la-fg-air11aug11,1,2200892.story?coll=la-news-a_section&ctrack=4&cset=true

Lagrosen, S. (2005). Customer involvement in NPD: A relationship marketing perspective. *European Journal of Innovation, 8*(4), 424-436.

Lambi, J.-J. (1995). *Marketing estratégico*. Madrid: McGraw-Hill.

Lapierre, J., & Medeiros, R. (2006). Information and community technology usage patterns: A case study. *Journal of Strategic Marketing, 14*(3), 229-244.

Larsen, G., & George, V. (2004, February). The social construction of destination image – A New Zeland firm example. *Working Papers*, 4/01.

Laudon, K., & Laudon, J. P. (2006). *Management information systems: Managing the digital firm*. Upper Saddle River, NJ: Prentice Hall.

Lawson, F., & Band-Bovy, M. (1977). *Tourism and recreational development.* London: Architectural Press.

Lazerson, M., & Lorenzoni, G. (1999). The firms that feed industrial districts: A return to the Italian source. *Industrial and Corporate Change*, *8*(2), 235-266.

Le Cam, F., Ruellan, D., & Cabedoche, B. (2006). *Public service information. The editorial identity of city councils' websites in Brittany.* Brest: Mole Armoricain de la Recherche sur la Societé de la Information (M@rsouin).

Le Galès, P. (1998). Regulations and governance in European cities. *International Journal of Urban and Regional Research*, *22*(3), 482-506.

Le Galès, P. (2001). Urban governance and policy networks: On the urban political boundedness of policy networks- A French case study. *Public Administration*, *79*(1), 167-184.

Le Galès, P. (2006). New state space in Western Europe? *International Journal of Urban and Regional Research*, *30*(3), 717-721.

Lee, G. M., Cai, L. A., & O'Leary, J. T. (2006). WWW. Branding.States.US: An analysis of brand-building elements in the U. S. state tourism Web sites. *Tourism Management*, *27*, 815-828.

Lee, G.-G., & Bai, R.-J. (2003). Organizational mechanisms for successful IS/IT strategic planning in the digital era. *Management Decision*, *41*(1/2), 32-42.

Leeds, K. (2005). *Technology: Fad or fixture? A study on students' perceptions of using Tablet PCs during class.* Retrieved June 6, 2006, from http://www.cccone.org/scholars/05-06/KelvinLeedsExecutiveSummary.pdf

Leeuwen, T. v. (2001). Semiotics and iconography. In T. V. Leeuwen & C. Jewitt (Eds.), *Handbook of visual analysis* (pp. 92-118). London, Thousand Oaks, New Delhi: Sage Publication.

Lento T., Welser, H. T., Gu, L., & Smith, M. (2006). The ties that blog: Relationship between social ties and continued participation in blogs. *Workshop on Weblogging* Edinburgh.

Leverick, F., Littler, D., Bruce, M., & Wilson, D. (1998). Using information technology effectively: A study of marketing installations. *Journal of Marketing Management, 14*(8), 927-962.

Li, C., & Stromberg, C. (2007). *The ROI of blogging.* Cambridge, MA: Forrester Research.

Lin, Y., & Huang, J. (2006). Internet blogs as a tourism marketing medium: A case study. *Journal of Business Research, 59*, 1201-1205.

Lin, Y., Su, H. Y., & Chien, S. (2006). Knowledge-enabled procedure for customer relationship management. *Industrial Marketing Management, 35*, 446-456.

Liu, Y. & Shrum, L. J. (2002). What is interactivity and is it always such a good thing? Implications of definition, person and situation for the influence of interactivity on advertising effectiveness. *Journal of Advertising*, *31*(4), 53-64.

Liu, Y. (2003, June). Developing a scale to measure the interactivity of Web sites. *Journal of Advertising Research*, 207-216.

Löfgren, O. (2001). Urbana koreografier—rum och rörelse i 1900-talets städer. In R. Solli & B. Czarniawska (Eds.), *Modernisering av storstaden: Marknad och management i stora städer vid sekelskiftet* (1. ed., pp. 15-34). Malmö: Liber.

López Lita, R., & Benlloch, M. T. (2005). La marca territorio. El marketing de ciudad, una herramienta al sevicio de las marcas territorio. *99% com*, *2*, 8.

Lowndes, V. (2005). Something old, something new, something borrowed . . . How institutions change (and stay the same) in local governance. *Policy Studies, 26*(3/4), 291-309.

Luley, P. M., Almer, A., Schnabel, T, Massimo, R., & Herpolsheimer, W. (2004). Geo-Data presentation on mobile devices for tourism application. *Symposium on Location Based Services & TeleCartography, University of Technology*, Wien, Austria. Retrieved June 20, 2008, from http://www.joanneum.at/uploads/tx_publicationlibrary/img1999.pdf

Luna-Reyes, L. F., Gil-Garcia, J. R., & Cruz, C. B. (2007). Collaborative digital government in Mexico: Some lessons from federal web-based interorganizational information integration initiatives. *Government Information Quarterly, 24*(4), 808-826.

Luna-Reyes, L. F., Gil-Garcia, J. R., & Estrada-Marroquín, M. (2008). The impact of institutions on interorganizational IT projects in the Mexican federal government. *International Journal for Electronic Government Research, 4*(2), 27-42.

Lundeval, K., & Ozan, H. (2006). *Report Mcity. Mobile projects in the city of Stockholm to make internal work more efficient and to increase the use of self service by city citizen.*

Lundvall, B. A. (Ed.) (1992). *National systems of innovation. Towards a theory of innovation and interactive learning.* London: Pinter Publishers.

Lury, C. (1998). *Prosthetic culture: Photography, memory, and identity.* London: Routledge.

Lynch, P. J., & Horton, S. (2004). *Manual de estilo web. Principios de diseño básico para la creación de sitios Web.* Barcelona: Gustavo Gili, SA.

Madon, S. (1997). Information-based global economy and socioeconomic development: The case of Bangalore. *The Information Society, 13*, 227-243.

Maghroori, R. &, Rolland, E. (1997). Strategic leadership: The art of balancing vision with policy, procedures, and external environment. *The Journal of Leadership Studies, 4*(2), 62-81.

Mahmood, T., Ricci, F., Venturini, A., & Höpken, W. (2008). Adaptative recommender systems for travel planning. In P. O'Connors, W. Höpken & U. Gretzel (Eds.), *Information and communication technologies in tourism* (pp. 1-11). Innsbruck, Austria: Springer Verlag.

Maitlamo (2005). *National policy for ICT development.* Ministry of Communication, Science and Technology. Government of Botswana. Retrieved June 10, 2008, from http://www.maitlamo.gov.bw/ict-vision.asp

Malnight, T., & Keys, T. (2007). Surfing the storm: Translating long-term global trends into today's decisions. *Perspectives for Managers, 145*, 1.

Mansell, R. (Ed.). (2002). *Inside the communication revolution: Evolving patterns of social and technical interaction* (1st ed.). Oxford: Oxford University Press.

Mansell, R., & Steinmueller, W. (2000). *Mobilizing the information society: strategies for growth and opportunity.* Oxford: Oxford University Press.

Markus, M., & Lassing, M. (2008). Some critical remarks on dynamic packaging from the perspective of SMEs and small tourism destination. . In P. O`Connors, W. Höpken & U. Gretzel (Eds.), *Information and communication technologies in tourism* Innsbruck, Austria: Springer Wien.

Marquardt, M. J. (2005). *Leading with questions.* San Francisco, CA: Jossey-Bass.

Martín Barbero, S., & Sandulli, F. (2005). *Marcating en la Web: BIWAM. Identidad desnuda.* Madrid: Dossat 2000.

Mason, R. B. (2007). The external environment's effect on management and strategy. *Management Decision, 45*(1), 10-28.

Matherly, L. L. (2007, August). *Mission: The impact of employee well being on processes, customer satisfaction and financial performance.* Paper presented at the meeting of the National Academy of Management, Philadelphia, PA.

Matherly, L. L., El-Saidi, M. A. & Martin, D. (2008, March). *Using project management to implement strategic planning and a strategic scorecard in a university setting.* Paper presented at the meeting of the Southwest Academy of Management, Houston, TX.

Maundeni, Z. (2004). *Mapping local democracy in Gaborone city.* Gaborone, Botswana, Botswana Association of Local Authorities – BALA.

Mayor Bloomberg, Mayor of New York City in a Keynote Address at the C40 Large Cities Climate Summit (2007, May). New York. Retrieved June 10, 2008, from http://www.nycclimatesummit.com

Mazanec J., & Schweiger, G. (1981) Improved marketing efficiency through multi-product brand names? An empirical investigation of image transfer. *European Research, 9*(1), 32-44.

McAfee, A. (2006). Mastering the three worlds of information technology. *Harvard Business Review, 84*(11), 141-149.

McGuire, S. (2000, February 7th). Shining Stockholm. *Newsweek,* 52-59.

McIntosh, R. W., Goeldner, C. R., & Ritchie J. R. B. (2000). *Turismo: Planeación, administración y perspectivas.* Mexico D.F.: Limusa Willey

McMillan, S. (2004). Internet advertising: One face or many? In D. Schumann and E. Thorson (Eds.), *Internet advertising: Theory and research (2nd edition).* New York, NY: Lawrence Erlbaum.

McMillan, S., Hwang, J-S., & Lee, G. (2003, December). Effects of structural and perceptual factors on attitudes toward the website. *Journal of Advertising Research,* 400-409.

Mecklenburg County, NC. (2007). 2007 Community Survey. Retrieved May 15, 2008 from http://www.charmeck.org/NR/rdonlyres/eknrcrfdromsgtehy7hr-cr34rvj2mdtqswbdnojx4mb7zktgglff7wsmusbwplng-u3xdkhlrzpmxibrzdcc6bmojpmc/2007Community-Survey091707.pdf.

Mecklenburg County, NC. (2007). 2007 Performance Report. Retrieved May 15, 2008, from http://www.charmeck.org/NR/rdonlyres/ea3bzxlsasj7kumjwyegasrdgjfhxl6m67gl2smozkijpki7wocaqqfnpnfl7mphydec7t6rqa5c4spt2bf3gpo7x4g/2007AnnualPerfRptfinal.pdf.

Mehta, D. Dewang Mehta. Retrieved August 4, 2007, from http://www.dewangmehta.com/main.htm.

Mellor, N. (2006). E-Citizen: Developing research-based marketing communications to increase awareness and take-up of local authority E-Channels. *Aslib Proceedings*, *58*(5), 437-446.

Metrex (2006, May). *International Conference on Planning and Sustainable Development of Metropolitan Capital Regions*, Madrid.

Miles, M., Hall, T., & Borden, J. (Eds.). (2000). *The city cultures reader*. London: Routledge.

Miller, D. (1998). *A theory of shopping*. Oxford: Polity Press.

Miller, D. (2001). *The dialectics of shopping*. Chicago; London: University of Chicago Press.

Mingers, J. (2001). Combining IS research methods: Towards a pluralist methodology. *Information Systems Research, 12*(3), 240-259.

Mingers, J., & Willcocks, L. (Eds.). (2004). *Social theory and philosophy for information systems*. Chichester, UK: John Wiley and Sons.

Mintzberg, H., & Waters, J. (1985). Of strategies, deliberate and emergent. *Strategic Management Journal, 6*(3), 257-272.

Miossec, J. M. (1977). L'image touristique comme introduccion à la géographie du tourisme. *Annales de Géographie*, 55-70.

Mitchell, W. (2007). *Ciudades inteligentes*. UOC Papers, num. 5. Retrieved July 10, 2008, from http://www.uoc.edu/uocpapers/5/dt/esp/mitchell.pdf.

Mithas, S., Krishnan, M. S., & Fornell, C. (2005). Why do customer relationship management applications affect customer satisfaction? *Journal of Marketing, 69*, 155-166.

Mobilkom (2004). *Mobilkom austria macht Bregenz mobil - Neue Massstaebe fuer Buerger- und Kundennaehe*. Retrieved June 20, 2008, from http://www.pte.at/pte.mc?pte=040708010

MobiSolutions (n. d.). *M-Governance Solutions*. Retrieved June 20, 2008, from http://www.mobisolutions.com/en/mgovt.php

Mohammed, R. A., Fisher, R. J., Jaworski, B. J., & Cahill, A. M. (2002). *Internet marketing: Building advantage in a networked economy*. New York, NY: McGraw-Hill.

Molebatsi, C. O. (1996). Towards a sustainable city. Gaborone, Botswana. *Ambio, 25*(2), 126-133.

Molina, F. X. (2001) European industrial districts: Influence of geographic concentration on performance of the firm. *Journal of International Management, 7*(4), 277-294.

MOMA –Museum of Modern Arts- New York (2006). On Site; new arquitecture in Spain.

Moon, M. J., & Norris, D. F. (2005). Does managerial orientation matter? The adoption of reinventing government and e-government at the municipal level. *Information Systems Journal, 15*(1), 43-60.

Morgan, N., Pritchard, A., & Piggott, R. (2002). New Zeland, 100% Pure: The creation of a powerful niche destination brand. *Journal of Brand Management, 9*(4/5), 335-354.

Morgan, N., Pritchard, A., & Pride, R. (2004). *Destination branding. Creating the unique destination proposition*. Oxford: Elsevier.

Morrison, A. (2002). *Hospitality and tourism marketing* (3rd ed.). Albany, NY: Delmar

Morton, P. (2006). Using critical realism to explain strategic information systems planning. *Journal of Information Technology Theory and Application, 8*(1), 1-48.

Mosha, A. C. (2004, September). *Challenges of municipal finance in Africa. Gaborone City, Botswana*. Paper presented at Habitat's 2nd World Urban Forum for Professionals on Land and Urban Poverty: The Challenges of City Financing: Organized by the Habitat Professionals Forum and the UN-HABITAT. Barcelona, Spain.

Mosha, A. C., & Cavric, B. (2001). Incorporating urban agriculture in Gaborone city planning. *Urban Agriculture, 4*, 25–27.

Mosha, A.C. (1996). The city of Gaborone, Botswana – Planning and management. *Ambio, 25*(2), 118-125.

Mumford, E. (1983). *Designing human systems for new technology: The ETHICS method*. Manchester: Manchester Business School.

Murphy, L., Moscardo, G., & Benckendorff, P. (2007). Exploring word-of-mouth influences on travel decisions: friends and relatives vs. other travelers. *International Journal of Consumer Studies, 31*(5), 517-527.

Myerson, G. (2001). *Heidegger, Habermas and the mobile phone*. Cambridge: Icon books & Totem books.

National League of Cities (NLC). (2007). City practices. Retrieved August 22, 2007, from http://www2.nlc.org/dbtw-wpd/exec/dbtwpub.dll.

National League of Cities (NLC). (2008). City practices. Retrieved June 1, 2008, from http://www2.nlc.org/dbtw-wpd/exec/dbtwpub.dll?AC=NEXT_BLOCK&XC=/dbtw-wpd/exec/dbtwpub.dll&BU=http%3A%2F%2Fwww2.nlc.org%2Fexamples%2Fcknsearchtest.htm&TN=EXAMPLES&SN=AUTO19263&SE=164&RN=40&MR=20&TR=0&TX=1000&ES=0&CS=0&XP=&RF=New+Web+Results&EF=&DF=New+Web+Details&RL=0&EL=0&DL=0&NP=3&ID=&MF=&MQ=&TI=0&DT=&ST=0&IR=3509&NR=0&NB=2&SV=0&SS=0&BG=&FG=&QS=&OEX=ISO-8859-1&OEH=ISO-8859-1.

Nedovic-Budic, Z. & Godschalk, D.R. (1996). Human factors in adoption of geographic information systems: A local government case study. *Public Administration Review, 56*(6), 554-567.

Nelson, R. (1993). *National innovation systems*. New York, NY: Oxford University Press.

Neus, A., & Scherf, P. (2005). Opening minds: Cultural change with the introduction of open-source collaboration methods. *IBM Systems Journal, 44*(2), 215-225.

Newsweek (2006, July 3). *The world's hottest cities, Special Issue*, (pp. 48-79).

Nielsen, J. (2002). *Usabilidad. Diseño de sitios web*. Madrid: Prentice Hall.

Nijkamp, P., & Jonkhoff, W. (2001). The city in the information and communication technology age: A comparative study on path dependency. *International Journal of Technology Policy and Management, 1*(1), 78-99.

Noisette, P., & Vallerugo, F. (1996). *Marketing des villes*. Paris: ESSIC.

Norris, D. F., & Moon, M. J. (2005). Advancing e-government at the grassroots: Tortoise or Hare? *Public Administration Review, 65*(1), 64-75.

North, D. C. (1999). *Institutions, institutional change, and economic performance*. New York, NY: Cambridge University Press.

Obaid, T. A. (2007, December). *Population and human development: The key connections*. The Millennium Development Goals Report 2007, UN.

OECD E-Government Studies The E-Government Imperative. (2003). *Organisation for Economic Co-operation and Development, 2003*(15), 1-199.

OECD. (1992). *Cities and new technologies*. Paris: OECD.

Ohmae, K. (1982). *The mind of the strategist: The art of Japanese business*. New York, NY: McGraw Hill.

Olshavsky, R.W., & Granbois, D.H. (1979). Consumer decision making: Fact or Fiction? *Journal of Consumer Research, 6*, 93-100.

OMT (1999). *Desarrollo turístico sostenible: Guia para administradores locales*. Madrid: OMT.

Orlikowski, W. (1992). The duality of technology: Rethinking the concept of technology in organizations. *Organization Science, 3*(3), 398-427.

Orlikowski, W. (2000). Using technology and constituting structures: A practice lens for studying technology in organizations. *Organization Science, 11*(4), 404-428.

Paddison, R. (1993). City marketing, image reconstruction and urban regeneration. *Urban Studies, 30*(2), 339-349.

Padín Fabeiro, C. (2004, April). *La formación de la imagen de un nuevo destino*. Paper presented at the XIII Simposi Internacional de Turisme i Lleure ESADE-Fira de Barcelona. Paper presented at the XIII Simposi Internacional de Turisme i Lleure ESADE-Fira de Barcelona. Retrieved April 30, 2007, from www.esade.es/cedit2004/cat/est_estudios.php

Pan, B., MacLaurin, T., & Crotts, J. C. (2007). Travel blogs and the implications for destination marketing. *Journal of Travel Research, 46*, 35-45.

Pan, S.-L., Tan, C.-W., & Lim, E. T. K. (2006). Customer relationship management (CRM) in e-government: A relational perspective. *Decision Support Systems, 42*(1), 237-250.

Paolillo, J. C., & Penumarthy, S. (2007, January). *The social structure of tagging internet video on del.icio.us*. Paper presented at the 40th HICSS, Waikoloa, Big Island, Hawaii.

Papazoglou, M. P., & Ribbers, P. M. A. (2006). *e-Business: Organizational and technical foundations*. Chichester: Wiley.

Parasuraman, A., Zeithaml, V. A., &, Berry, L. L. (1988). A multiple-item scale for measuring consumer perceptions of service quality. *Journal of Retailing, 64*(1), 12-40.

Park, C. W., Jaworski, B. J., & MacInnis, D. J. (1986, October). Strategic brand concept management, *Journal of Marketing, 50*, 135-145.

Park, D. H., Lee, J., & Han, J. (2007). The effect of online consumer reviews on consumer purchasing intention: The moderating role of involvement. *International Journal of Electronic Commerce, 11*(4), 125-148.

Parsons, A., Zeisser, M., & Waitman, R. (1998). Organizing today for the digital marketing of tomorrow, *Journal of Interactive Marketing, 12(1)*, 31-46.

Partnership for the Heart (2006). Retrieved August 13, 2007, from http://www.partnership-for-the-heart.de

Payne, A., & Frow, P. (2005). A strategic framework for customer relationship management. *Journal of Marketing, 69*, 167-176.

Peak, D., & Guynes, C. S. (2003). The IT alignment planning process. *The Journal of Computer Information Systems, 44*(1), 9.

Pereira, C. S., & Soares, A. L. (2007). Improving the quality of collaboration requirements for IM through social networks analysis. *International Journal of Information Management, 27*, 86–103.

Perryman, M.R. (2006). The market for prosperity: Understanding the economic development process. Retrieved July 22, 2007, from http://www.texasedc.org/files/File/ED%20Publications/Perryman%20report_color.pdf.

Pitta, D., & Fowler, D. (2005). Online consumer communities and their value to new product developers. *Journal of Product & Brand Management, 14*(5), 283-291.

Poon, A. (1993). *Tourism, technology and competitive strategies*. Oxon, United Kingdom: Cab International.

Porsander, L. (2000). *Titt-skåp för alla, en berättelse om hur Stockholm blev kulturhuvudstad*. Göteborg: BAS.

Porsander, L. (2000). Translating a dream of immortality in a (con)temporary order. *Journal of Organizational Change Management, 13*(1), 14-29.

Porter, M. (1980). *Competitive strategy: Techniques for analyzing industries and competitors*. New York, NY: Free Press.

Porter, M. (1985). *Competitive advantage*. New York, NY: Free Press.

Porter, M. (1985, Winter). Technology and competitive advantage. *The Journal of Business Strategy*, 60-70.

Porter, M. (1999). *On competition*. Boston, MA: Harvard Business School Press.

Porter, M. E. (1980). *Competitive strategy: Techniques for analyzing industries and competitors*. New York, NY: Free Press.

Porter, M. E. (1985). *Competitive advantage: Creating and sustaining superior performance*. New York, NY: Free Press.

Porter, M. E. (1990). *The competitive advantage of nations*. London: The McMillan Press.

Prats, L., & Balagué, J. (2005). Cohesión y sostenibilidad, elementos clave en la competitividad del territorio turístico: El caso de la Costa Brava. *Retos Turísticos, 3*(3), 8-15.

Prats, L., & Guia, J. (2005). The destination as a local system of innovation. In C. Petrillo & J. Swarbrooke (Eds.), *Networking and partnership in destinations and development management* (pp. 121-136). Arnhem: ATLAS.

Prats, L., & Guia, J., Molina, F. X. (2007). Tourism local innovation systems or how tourism destinations evolve. In M. Smith & L. Onderwater (Eds.), *Destinations revisited. Perspectives on developing and managing tourist areas* (pp. 35-53). *Arnhem: ATLAS*

Prats, L., Camprubí, R., & Comas, J. (2005). Network ties relevance on the destination business relationships. In Universidade do Algarve (Eds.), *Recent developments in tourism research*. Faro: University of Algarve.

Prebensen, N. K. (2007). Exploring tourists' images of a distant destination. *Tourism Management, 28*, 747-756.

Press Conference (2007, July 11), ITU Secretary-General Dr. Hamadou Touré for the *Connect Africa* Summit held in Kigali, Rwanda, 29-30 October 2007 held jointly with the UN Global Alliance for ICT and Development (GAID).

Prud'homme, R. (1996). Size, sprawl, speed and the efficiency of cities. *Urban Studies Journal, 36*(11), 1849-1858.

Rainisto, S. (2003). *Success factors of place marketing. A study of place marketing practices in northern Europe and the United States*. Helsinki: Helsinki University of Technology.

Rakkolainen, I., & Vainios, T. (2000, November). *A 3-D city info for mobile users*. Paper presented at the 3rd International Workshop in Intelligent Interactive Assistance and Mobile Multimedia Computing, Rostock, Deutschland, S.115-212.

Rakodi, C. (1997) (Ed.). *The urban challenge in Africa: Growth and management of its large cities*. New York, NY: United Nations University Press.

Rakodi, C. (1998). *Globalization: Trends and Sub-Saharan African cities*. In F.C. Lo and

Rallet, A., & Torre, A. (2004). Proximité et localisation. *Economie Rurale, 280*.

Ramachandran, K., & Sougata, R.(2005). Creating information technology industrial clusters: Learning from strategies of the early and late movers. In T. Thatchenkery & R. Stough (Eds.), *Information communication technology and economic development: Learning from the Indian experience* (pp. 149-166). Northampton: Edward Elgar Publishing.

Ramos, A. J., Nangit, G., Ranga, A. I., & Triñona, J. (2007). ICT-enabled distance education in community development in the Philippines. *Distance Education, 28*(2), 213.

Rannu, R. (2003). *Mobile serivces in Estonia*. Praxis Working Paper No. 8.

Rannu, R. (2003). M-services in Estonia. *Baltic IT&T Review, 32*, 28-33.

Rannu, R. (2004). M-Governance from vision to reality. *Baltic IT&T Review, 35*, 35-40.

Rannu, R., & Semevsky, M. (2005). *Mobile services in Tartu - Existing services, best practices, methodology and suggestions for future development*. Retrieved June 20, 2008, from http://www.ega.ee/public/Mobile_services_in_Tartu_FINAL1.pdf

Raymond, C., & Brown, G. (2007). A spatial method for assessing resident and visitor attitudes towards tourism growth and development. *Journal of Sustainable Tourism, 15*(5), 520-540.

Reime, M., & Hawkins, C. (1979). Tourism development: A model for growth. *Hotel and Restaurant Administration Quarterly, 20*, 67-74.

Ricci, F., & Werthner, H. (2002). Cased-based querying for travel planning recommendation. *Information Technology and Tourism, 4(3/4)*, 215-226.

Ricci, F., Cavada, D., Mirzadeh, N., & Venturini, A. (2006). *Case based travel recommendations*. Paper presented at the Destination Recommendation Systems: Behavioral Foundations and Applications, Oxford.

Riezebos, R. (2003). *Brand management. A theoretical and practical approach*. Harlow, UK: Prentice-Hall.

Rimmon-Kenan, S. (1983). *Narrative Ficion: Contemporary poetics*. London and New York: Routledge.

Ritchie, B., & Ritchie, R. (1998). *The branding of tourism destination: Past achievements and future trends*. Paper presented at the 48th Congress, AIEST, St-Gall.

Ritchie, J. R. (1993). Crafting a destination vision: Putting de concept of residence responsive tourism into practice. *Tourism Management, 14*, 279-289.

Roberts, J. (2001). Challenges facing service enterprises in a global knowledge-based economy: Lessons from the business services sector. *International Journal of Services Technology and Management, 2*(3), 402-433.

Roberts, M. L. (2003). *Internet marketing: Integrating online and offline strategies*. New York, NY: McGraw-Hill Irwin.

Robey, D., & Boudreau, M. (1999). Accounting for the contradictory organizational consequences of information technology: Theoretical directions and methodological implications. *Information Systems Research, 10*(2), 167-185.

Rocheleau, B. (2006). *Public management information systems*. Hershey, PA: Idea Group Publishing.

Roettger-Gerigk, S. (2002). *Handbuch Mobile Commerce: technische Grundlagen, Marktchancen und Einsatzmoeglichkeite (pp. 419-426)*. Berlin/Heidelberg, Germany: Springer.

Roge, J. N., & Chakrabarty, S. (2003). Waiting for the other shoe to drop: Has information technology integrated marketing operations with marketing strategy? *The Journal of Computer Information Systems, 43*(2), 16.

Rolland, E., Patterson, R. A. &, Ward, K. F. (2008). Dynamic capabilities and E-Service. Working paper, University of California at Riverside.

Rosemberg, M. (2000). *Le marketing urbain en question*. Paris: Anthropos.

Rosenbloom, A. (2004). The blogosphere. *Communications of the ACM, 47*(12), 31–33.

Rowley, J. (2004). Online branding. *Online Information Review, 28*(2), 131-138.

Rowley, J. (2004). Online branding. The case of McDonald's. *British Food Journal, 106*(3), 228-237.

Rowley, J., Teahan, B., & Leeming, E. (2007). Customer community and co-creation: A case study. *Marketing Intelligence & Planning, 25*(2), 136-146.

Rust, R. T., & Lemon, K. N. (2001). E-Service and the consumer. *International Journal of Electronic Commerce*, *5*(3), 85-101.

Sabherwal, R., Hirschheim, R., & Goles, T. (2001). The dynamics of alignment: Insights from a punctuated equilibrium model. *Organization Science, 12*(2), 179-197.

Sadler, D. (1993). Place marketing, competitive places and the construction of hegemony in Britain in the 1980s. In G. Kearns & C. Philo (Eds.), *Selling places. The city as cultural capital. Past and present* (pp. 175-192). Oxford: Pergamon Press.

Sairosse, M. T. (2003). *Economic and social impact of the Internet: A study of cybercafés in Gaborone, Botswana*. University of Botswana, Gaborone. MLIS dissertation.

Sandholz, J. H., Ringstaff, C., & Dwyer, D. C. (1991). *Teaching with technology: Creating pupil centred classrooms*. New York, NY: Teachers College Press.

Sassen, S. (2001). Cities in the global economy. In R. Paddison (Ed.), *Handbook of urban studies* (pp. 256-272). London: Sage Publications.

Satterthwaite, D. (1996). Toward healthy cities. *People and the Planet, 5*(2), 12-15.

Satterthwaite, D. (2002). *Why the cities have grown*: Commissioned Paper by the International Institute for Environment and Development (IIED). London

Satterthwaite, D. (2002, December) *City Governance for and with Children*. Paper presented at the Conference on Children and the City; Learning from international experiences, creating local solutions. Patronage of Her Majesty Queen Rania Al-Abdullah. Amman, Jordan.

Schermerhorn, J. S. (1999). *Management (*2nd edition), New York, NY: Wiley.

Schmallegger, D., & Carson, D. (2008). Blogs in tourism: Changing approaches to information exchange. *Journal of Vacation Marketing 14(2)*, 99-110.

Schneider, J., & Schroeder, F. (2003). *The M-Toguide Project—Development and deployment of an European Mobile Tourism Tool*. Eurescom Summit 2003.

Schroeder, J. (2003). Att marknadsföra identiteter och konsumera skillnader. In Á. Sverrisson, P. Aspers & P. Fuehrer (Eds.), *Bilderna i samhällsanalysen*. Lund: Studentlitteratur.

Schroeder, J. E. (2004). Produktion och konsumtion av reklambilder. In P. Aspers, P. Fuehrer & Á. Sverrisson (Eds.), *Bild och samhälle. Visuell analys som vetenskaplig metod.* (pp. 75-96). Lund: Studentlitteratur.

Schroeder, J., & Borgerson, J. L. (2003). Dark desires: Fetishism, ontology, and representation in contemporary advertising. In T. Reichert & J. Lambiase (Eds.), *Sex in advertising. perspectives on the erotic appeal* (pp. 65-87). Mahwah, N.J & London: Lawrence Erlbaum Associates.

Schuler, M. (2004). Management of the organizational image: A method for organizational image configuration. *Corporate Reputation Review, 7*(1), 37-53.

Schumpeter, J. A. (1934). *Theory of economic development: An inquiry into profits, capital, credit, interest and the business cycle*. Cambridge, MA: Harvard University Press.

Schware, R. (Ed.) (2005, November). *E-Development: From excitement to effectiveness.* Paper presented at the World Summit on the Information Society. Tunis, Tunisian Republic.

Schware, R. (Ed.) (2005, November). *Global forces, urban change and urban management in Africa:* The urban Challenge. In Africa: Growth and Management of its Large Cities. Paper presented at the World Summit on the Information Society. Tunis, Tunisian Republic.

Scott, D. M. (2007). *The new rules of marketing & PR*. Hoboken, NJ: John Wiley & Sons.

Scott, W. R. (2001). *Institutions and organizations* (2nd ed.). Thousand Oaks, CA: Sage.

Selby, M., & Morgan, N. G. (1996). Reconstructing place image: A case study of its role in destination market research. *Tourism Management, 17*, 287-294.

Selden, R., & Widdowson, P. (1993). *A reader's guide to contemporary literary theory* (3rd ed.). New York & London: Harvester Wheatsheaf.

Selinger, M. (2002). Learning computer technology skills in the subject context of learning. *Journal of Information Technology for Teacher Education, 10*(1-2), 143-157.

Shark, A. (2007, August). *Revitalizing Rural America.* Paper presented at the Texas Municipal League Technology Summit, Lewisville, TX.

Shaw, D. (2001). The post-industrial city. In R. Paddison (Ed.), *Handbook of urban studies* (pp. 284-295). London: Sage Publications.

Sherry, J., & Salvador, T. (2002). Running and grimacing: The struggle for balance in mobile work. In B. Brown, N. Green & R. Harper (Eds.), *Wireless world. Social and interactional aspects of the mobile age* (pp. 108-120). London: Springer.

Shih, E., Kraemer, K. L., & Dedrick, J. (2008). IT diffusion in developing countries. *Communications of the ACM, 51*(2), 43-48.

Shlomo, A., Stephen, C., Sheppard, S. C., Civco, D. L., Buckley, R., Chabaeva, A., Gitlin L., Kraley, A., Parent, J., & Perlin, M. (2005). *Transport and urban development department.* Washington, D.C. International Bank for Reconstruction and Development. World Bank Group.

Short, J. R., & Benton, L. M. (1993). Reconstructing the image of an industrial city. *Annals of the Association of American Geographers, 83*(2), 207-224.

Short, J. R., & Kim, Y. H. (1998). Urban crisis / urban representations: Selling the city in difficult times. In T. Hall & P. Hubbard (Eds.), *The entrepreneurial city: Geographies of politics, regime and representation* (pp. 55-75). Chichester, UK: John Wiley and Sons.

Shostak, A. (1982). Seven scenarios of urban change. In G. Gappert & R. V. Knight (Eds.), *Cities in the 21st century* (pp. 69-93). Beverly Hills: Sage.

Sicilia, M., Ruiz, S., & Munuera, J. L. (2005). Effects of interactivity in a Web site. *Journal of Advertising, 34*(3), 31-45.

Sigala, M. (2003). Developing and benchmarking internet marketing strategies in the hotel sector in Greece. *Journal of Hospitality & Tourism Research, 27*(4), 375–401.

Sigala, M. (2005). In search of online post-modern authenticity: Assessing the quality of learning experiences at eternalegypt.org. In M. SIGALA & D. Leslie (Eds.), *International Cultural Tourism: Management, implications and cases* (pp. 123–136). Oxford, UK: Butterworth Heinemann, Elsevier.

Sigala, M. (2005). Integrating customer relationship management in hotel operations: Managerial and operational implications. *International Journal of Hospitality Management, 24*(3), 391–413.

Sigala, M. (2006). e-Customer relationship management in the hotel sector: Guests' perceptions of perceived e-service quality levels. *Tourism: An International Interdisciplinary Journal, 54*(4), 333-344.

Sigala, M. (2008, January). Developing and implementing an eCRM 2.0 strategy: Usage and readiness of Greek tourism firms. *ENTER 2008 conference*, Innsbruck, Austria.

Simon, S. J. (2001). The Impact of culture and gender on web sites: An empirical study, *Database for Advances in Information Systems, 3*(1), 18-37.

Singer, J. G. (2006). Systems marketing for the information age. *MIT Sloan Management Review, 48*(1), 95.

Smith, D., Menon, S., & Sivakumar, K. (2005). Online peer and editorial recommendations, trust, and choice in virtual markets. *Journal of Interactive Marketing, 19*(3), 15-37.

SMS Cell Broadcast (n. d.). Retrieved May 23, 2007, from http://www.gsm-modem.de/sms-cell-broadcast.html

Smyth, H. (1994). *Marketing the city. The role of flagship developments in urban regeneration.* London: E and FN Spon.

Sorensen, F. (2004). *Tourism experience innovation networks.* Doctoral dissertation. Roskilde University, Denmark.

Spool, J. M. (1996, January 1). Branding and usability. *User Interface Engineering.* Retrieved June 1, 2008, from www.uie.com/articles/branding_usability.

Stavrou, A. (2005, October). *Productive cities: Information needs of the urban poor.* Paper presented at the 4th International Forum on Urban Poverty Conference, Marrakech, Morocco.

Steiner, G. A., & Miner, J. (1982). *Management policy and strategy.* New York, NY: MacMillan.

Stewart, M. (1996). Competition and competitiveness in urban policy. *Public Money and Management, 16*(3), 21-26.

Stockholm Business Region. (2007). *Marknadsplan. Investment Promotion.* Stockholm: City of Stockholm.

Stockholm Business Region. (2007). *Verksamhetsprogram och budget Stockholm Business Alliance 2007.* Stockholm: City of Stockholm.

Stough, R. R., Haynes, K. E., & Salazar, M. E. (2005). Economic Development Theory and practice: the Indian development experience. In T. Thatchenkery & R. Stough (Eds.), *Information communication technology and economic development: Learning from the Indian experience* (pp. 11-28). Northampton: Edward Elgar Publishing.

Strannegård, L., & Friberg, M. (2001). *Already elsewhere—Play, identity and speed in the business world.* Stockholm: Raster förlag.

Strauss, J., & Frost, R. (1999). *Marketing on the Internet. Principles of online marketing.* New Jersey: Prentice Hall.

Struse, D. (2000). Marketing research's top 25 influences. *Marketing Research, 11*(4), 4-10.

Suh, T., & Amine, L. S. (2007). Defining and managing reputational capital in global markets. *Journal of Marketing Theory and Practice*, *15*(3), 205-217.

Sullivan, B., & Estes, C. (2007). Measuring customer service quality in local government: Fulton County Human Services shares its experience with surveys and recommends ways to use them to improve customer satisfaction. *The Public Manager, 36*(1), 37-39.

Sundbo, J. (1998). *The theory of innovation: Entrepreneurs, technology and strategy.* Cheltenham, UK: Edward Elgar.

Sussmann, S., & Baker, M. (1996). Responding to the electronic marketplace: Lessons from destination management systems. *International Journal of Hospitality Management, 15*, 99-112.

Swope, C. (2007, May). Working without wires. Municipal WiFi is coming. Government employees, not citizens, may be the biggest users. *Governing,* 29-34.

Syfert, P., & Elliott, N. (1998). Charlotte adapts the balanced scorecard. *American City & County, 113*, 32.

Tarlow, M., & Tarlow, P. (2002). *Digital aboriginal / the direction of business now: Instinctive, nomadic and ever-changing.* New York, NY: Warner Books, Inc.

Taylor, G. D. (1980). How to match plan with demand: A matrix for marketing. *International Journal of Tourism Management, 1*, 56-60.

Technology Parks of India Bangalore. Retrieved August 4, 2007, from http://www.blr.stpi.in

Technology Parks of India Hyderabad. Retrieved August 4, 2007, from http://www.hyd.stpi.in

Telisman-Kosuta, N. (1989). Tourism destination image. In S. F. Witt & L. Moutinho (Eds.), *Tourism marketing and management handbook* (pp. 557-561). Cambridge: Prentice Hall.

Thorngate, W. (1976). "In general" vs. "it depends": Some comments on the Gergen-Schlenker debate. *Personality and Social Psychology Bulletin*, *2*, 404-410.

Thorogood, A., Yetton, P., Vlasic, A., & Spiller, J. (2004). Raise your glasses - the water's magic! Strategic IT at SA Water: a case study in alignment, outsourcing and governance. *Journal of Information Technology, 19*(2), 130-139.

Tietze, S., Cohen, L., & Musson, G. (2003). *Understanding organizations through language*. London, Thousand Oaks, New Delhi: Sage.

Tjostheim, I., & Tronvoll, B. (2002). *The Internet and city tourist: A study of preferences for information sources in travel planning.* Paper presented at the Cities Tourism 2002, Vienna, Austria.

Toshiba (2007). *The Classroom and beyond*, Retrieved September 28, 2007, from http://eu.computers.toshiba-europe.com/cgi-bin/ToshibaCSG/download_whitepaper.jsp?service=EU&WHITEPAPER_ID=2007-09-Classroom-and-Beyond-EN

Toshiba (2007). *Trends and technology 2007.* Retrieved March 15, 2007, from

Townsend, A. M. (2002). Mobile communications in the Twenty-first Century City. In B. Brown, N. Green & R. Harper (Eds.), *Wireless world. Social and interactional aspects of the mobile age* (pp. 62-77). London: Springer.

Trueman, M., Klemm, M., & Giroud, A. (2004). Can a city communicate? Bradford as a corporate brand. *Corporate Communications: An International Journal, 9*(4), 317-330w.

Tschirhart, M. (2008). Evaluation of brand use on city government websites: A search for brand type, repetition and consistency. *Journal of Nonprofit & Public Sector Marketing, 19*(1), 35-53.

Tsikriktsis, N. (2002). Does culture influence web site quality expectations? An empirical study. *Journal of Service Research*, *5*(2), 101-112. AHRQ (2008). Consumer assessment of healthcare providers and systems. Retrieved February 21, 2008, from https://www.cahps.ahrq.gov/default.asp

Turowski, K., & Pousttchi, K. (2004). *Mobile Commerce—Grundlagen und Techniken*. Berlin/Heidelberg, Germany: Springer.

UNEP/UNESCO (2006). *Facts and figures.* New York, NY: United Nations.

UNFPA, (2007). World population Report: Unleashing the potential of Urban Growth. New York, NY: United Nations. http://www.unfpa.org/swp/2007/english/chapter_1/smaller_cities.htm. Retrieved February 16, 2008.

United Nations Development Programme's Regional Project on Local Governance for Latin America (2005). New York, NY: United Nations.

Urban, F. (2002). Small town, big website? Cities and their representation on the internet. *Cities, 19*(1), 49-59.

Urry, J. (1990). *The tourist gaze*. London: Sage:

Urry, J. (2000). *Sociology beyond societies, mobilities for the Twenty-first Century*. London & New York, NY: Routledge.

Uyanga, S. (2005). The usage of ICT for secondary education in Mongolia. *International Journal of Education and Development using Information and Communication Technology, 1*(4), 101-118.

Van der Meer, A., & Van Winden, W. (2003). E-governance in cities: A comparison of urban ICT policies. *Regional Studies, 37*, 407-419.

van Limburg, B. (1998). City marketing: A multi-attribute approach. *Tourism Management, 19*(5), 475-477.

Vegara, A., & De Las Rivas, J.L. (2004). *Territorios inteligentes*. Madrid: Fundación Metrópoli.

Verdú, V. (2001, April 13). La sexy-ciudad. El País, (p. 20).

Verdú, V. (2003). *El estilo del mundo. La vida en el capitalismo de ficción*. Barcelona: Anagrama.

Verdú, V. (2005). *Yo y tú, objetos de lujo. El personismo: La primera revolución cultural del siglo XXI*. Barcelona: Debate.

Voutsis, N., & Zimmermann, F. (2005, July). *Anonymous code lists for secure electronic voting over insecure mobile channels*. Paper presented at the International Conference on Mobile Government, Sydney, Australia, (pp. 434-444).

Waldock, L. (2006). Travel light with a tablet [Tablet PC review]. *Personal Computer World, 29*(4), 94-96.

Wang, Y., & Fesenmaier, D. R. (2004). Modeling participation in an online travel community. *Journal of Travel Research, 42*, 261-270.

Wang, Y., & Fesenmaier, D. R. (2006). Identifying the success factors of web-based marketing strategy: An investigation of convention and visitors bureaus in the United States. *Journal of Travel Research, 27*, 326-341.

Wang, Y., Yu, Q., & Fesemaier, D. R. (2002). Defining the virtual tourist community: Implications for tourism marketing. *Tourism Management, 23(4)*, 407-417.

Ward, J., & Griffiths, P. (2002). *Strategic planning for information systems*. Chichester, UK: John Wiley and Sons.

Ward, K., Rolland, E., & Patterson, R. (2005). Improving outpatient healthcare quality: Understanding the quality dimensions. *Health Care Management Review, 30*(4), 361-371.

Ward, S. V. (1998). *Selling places. The marketing and promotion of towns and cities, 1850-2000*. London: E & FN Spon.

Ward, S. V., & Gold, J. R. (Eds.). (1994). *Place promotion. The use of publicity and marketing to sell towns and regions*. Chichester: John Wiley.

Warnaby, G., & Bennison, D. (2006). Reciprocal urban place marketing and co-branding? Retail applications. *Place Branding, 2*(4), 297-310.

Weber, L. (2007). *Marketing to the social web*. Hoboken, NJ: John Wiley & Sons.

Weber, S. (2007). *Plug your business*. Falls Church, VA: Weber Books.

Webster, F. E. (1996). The future is interactive marketing. *Harvard Business Review, 74*(6), 156-157.

Webster, F. E. (1998). Interactivity and marketing paradigm shifts. *Journal of Interactive Marketing, 12*(1), 54-55.

Webster, F. E. (2005). A perspective on the evolution of marketing management. *Journal of Public Policy and Marketing, 24*(1), 121-126.

Weick, K. E. (1979). *The social psychology of organizing (2nd ed.)*. New York, NY: McGraw-Hill, Inc.

Weick, K. E. (1995). *Sensemaking in organizations*. Thousand Oaks, CA: Sage.

Wellman, B. (1997). An electronic group of virtually a social network. In S. Kiesler (Ed.), *Culture of the internet*. Mahwah: Lawrence Erlbaum.

Wells Jr, L. T., & Wint, A. G. (2000). *Marketing a country. Promotion as a tool for attracting foreign investment* (revised edition ed.). Washington: Foreign Investment Advisory Service.

Wen, H. J., & Shih, S. C. (2006). Strategic information technology prioritization. *The Journal of Computer Information Systems, 46*(4), 54-63.

Werthner, H., & Klein, S. (1999). *Information technology and tourism, a challenging relationship*. Vienna: Springer Verlag.

Werthner, H., & Ricci, F. (2004). E-Commerce and tourism. *Communications of the ACM, 47(12),* 101-115.

West, J., & Berman, E. M. (2001). The impact of revitalized management practices on the adoption of information technology: A national survey of local governments. *Public Performance and Management Review, 24*(3), 233-253.

Williams, P. (2002, December 13). C&W Lowers Rates at MoBay Digiport. *The Jamaica Observer.* Retrieved August 5, 2007, from http://www.jamaicaobserver.com/news/html/20021213T000000-0500_36551_OBS_C_W_LOWERS_RATES_AT_MOBAY_DIGIPORT.asp.

Winer, D. (2005). What is a .River of News. style aggregator? *Really Simple Syndication.* Retrieved June 22, 2005, from http://www.reallysimplesyndication.com/riverOfNews

Wirtschaftswoche. (2002). Europas Städte im Vergleich. *Wirtschaftswoche, 33,* 18-24. Ambikairajah, E. (2007). Tablet PC and electronic whiteboard use in signal processing education. *IEEE Signal Processing Magazine,* 24(1), 130-133.

Wöber, K. W. (Ed.). (2002). *City tourism 2002.* Vienna: Springer Wien.

Wolf, M. (1999). *The entertainment economy. The mega-media forces that are re-shaping our lives.* New York, NY: Penguin Books. Bakhtin, M. (1934-35/1981). Discourse in the novel (C. Emerson & M. Holquist, Trans.). In M. Holquist (Ed.), *The dialogic imagination: Four essays* (p. 259-422). Austin: Univ. of Texas P.

Wong, C. (2001). The relationship between quality of life and local economic development: An empirical study of local authority areas in England. *Cities, 18*(1), 25–32.

Wong, C. Y. L., Millar, C. J. M., & Choi, C. J. (2006). Singapore in transition: From technology to culture hub. *Journal of Knowledge Management, 10*(5), 79-91.

Wonglimpiyarat, J. (2007). National foresight in science and technology strategy development. *Futures, 39*(6), 718-728.

Wood, P., & Taylor, C. (2004). Big ideas for a small town: The Huddersfield creative town initiative. *Local Economy, 19*(4), 380-395.

World Bank (1998). *Reports of findings on ongoing operational, economic and sector work carried out by the World Bank and its member governments in the Africa Region* Published periodically by the Africa Technical Department on behalf of the Region. A report of findings on the Africa Region. Number 7.

World Bank (2006). *Cities in transition: A strategic view of urban and local government issues.* International Bank for Reconstruction and Development, Washington, D.C.: World Bank Group.

World Bank (2006). *Information and communications for development (IC4D): Global trends and policies.* Washington D.C. Global Information and Communication Technologies Department. World Bank Group.

World Tourism Organisation (1999). *Promoción de destinos turísticos en el Ciberespacio.* Madrid: WTO.

World Tourism Organisation (2001). *Comercio electrónico y turismo.* Madrid: WTO.

Wright, K. B. (2005). Researching Internet-based populations: Advantages and disadvantages of online survey research, online questionnaire authoring software packages, and web survey services. [Electronic Version]. *Journal of Computer-Mediated Communication, 10(3).* Retrieved from http://jcmc.indiana.edu/vol10/issue3/wright.html

WTO. (1999). *Marketing tourism destinations online: Strategies for the information age.* Madrid: World Tourism Organization. .

WTO. (2001). *eBusiness for tourism: Practical guidelines for destinations and business.* Madrid: World Tourism Organization

Wu, F., & Zhang, J. (2007). Planning the competitive city-region: The emergence of strategic development plan in China. *Urban Affairs Review, 42*(5), 714-740.

Y. Yeung (Eds.), *Globalization and the world of large cities (pp. 314-351).* Tokyo, New York; Paris: UNU Press.

Ying Z., & Davis, J. (2007, January). *Web communities in blogspace.* Paper presented at the 40th HICSS, Waikoloa, Big Island, Hawaii.

Yuan, Y., Gretzel, U., & Fesenmaier, D. R. (2003). Internet technology use by American Convention and Visitors Bureaus. *Journal of Travel Research, 41(3),* 240-255.

Yuan, Y., Gretzel, U., & Fesenmaier, D. R. (2006). The role of information technology use in American convention and visitors bureaus. *Tourism Management, 27*(2), 326-341.

Yuankai, T. (2007). Shanghai shows its heart. *Bejing Review, 50*(42), 27.

Yunus, M. (2007). The Nobel Peace Prize 2006 - Nobel Lecture. *Law and Business Review of the Americas, 13*(2), 267. Ahlstrand, K. I., Asperen, A., Gardlund, M., & Leandri, P. (2004). *explorestockholm.com—A tourist service from .tourism.* Retrieved June 20, 2008, from www.tslab.ssvl.kth.se/csd/projects/0412/tourism%20final%20report.pdf

Zaera-Polo, A. (2004, September 10). La era urbana. El País, special issue on the World Urban Forum, (p. 12).

Zimmermann, J. B. (2002). Des «clusters» aux «small-worlds» une approche en termes de proximité. *Géographie, Économie, Société, 4*, 3-17.tecedents of executive information system success: A path analytic approach. *Decision Support System, 22*(1), 31-43.

Barrow, C. (1990). Implementing an executive information system: Seven steps for success. *Journal of Information Systems Management, 7*(2), 41-46.

Basu, C., Poindexter, S., Drosen, J., & Addo, T. (2000). Diffusion of executive information systems in organisations and the shift to Web technologies. *Industrial Management & Data Systems, 100*(6), 271-276.

Beer, S. (1979). *The heart of enterprise*. Chichester: Wiley.

Belkin, N. J., Cool, C., Koenemann, J., Bor Ng, K., & Park, S. Y. (1996). Using relevance feedback and ranking in interactive searching. In *Proceedings of the Fourth Text Retrieval Conference,* 181-210.

Belkin, N. J., & Croft, W. B. (1992). Information filtering and information retrieval: Two sides of the same coin? *Communications of the ACM, 35*(12), 29-38.

Bernhardt, D. (1994). I want it fast, factual, actionable: Tailoring competitive intelligence to executive's needs. *Long Range Planning, 27*(1), 12-24.

Biernacki, P., & Waldorf, D. (1981). Snowball sampling. Problems and techniques of chain referral sampling. *Sociological Methods & Research, 10*, 141-163.

Boyd, B. K., & Fulk, J. (1996). Executive scanning and perceived uncertainty: A multidimensional model. *Journal of Management, 22*(1), 1-21.

Brittin, M. (1991). *Business research guide: How to develop your competitor intelligence system: Five case studies*. Cleveland, UK: Headland Press.

Brackett, M. H. (1999, March). Business intelligence value chain. *DM Review Magazine*.

Breeding, B. (2001). CI and KM converge: A case study at Shell Services International. In J.E. Prescott & S. H. Miller (Eds.), *Proven strategies in competitive intelligence: Lessons from the trenches* (pp. 45-68). New York: John Wiley & Sons.

Brunson, D. (2005). Top *10 trends in business intelligence and data warehousing for 2005 revisited.* Retrieved December 29, 2006, from http://www.b-eye-network.com/view/969

Budzik, J., Bradshaw, S., Fu, X., & Hammond, K. (2002). Supporting online resource Dsiscovery in the context of ongoing tasks with proactive software assistants. *International Journal of Human-Computer Studies, 56*(1), 47-74.

Burkan, W. C. (1991). *Executive information systems. From proposal through implementation.* New York: Van Nostrand Reinhold.

Burke, M. E. (1995). Sources of UK business information: Knowing where to find relevant information is vital for today's business manager. *Management Decision, 33*(5), 33-43.

Bussen, W., & Myers, M. D. (1997). Executive information system failure: A New Zealand case study. *Journal of Information Technology, 12*(2), 145-153.

Cady, J. F. (1984). *Strategic marketing management: The course* (Paper No. 9-584-076). Boston: Harvard Business School.

Carvalho, J. A. (1998). Using the viable system model to describe the role of computer-based systems in organizations. In N. Callaos, L. Holmes, & R. Osers (Eds.), In *Proceedings of the World Multiconference on Systems, Cybernetics and Informatics* (pp. 497-502), Orlando, FL.

Cavalcanti, E. P. (2005). The relationship between business intelligence and business success. *Journal of Competitive Intelligence and Management, 3*(1).

Chen, H., Chau, M., & Zeng, D. (2002). CI spider: A tool for competitive intelligence on the Web. *Decision Support Systems, 34*, 1-17.

Christiansen, J. A. (2000). *Competitive innovation management: Techniques to improve innovation performance.* Basingstoke: Macmillan Press Ltd.

Claraview, LLC. (2002). *Business intelligence applications for the public sector.* Retrieved December 29, 2006, from http://www.dmreview.com/whitepaper/WID592.pdf

Cottrell, N. & Rapley, K. (1991). Factors critical to the success of executive information systems in British airways. *European Journal of Information Systems, 1*(1), 65-71.

Crane, A. (2004). In the company of spies: When competitive intelligence gathering becomes industrial espionage. *Business Horizons*, *48*(3), 233-240.

Crowcombe, P. (2002, December 9). *Scalable security solutions*. Paper presented at the Business Continuity Briefing Seminar, London, UK.

Crowley, E. (2004, March/April). A market intelligence primer. *Productmarketing.com*, pp. 3-6.

Curtis, G., & Cobham, D. (2005). *Business information systems: Analysis, design and practice* (5th ed.). FT: Prentice Hall

D'Aveni, R., & MacMillan, I. (1990). Crisis and the content of managerial communications: A study of the focus of attention of top managers in surviving and failing firms. *Administrative Science Quarterly*, *35*, 634-657.

Daft, R., Sormunen, J., & Parks, D. (1988). Chief executive scanning, environmental characteristics, and company performance: An empirical study. *Strategic Management Journal*, *9*(2), 123-139.

About the Contributors

Mila Gascó-Hernández holds a MBA and a PhD in public policy evaluation (Award Enric Prat de la Riba granted to the best PhD thesis on public management and administration, given by the Escola d'Administració Pública de Catalunya in Barcelona, Spain). She is co-founder of Estratic as well as an associate professor at both the Open University of Catalonia and the Pompeu Fabra University, both in Spain. For seven years, she was a senior analyst at the International Institute on Governance of Catalonia. She has a wide teaching experience (she worked as a full professor in the Rovira i Virgili University in Tarragona, Spain) as well as a broad researching experience. She has taken part in numerous national and international seminars, she has published both in Spanish and English and she has supervised some PhD thesis. She has collaborated with several institutions such as both the provincial and city government of Barcelona, the World Bank Development Gateway, the United Nations Development Program, the University of Hull in United Kingdom, the Mayor's Office in Valencia (Venezuela) or the Governments of Brazil and Dominican Republic. Her main interests are related to public policies that allow the transition of a society to the so-called knowledge era (in particular she is interested in e-government and e-governance), to the use of ICTs for human development and to public policy evaluation.

Teresa Torres-Coronas, has a bachelor's degree in economics (Barcelona University) and a PhD in management (Rovira i Virgili University). She won first prize in the 2000 edition of EADA related management research. She is the author of the book *Valuing Brands* (Ediciones Gestión 2000, Spain), co-author of the book *Retrieve Your Creativity* (Septem Ediciones, Spain), and co-editor of the books *Changing the way you teach: Creative tools for management education* (Septem Ediciones, Spain), *e-HRM: Managing knowledge people* (Idea Group, USA), *Higher creativity for virtual teams: Developing platforms for co-creation* (Information Science Reference) and, *The Encyclopedia of HRIS: Challenges in e-HRM* (Information Science Reference). She is author of many articles and conference papers about intangible management, management education, and applied creativity and IT. She is management professor at the Universitat Rovira i Virgili. She is one of the researchers of the ELIS group: E-government for Local Integration with Sustainability (Hull University). She is an active member of the Management Education and Development Division (Academy of Management) and the Information Resources Management Association (IRMA).

***

Miguel Cervantes Blanco, PhD, is professor of marketing at the Universidad de León (Spain) and he is doctor of business sciences. He is authors of several papers and publications in market research and marketing. He also works as consultant. His research interests include the areas of marketing research, brands, city marketing.

Michael Callaghan is an expert in multimedia technologies teaching and researching in areas such as multimedia, wireless technologies, technology in education, computer gaming, remote experimentation, machine vision and hybrid intelligent systems. Michael is a reviewer for several journals and international conferences and an editor for the *International Journal of Online Engineering*. He is on the program committee of Remote Engineering and Virtual Instrumentation conference and serves on the Scientific Advisory Board of the IAOE (International Association of Online Engineering). He has published 20+ articles in peer-reviewed journals and conferences.

Raquel Camprubí is research fellow at Universitat de Girona. She has been teaching at the Faculty of Tourism since 2004. Her fields of interest include destination marketing and business organization. Currently, she is working on her dissertation which is focused on tourism image formation.

José Fernández-Cavia, PhD, is a professor of advertising and public relations at the Universitat Pompeu Fabra, Barcelona, Spain. He has published several papers, chapters and books on communication issues. He is a member of the governing body of the Catalonian Advertisers and Public Relations Association. He is also member of the editorial board of the *Catalan Journal of Communication & Cultural Studies*.

Juliane Chudalla is a master student at the University of Augsburg and works in the research project *Mobile Services in City Marketing*. From 2006 to 2007 she was a project assistant at the wi-mobile Research Group and was involved in mobile payment and mobile marketing research. Since 2007 she works in the network and company support department of Augsburg AG, a public private partnership for business development of the Augsburg region.

José-Rodrigo Córdoba, PhD, is lecturer of management systems at the business school, University of Hull. During his previous career as an IT project management in Colombia he became involved in the formulation of policies to bring the information society to reality in the country. He researches on the use of systems thinking methodologies to bring about positive chance in the implementation of information technologies. He is a founding member of the ELIS group which explores implementation of e-government across countries. He has written several papers in internationally recognized journals in information systems, systems thinking and management.

Pablo Díaz Luque, PhD, has been a lecturer in Tourism and Electronic Marketing since 2004 at the Department of Business Administration, Universidad Pablo de Olavide (Sevilla, Spain). He was a visiting Researcher in the School of Services Management, Bournemouth University (United Kingdom) during 2007-2008 academic year. He is a research member of the Centro de Sociología y Políticas Locales (Sociology and Local Policias Research Center), Pablo de Olavide University (Sevilla, Spain). He holds a PhD in Economic and Business Sciences (January 2006, Department of Applied Economy, Universidad de Málaga (Spain). His PhD thesis was about the development of Spanish tourism websites and the Destination Marketing Organization and it was funded by the Spanish Ministry of Economic. He has done research stages at Universidad de Málaga, Andalusian Institute of Statistics and, Regional Institute of Development to conduct regional, national and international projects.

Peter Dobers takes interest in how ideas of broadband, city images, corporate (social) responsibility or sustainable development travel the world, are enabled or disabled and has published widely in these areas. He currently holds a chair in management and sustainable development at Mälardalen University (Västerås, Sweden) and is associate dean of the Faculty for Humanities, Social and Caring Sciences. Between 2006-2008 he was visiting professor at the Umeå School of Business, Sweden. He is board member of the International Sustainable Development Research Society, on the editorial board of *Business Strategy and the Environment*, and *Corporate Social Responsibility and Environmental Management*.

Prescott C. Ensign, PhD, is an assistant professor in the Telfer School of Management at the University of Ottawa, Ontario, Canada. He is a fulbright scholar and has been recognized for excellence in teaching and research. Ensign has written articles on the strategy and structure of multinational enterprises, and his current work investigates innovative efforts of entrepreneurs in high-growth, technology-based firms. This is his second article to appear in *Thunderbird International Business Review*.

Anette Hallin is currently writing up her PhD-thesis at the Department of Industrial Management and Organization which is about the size of organizations and how organizations can be perceived bigger than they are in factual numbers. Her other research interests include the concept of "cities" and "organizations", the non-managerial emerging of organizations as well as the management of these, and the social and cultural aspects of sustainable development in organizations and societies. Together with Tina Karrbom-Gustavsson she is editing a book called *Organizational Communication and Sustainable Development: ICTs for mobility* (Idea Group Publishing).

Jim Harkin, PhD, holds a BTech, MSc and PhD in electronic engineering. His research interests relate to the creation of intelligent and adaptive reconfigurable systems, and distance learning through remote experimentation. He is an associate member of the Research Institute for Computer Science with the Faculty of Engineering and the author of 40+ publications.

Assumpció Huertas, PhD, is an assistant professor of public relations at Universitat Rovira i Virgili, Spain. She is also a professor of advertising at Universitat Oberta de Catalunya. She has a background in sociology and advertising and public relations. Her research spans tourism marketing, destination branding, new technologies applied to tourism, e-commerce, e-marketing and tourism communication. She is working in a project about new technologies and tourism at Universitat Rovira i Virgili and she is also doing research on identity and destination branding with Annette Pritchard and Nigel Morgan from the University of Wales Institute, Cardiff. She has published several papers in international journals. She is a member of AIRP, an association of public relation researchers.

María I. Huerta-Carvajal is a professor of business at the Universidad de las Américas, Puebla in México. She holds an MBA in business administration with specialty in service marketing from the Universidad de las Américas, Puebla. Her research interests lie on marketing of cities and towns, cultural marketing and marketing research.

Maureen Jouett, MBA is an adjunct professor at Tarleton State University where she teaches marketing and management. Maureen has a wide range of business experience in management and supervision, strategic planning, marketing, finance and accounting and computer programming. Additionally, she served six years as a council member and six years as mayor of Killeen, TX during a time of unprecedented growth. As chair of the city's information management technology committee she was responsible for the allocation of millions of dollars in city financial resources for the long-term development of the city's technology infrastructure. Her community leadership and passion for excellence is without parallel.

Nicolas Jullien, PhD, is assistant professor at GET school ENST Bretagne. He holds a PhD in economics from ENST Bretagne and is director of M@rsouin (Môle Armorcain de la Recherche sur la Société de l'Information et l'Usages d'Internet) in France. He has published several papers in international journals and books about information technologies and more specifically open source software. He has also worked in several European projects and has organized European conferences on software and information technology, including the M@rsouin annual conference in June every year. Nicolas' expertise lies in the design of local activities to assess perceptions of groups of citizens when using information technologies

Luis F. Luna-Reyes is a professor of business at the Universidad de las Américas in México. He holds a PhD in information science from the University at Albany. Luna-Reyes is also a member of the Mexican National Research System. His research focuses on electronic government and on modeling collaboration processes in the development of information technologies across functional and organizational boundaries.

Laura L. Matherly, PhD, MBA is an assistant professor at Tarleton State University where she teaches strategy, statistics and research methods in the MBA program. Prior to this, she was an assistant professor at Pennsylvania State University. She has extensive executive and consulting experience in training and developing strategic plans to improve organizational effectiveness and efficiency to meet and exceed stakeholder expectations in Fortune 100 companies such as Boeing Commercial Aircraft Group, Weyerhaeuser Paper Company, R. R. Donnelley and Microsoft. She is a member of the Academy of Management and the American Society for Quality.

Al D. McCready, PhD, is an experienced executive who been a partner in two major international professional service firms, has worked in the technology sector at the federal and state government levels in the United States, and has conducted consulting assignments for companies in the consumer marketing segment in the U.S.A. and other countries for more than 25 years. He has a master's in business administration from the University of Utah, and doctorate from The George Washington University in Washington, D.C. Dr. McCready was a partner at both Arthur Young & Co. (now Ernst and Young) and then Deloitte & Touche, and since 1991 has been CEO of McCready Manigold Ray & Co., Inc., a strategic technology consulting firm.

Martin McGinnity, PhD, holds the post of professor of intelligent systems engineering. He is the author or co-author of over 175 research papers with research interests focused on computational intelligence and the creation of bio-inspired intelligent computational systems in general, particularly in relation to hardware and software implementations of biologically plausible artificial neural networks, fuzzy systems, genetic algorithms, embedded intelligent systems utilizing re-configurable logic devices, brain computer interfacing and intelligent systems in robotics. He is a core member of the Research Institute for Computer Science with the Faculty of Engineering.

Barry Mishra, PhD, is an associate professor in the Department of Accounting and Information Systems at the Anderson Graduate School of Management, University of California at Riverside. He has served as department chair and associate dean in the Anderson Graduate School of Management at UC Riverside. Barry's research is in the area of accounting and information management. His research interests are diverse and include the areas of accounting disclosure, management and control, information systems security, and open source software. He has been published in the *Journal of Accounting Research, The Accounting Review, Management Science, Marketing Science, Information Systems Research*, and others.

Sandra Moffett, PhD, has been researching the field of knowledge management for 10 years, completing her PhD in this area. As part of her work she created the MeCTIP model and Benchmarking KM assessment tool, winning several awards and research grants for her work. She has 40+ publications in the areas of knowledge management, innovation, strategic business improvement and Semantic Web. Before taking up academic post in 2000, Sandra was a secondary school teacher, hence her interest in pedagogy and technical application to enhance the learning environment. She is a core member of the Research Institute for Business and Management.

Bantu L. Morolong, PhD, is a lecturer at the University of Botswana. She holds a PhD in international/intercultural educational policy studies from the University of Alberta, Canada and an MA in sociology from Michigan State University, USA. Dr. Morolong's career spans more than twenty five years of university teaching, administration and outreach and development. She has served as a technical advisor and a training and evaluative research consultant for programmes in the local and regional public and private sectors. Dr Morolong has, researched, published and presented papers at regional and international meetings, covering her areas of interest which are; rural development, literacy, sociology of education, gender, education and sustainable development, women and the law, the potential role of ICTs in poverty reduction and inclusive development, gender, energy and the environment, the gendered project cycle and gender dynamics in the changing family with special reference to access to and control over resources.

Norberto Muñiz-Martínez, PhD, is doctor of economics and business administration, professor of marketing and market research at the Universidad de León, Spain. He holds a degree in foreign trade and European communities (Polytechnic University, Madrid) and a Master of science transport & distribution management (University of Central England). He has published articles in national and international journals and presented papers at conferences. He has been teaching in various master and postgraduate courses as well as conducting research stays at universities in Europe, the Americas and Asia. His research interests include the areas of city marketing, international retailing, new trends in tourism, and strategic marketing.

Maeve Paris is an expert in the areas of natural language processing and object-oriented systems development with specific research focused on computer-based assessment of diagrams, e-government, accessibility, and mobile learning. She was previously employed in Liverpool John Moores University, and more recently, in the Humanities Computing Unit of Oxford University, and she is a member of the European Platform of Women Scientists.

Lluís Prats Planagumà is associate professor at the Faculty of Tourism at Universitat de Girona. He holds a PhD in tourism economics and he is member of the Research Center on Tourism Innovation (CRIIT). His research fields include innovation management and tourism destination management and planning.

Key Pousttchi, PhD, is an assistant professor and heads the wi-mobile Research Group at the University of Augsburg since 2001. The research focus of his group comprises mobile-integrated business processes, the development of mobile markets, mobile financial services and mobile marketing. Prior to his academic career he has been a regular officer of the German army for 12 years, lastly being responsible for integrated information processing and simulation systems in an R&D department.

Nicholas P. Robinson is a member of the Canadian Bar Association and a graduate of the Faculty of Law at McGill University, Montreal, Quebec, Canada. He has been active in numerous areas of business and academia. Robinson has worked in the recording industry, run an IT-based business, conducted tax policy research for government, and worked in the insurance industry. He has written a number of case studies and research papers on international management, marketing, public policy, and law.

Erik Rolland, PhD, is an associate professor in the Department of Accounting and Information Systems at the Anderson Graduate School of Management, University of California at Riverside. He has served as department chair and associate dean in the Anderson Graduate School of Management at UC Riverside. Erik's research is in the area of strategic use of management information systems and technology, dealing with a wide array of topics such as IT risk, information & service quality, pricing, telecommunications, customer relationship management, e-commerce, and the development of dynamic (strategic) management capabilities. His papers have appeared in such journals as *Operations Research, The European Journal of Operational Research, Decision Sciences.*

Marianna Sigala is a lecturer of operations and production management at the University of the Aegean, Greece. Before joining the University of the Aegean, she had been lecturing at the Universities of Strathclyde and Westminster in the UK. Her interests include service operations management, information and communication technologies (ICT) in tourism and hospitality, CRM and e-learning. She has professional experience from the hospitality industry in Greece, while she has also contributed to several international research and consultancy projects. She has published three books, and numerous research papers at academic journals and international conferences. She had served as president of Euro-CHRIE (2004 – 2005) and she currently serves at the board of directors of International Federation of IT and Travel (IFITT) and Hellenic Association of Information Systems (HeAIS).

Jocelyne Trémenbert holds degrees on statistics from Dauphine and Paris VII universities. She is working with Nicolas Jullien in M@rsouin, a Breton association of social science research centers for the study of information society. She is responsible for an ICT Observatory named OPSIS used to conduct its own surveys (among individuals, firms, city halls …) to inform policy making in Brittany. Jocelyn's expertise lies in the preparation and control of data collection and interpretation. She has published several papers about the measure of ICT uses. Currently she supervises research students in a French school of statistics.

Derek Woods is a senior lecturer in SCIS. His teaching and research focuses on pervasive computing, mobile technologies and wireless capability.

Index

U

V

W